MILK PROTEINS
chemistry
and
molecular biology

VOLUME II

MILK PROTEINS
chemistry and molecular biology

Edited by

Hugh A. McKenzie

Department of Physical Biochemistry
Institute of Advanced Studies
Australian National University
Canberra, Australia

VOLUME II

 1971

ACADEMIC PRESS   New York and London

COPYRIGHT © 1971, BY ACADEMIC PRESS, INC.
ALL RIGHTS RESERVED
NO PART OF THIS BOOK MAY BE REPRODUCED IN ANY FORM,
BY PHOTOSTAT, MICROFILM, RETRIEVAL SYSTEM, OR ANY
OTHER MEANS, WITHOUT WRITTEN PERMISSION FROM
THE PUBLISHERS.

ACADEMIC PRESS, INC.
111 Fifth Avenue, New York, New York 10003

United Kingdom Edition published by
ACADEMIC PRESS, INC. (LONDON) LTD.
Berkeley Square House, London W1X 6BA

LIBRARY OF CONGRESS CATALOG CARD NUMBER: 78-86363

PRINTED IN THE UNITED STATES OF AMERICA

Contents

List of Contributors ix
Preface xi
Contents of Volume I xiii

PART D CASEINS AND RENNIN (CHYMOSIN) 1

General Introduction
H. A. McKenzie

9 ☐ Formation and Structure of Casein Micelles . . . 3
D. F. Waugh

I.	Introduction	4
II.	Micelle Types	6
III.	The Natural Casein Micelle	10
IV.	Properties of Monomer Caseins	13
V.	The Structure of Casein Micelles	23
VI.	The Micelle Core	35
VII.	The Micelle Coat	47
VIII.	Equilibrium Casein Micelle Systems	50
IX.	Natural Micelles, Other Colloid Particles, and the Aqueous Phase of Milk	58
X.	Rennin (Chymosin) Coagulation	75
	References	79

10 ☐ Whole Casein: Isolation, Properties, and Zone Electrophoresis 87

H. A. McKenzie

I. Introduction 87
II. Methods of Isolation 90
III. Properties 94
IV. Controversial Components 101
V. Zone Electrophoresis in Casein Typing 109
VI. Summary 113
References 114

11 ☐ α_s- and β-Caseins 117

M. P. Thompson

I. Introduction 117
II. α_s-Caseins 120
III. β-Caseins 155
IV. General Considerations 168
References 169

12 ☐ κ-Casein and Its Attack by Rennin (Chymosin) . . . 175

A. G. Mackinlay and R. G. Wake

I. The General Properties of κ-Casein 175
II. The Action of Rennin (Chymosin) on κ-Casein . . 198
References 212

13 ☐ The Biochemistry of Prorennin (Prochymosin) and Rennin (Chymosin) 217

B. Foltmann

I. Introduction 217
II. Assay of Rennin 218
III. Preparation of Prorennin and Rennin 222
IV. Formation of Rennin from Prorennin 230
V. Physical and Chemical Properties 236
VI. Proteolytic Activity 246
VII. A Comparison of Rennin and Pepsin 249
References 251

PART E WHEY PROTEINS AND MINOR PROTEINS 255

General Introduction
H. A. McKenzie

14 ☐ β-Lactoglobulins 257
H. A. McKenzie

I. Introduction 258
II. Isolation of β-Lactoglobulins 259
III. Methods of Zone Electrophoresis of Whey Proteins . 271
IV. Species Differences and Genetic Variants 274
V. Amino Acid Composition 277
VI. Electrochemical Properties 294
VII. Molecular Size and Conformation 304
VIII. Denaturation of β-Lactoglobulins 316
IX. Interaction of β-Lactoglobulins and κ-Casein . . . 321
X. X-Ray Crystallographic Studies 323
XI. Summary and Conclusions 324
References 325

15 ☐ α-Lactalbumin 331
W. G. Gordon

I. Reports of the Isolation of Crystalline "Albumins" from Cow Milk 332
II. Preparation and Purification of Bovine α-Lactalbumin 334
III. Composition and Structure 339
IV. Physico-Chemical Properties 345
V. Genetic Polymorphism 351
VI. α-Lactalbumins in the Milk of Other Mammals . . 352
VII. Crystallography and X-Ray Diffraction 356
VIII. The Biological Function of α-Lactalbumin . . . 356
References 361

16 ☐ Minor Milk Proteins and Enzymes 367
M. L. Groves

I. Introduction 367
II. Minor Milk Proteins 368
III. Milk Enzymes 385
References 411

PART F   MILK PROTEINS AND TECHNOLOGY 419

General Introduction
H. A. McKenzie

17 ☐ Milk Protein Research and Milk Technology 421
R. Beeby, R. D. Hill, and N. S. Snow

    I. Introduction 422
   II. Cheese Manufacture 422
  III. Concentrated Milk and Milk Powder 434
  IV. The Manufacture of Casein, Coprecipitate, and
       Whey Proteins 446
   V. Specialized Products 452
  VI. Conclusion 457
     References 459

PART G   THE FUTURE 467

General Introduction
H. A. McKenzie

18 ☐ Milk Proteins in Prospect 469
H. A. McKenzie

    I. Introduction 469
   II. Prospects 470
     References 480

Appendix 483
H. A. McKenzie

   Methods for Zone Electrophoresis of Milk Proteins . 483
   Summary of Zone Electrophoresis Methods . . . 487
   References 508

Author Index 509
Subject Index 529

List of Contributors

Numbers in parentheses indicate the pages on which the authors' contributions begin.

RALPH BEEBY   C.S.I.R.O., Division of Dairy Research, Highett, Victoria, Australia (421)

BENT FOLTMANN   The Protein Laboratory, Faculty of Medicine, University of Copenhagen, Copenhagen, Denmark (217)

WILLIAM G. GORDON   Eastern Utilization Research and Development Division, Agricultural Research Service, U.S. Department of Agriculture, Philadelphia, Pennsylvania (331)

MERTON L. GROVES   Eastern Utilization Research and Development Division, Agricultural Research Service, U.S. Department of Agriculture, Philadelphia, Pennsylvania (367)

RONALD D. HILL   C.S.I.R.O., Division of Dairy Research, Highett, Victoria, Australia (421)

HUGH A. McKENZIE   Department of Physical Biochemistry, Institute of Advanced Studies, Australian National University, Canberra, Australia (87, 257, 469, 483)

ANTHONY G. MACKINLAY*   Department of Biochemistry, University of Sydney, N.S.W., Australia (175)

NORMAN S. SNOW†   C.S.I.R.O., Division of Dairy Research, Highett, Victoria, Australia (421)

MARVIN P. THOMPSON   Eastern Utilization Research and Development Division, Agricultural Research Service, U.S. Department of Agriculture, Philadelphia, Pennsylvania (117)

ROBERT G. WAKE   Department of Biochemistry, University of Sydney, N.S.W., Australia (175)

DAVID F. WAUGH   Department of Biology, Massachusetts Institute of Technology, Cambridge, Massachusetts (4)

* *Present address*: Department of Biochemistry, University of New South Wales, Kensington, N.S.W., Australia.
† *Present address*: Australian Dairy Produce Board, Melbourne, Victoria, Australia.

Preface

The general objectives of this book have been given in the Preface to the first volume and the reader is referred to it for details. In brief, Volume I is concerned with giving a general perspective of the history, occurrence, and properties of milk proteins and a detailed discussion of the strategy for their study. No apology is made for the heavy emphasis on physicochemical approaches. Indeed, the complexity of milk proteins and the need for sophisticated methods in their study should be apparent at the end of Volume I. With firm foundations laid, the reader should be well prepared for the detailed discussion of the individual milk proteins which is the concern of this volume.

The caseins constitute the major group of milk proteins; they are the most complex and, in many ways, the least understood. The first part of Volume II is concerned with the caseins. The central problem is to gain an understanding of the formation and structure of casein micelles. Thus, the opening chapter is a critical discussion of these problems by a distinguished worker, David Waugh. This chapter will be of value also to those who are interested in other classes of micelles. An important feature is that Waugh stresses the differences in structure of various classes of micelles and integrates our knowledge of the casein micelles into this general picture. At the same time we must learn more about the individual caseins. Before discussing them it is necessary to understand what we mean by whole casein and to be able to identify caseins. In Chapter 10 I have discussed some of the problems involved. Then the α_s- and β-caseins are discussed thoroughly in Chapter 11 by Marvin Thompson who has contributed so diligently to our knowledge of them. κ-Casein and its attack by rennin are discussed by Anthony Mackinlay and Robert Wake in Chapter 12. This problem is of great practical as well as of theoretical importance in protein chemistry. Thus, Bent Foltmann's review of the biochemistry of rennin in Chapter 13 is especially pertinent.

We then proceed to consider the "whey" proteins and enzymes: both those synthesized in the mammary gland and those transported from the blood. I discuss the β-lactoglobulins in Chapter 14. The remarkable array of their physico-chemical properties is stressed and at the same time they are discussed as an example of the amount of information that can be obtained about a protein whose X-ray crystal structure is unknown. The veteran investigator, William Gordon, deals with the α-lactalbumins in Chapter 15. The biological role of the α-lactalbumins, unlike the β-lactoglobulins, is known, and it has a remarkable evolutionary relationship with lysozyme. The comparison of the properties of the two proteins in this chapter is of great interest. Merton Groves assesses critically and authoritatively the minor milk proteins and enzymes in Chapter 16. This is an area in which there has been a great upsurge in interest recently. In all these chapters an effort is made to consider the evolution and origin of the proteins.

Milk technology plays a role in our daily lives, and recent developments in food technology and immunology have important implications for us. In Chapter 17 Ralph Beeby, Ronald Hill, and Norman Snow assess the relationship of recent advances in milk protein chemistry to problems of milk technology.

It will be seen by the end of these chapters that we have reached an important stage of development in our basic knowledge of milk proteins, but important problems lie ahead. These are important not only in their own right but have considerable implications in future problems of nutrition. Thus, the future is assessed in Chapter 18 in the light of recent developments.

In addition to the acknowledgments given in Volume I, appreciation is expressed to the Australian Dairy Produce Board for grants that have greatly assisted me in the completion of this book. Also, Doreen McLeod and her staff of the library of the John Curtin School of Medical Research have been unstinting in their assistance. The subject indexes have been prepared with the able assistance of Margaret, Judith, and Ross McKenzie.

H. A. McKenzie

Contents of Volume I

Milk Proteins in Retrospect
  T. L. McMeekin

Protein Composition of Milk
  R. Jenness

Immunological Studies of Milk
  L. Å. Hanson and B. G. Johansson

General Methods and Elemental Analysis
  H. A. McKenzie and W. H. Murphy

Amino Acid, Peptide, and Functional Group Analysis
  H. A. McKenzie

Analysis and Structural Chemistry of the Carbohydrate of Glycoproteins
  E. R. B. Graham, H. A. McKenzie, and W. H. Murphy

The Elucidation of Interacting Systems in Terms of Physical Parameters
  H. A. McKenzie and L. W. Nichol

Effects of Changes in Environmental Conditions on the State of Association, Conformation, and Structure
  H. A. McKenzie

Part D
Caseins and Rennin (Chymosin)

General Introduction

In Part C of Volume I it was emphasized that a major problem in the study of milk proteins is their interaction with one another and with various small molecules and ions. This creates difficulties not only in isolating them in a pure state for characterization but also in interpreting their physical properties. No class of milk proteins better exemplifies these difficulties than the major protein group of mature milk, the phosphoproteins or, as they are commonly termed, the caseins.

The caseins interact with themselves, with calcium(II) and phosphate to form micelles or colloidal aggregates. Micelles are essentially spherical in shape and may (in cow milk) be of the order of 40–300 nm in diameter, having particle weights in the range 10^6–3×10^9 daltons. Smaller aggregates and casein monomers have molecular weights as low as 20,000 daltons. They and the micelles may interact with the whey proteins, such as β-lactoglobulins, and this can have a modifying effect on the properties of the micelles. It can be seen that the behavior of the caseins is a complex problem of colloidal science. In Chapter 9 David F. Waugh gives a penetrating analysis of the formation and structure of casein micelles.

It is not surprising that even the definition of a casein is a matter for controversy. In Chapter 10 the isolation, composition and properties of whole casein are discussed by Hugh A. McKenzie. Considerble emphasis is placed in this chapter on gentle methods of isolation and on the so-called minor casein components. Some of these may be minor in amount but nevertheless may exert an important influence on properties.

In Chapter 11 Marvin P. Thompson discusses the calcium-sensitive caseins, α_s- and β-caseins. They are important components of the micelles, show some very interesting association-dissociation reactions, and have important genetic relationships.

κ-Casein is now recognized as the micelle-stabilizing casein, first proposed in the theory of casein micelles of the Danish chemists, Linderstrøm-Lang and Kodama. It is unique among the caseins in that it contains cystine (or cysteine), carries varying amounts of a carbohydrate moiety and is the casein upon which the enzyme rennin (or chymosin)* acts immediately. κ-Casein and the action of rennin are discussed by Mackinlay and Wake in Chapter 12. The biochemistry of rennin and its precursor prorennin (prochymosin)* is thus of great importance and is gradually being unraveled, as shown by Foltmann in Chapter 13. This enzyme provides an interesting example of limited proteolysis.

H. A. McKenzie

---

* Foltmann (1969) has proposed that the name *chymosin* be used in place of *rennin* (E.C. 3.4.4.3), from calf stomach, to avoid the confusion that often occurs with renin (E.C. 3.4.4.15), from the kidney. The term chymosin was used originally (for example Deschamps, 1840; Lea and Dickinson, 1890). In this book chymosin is given at the head of each relevant chapter as an alternative to rennin, and *prochymosin* is given as an alternative to *prorennin*.

9 ☐ Formation and Structure of Casein Micelles

D. F. WAUGH*

I.	Introduction	4
II.	Micelle Types	6
III.	The Natural Casein Micelle	10
IV.	Properties of Monomer Caseins	13
	A. α_s-Casein	15
	B. β-Casein	18
	C. κ-Casein	19
	D. General Remarks	21
V.	The Structure of Casein Micelles	23
	A. Micelles from Mixtures of α_s- and κ-Caseins	23
	B. Micelles from Mixtures of β- and κ-Caseins	34
VI.	The Micelle Core	35
	A. Core Polymers	35
	B. Ion Binding	38
VII.	The Micelle Coat	47
VIII.	Equilibrium Casein Micelle Systems	50
IX.	Natural Micelles, Other Colloid Particles, and the Aqueous Phase of Milk	58
	A. The Ionic Environment and Inorganic Colloid	58
	B. Cooperative Effects	63
	C. Hysteresis Effects	69
	D. Natural Micelles	73
	E. Metastable Colloid	73
	F. Minor Casein Components	74
X.	Rennin (Chymosin) Coagulation	75
	References	79

* Supported by Research Grant GM 05410 from the National Institute of General Medical Science of the National Institutes of Health, Bethesda, Maryland.

I. Introduction

The long history of casein studies, including the work of Braconnot, Mulder and Hammarsten, for example, on casein micelles, has been considered by McMeekin in Chapter 1, Volume I. For present purposes, the two most important themes running through this early work are the heterogeneity or, as we would now wish to state it, the paucidispersity of casein, and the gradual recognition that some special mechanism for stabilization appeared to be required.

One of the chief attractions in discussing the formation and structure of casein micelles is the opportunity to show that structure, and the reactions which lead to structure, are based on the most fundamental subtleties of biostructure formation—the highly specific sequences of protein–protein and protein–ion interactions.

In view of the strong interaction properties of the caseins, even in the absence of divalent cations (to be described), it is not surprising that the dispersity of casein remained a problem which was not solved for a long time. Understanding of this problem could only be gained by the development and proper application of adequate physical techniques. Our present considerations start with the pioneering work of Linderstrøm-Lang and his colleagues (Linderstrøm-Lang, 1925, 1929; Linderstrøm-Lang and Kodama, 1925), who demonstrated clearly for the first time that the casein of Hammarsten contained more than one casein component. Shortly thereafter, Pedersen (1936) demonstrated possible heterogeneity by ultracentrifugal analysis. In the work of Linderstrøm-Lang, fractions obtained from the acid precipitate by acid–alcohol extraction varied in their response to the addition of calcium. They ranged from certain minor fractions of low phosphorus content which remained in clear solution to major fractions which formed precipitates. On the basis of this behavior pattern Linderstrøm-Lang clearly stated a fundamental postulate; namely, that soluble components interact with and stabilize less soluble components. A direct demonstration of the protective effect was not made, partly because pure components were not available but mainly, as we now know, because the relative stabilities of mixtures of available fractions were not examined. Linderstrøm-Lang did, however, predict that a component or components insoluble in the presence of calcium ion would eventually be isolated. He predicted further that the clotting enzyme rennin* would be found to split off a stabilizing component.

---

* The term *chymosin* may be used as an alternative to *rennin* throughout this chapter, as explained in the introduction to Part D.

In his fractionations, Linderstrøm-Lang used techniques involving precipitation at low pH, and it was known that low pH could alter other proteins. In the meantime, electrophoretic procedures were applied by Mellander (1939) to whole casein prepared by acid precipitation, with the result that three peaks were observed: two main peaks designated α-casein and β-casein and a minor peak designated γ-casein, in descending order of anodal mobility at alkaline pH. With this information as a guide, Warner (1944) developed more efficient and less objectionable fractionation procedures than those of Linderstrøm-Lang and isolated materials corresponding to the α- and β-caseins of Mellander. The question of protective action was again taken up by Cherbuliez and Baudet (1950a,b), who obtained the caseins by the methods of Warner. By mixing experiments at 37°C, they showed that α-casein formed a colloid in the presence of calcium and that α-casein was capable of stablizing β-casein. Neither α- nor β-caseins had the calcium insensitivity, that is, absence of precipitation or colloid formation, of some of the minor fractions of Linderstrøm-Lang. It is also interesting to note that Cherbuliez and Baudet were unaware of the extraordinary temperature dependence of β-casein solubility in the presence of calcium (Waugh, 1958).

Additional insight into the structure of the system came when a new procedure for the preparation of caseins was worked out by von Hippel and Waugh (1955) and Waugh and von Hippel (1956). The procedure, in essence, was to harvest micelles directly by centrifugation; to bring them into solution by sequestering calcium; to dialyze against monovalent salts; and to precipitate out a major protein fraction by re-adding calcium, at constant pH, to a high concentration, for example, 0.25 M. We now know that this operation, termed *splitting*, can be carried out under a wide variety of conditions and that the result is essentially the same. The precipitate formed at 37°C is rich in protein having the electrophoretic characteristics of α- and β-caseins. The supernatant, fraction S, while containing components with mobilities like that of β-casein, has a new component which is distinguished mainly by its large sedimentation coefficient; its electrophoretic mobility is close to that of α-casein. The important observations are that (a) the precipitated proteins, freed of calcium and thus brought into solution, reprecipitate in the presence of small amounts of calcium at 37°C; (b) a solution of α- and β-caseins can be fractionated at 0°C on the addition of small amounts of calcium, the precipitate being largely Ca α-caseinate; (c) fraction S proteins remain clear on the addition of calcium; and (d) if all proteins are mixed in proper proportions, the addition of calcium leads to the formation of stable micelles, and precipitate is absent. The calcium-insensitive component is considered to be necessary for stabilization. Least abundant of the three, it is termed κ-casein. It is now apparent that the materials obtained by Warner (1944) are to be equated directly with those obtained by von Hippel and Waugh (1955) and Waugh and von Hippel (1956) on the basis that the α-casein of Warner, from its electrophoretic properties and stability characteristics as found by Cherbuliez and Baudet, is an interaction product which contains κ-casein. In order to preserve continuity the major related components, calcium precipitable at low temperatures, are termed α_s-caseins (Waugh, 1958; Waugh et al., 1962).

Certain terms and nonstandard abbreviations will be used extensively throughout this chapter:

Solubilized skim milk is a product obtained by adding sodium citrate to skim milk and dialyzing the resulting protein solution against sodium chloride, usually 0.07 or 0.05 M, at pH 7.

First-cycle casein is a solution which contains essentially all of the casein components and little whey protein. Calcium is essentially absent.

Nonpolar amino acid side chains are those of valine, leucine, isoleucine, phenylalanine, tryptophan, and proline.

Calcium is used instead of the more accurate term calcium (II). When calcium is added it is usually as calcium chloride.

I indicates ionic strength due to mono-monovalent salts. It does not include calcium chloride, etc., which are specified separately.

R is the weight ratio in a system of (α_s-casein + β-casein)/κ-casein.

R_i is the initial weight ratio established before micelle formation is initiated, by the addition of calcium, for example.

R_s is the final weight ratio established after calcium has been introduced at some point in an experiment, for example, if two micelle systems established at different R_i are mixed.

G is the total solvent in precipitates or micelles expressed as g water/g protein.

U is tightly bound solvent, expressed as g water/g protein. It is taken as 0.5.

II. Micelle Types

A considerable amount of space could be devoted to the question of the meaning of the word "micelle." In most cases micelle has been used to designate what obviously are colloidal association products. However, the designation has most frequently been used before even a general understanding of structure was available. It is not surprising, at this time, that different micelle types have different structures: The cellulose micelles of von Nägeli (see von Nägeli and Schwendener, 1877) are different in structure from soap micelles* (see Shinoda et al., 1963) and from casein micelles. The micelles of milk are customarily defined as the colloidal association products of the caseins. To write of them as colloidal particles is permissible on the basis that they are large with respect to constituent monomers but stable with respect to each other and to the earth's gravitational field.

What might be anticipated, from the properties of other systems, as permissible plans for the structure of the casein micelle? Micelles are expected to fall into one or more of the categories† of (a) single-phase particles, (b) large chemical compounds or (c) structures having a composition which changes in going from the surface to the center.

If all the dimensions of a colloidal particle are greater than the longest dimension of its constituent subunit or subunits, the average composition could be kept uniform throughout. However, some subunits must be entirely surrounded by others, and to give a minimum surface energy (zero or positive), surface orientation would be expected. Verwey and Overbeek (1948) considered such situations for positive surface energy. Surface-to-surface interaction must be retarded by an energy barrier, probably a combination of surface charge and hydration. The system would be expected to be metastable but to exhibit stability so long as the rate of exchange of subunits through the solution phase is slow compared to the duration of an experiment.‡ Lowering the energy barrier to close approach or artificially bringing particles into surface contact should lead to irreversible aggregation, and thus metastability should be revealed. It is noted that if colloid systems are metastable, they are not at equilibrium and thus will not form spontaneously.

Systems of particular interest are those formed from small polar organic molecules, such as soaps, detergents and a variety of important molecules of biological origin, such as the bile salts as studied by Borgstrom (1965) and Laurent and Persson (1965). Micelles can form from a variety of mixtures of molecules. In these systems micelles form spontaneously, and the systems are in equilibrium. Interaction is driven by the decrease in

---

* *Editor's note:* The colloid chemist, McBain, did not use the term micelle originally in connection with soap micelles. In a discussion at the Faraday Society Meeting in 1913 on colloids McBain remarked, "Now take some of these highly charged colloidal aggregates, micelles, or 'colloidal ions' we are discussing." In a later paper (McBain and Salmon, 1920) he pointed out that he had introduced the concept of "micelle" at the 1913 discussion "to remove one of the chief difficulties in interpreting the properties of acid and alkali albumin."

† Several of these categories are discussed briefly by Waugh and Noble (1965).

‡ The micellar system is discussed in Chapter 18.

free energy resulting from hydrophobic bonds, which are formed by an association of nonpolar portions of molecules, and it is opposed by electrostatic interactions of charged groups, when these are present. Micelles are most often spherical but may be rod shaped (Debye and Anacker, 1951) or the subunits may be arranged in sheets. The structure and formation of soaplike micelles is reviewed by Shinoda et al. (1963). With respect to the various shapes of the soaplike micelles it is observed that the diameter of the sphere, the diameter of the rod or the thickness of the sheet is about twice the largest dimension of the constituent subunits. The significance of double-layer structures is that all subunits (which may be of more than one type and/or consist of more than one type of molecule) are at the surface and, on the average, can be exchanged without changing immediate environment; essentially none are entirely surrounded by others. Micelles of this type have been treated as equilibrium single-phase systems by Shinoda and Hutchinson (1962).

There exists a variety of colloidal association products whose structures suggest that they be classified as chemical compounds. To satisfy this specification rigorously, each kind of colloidal particle must contain the same definite types, and numbers of each type, of interacting molecules. To accommodate a description of experimental observation, it may be found convenient to group particular combinations of interacting molecules into subunits and to consider these as dimers, trimers, etc. Examples of association products which can be classified as chemical compounds are hemoglobin (Cullis et al., 1962), a number of complex enzyme systems (Monod et al., 1965; Koshland et al., 1966), hemocyanin (van Bruggen et al., 1962) and tobacco mosaic virus and bacteriophage (Caspar and Klug, 1962). In these systems the increase in particle size which accompanies molecular or subunit association appears to be limited by the formation of a single kind of association product in which specificity of interaction plays the important role of dictating a closed geometry. On closure the complementary interacting surfaces of constituent monomers are fully satisfied, and the association product presents, externally, a surface which has no capacity for additional interaction. Another mechanism for introducing stability and limiting association is that which apparently determines the structure of tobacco mosaic virus: One of the subunits (RNA) is a single molecule, and although protein subunits may interact in the absence of RNA to give helical association products, in the complete system association is promoted to the extent that the single RNA molecule can interact with protein. It should be noted that the tobacco mosaic rodlet bears a resemblance to a soap micelle in that the radius of the rodlet is determined by interaction characteristics and the length of a single subunit.

Many colloidal systems are known in which particle composition changes in going from the surface to the center. The most common of these are dispersions of liquid in liquid, or solid in liquid. The dispersed phase has a low solubility in the dispersion medium, thus slowing monomer exchange between particles through the aqueous phase, and the surface of the dispersed phase adsorbs molecules of appropriate structure. Adsorption is positive and the adsorbed layer lowers interfacial tension and introduces an energy barrier to close approach. Examples are emulsions of fat in water stabilized by the adsorption of phospholipid and/or protein (see Shulman, 1957). Systems of this type usually appear to be metastable: They will not form spontaneously; i.e., homogenization is required to decrease the particle size of the disperse phase, and in many instances reducing the relative amount of dispersion medium by centrifugation or drying leads to irreversible aggregation. In some systems, however, the coat layer is so tightly bound that the disperse phase may occupy the major fraction of the total volume. These systems appear to be essentially dry.

There are some interesting situations in which emulsions form spontaneously. Different surface-active compounds are introduced individually into the oil phase and the aqueous phase. On contact of the two phases, mutual adsorption at the interface may be sufficiently energetic to overcome interfacial tension, which may be as low as 1 erg/cm^2, and the oil phase spontaneously fragments to give a size distribution of droplets of the order of microns in diameter (see Shulman, 1957). Such spontaneous formation of droplets does not mean, of course, that for these systems the dispersed state is the equilibrium state.

The possibility that a colloidal state may be the equilibrium state is of great interest from the standpoint of structural development in living systems. Clearly no particular problems are expected if the colloid has the structure of a soaplike micelle or a large chemical compound. The situation is otherwise when the colloid particle is of nonuniform composition. Consider an emulsion as the equilibrium state. An approach to the equilibrium state can be attempted from several directions. For example, in principle it should be possible to develop a colloid by allowing the emulsifier in solution to act on a large droplet of pure disperse phase; a progressive, spontaneous decrease in the size of preexisting emulsion particles should be observed as the ambient concentration of surface-active agent is increased; a preexisting emulsion might be able to emulsify a droplet of disperse phase, etc. Whether such transformations will be observed will depend on the molecular mechanics, particularly the energy barriers, associated with the reactions by which an increased surface area is to be generated.

In the casein micelle system, as will be seen, the micelle state may be the lowest free-energy state of the system. Of particular interest will be micelle structure and the mechanisms which operate in determining micelle size.

III. The Natural Casein Micelle

The standard against which altered or reconstituted systems are to be compared is the micelle system of fresh skim milk. One might then expect that the natural system has been studied in great detail. The contrary is the case; to capture the micelle before the conditions of the experiment have altered it in structure has proven to be a difficult task, and the information which follows should be taken on the understanding that a particular kind of artifact may be introduced in each approach.

In gross aspect, natural micelles appear to be nearly spherical, to have a broad size distribution and to be highly solvated. That micelles could not be particularly anisometric has been appreciated for some time. The viscosity of skim milk, according to Whitaker et al. (1927), Whitnah and Rutz (1959) and Eilers et al. (1947), is low and approximately 1.04 cp* at 37°C; thus it is only slightly higher than that of skim milk serum, which can be taken as 0.87 cp. If these values are inserted into Einstein's equation for the viscosity of a population of spheres relative to the viscosity of the suspensions medium

$$(\eta_{\text{rel}} - 1) = 2.5\Phi \tag{1}$$

the volume fraction, Φ, of micelles is predicted to be 0.06. On the basis of the average casein content of skim milk, it is then required that 1 g of micellar casein occupy a volume of about 2.7 ml, thus, that the solvation be nearly 2 g solvent/g protein. The changes in viscosity with micellar protein content observed by Whitnah and Rutz (1959) yield a similar value. A common observation after ultracentrifugation of natural micelles prior to the preparation of the caseins (von Hippel and Waugh, 1955; Waugh et al., 1962) is that the pellet volume is several times greater than the volume of its protein content. Using the procedure of Waugh and Noble (1965), it is observed that pellets obtained from skim milk centrifuged for 35 min at 67,000 g and 37°C have a solvation of 1.9 g water/g protein. None of these data is sufficiently accurate to reveal small deviations from sphericity, as can be appreciated from an examination of Simha's equation (Cohn and Edsall, 1943) relating the viscosity increment to the asymmetry of kinetic units.

It is not surprising that when electron micrographs of micelles were first made, they appeared to have a natural spherical shape (Nitschmann, 1949; Hostettler and Imhof, 1951). These observations have been subsequently confirmed in many laboratories. Nitschmann (1949) found an important characteristic of micelle systems; namely, that the natural micelles have a size distribution ranging from 21% with a diameter of approximately 400 Å for the smallest, to ~2% with a diameter of 2800 Å for the largest, the

---

* cp, Centipoise.

average being about 1200 Å. Although the volume spread, a factor of 350, is impressive, an initial critical comparison is appreciated to be the ratios of micelle diameters to the dimensions of the casein monomers, particularly the longest dimension of the latter. A critical comparison will reveal that some monomers must be entirely surrounded by others, a circumstance which has a direct bearing on the choice of model. This will be discussed below.

Shimmin and Hill (1964, 1965) have examined fresh skim milk fixed in the presence of osmium tetroxide. Micelles were either deposited on a carbon film and shadowed with platinum–palladium or fixed by vapor-phase perfusion of osmium tetroxide, sedimented by ultracentrifugation, dispersed in distilled water, stabilized by propylene oxide, dehydrated in an ascending sequence of ethanol concentrations, embedded in plastic and cut into sections about 100 Å thick. Although the possibility of artifact production clearly exists, two important aspects of micelle structure are offered for consideration. The first of these is that, internally, the osmium stain accumulates in regions about 90 Å in diameter, with an irregular distance between stain centers of 140–150 Å. The second is that the surface of the micelle appears to have local crenations of the order of magnitude of 150 Å in diameter; thus, it appears not to be smooth. The authors suggest that the micelle is built up of subunits that appear to be spherical and about 80–100 Å in diameter. Rose and Colvin (1966a, b), using a more difficult replication technique, have obtained electron micrographs which they interpret as confirming the observations of Shimmin and Hill.* Rose and Colvin are concerned with the possibility that the electron-dense regions are due to granules of calcium phosphate. This is considered unlikely in view of the fact that small volumes of skim milk are either suspended in large volumes of osmium tetroxide solution or washed with distilled water.

It appears safe to assume that the electron micrographs reveal asymmetries in the distribution of protein material, internally and at the surface, whose dimensions are comparable to those of the casein monomers. An electron micrograph of casein micelles is shown in Fig. 1.

Adachi (1963) has reconstituted micelles from casein, prepared according to the method of Warner (1944), by adding calcium and other divalent cations. He obtained typical micelles, including the appearance of a size distribution. We have obtained the same result on many occasions.

One of the striking properties of natural casein micelle systems is their extraordinary stability; indeed, were it not for the occurrence of proteolysis in milk one suspects that micelles would survive indefinitely at 37°C. Certainly stability has limits with respect to environmental conditions—acidification produces the well-known precipitation of caseins near pH 4.7, micelles can be destabilized by the addition of salt, the caseins can be salted

---

* Observation of spherical substructure is now frequent.

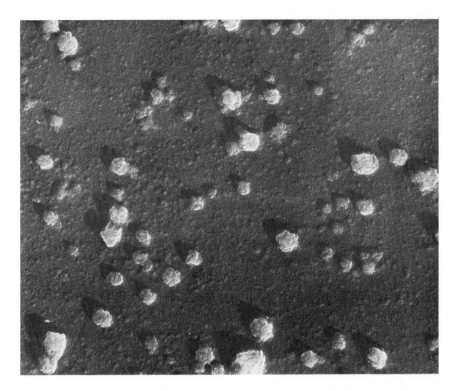

FIGURE 1. Electron micrograph of the micelles of fresh milk fixed in 3.64% glutaraldehyde solution for 15–30 min. Shadowed with platinum–palladium. Average micelle diameter 1350 Å. (Courtesy of Dr. Robert J. Carroll, Eastern Utilization Research and Development Division, U. S. Department of Agriculture, Philadelphia, Pennsylvania.)

out at pH 7, the addition of sufficient divalent cation can lead to the deposition of a precipitate enriched in Ca α_s- and β-caseinates and so on. The point is the extent to which the environment can be changed without ensuing micelle degeneration. Three examples, of many, will suffice. The first two are so familiar as sometimes to be overlooked. They are that milk may be cooled to 0°C or boiled for considerable periods without obvious micelle degeneration and that micelle systems can be appropriately dried and reconstituted simply by the addition of water (see Chapter 17). Stability to close approach in solution is the third example. This has been a particularly useful test for stability as developed by Waugh and Noble (1965). Skim milk is centrifuged at 37°C and 97,000 g for 30 min. Pellets are placed in their supernatants under conditions where the supernatants are either slowly stirred or held quiescent. During 8 hr of slow stirring the pellet essentially disappears and a micelle system is established. When the pellet or its grainy lowest layer is suspended from a platinum hook in the quiescent supernatant, a micelle suspension streams off the surface and, because of its increased density, accumulates transiently in the bottom of the vessel. Similar results are obtained with skim milk made 0.07 M in added calcium chloride. Natural micelle surfaces thus show little or no tendency to interact.

The surface charge of the micelle is not zero. Whitney (1961) has shown that micelles move in an electric field; although it is not stated in his publication, they move toward the positive electrode.

Choate et al. (1959) have differentially centrifuged micelles from milk. They find constant proportions of α- and β-caseins, independent of micelle size. Rose (1965), using starch-gel electrophoretic analysis in the presence of mercaptoethanol or cysteine by the method of Neelin (1964), reports that the larger micelles contain a higher relative amount of β-casein. Sullivan et al. (1959) discovered the important fact that the relative κ-casein content of small micelles is greater than that of large micelles. They obtained samples by differential centrifugation of skim milk and found κ-casein content to be inversely proportional to the logarithm of micelle diameter. McGann and Pyne (1960) and Rose (1965) have also shown that the smaller micelles contain larger relative amounts of κ-casein. Certain other aspects of natural micelles—cooperative protein–ion interactions, the distribution of protein between micellar and solution forms, hysteresis effects and equilibrium aspects—are more conveniently considered in other sections.

IV. Properties of Monomer Caseins

Since fractionation and the properties of the casein monomers are dealt with more extensively in Chapters 10, 11, and 12, only those aspects of importance to a consideration of micelle formation and structure will be outlined here.

There appear to be many casein components; the number has been set as high as 20 by Wake and Baldwin (1961). Similar results have been obtained by Neelin et al. (1962). How many of these casein components are produced by the synthetic mechanisms of the cell (primary caseins) and how many are derived as a result of subsequent proteolysis remains unknown. A formulation of the problem can be made by noting that κ-casein, which constitutes only about 15% of the total casein, plays an obviously important role in micelle formation and structure. Other components are also present to the extent of a few percent. Do they play an important role in micelle formation? This will be considered in Section IX.

It is clear that all casein components *may* contribute to the structure and properties of natural micelles. However, in what follows, we shall assume that most of the important micelle characteristics can be obtained by examining appropriate combinations of α_s-, β-, and κ-caseins.

Fractionation, in particular, has been complicated not only by the number of possibly important components, but by the fact that some, possibly all at 37°C, interact strongly with each other even in the absence of calcium. Thus, at pH values near 7.0 and ionic strength values near 0.1, a wide variety of interaction products have been observed for α_s-, β-, and κ-caseins, alone or in all possible combinations. Only β-casein appears to yield monomers at a low temperature (Sullivan *et al.*, 1955). Evidence of extensive interaction in mixtures is provided by the work of Krecji *et al.* (1941, 1942); Warner (1944); Nitschmann and Zurcher (1950); Slatter and van Winkle (1952); Halwer (1954); Sullivan *et al.* (1955); von Hippel and Waugh (1955); Waugh and von Hippel (1956); McKenzie and Wake (1959a); Payens (1961); Driezen *et al.* (1962); Cheesman (1962); Swaisgood and Brunner (1962); Payens and van Markwijk (1963); Pepper and Thompson (1963); Swaisgood *et al.* (1964); Garnier *et al.* (1964); Kresheck *et al.* (1964); Dresdner (1965); Ashworth (1964); Payens and Schmidt (1965); Noble and Waugh (1965); Garnier (1966); and Payens and Schmidt (1966).

Such ubiquitous interaction would have made fractionation difficult indeed without the introduction of dissociating agents. The effectiveness of dissociating agents was apparent in the early work of Burk and Greenberg (1930) (confirmed by Nielsen and Lillevik, 1957) on casein fractions and components. Burk and Greenberg found that in 6.6 M urea, in the absence of calcium, the number-average molecular weight (M_n) of whole casein components is near 30,000. Subsequent work has revealed that κ-casein can polymerize to give covalently bonded structures having molecular weights far in excess of this value. The fact is that the dissociating action of urea has been made an important part of several fractionation procedures, for example, the differential precipitation procedures of Hipp *et al.* (1952) and Waugh *et al.* (1962) to permit fractionation on DEAE-cellulose columns. Urea has become a standard additive in most fractionation procedures for α_s- and β-caseins but not in the effective procedure of McKenzie and Wake (1961) for preparing κ-casein.

Many of the studies of interaction in the absence of divalent cation have intrinsic interest. However, their relationship to micelle formation is obscure for, as indicated by Noble and Waugh (1965), the addition of calcium (or other divalent cation) initiates a particular interaction sequence. Discussion of interaction in the absence of calcium will not be made, except as such interaction appears to bridge calcium-free and calcium-containing systems.

Properties of the monomers of α_s-, β-, and κ-caseins are summarized in Table I. Although there is general agreement as to the properties of α_s- and β-caseins, in several respects κ-casein presents difficulties and these must receive special attention. Also, not recognized completely in Table I are the genetic variants of the caseins. These are discussed by Thompson in Chapter 11 and by Mackinlay and Wake in Chapter 12.

A. α_s-CASEIN

The molecular weight of α_s-casein obtained by physical measurements is 27,300 daltons (Table I); there are nine phosphorus atoms per molecule; it has C-terminal Trp and N-terminal Arg; and it is free of carbohydrate and disulfide or sulfhydryl. There is a high fractional content of large nonpolar side chains of 0.34 (Val, Leu, Ile, Phe, Trp, Pro) and the protein is insoluble in the presence of sufficient calcium to produce micelle formation ($\sim 0.02\ M$) over a wide temperature range at pH 7. The occurrence frequency for proline is 0.08.

One of the most important developments with respect to structure is the demonstration of a limited distribution of phosphate groups. The most interesting results have been gained through peptic or tryptic hydrolysis of caseins, followed by fractionation and examination of the resulting phosphopeptides. The presence of phosphorus has made such peptides relatively easy to isolate. This fact and a long-standing interest in the effects of rennet and rennin and other enzymatic actions on casein have made the list of peptide examinations quite extensive. These start with early isolations of a peptide or peptides from whole casein, for example, by Damodaran and Ramachandran (1940), Rimington (1941) and Lowndes et al. (1941). Many additional isolations from whole casein have taken place more recently. Isolation of peptides from α_s-casein and β-casein have been carried out, for example, by Groves et al. (1958), Pantlitschko and Grundig (1958) and Bennich et al. (1959). Phosphopeptides from α_s-casein have been isolated and studied by Österberg (1961, 1964, 1966) and from β-casein by Peterson et al. (1958) and Peterson (1969).

The important observation for α_s-casein is that most of the phosphate groups are present in a short nonterminal segment of 35 amino acid residues, which includes 7 phosphate, 11 carboxylate and 2 amino groups. This segment contains only 7 nonpolar side-chain groups. A partial structural for-

TABLE I

PROPERTIES OF BOVINE CASEINS

Properties	α_s-Casein	Reference	β-Casein	Reference	κ-Casein	Reference
Molecular weight (daltons)	27,300	Dreizen et al. (1962); Swaisgood and Brunner (1963)	24,200	Gordon et al. (1949)	50,000	Beeby (1963)
	31,000	Kalan et al. (1964); Manson (1961)	19,800	McKenzie and Wake (1959b)	26,000	McKenzie and Wake (1959b)
			17,300	McKenzie and Wake (1959b)	24,000	Swaisgood and Brunner (1962)
			25,000	Payens and van Markwijk (1963)	16,000	Waugh (1958)
	24,300	McKenzie and Wake (1959b)	24,100	Sullivan et al. (1955)	56,000–135,000	Swaisgood et al. (1964)
	25,500	McKenzie and Wake (1959b)			19,000[a]	Woychick et al. (1966)
	26,700	Noelken (1967)			17,000–23,500[a]	Pujolle et al. (1966)
	16,500	Schmidt and Payens (1963)			18,000–20,000[a]	Swaisgood and Brunner (1963)
	27,600	McKenzie and Wake (1959b)				
	23,600[b]	Grosclaude et al. (1970)				
Solubility in 0.03 M CaCl$_2$, pH 7 (g/liter)						
4°C	0.17	Noble and Waugh (1965); Waugh (1958)	Soluble	Waugh (1958)	Soluble	Waugh (1958)
37°C	0.04	Noble and Waugh (1965); Waugh (1958)	0.2	Waugh (1958)	Soluble	Waugh (1958)
Phosphorus content (g/100 g)	1.01	Pepper and Thompson (1963); Thompson and Kiddy (1964); Thompson and Pepper (1964a); Zittle and Custer (1963)	0.61	Gordon et al. (1949); Warner (1944)	0.35	Hipp et al. (1961b); Swaisgood and Brunner (1962)
	1.12	Schmidt and Payens (1963)	0.55	Hipp et al. (1950, 1952)	0.17	Mackinlay and Wake (1965)
	1.10	Waugh (1958)	0.64	Hipp et al. (1952)	0.22	Thompson and Pepper (1962); Swaisgood et al. (1964); Jollès et al. (1962)
	1.03	Waugh et al. (1962)	0.58 (variant A)	Pion et al. (1965)	0.19	Schmidt et al. (1966); Waugh (1958)
			0.59 (variant A)	Thompson and Pepper (1964a)		

Property	Value	Reference		Reference
Carbohydrate	0.56 (variant B)	Pion et al. (1965)	0.3	Zittle and Custer (1963)
	0.57 (variant B)	Thompson and Pepper (1964a)	0.19 (variant A)	Schmidt et al. (1966)
	0.50 (variant C)	Thompson and Pepper (1964a)	0.14 (variant B)	Schmidt et al. (1966)
	0.45 (variant C)	Dresdner (1965)		
	0.55 (variant C)	Pion et al. (1965)		
	0.57 (variant AC)	Dresdner (1965)		
	0	Dresdner (1965); Waugh (1961)	N-Acetylneuraminic acid, galactosamine, galactose	
	0	Waugh et al. (1962)		
N-Terminal residues	Arg.	Kalan et al. (1964); Mellon Manson (1961); Mellon et al. (1953); Schmidt and Payens (1963); Thompson and Kiddy (1964)	Gln(?)	Jollès (1966)
	Arg	Kalan et al. (1965)		Jollès et al. (1969)
	Arg, Lys	Mellon et al. (1953)		
C-Terminal residues	.Leu, Trp	Kalan et al. (1964); Waugh et al. (1962)	Ile$_2$, Val	Dresdner (1965); Kalan et al. (1965)
	.Trp	Thompson and Kiddy (1964)	Ala, Val	de Koning et al. (1966); Pujolle et al. (1966)
			Ser, Thr, Ala	Jollès (1966)
Cystine content as cysteine (mole/10⁵ g)	0	de Koning and van Rooijen (1965); Gordon et al. (1965); Hipp et al. (1961a); Ho and Waugh (1965a); Waugh et al. (1962)	8.9	de Koning et al. (1966)
	0	Dresdner (1965); Gordon et al. (1949); Pion et al. (1965)	6.9	Hipp et al. (1961a, b)
			11.7	Jollès et al. (1962)
			5.4	Schmidt et al. (1966)
			3.3	Swaisgood and Brunner (1963)
			10	Woychick et al. (1966)
			5–8.4	Hill and Hansen (1963)
			7.1–7.7	Swaisgood et al. (1964)
Absorbancy ($A_{1\,cm}^{1\%}$ at 276–280 nm)	10.1	Thompson and Kiddy (1964); Waugh et al. (1962)	10.5	Garnier (1963)
	10.2	Zittle and Custer (1963)	9.2	Noble and Waugh (1965)
			12.2	Zittle and Custer (1963)
	4.6	Dresdner (1965); Thompson and Pepper (1964b)		
	4.7	Hipp et al. (1950)		
	4.8	Pion et al. (1965)		

^{a} Reduced κ-casein.

^{b} This value is considered to be the correct one and is based on peptide studies.

mula for the phosphopeptide is shown in Structure I. The arrows in the structure indicate major points of peptide hydrolysis.

$$\text{Asp—(SerP, Ile)—(Asp, ThrP, Gly, Glu}_2\text{)—(SerP, Glu)} \overset{P1}{\underset{\downarrow}{}} \text{Glu—}$$

$$\overset{P2}{\text{—Ala—SerP—SerP—(SerP}_2\text{, Ile}_2\text{, Glu}_2\text{, Asp, Val}_2\text{, Lys, Pro)—}}$$

$$\text{—GluNH}_2\overset{P3}{\underset{\downarrow}{}}\text{GluNH}_2\text{—(Ala, Met)}\overset{P4}{\underset{\downarrow}{}}\text{Asp—Glu—(GluNH}_2\text{, Met)—Ile—Lys}$$

STRUCTURE I

Perlmann (1955), on the basis of phosphatase action, suggested that both mono- and diester forms of phosphorus are present. This possibility was raised again by Ho and Kurland (1966) on the basis of nuclear magnetic resonance studies. The conclusion that monoesters alone are present is drawn from the titration data of Österberg (1961); the composition studies of Hofman (1958), Kalan and Telka (1959) and Anderson and Kelley (1959); the nuclear magnetic resonance studies of Ho et al. (1969); and binding studies discussed in Section VI.

Mercier et al. (1970) have determined the sequence of the 48 residues at the COOH end of the polypeptide chain of the B variant. The partial sequence of 36 residues at the NH_2 terminal is given by Thompson in Chapter 11. Grosclaude et al. (1970) have confirmed that 8 of the 9 phosphate groups are in an acidic peptide 36 residues long and located between the terminal peptides.

B. β-CASEIN

The molecular weight of β-casein is 24,100 daltons; there are 4–5 phosphorus atoms per molecule; it has C-terminal Val and N-terminal Arg; and it is free of carbohydrate and disulfide or sulfhydryl. The nonpolar sidechain frequency from amino acid composition is 0.45, and the occurrence frequency for proline is 0.18.

The work of Peterson et al. (1958) has shown that all of the phosphate groups of β-casein are in a terminal segment (Peterson, 1969) of 24 amino acids of molecular weight ~3100 and that this segment is highly acidic since it also contains five free aspartic and glutamic acid carboxyl groups but only two (terminal) arginines. The peptide contains only eight nonpolar side chains. Manson and Annan (1970) have determined the sequence of the first 25 residues of β-casein A-1. Four of the five phosphate groups of this variant are in monoester linkages on adjacent serine residues and are located in the hydrophilic region near the NH_2 terminal end.

One of the extraordinary characteristics of Ca β-caseinate is the dependence of its solubility on temperature. The temperature effect is dependent on protein concentration, ionic strength and concentration of calcium. The general result is illustrated by examining a system containing 0.5% β-casein, 0.015 M calcium at I = 0.05 (sodium chloride plus 0.002 M cacodylate at pH 6.6). At 26°C the protein is in solution and at 27°C it is precipitated; precipitation and resolution are reversible. As the calcium concentration increases, the precipitation temperature decreases; at 0.03 M calcium it is near 20°C. The temperature characteristics of the solubility of β-casein in the presence of calcium have been accepted by the Committee on Nomenclature as part of the characterization of β-casein (see Thompson et al., 1965 and Chapter 2, Volume I).

The temperature dependence of β-casein solubility in the presence of calcium may well be related to a temperature-dependent conformational change observed in the absence of calcium. By difference spectrophotometry and changes in optical rotation, Garnier (1966) has observed a conformational change between 5° and 40°C. The transition is endothermic, ΔH = 30 ± 3 kcal/mole, and at the half-transition temperature of 23°–24°C, ΔS = 100 ± 10 eu. As the temperature is increased, the absorption bands of Trp, Tyr, and Phe are shifted toward longer wavelengths and show hyperchromicity. Other optical changes are described. Garnier interprets the change in conformation, with increasing temperature, as a decrease in content of poly-L-proline left-handed helix and an increase in α helical conformation, the proportion of aperiodic structure remaining constant.

In the sections which follow, several situations will be discussed in which the presence of β-casein, through its interactions with α_s- and κ-caseins, modifies the system so as to introduce important temperature-dependent effects.

C. κ-Casein

The characteristics of κ-casein appear to be more complex than those of α_s- or β-casein. The problems to be resolved have two sources: (a) As shown by Swaisgood and Brunner (1963) and Swaisgood et al. (1964), a native κ-casein preparation appears to consist of disulfide-linked (covalent) polymers in a distribution of sizes. (b) The constituent monomers are of several different compositions.

After it was reported by Waugh et al. (1960) that κ-casein is the only major primary casein to contain disulfide, it was shown by Waugh (1962) that reduction of disulfide gives rise on starch-gel electrophoresis to several bands, four of which are prominent. Later Beeby (1964) suggested that κ-casein contains masked sulfhydryl groups, but this has not been confirmed. The effects of reduction were also observed by Schmidt (1964), Mackinlay and Wake (1964), Woychik (1964), and Neelin (1964), and since then, components of reduced κ-casein have been isolated and studied extensively. The average molecular weight of reduced κ-casein can be taken as about 21,000 daltons (Table I).

There are important differences between the components which appear on κ-casein reduction. Stabilized components have been isolated by Mackinlay and Wake (1965), Woychik et al. (1966), Schmidt et al. (1966) and Pujolle et al. (1966). As the mobility of the κ-casein component increases, its carbohydrate content increases (Mackinlay and Wake, 1965; Woychik et al., 1966; Schmidt et al., 1966). The component of lowest mobility, apparently the major component, contains a negligible amount of carbohydrate. The component of highest mobility may contain as much as 10% carbohydrate, including N-acetylneuraminic acid (Mackinlay and Wake, 1965; Schmidt et al., 1966). Apparently, the components have negligible differences in cystine and phosphorus. Rennin splits κ-casein into an insoluble para-κ-casein (Waugh, 1958) and soluble peptides which retain the C-terminal Val (Jollès et al., 1962). The peptides are discussed by Jollès (1966) and by Mackinlay and Wake in Chapter 12. Whole reduced casein has been considered by some workers to consist of at least two components (see, for example, Mackinlay et al., 1966; de Konig et al., 1966; Hill and Wake, 1969b). However, strong evidence against there being more than one para-κ-casein component has been presented by Kim et al. (1969).

Each κ-casein component appears to give rise on rennin treatment to soluble peptides having the same amino acid composition (Mackinlay and Wake, 1965; de Koning et al., 1966). The differences between peptides, then, are due to differences in carbohydrate content. The average weight of the glycomacropeptide may be taken as 7500 daltons from ultracentrifuge-diffusion measurements of 6000–8000 by Nitschmann et al. (1957) and as 8000 daltons from amino acid composition according to Jollès et al. (1961). The phosphorus content of κ-casein is 0.22%. If all phosphorus were in the macropeptides, peptide content should be about 0.7% on the basis that the average molecular weight of reduced κ-casein is 21,000 daltons. Since the phosphorus content of the glycomacropeptides averages 0.4% (Alais and Jollès, 1961), it is apparent that phosphorus is probably distributed between para-κ-casein and macropeptides. Since the average reduced κ-casein monomer contains 1.5 phosphorus atoms per molecule, there is the possiblity that κ-casein monomers can be phosphorylated differently with respect to their para-κ-casein and macropeptide portions.

According to Beeby and Nitschmann (1963) and Beeby (1965), products similar to the glycomacropeptide and para-κ-casein can be produced by actions other than rennin, such as acid precipitation or urea treatment. Jollès (1969) has shown that the glycopeptide has a glycosidic linkage between threonine and galactosamine [contrast the linkage in α_1-acid glycoproteins (Kamiyama and Schmid, 1962)].

With the exception of Beeby (1964), it is reported that κ-casein obtained from milk does not contain SH groups (Waugh et al., 1960; Jollès et al., 1962; Mackinlay and Wake, 1964; Swaisgood et al., 1964; Nakai et al., 1965; Woychik et al., 1966). It is also noted that an SH oxidizing enzyme has been found in milk (Kiermeier and Petz, 1967) and that normal milk serum appears to be free of small-molecule SH compounds (Hutton and Patton, 1952; Zweig and Block, 1953) which might promote SS reduction or exchange. It seems likely that there is little SH-κ-casein in milk and that the κ-casein distribution in normal milk is stable. In this respect, then, disulfide exchange must account for the heat-induced interaction of κ-casein and β-lactoglobulin observed by Zittle et al. (1962) and Long et al. (1963).

The present knowledge of the structure of κ-casein is summarized in Chapter 12 (see especially Fig. 10). It should be noted that the nonpolar side-chain frequency is 0.35, and the occurrence frequency for proline is 0.10.

D. GENERAL REMARKS

The abundances of the caseins are difficult to estimate. The problems are the large number of casein components and the lack of a sufficiently extensive set of distinctive characteristics. This is particularly the case for the minor caseins. Fractionation has so far not been developed into a quantitative tool. Waugh and von Hippel (1956) estimate the abundance of κ-casein as 15% by weight, as do Sullivan et al. (1959). An abundance of 11–26% is given by Marier et al. (1963), and yields from whole casein of 7–12% have been obtained by Zittle and Custer (1963). We may take κ-casein as 15%. Waugh et al. (1962) estimate the abundance of α_s-casein by C-terminal Trp analysis as 47% of the total absorbancy of first-cycle casein. Taking this information and the absorbance indices of α_s-, β- and κ-caseins, we estimate the abundances by weight to be 40% α_s-, 35% β-, 15% κ- and 10% minor casein components. McKenzie and Wake (1959a) estimate α_s-casein plus κ-casein as 77%. This value, based on moving-boundary electrophoresis, is approximate. The abundances given by Rose (1965) are in reasonable agreement with those given above.

With respect to the determination of absorbance indices, attention is drawn to the necessity of correcting apparent absorption at 280 nm for scattering. This is usually carried out by determining apparent absorbance at some wavelength (usually 320 nm) where intrinsic absorption is negligible. Scattering at 280 nm is calculated on the assumption that scattering increases with the inverse fourth power of the wavelength. A careful study of β-casein has shown that this is a necessary procedure (Dresdner, 1965). Scattering may well account for the spread in absorbance indices observed for κ-casein.

All of the caseins contain unusually large amounts of proline; the frequencies are $\sim$0.08 for α_s-casein, $\sim$0.18 for β-casein and $\sim$0.1 for κ-casein. Cohen and Szent-Györgyi (1957), Herskovits and Mescanti (1965), Kresheck (1965) and Herskovits (1966) have shown that the caseins have low optical rotatory dispersions. It has been inferred from these observations that the caseins are random coils. Interpreting these observations in the usual way, as did these authors, one would certainly conclude that the caseins contain only small amounts of α-helix. However, it is desirable to point out that a random coil structure does not require that the peptide chain be extended and all parts of it readily available to the solvent. As will be discussed in more detail in Section VI, a consideration of the high nonpolar side-chain frequencies and the particular distributions of ionic groups for α_s- and β-caseins suggests that most of the monomeric structure of each is a compact body and that the highly acidic portions are external to this body and occupy special positions.

Urea is frequently used as a dissociating agent for preparation and studies on the caseins (see Chapters 10, 11, 12). Commercial urea, even the reagent grade, contains heavy-metal ions and cyanate as contaminants. The cyanate can produce chemical alterations as shown by Stark *et al.* (1960). A procedure for removing both cyanate and heavy-metal ions is given by Waugh *et al.* (1962). Manson (1962) has demonstrated the effects of cyanate on the caseins (see also Chapter 10).

Some time ago, Burnett and Kennedy (1954) showed that rat liver mitochondria could phosphorylate serine in α- and β-caseins, or at least mediate a transfer of phosphate between ATP and phosphoserine, since alkali-dephosphorylated casein was not phosphorylated. Free serine was not phosphorylated, and the authors suggested that the conformation of the intact protein is required. Since caseins were found to be unique in phosphorylation, the authors further suggested that the groups to be phosphorylated sterically are more available in the caseins than in other proteins. According to Turkington and Topper (1966), explants of mouse mammary glands produce a pool of casein which, to a major extent, must be phosphorylated after synthesis.

V. The Structure of Casein Micelles

A. MICELLES FROM MIXTURES OF α_s- AND κ-CASEINS

The α_s-κ-casein system is an attractive system to be used in initiating a discussion of micelle structure. In appropriate mixtures the two proteins can form micelles which resemble natural micelles in several important respects, and temperature effects on the individual monomers are small. At any calcium concentration consistent with micelle formation, calcium κ-caseinate is soluble and calcium α_s-caseinate is essentially insoluble over the temperature range 0°–37°C. When research on the α_s-κ-casein system was initiated by Noble and Waugh (1965), previous work had suggested the existence, in mixtures of α_s-, β- and κ-caseins and in the absence of divalent cation, of a preferential combination of α_s- and κ-caseins to form a stoichiometric complex. The inference (Waugh, 1958) was that complex formation is the first and most important interaction leading to micelle formation. A second assumption, drawn from observations that calcium caseinate precipitates transformed, at times, into micelle suspensions (to be discussed), was that the natural micelle system and a variety of reconstituted systems are equilibrium systems. The general assumption of stoichiometry in the micelle, as will be shown in Section VIII, is untenable. The assumption of equilibrium requires an understanding of the effects of the environment. Only under particular conditions is evidence of equilibrium obtained.

The studies of Noble and Waugh (1965) and Waugh and Noble (1965) were designed to arrive at critical properties of micelle systems and, from these, to derive a model. Studies of the α_s-κ system now merit critical examination, since the model arrived at can be applied in fundamental aspect to all casein micelle systems with which this author is familiar. However, the specification of some important details of the model will depend on environmental conditions such as pH, ionic strength, ion composition and temperature.

A study of micelle formation with α_s-κ-casein mixtures was made at 37°C, pH 7.1 and $I = 0.075$ (0.07 M NaCl and 0.01 M imidazole); see p. 25. Once a set of weight ratios of α_s-κ-caseins = R_i had been selected, calcium addition was carried out at a constant pH of 7.1 by two different procedures: by single-aliquot addition using a rapid-mixing technique and by incremental addition in which the same amount of calcium was added but in a series of small aliquots (to give 0.002 M increments) with time elapsing between each addition. After 75 min of equilibration, aliquots were centrifuged for 1 min at 1200 g, and the concentration of supernatant protein was then determined. The supernatant contains soluble protein and micelles small enough to remain suspended during application of the assay field; the pellet contains calcium α_s-caseinate precipitate and/or large micelles. When calcium α_s-caseinate (or calcium β-caseinate) alone is present, particles spontaneously coalesce, and on assay centrifugation form a pellet of bulk phase. In contrast, pellets obtained by centrifuging micelle systems are particulate. They redisperse spontaneously when gently agitated or when suspended in their supernatants. To make a significant contribution to the pellet, a micelle size must have a significant abundance, and a significant fraction must be removed by assay centrifugation. It is estimated that assay centrifugation removes all micelles having radii greater than 545 nm and 0.65 of micelles having radii of 440 nm. Micelles having radii of 150 nm are not removed significantly.

Estimates are made using the centrifuge equation

$$\phi_i = \frac{x_m}{x_b - x_m} [\exp(\omega^2 s_i t) - 1] \qquad (2)$$

where ϕ_i is the fraction of the ith size micelle removed in time t, x_m and x_b are radii to the meniscus and base of the cell, and s_i is the micelle or particle sedimentation coefficient.

The relation between micelle radius and sedimentation coefficient is given by

$$s_i = \frac{2\rho_p (1 - \bar{v}\rho_s)}{9\eta r_i} \left[\frac{(r_i - \Delta)^3}{1 + Q_\alpha} + \frac{r_i^3 - (r_i - \Delta)^3}{1 + Q_\kappa} \right] \qquad (3)$$

and the ratio of the total micelle population surface area to the total κ-casein is given by

$$\frac{\text{total surface area}}{\text{total } \kappa\text{-casein}} = \frac{3(1 + R)}{\rho_p}$$

$$\times \sum_{i=1}^{n} \frac{f_i (r_i - \Delta/2)^2}{[(r_i - \Delta)^3/(1 + Q_\alpha)] + \{[r_i^3 - (r_i - \Delta)^3]/(1 + Q_\kappa)\}} \qquad (4)$$

In (3) and (4), ρ_p is the protein density, ρ_s is the solvent density, $\bar{v}$ is the protein partial specific volume, η is the solution viscosity, r_i is the radius of the ith size micelle, Δ is the thickness of κ-casein in the coat, Q_α and Q_κ are the solvations of α_s- and κ-caseinates in ml solvent/ml protein, and R is the system weight ratio of (α_s-casein + β-casein)/κ-casein. By application of these equations to appropriate experimental data, a micelle distribution can be characterized as containing mass fractions of its micelles in a set of chosen sizes (classes), and the micelle population surface area can be calculated. In publications from this laboratory the assay centrifugal field has been given as 400 g. This should have been 1200 g.

An earlier stability test used in several laboratories is that developed by Zittle (1961). In this test the pH is 6.7, the temperature 30°C, and the α_s-/κ-casein ratio is varied at constant 0.02 M calcium. No ionic strength is specified. In the studies of Noble and Waugh (1965) and Waugh and Noble (1965) the α_s-/κ-casein ratio is varied, and examinations are carried out with respect to calcium concentration and protein concentration at pH 7, $I = 0.075$ and 37°C. As will be seen (Section VIII), the sytems are particularly sensitive to ionic strength.

The main curve of Fig. 2, which is illustrative of typical results, shows supernatant protein as a function of total calcium concentration for a solution containing initially 10 mg/ml of α_s-casein and 1 mg/ml of κ-casein. At very low calcium concentrations (0–0.004 M) no precipitate forms and the turbidity of the solution does not visibly increase. From 0.004 to 0.006 M calcium there is a sharp decrease in supernatant protein (descending limb of the dip) to about 5 mg/ml at 0.006 M calcium. Pellets contain Ca α_s-caseinate. From 0.006 to approximately 0.015 M calcium, supernatant protein increases (ascending limb of the peak) until pellets are absent.

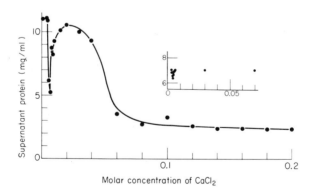

FIGURE 2. Supernatant protein, resulting from single-aliquot addition of calcium, plotted as a function of CaCl$_2$ concentration for two α_s-κ-casein mixtures in standard KCl buffer at 37°C. The main curve represents data for a solution initially containing 10 mg/ml of α_s-casein and 1 mg/ml of κ-casein. The inset represents data for a solution initially containing 5 mg/ml of α_s-casein and 2 mg/ml of κ-casein. The designations of the segments of the curves, in order of increasing CaCl$_2$ concentration, are the dip, the peak, and the pseudoplateau.

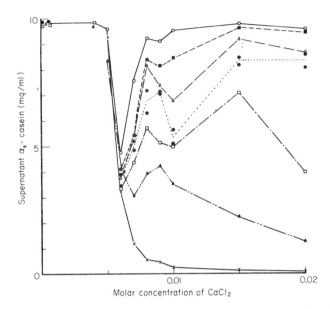

FIGURE 3. Supernatant protein, resulting from single-aliquot addition of calcium, plotted as a function of $CaCl_2$ concentration for a series of α_s-κ-casein mixtures in standard KCl buffer at 37°C. Each solution contained initially 10 mg/ml of α_s-casein, but the initial κ-casein concentration varied. The different κ-casein concentrations are represented as follows: ○—1 mg/ml; ■—0.9 mg/ml; △—0.8 mg/ml; ●—0.7 mg/ml; □—0.6 mg/ml; ▲—0.5 mg/ml; and X—0.0 mg/ml of κ-casein, i.e. pure α_s-casein.

Micelles first appear at the start of the ascending limb of the peak. Above 0.02 M calcium, supernatant protein decreases rapidly (descending limb of the peak) up to 0.06 M calcium, after which it decreases slowly (pseudoplateau). Micelle formation is a maximum at the top of the peak in the sense that the micelle distribution has the smallest average size. The average size increases progressively over the descending limb of the peak and the pseudoplateau.

Using single-aliquot addition of calcium, the dip region for constant α_s-casein concentration can be compared with Ca α_s-caseinate solubility as the amount of κ-casein is increased, therefore as the initial α_s-/κ-casein weight ratio, R_i, is decreased. The lower solid line of Fig. 3 refers to the solubility of Ca α_s-caseinate, starting with 10 mg of protein per milliliter. The others refer to systems containing increasing amounts of κ-casein, from 0.5 to 1.0 mg/ml (R_i of 20–10). The descending limb of the dip is nearly independent of the κ-casein concentration; the ascending limb of the peak is not. The precipitates which form during the descending limb do not contain κ-

casein and coalesce readily to form single phases. Prior to precipitate formation in this region, the solution becomes turbid with a characteristic gray appearance, in contrast to the white opacity associated with micelle formation. Precipitation of Ca α_s-caseinate is progressively retarded by the presence of κ-casein; retardation is negligible at an initial ratio of 20 and obvious at an initial ratio of 10. Above about 0.007 M calcium there occurs a dramatic increase in supernatant protein which is far above the solubility of Ca α_s-caseinate. As the peak is approached, pellets become progressively more particulate (micellar). Maximum micelle stabilizations are achieved between 0.01 and 0.02 M calcium. Complete stabilization occurs at, or below $R_i = 10$, and stabilization against assay centrifugation decreases as R_i increases. Evidently, at peak micelle formation the average micelle size increases as R_i increases. At the peaks of the highest R_i, the largest of the micelles eventually sediment under unit gravitation. It is important to note that the dip, although reduced, can be obtained using first-cycle casein and solubilized skim milk and that the former behaves like an α_s-κ-casein mixture at $R_i \simeq 6$.

The response of a system on incremental addition of calcium chloride is quite different from that on single-aliquot addition. Typical results are shown in Fig. 4. Here, supernatant α_s-casein concentration is plotted against

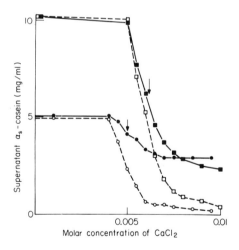

FIGURE 4. Supernatant α_s-casein, resulting from incremental addition of calcium, plotted as a function of CaCl$_2$ concentration for solutions containing, in standard KCl buffer, 1 mg/ml of κ-casein and 10 mg/ml of α_s-casein (■); 1 mg/ml of κ-casein and 5 mg/ml of α_s-casein (●); 10 mg/ml of α_s-casein (□); and 5 mg/ml of α_s-casein (○). The arrows indicate calcium concentrations for the first appearance of micelles and for the bottom of the dip using single-aliquot addition.

total calcium concentration for solutions containing 10 and 5 mg/ml of α_s-casein and 1 mg/ml of κ-casein (solid lines), and the corresponding calcium α_s-caseinate solubility curves (dotted lines) are shown. As observed with single-aliquot addition, the calcium concentration at which a precipitate first appears is nearly the same whether or not κ-casein is present. The descending limb of the dip is also observed. The vertical arrows indicate the location of the bottom of the dip, the first appearance of micelles on single-aliquot addition, and the point where micelles first become apparent on incremental addition. Beyond the points indicated by the arrows, while the systems for single-aliquot addition rise to complete stabilization, incremental addition gives increasing amounts of assay centrifugate, extending the excursion of the descending limb. This limb eventually levels off well above calcium α_s-caseinate solubility at stabilization ratios between 2 and 3. The supernatant protein does not increase either with extended time at a particular calcium concentration or when the calcium concentration is increased above 0.01 M.

Using single-aliquot addition of calcium to systems having different initial ratios of α_s-/κ-caseins at 37°C, I = 0.075 and pH 7, Waugh and Noble (1965) examined peak and postpeak behavior. Assay supernatant protein was determined for various total calcium concentrations using systems containing 1 mg/ml of κ-casein and varying concentrations of α_s-casein (initial ratios between 20 and 4). As the initial ratio is varied, the systems exhibit two progressive shifts. First, for R_i = 8, the calcium range at the top of the peak over which little or no pellet forms is from 0.015 to 0.04 M; for R_i = 6, this range is from 0.01 to $\sim$0.05 M; and for R_i = 5, the range extends to 0.07 M calcium. When R_i is increased to 15 or 20, less stabilization is accomplished, although there is still a small peak in both curves. For all points, subtracting the initial κ-casein concentration from supernatant protein leaves an amount that greatly exceeds the calcium α_s-caseinate solubility (Fig. 3). Second, in the region of the pseudoplateau, stabilization at R_i of 5 or below is complete. As R_i is increased above 5, the supernatant protein progressively decreases, leading obviously to a decrease in the apparent stabilization ratio.

Increasing the initial protein concentration accentuates the region of the dip. In all cases, in the region of the pseudoplateau, supernatant protein increases with increasing protein concentration. At the same time, the fraction of the initial protein remaining in the supernatant decreases.

Whatever the result of single-aliquot addition, each system rapidly comes to a state which is stable, as far as assay centrifugation is concerned, for as long as the experiment can be conducted. Experiments are eventually terminated after a few days as a result of enzymatic or microbial action.

Useful information has been gained for the α_s-κ-casein micelle system under the given environmental conditions by examining the stability of final states. The addition of κ-casein to a micelle system produces a decrease in turbidity, most of which occurs within a few minutes. Differential centrifugation shows that there is a concomitant decrease in average micelle size. In these experiments the α_s/κ ratio was reduced, by adding κ-casein, from $R_i = 5.0$ to levels of $R_s = 2.5$, 1.25 and 0.62. An important consideration is the mechanism by which reduction in size takes place. This will be considered in Section VIII. It was also observed that the addition of κ-casein to calcium α_s-caseinate precipitates alters their surface characteristics from being extraordinarily sticky to being nonadherent. Some evidence indicates that homogenization of these systems would give smaller stable micelles.

Another result of importance in deriving a model of micelle structure was obtained from studies of micelle stability to close approach. These studies revealed the extraordinary nonreactivity of the micelle surface. Close approach was accomplished by ultracentrifugation in a swinging bucket rotor, Spinco SW 39 (Beckman Instruments, Inc.), at 37°C for 30 min at 97,000 g. The resulting micelle pellets varied from soft gelatinous (top) to hard grainy (bottom); the amounts present depended on the nature of the system being centrifuged. Micelle stability was tested by determining the extent to which a micelle suspension reformed, under conditions of gentle stirring or convection (to avoid diffusion as the rate-limiting step), when pellets were placed in contact with their corresponding supernatants. For example, a solution containing 10 mg/ml of α_s-casein and 1 mg/ml of κ-casein ($R_i = 10$) was converted to micelles at 0.02 M calcium and ultracentrifuged as described. Pellet and supernatant were placed in a jacketed beaker at 37°C. Within 1 hr, supernatant protein stable to assay centrifugation increased from 1 to 7 mg/ml. After 24 hr, 8.5 mg/ml were present out of a possible 11 mg/ml. Thus, at the maximum initial ratio for complete stabilization, 75% of the pellet appeared as micelles within 24 hr. At an initial ratio of 4, complete resuspension of pellets obtained by ultracentrifugation required less than 1 hr. Resuspension of pellets obtained from skim milk is described in Section III.

These studies are summarized in the following, which give experimental results and conclusions considered important in developing a model of micelle formation and structure. (a) Prior to micelle formation, when all protein is still in solution at 37°C, there are present free calcium α_s-caseinate and a calcium α_s-κ-caseinate interaction product with a low weight ratio. (b) For stability, micelles require a minimum level of calcium which is greater than that required to precipitate calcium α_s-caseinate. (c) In confirmation of the results of Sullivan et al. (1959), the more centrifugable (larger) the micelle the lower its weight fractional content of κ-casein. (d) Large micelles formed in the presence of κ-casein, or calcium α_s-caseinate precipitates treated with κ-casein at calcium concentrations sufficient for micelle stability, differ from untreated calcium α_s-caseinate precipitates in being nonadherent. (e) Micelles are stable with respect to the close approach induced by ultracentrifugation into a pellet. (f) A micelle system has a size distribution which depends strongly on the initial ratio, the final calcium concentration on single-aliquot addition, and the initial protein concentration. The apparent final state of a given system is dependent on the path by which final conditions are approached. For example, different distributions are obtained by single-aliquot and incremental addition of calcium. (g) Complete stabilization in the region of the peak can be achieved by single-aliquot addition of calcium at all initial ratios up to 10, and micelles stable with respect to assay centrifugation may have weight ratios in excess of 10. (h) The size distribution of a micelle population can be altered by the addition of κ-casein and by the addition of calcium. Size transformations induced by κ-casein addition occur in minutes, while those following calcium chloride addition take hours. (i) Stabilizing capacity is evident in some micelle systems after the systems have come to apparent final states. (j) A micelle distribution, once formed, is relatively stable to dilution with a buffer containing an appropriate calcium concentration.

Waugh and Noble (1965) examined these experimental results using the general possibilities for micelle structure described in Section II and excluded the possibilities that micelles could be single-phase particles or large chemical compounds. The authors concluded that micelles have a composition which changes in going from the surface to the center and that the surface is rich in κ-casein and the core is rich in α_s-casein. The proposed structure has a pure calcium α_s-caseinate core covered by a uniform coat composed of a calcium α_s-κ-caseinate interaction product with a low weight

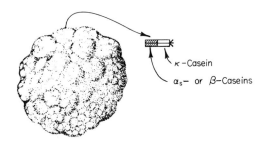

FIGURE 5. A sketch of the casein micelle. The surface crenations are due to the interaction of core polymers with κ-casein. This interaction leads to the formation of a coat unit, which is a near unit weight-ratio interaction product of either α_s-κ-caseins or β-κ-caseins.

ratio. This interaction product will be referred to here as a *coat unit*. Coat units can also be constructed of β- and κ-caseins (see below). That κ-casein contributes importantly to the surface properties of micelles has been suggested by Rose (1965) and Payens (1966). In this model (Fig. 5) the coat units are required to have certain interaction specificities. That portion of the unit exposed at the surface, the κ-casein macropeptide, has little tendency to interact with itself or other milk proteins. Internally, the units have a strong interaction with core calcium α_s-caseinate and a lateral specificity for each other. In the coat units, α_s- or β-caseins have a specificity of interaction with κ-casein.

The sequence of events following the addition of calcium to a solution of α_s- and κ-caseins is considered to be as follows. Immediately there occurs the redistribution of components to form, in solution, calcium α_s-caseinate and coat units. As calcium α_s-caseinate polymerizes, the polymers are covered and stabilized by coat. The average micelle size will be determined by the amount of κ-casein present. The development of a micelle size distribution under the conditions used is attributed to randomness in the precipitating and coating reactions.

Differential rates of coat and core formation readily account for the dependency, for example, of micelle Ca α_s-caseinate and micelle size distributions on initial ratio, initial protein concentration, and calcium concentration via single-aliquot addition. When, by a progressive change in a condition, calcium α_s-caseinate polymerization is favored over coating, there results a shift toward larger micelles, and an increasing fraction of the micelle population is removed by assay centrifugation. According to this model, there is no qualitative difference between micelles of widely different size. Additional evidence giving strong support for the coat-core model will be given in Section VIII where it will be shown that metastable colloid particles of calcium α_s-caseinate of 1 μ or more in diameter are converted to stable micelles on being coated by addition of κ-casein.

Particular calcium requirements are evident from the observation that calcium α_s-caseinate precipitation occurs when calcium is added to a protein mixture or when a preformed micelle system is diluted, in both cases to obtain a calcium concentration near the bottom of the dip. Upon dilution, the coat becomes unstable, goes into solution and cores then precipitate. A clear inference is that the calcium dependency of coat formation must be greater than that of calcium α_s-caseinate precipitation.

The addition of calcium to a peak micelle system increases average micelle size. This suggests that the coat has its maximum stability at calcium concentrations in the region of the peak (Figs. 2 and 4) and that the addition of calcium must cause an instability such that κ-casein is removed, and the core surface is revealed. Coalescence then follows until complete coverage is again attained. Assume that the micelle coat is constructed of identical units of calcium α_s-κ-caseinate. Coat instability would result if coat components react with calcium to produce noncoat material, by a reaction which has a higher calcium dependency than that of coat formation. Calcium concentration, equilibrium constants and calcium dependencies would determine the relative concentrations of coat and noncoat materials. A noncoat material might be a calcium α_s-κ-caseinate interaction product having a lower weight ratio than the coat unit; possibly this product is pure calcium κ-caseinate. Formation of a noncoat calcium κ-caseinate would account for the splitting step in calcium fractionation.

It was suggested by Waugh and von Hippel (1956) and Waugh (1958), largely on the basis of experimentally observed interactions of the caseins in the absence of calcium, that preexisting complexes of α_s- and κ-caseins at a weight ratio of 4/1 were the most important preliminary structures involved in the formation of casein micelles. This stoichiometry was expected to limit the stabilizing capacity of κ-casein. The occurrence of the dip at weight ratios below 4 precludes the formation of stable complexes under the conditions used, and the existence of micelles having weight ratios of 10, as observed also by Zittle (1961) and Zittle et al. (1962), and well in excess of 10, effectively excludes any fixed stoichiometry.

The experimental observations which led to a model based on stoichiometry and to the high-weight-ratio α_s-κ-casein complex are of more than passing interest; they indicate the extent to which rearrangement can take place on the addition of divalent cation to a system free of such ion. The complexes of Waugh and von Hippel (1956) were revealed by ultracentrifugation at 1°C (pH 7, $I = 0.15$, phosphate buffer) to decrease interaction with β-casein. In 4/1 α_s-κ-casein mixtures the typical α_s-casein peak ($s_{20} = 4.5$ S) and κ-casein peak (13.5 S) were found to be absent, and an α_s-κ-casein peak was present (7.5 S). Excess κ'-casein introduced a κ-casein peak. The sedimentation coefficient of the α_s-κ-casein peak was found to be sensitive to high pH treatment. The spontaneous occurrence of a complex of $s_{20} = 7.5$ S, associated with a skewed peak, was later reported by Pepper and Thompson (1963) and observed even for dephosphorylated proteins. Treatment at high pH, however, did not alter the s_{20} of the complex they obtained. Swaisgood and Brunner (1962) found that spontaneous complexing does not occur at room temperature but only after pretreatment with urea or high pH, when complexing again takes place at an α_s/κ weight ratio of 4. The sedimentation coefficients of the complexes obtained by these two treatments differ markedly: 6.2 S for the alkaline-treated and 4.7 S for the urea-treated materials. Garnier et al. (1964) reported spontaneous complexing at 25°C, but only at a weight ratio of unity, to give a product of $s_{20} = 14.7$ S. They reported further that the liberation of protons on rennin action (over a wide range of temperature) also supports the existence of an α_s-κ-casein interaction product of weight ratio near unity. All of their experiments were carried out in 0.1 M NaCl at pH 6.95. Garnier (1966) reported that neither α_s-casein nor β-casein forms complexes with κ-casein at a weight ratio of 4 in the absence of calcium at pH 7, $I = 0.1$ (sodium chloride) and 25°C.

A reexamination of complex formation by Noble and Waugh (1965) did not agree with the earlier results of Waugh and von Hippel (1956). The more recent data suggest that interaction occurs spontaneously at 37°C, requires pretreatment with urea or high pH in order to occur at 20°C, and occurs neither spontaneously nor after such pretreatment at temperatures near 5°C. When interaction occurs at 37°C, the interaction product has a weight ratio near unity and a sedimentation coefficient near 7 S.

Noble and Waugh (1965) went further to show, by sampling distributions after ultracentrifugation at 37°C and testing these by calcium addition, that even in first-cycle casein, κ-casein moves preferentially to the bottom of the tube and is thus not uniformly combined with α_s- and/or β-caseins.

Examination of the data presented in the group of publications under discussion, and the more extensive literature cited at the beginning of Section IV, reveals other differences too numerous to mention. It is concluded that the many interactions into which the caseins can enter, with themselves and with each other in the absence of calcium, make a particular interaction pattern which is temporarily reproducible but strongly dependent on unknown experimental conditions, including the past history of the sample. However, it appears that, whatever the structure of the system in the absence of calcium, the only α_s-κ-casein interaction product that can survive during the descending limb of the dip is one with a low weight ratio. The existence of a low weight ratio interaction product is also deduced from the progressive divergence of curves for calcium α_s-caseinate solubility and descending limbs for α_s-κ-casein mixtures (Fig. 4).

Although the coat-core model has been accepted by Rose (1965), Payens (1966), and McKenzie (1967), Garnier and Ribadeau-Dumas (1969) present an inherently different model for micelle structure. In their model, small nodes of κ-caseinate are interconnected by constant-diameter strands of α_s- and/or β-caseinates. This model is included in a group of models in which it is assumed that micelles are large chemical compounds. Waugh and Noble (1965) considered models of this type unlikely on the basis of the phase rule. It is also felt that, without highly restrictive specifications, the model of Garnier and Ribadeau-Dumas would have difficulty in accounting for the micelle-system. characteristics described in this chapter and the equilibrium characteristics summarized in Section VIII. In addition, the model predicts that micelle solvation will increase as R_i increases, from 10 g solvent/g protein at $R_i = 14$ to 180 g solvent/g protein at $R_i = 100$. Experimentally (Waugh and Noble, 1965), micelle solvation is found to decrease as R_i increases and to be less than 2.1 at $R_i = 20$.

Parry and Carroll (1969) have attempted to locate κ-casein in milk micelles with electron microscopy. Using κ-casein-combining antibodies, they obtained no evidence for a κ-casein stabilizing coat. They propose that 30% of κ-casein is free in serum and associated with small amounts of $α_s$- and β-caseins. High molecular weight aggregates of the remainder are located at the centers of micelles and are surrounded by insoluble calcium $α_s$- and calcium β-caseinates. The problems of accounting for stability with respect to close approach, and other micelle characteristics, are apparent. Nonreactivity of κ-casein at the micelle surface with combining antibodies is consistent with coat interaction specificity given in this section: To react and reveal coat, the antibody must be directed against the substituent macropeptides, which are exposed at the micelle surface and which are remarkably nonreactive.

B. Micelles from Mixtures of β- and κ-Caseins

Zittle and Walter (1963) have shown that κ-casein can stabilize β-casein against calcium precipitation at 30°C. In near agreement with the results of Waugh (1958), they found the solubility of β-casein in the presence of 0.03 M calcium to be 0.3 g/liter. No ionic strength was specified, a circumstance which, in view of strong ionic strength dependency, makes it difficult to compare these experiments with others. Zittle and Walter found, surprisingly, that above 0.008 M calcium, precipitation of solutions containing 0.15% β-casein was constant at 80% of the protein. At 0.15% β-casein, κ-casein produced complete stabilization at $R_i = 6.7$. At 0.01 M calcium, they found the maximum stabilization ratio to decrease, thus κ-casein to become less effective, as the protein concentration increased. However, the experiments were not carried to the point where complete stabilization was demonstrated either at 0.3 or 0.6% protein, which were the highest concentrations used. A pH dependency of stabilization was observed, the optimum being pH 7.2. Above pH 9 an increasing solubility of β-casein made stabilization results difficult to interpret. The authors concluded that the β-κ-casein system behaves qualitatively much like the $α_s$-κ-casein system. At 30°C it is apparent that the coat-core model for micelle structure should be retained for the β-κ-casein system. There is an important difference, however, in temperature dependence. As the temperature is decreased, the $α_s$-κ-casein system can degenerate to yield a calcium $α_s$-

caseinate precipitate. Due to solubility characteristics, this is not possible for the β-κ-casein system. We have checked this point and have found it to be correct, with the addition of an interesting qualification. As the temperature of a calcium β-κ-casein micelle system is decreased, the protein does not dissolve to give an entirely clear solution. Turbidity is present even after long standing at 4°C. Possibly there are cooperative interactions between calcium, β-, and κ-caseins which are not easily reversed by a decrease in temperature.

VI. The Micelle Core

In this section the assumption is made that certain important core properties can be determined from a characterization of the interaction of calcium and $α_s$-casein, β-casein or mixtures. Such a characterization, including the characteristics of precipitates, is obviously satisfactory for the cores of micelles made under conditions in which calcium is the driving ion and the other ions present are monovalent. It will be apparent in succeeding sections that the cores of natural micelles, and of micelles made with buffers which duplicate or simulate the skim milk environment, are more complex in that they involve the participation at least of inorganic phosphate and of citrate. The evidence to be presented supports the view that this greater complexity involves modification of the same fundamental structure of the micelle in the details of ion and protein interactions.

A. Core Polymers

Waugh et al. (1970), using $α_s$-casein, β-casein, and a unit weight ratio mixture of the two, examined, at pH 6.6, the degree of monomer association and the structure of the association product as affected by environmental conditions of ionic strength (NaCl), divalent cation concentration (Ca) and temperature. Sedimentation coefficients and reduced viscosities were obtained. For β-casein and the mixture it was observed that a progressive increase in an environmental condition increases association and leads to increasing $s_{20,w}$ and decreasing viscosity. However, prior to precipitation both have significant plateau values. This suggests that polymers arrive at limited average values of size, shape, and solvation. By applying the equations of Simha and Perrin it is determined that these limited polymers are spherical, have a maximum degree of association near 30, have radii near 100 Å, and have solvations ranging from 2.2–4.4 g water/g protein. The degrees of association calculated using $s_{20,w}$ and $[η]$ are in agreement with those obtained independently by Archibald analysis of polymer weights. Limited polymers (core polymers) interact at their surfaces in the formation of cores or precipitates.

A comparison of α_s- and β-caseins suggests that they have the same basis of physical structure. The striking features of both proteins are the accumulation of organic phosphate and carboxylate groups in short chains of amino acid residues and their high content of nonpolar side chains. Monomer models which have been helpful in accounting for the viscosities of solutions of monomers, the properties of core polymers, and the binding of cations during progressive association and precipitation follow.

For α_s-casein, the model is a compact, prolate ellipsoidal body having a length of about 100 Å and a mass of 22,300 daltons. The body is stabilized by hydrophobic and hydrogen bond interactions, but not all of those possible are accommodated within the body; the surface of the body also exposes groups capable of such interactions. Covalently bonded to the body are the ends of a peptide chain of approximately 5000 daltons, which forms a torus around the body end. The torus carries phosphate groups and additional groups as specified by Österberg (1964)—7 phosphate groups, 11 carboxylate groups, and 2 amino groups (Structure I). The torus has sufficient length, as shown in Fig. 6a, to encircle the end of the body in such a way as to permit access of solvent to all charged groups.

For β-casein, the body has a mass of approximately 21,000 daltons and a length of approximately 100 Å. The acidic peptide, which has a mass of approximately 3000 daltons, contains all five phosphate groups and, in

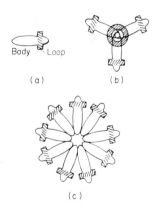

FIGURE 6. A sketch of α_s-casein association. (a) A compact body to which the acidic peptide is attached near one end. (b) A tetramer formed at low I in the absence of divaent cation. (c) A planar view of monomers as they are packed in the limited core polymer.

addition, five carboxylate groups and two arginine groups. It is covalently bonded to the body at one of its ends.

The main direct evidence to suggest that the body is compact is its unusually high content of nonpolar amino acid side chains, and, as will be shown, its low net charge. That the acidic peptides must be able to disengage from the body and become surrounded by solvent can be seen by considering electrostatic and hydrodynamic properties.

According to Waugh et al. (1970), these structures can account for the high intrinsic viscosities of solutions of monomers as observed by Sullivan et al. (1955) and Ho and Chen (1967), and for the observation that $[\eta]$ does not increase significantly when proteins are placed in Ga·HCl (Ho and Chen, 1967; Noelken and Reibstein, 1968). Core polymers are considered to contain radially arranged anisometric monomers, as shown in Fig. 6 for α_s-casein (Waugh et al., 1970). In the polymer, electrostatic free energy is minimized by placing the acidic peptides near the polymer surface. When the acidic peptides are sufficiently discharged by ion shielding or binding, interpolymer interactions allow precipitation (core formation) to occur. A comparison of the behavior of individual components and the mixture shows that α_s- and β-caseins commingle at the molecular level in the mixture and that polymers of α_s-casein have a larger interpolymer attractive energy than those of β-casein. Core polymer structure and interaction automatically lead to a crenated micelle.

Commingling of α_s- and β-casein monomers is also indicated for a variety of systems, including the natural micelle system. That interactions observed in the presence of calcium are initiated in the absence of calcium is indicated in the results of von Hippel and Waugh (1955). Between 4° and 40°C in mixtures, there is a progressive reduction in the area of the slow ultracentrifuge peak associated with β-casein and a progressive increase in the area and sedimentation coefficient of a faster polymer peak.

Metastable colloid* has been prepared by dialyzing a mixture of α_s- and β-caseins, 12.5 mg/ml each, against type IVa or type IVc buffer (Table V) at 37°C for 16 hr, or for 20 hr against 0.007 M calcium chloride con-

---

* Frequent use is made of metastable colloids which can be produced from solutions of α_s-casein, β-casein, or mixtures of the two. The average size of colloid particles is condition dependent—on pH, ionic strength, the particular ions present, calcium concentration, temperature and protein concentration. By electron microscopy, colloid particles are observed to be spherical. They can be rendered unstable: The addition of calcium, to give a concentration near that required to form micelles, leads to immediate precipitation, and on being brought into contact by centrifugation the colloid particles irreversibly coalesce to form a pellet of bulk phase. Colloids are frequently formed at low ionic strengths and calcium concentrations. A size distribution is not highly reproducible. Conditions for metastable colloid formation will be specified in the text.

taining 0.03 M sodium chloride at pH 6.6 and 37°C. Each metastable colloid was centrifuged so that half of the protein present was removed. Pellet and supernatant proteins were harvested, brought into solution and examined by gel electrophoresis. No significant differences were observed in the distributions of α_s- and β-caseins among the starting solutions, pellets or supernatants.

Using skim milk systems and differential centrifugation Choate et al. (1959) found a constant proportion of α- and β-caseins, independent of micelle size. That the interaction ratio might not be constant is indicated by the results of Rose (1965), who reported that the larger micelles contain a somewhat greater relative amount of β-casein. Rose and Colvin (1966b) reported that micelles of different size contain the same fractional content of β-casein. Some additional evidence is provided by the centrifugation of skim milk which has been allowed to stand at 0°–4°C for 24 hr or more: The centrifugate contains both α_s- and β-caseins in about equal amounts; the supernatant contains excess β-casein, but also a significant amount of α_s-casein. To a first approximation we have found that β- and α_s-casein can replace each other in micelle formation. For example, a mixture containing 8, 6 and 2 mg/ml of α_s-, β- and κ-caseins, respectively, readily forms micelles under conditions in which an α_s-κ-casein mixture at 16 mg/ml and $R_i = 7$ forms micelles.

A precipitate of core polymers should have little regularity in structure over distances greater than that of the core polymer diameter. Available evidence on Ca caseinate precipitates shows that this is the case: Extended structural regularity is contraindicated by the absence of appropriate X-ray diffraction spacings. Tuckey et al. (1938) found this to be true for natural casein micelles.

Payens (1966) suggests that the micelle core consists of a network of β-casein molecules, which consist of open coils, and that compact α_s-casein molecules are attached to these by hydrophobic bonds. Rose (1969) accepts the core model of Payens, with the addition of colloidal calcium apatite (Rose, 1965) as a stabilizing factor. Colloidal calcium apatite is examined in Section IX.

B. Ion Binding

The degree of monomer association increases with increasing ionic strength and divalent cation binding. Studies of the interactions of the caseins with calcium and monovalent cation go back at least to the extensive studies of Chanutin et al. (1942), who used whole casein and interpreted their binding data in terms of mass-action relationships. Calcium binding was found to increase with calcium concentration, to decrease with increasing sodium chloride concentration, and to be relatively independent of pH between pH 6.3 and 8.5. Carr and Topol (1950) and Carr and Engelstadt (1958) reported that α_s-casein binds 4.8 sodium ions/27,300 g protein in 0.01 M NaCl at pH 7.4. In the latter paper, binding was reported to decrease as the pH decreases, becoming zero at pH 5.7. The total binding, but surprisingly not the pH dependency, was confirmed for α_s-casein by Ho and Waugh (1965a). Carr and Topol (1950) reported that whole casein does not bind significant chloride ion at pH 5 and 0.012 M sodium chloride. Carr (1955) concluded that phosphate groups are responsible for sodium binding.

To return to calcium binding, Zittle et al. (1958), using whole casein, found a maximum binding of ~ 25 mole Ca/10^5 g protein. Ionic strengths were not given. Binding was found to decrease with decreasing pH. There has been a brief account of the interaction of calcium with α_s-casein by Ho and Waugh (1965b). In this work, infrared spectroscopy was used to show that the first calcium ions to bind do so by interacting with organic phosphate groups. The same conclusion is drawn from the work of Reeves and Latour (1958), in which calcium is found to bind extensively to mixed acidic peptides obtained from whole casein. Dickson and Perkins (1969) reported that the relative order of ion-binding constants to α_s-, β-, and κ-caseins is Ca > Ba > Sr and that the relative order of binding capacities is α_s- > β- > κ-casein.

Waugh et al. (1971) have examined the binding of cations to α_s-casein, β-casein and a unit weight ratio mixture at pH 6.6. Starting with isoionic proteins, they determined proton release on bringing proteins into solution at pH 6.6. Thereafter, proton release was determined at sodium chloride concentrations, I, of 0.04, 0.08 and 0.16 M. For each of these, proton release was determined for progressive additions of calcium chloride up to 0.08 M. The apparent molar binding of calcium, $\bar{\nu}_{Ca,A}$, was determined after equilibrium dialysis. When precipitates appeared, their solvations were measured. Table II summarizes the results. Column 1 gives the protein examined and Column 2 the concentration of sodium. The first row for each protein gives the sodium hydroxide concentration required to bring the isoionic protein into solution at pH 6.6. Little chloride ion is present and the ionic strength is uncertain. The following three rows record for each protein the three added concentrations of sodium chloride: 0.04, 0.08 and 0.16 M. Column 3 gives increments in proton release, δH_I^+ per mole protein. The first row gives δH^+ between isoionic protein and protein just at pH 6.6, and the remaining rows give the increments due to changes in I between values of column 2. Column 4 gives $\bar{Z}_I$, the charge per monomer, exclusive of site-bound sodium. Column 5 gives ξ, the average incre-

TABLE II

PROTON RELEASE OF THE CASEINS[a]

Protein	I^b	δH_I^{+c}	$\bar{Z}_I^d$	ξ^e	Total δH^{+f}
α_s-Casein	6×10^{-4}	22.0	−22.0		
	0.04	2.8	−24.8	0.21	6.58
	0.08	0.65	−25.4	0.18	6.69
	0.16	0.85	−26.3	0.13	6.64
β-Casein	3.7×10^{-4}	13.0	−13.0		
	0.04	1.25	−14.2	0.37	4.95
	0.08	0.67	−15.9	0.30	4.92
	0.16	0.48	−16.3	0.26	4.90
α_s- + β-Caseins	5×10^{-4}	17.2	−17.2		
	0.04	1.95	−19.2	0.30	6.3
	0.08	0.75	−19.9	0.25	6.3
	0.16	0.60	−20.5	0.21	6.3

[a] The reference materials are the isoionic proteins.
[b] Sodium concentrations; the first for each protein does not accurately represent the ionic strength.
[c] The increments in proton release between sodium concentrations.
[d] The monomer net charge for each sodium concentration.
[e] The average proton release per apparent bound calcium.
[f] The total proton release at 0.08 M calcium.

ment in δH^+ per molecule per apparent bound calcium, and column 6 gives the total proton release between protein just at pH 6.6 and 0.08 M calcium chloride. Total δH^+ per molecule is essentially constant at 6.6 protons for α_s-casein, 4.9 protons for β-casein, and 6.3 protons for the α_s-β-casein mixture.

Figure 7 refers to α_s-casein and shows, for the three values of I, the relation between apparent calcium binding, $\bar{\nu}_{Ca,A}$, and the logarithm of the solution calcium concentration, $[Ca^{2+}]_s$. Curves for β-casein and the α_s-β-casein mixture are similar to these. It is apparent from Fig. 7 that at constant $[Ca^{2+}]_s$, increasing the concentration of sodium ion results in lowered binding, with the effect diminishing as $[Ca^{2+}]_s$ increases.

In Fig. 8, G, in g water/g protein, is plotted against site-bound calcium, $\bar{\nu}_{Ca,S}$, which was obtained from $\bar{\nu}_{Ca,A}$ by a process to be described. Figure 8 refers only to precipitates where G is measured directly. Evidently, for the three protein systems, as $\bar{\nu}_{Ca,S}$ increases, G decreases essentially to plateau levels of 1.6–1.7 for both α_s-casein and the α_s-β-casein mixture and to 1.9 for β-casein. There is an indication that if $\bar{\nu}_{Ca,S}$ were to be increased sufficiently, G would increase.

9. FORMATION AND STRUCTURE OF CASEIN MICELLES 41

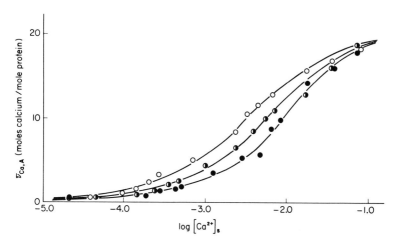

FIGURE 7. Apparent binding of calcium, $\bar{v}_{Ca,A}$, in moles calcium per mole protein for α_s-casein vs. the logarithm of the solution calcium concentration, $[Ca^{2+}]_s$. Total sodium concentrations are 0.04 M (○); 0.08 M (◐) and 0.16 M (●).

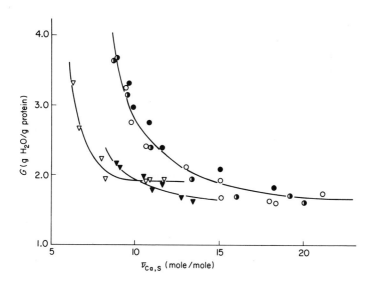

FIGURE 8. Precipitate solvation, G, vs. site-bound calcium, $\bar{v}_{Ca,S}$. Sodium concentrations for α_s-casein (circles) are 0.048 M (○); 0.085 M (◐); and 0.154 M (●). Sodium concentration for β-casein (▽) and a unit weight-ratio mixture (▼) are 0.048 M.

By taking the difference in total calcium between protein solution and diffusate, $\bar{\nu}_{Ca,A}$ is obtained. The problem is to divide this between Donnan bound, $\bar{\nu}_{Ca,D}$, and site bound, $\bar{\nu}_{Ca,S}$, taking monomer net charge, $\bar{Z}$, into account. Site binding of sodium is considered neglibible; $\bar{\nu}_{Ca,S}$ are corrected for Donnan bindings, $\bar{\nu}_{Ca,D}$, which are calculated using equations, for protein and diffusate, of mass and charge balance and equivalence of salt activities in accessible regions.

Simple calculations, using any reasonable hydrogen ion binding constants for phosphate and carboxylate groups, show that (exclusive of sodium or chloride binding) the acidic peptides must carry essentially all monomer net charge (the body is -2 to -3). The peptides, then, give rise to regions of high local fixed-charge density.

Overbeek (1956) has shown that when fixed charge is located so that the charge density on some portion of a polyelectrolyte is high, the classical Donnan theory does not apply. In particular, the mobilities of the counter ions decrease as local charge density is increased, and an increasing fraction of fixed charge is compensated for by the accumulation of counter ions and less by the expulsion of co-ions. Since monomer net charge is concentrated essentially on short acidic peptides, Donnan volumes for protein in solution are calculated using apparent $\bar{Z}$ and standard ratios of solvation to apparent $\bar{Z}$, obtained at minimum conditions for precipitation. Measured solvations are used for precipitates. Solvations, total amounts, equilibrium concentrations, apparent $\bar{Z}$, and $\bar{\nu}_{Ca,A}$ are then used to obtain $\bar{\nu}_{Ca,D}$ and $\bar{Z}$. Site-bound calcium, $\bar{\nu}_{Ca,S}$, is then obtained from $\bar{\nu}_{Ca,A}$. For progressive calcium addition, $\bar{\nu}_{Ca,D}$ are small at low $\bar{\nu}_{Ca,S}$ ($\sim 15\%$ of $\bar{\nu}_{Ca,A}$), and they decrease as I or $\bar{\nu}_{Ca,S}$ increases, thus as $|\bar{Z}|$ decreases.

An analysis of binding and solvation data indicate the following. Each protein initiates precipitation, independent of I, at about the same values of $\bar{\nu}_{Ca,S}$ and $\bar{Z}$. These are, respectively, for α_s-casein, β-casein and the α_s-β-casein mixture $\bar{\nu}_{Ca,S} = 9.3$, 5.4, and 10.4, and $\bar{Z} = -9$, -6, and -1.5. The mixture is not intermediate but has the largest $\bar{\nu}_{Ca,S}$ and smallest $|\bar{Z}|$. From Fig. 8 the same is evident for $(G - U)$: At precipitate initiation, α_s-casein has the largest value (~ 3), β-casein is intermediate (~ 2.1) and the α_s-β-casein mixture is least (~ 1.9); U is tightly bound solvent and is taken as 0.5.

At high $\bar{\nu}_{Ca,S}$ there is apparently a large reversal of monomer net charge for all systems. However, $(G - U)$ and solubility do not reverse significantly. Since ionic strength alone could not cause precipitation using β-casein or the α_s-β-casein mixture (Waugh et al., 1970), it is proposed that an important source of attractive energy for precipitation is cross-linking of polymers by calcium. Other interaction bond types are probably also involved.

Interactions between α_s- and β-caseins appear to alter significantly inter- and intracore polymer interactions. For example, while both α_s- and β-caseins alone initiate precipitation at relatively high $\bar{Z}$, the mixture requires a low $\bar{Z}$. It has been shown (Section VI.A) that α_s- and β-caseins commingle at the monomer level. The present results suggest that when either protein alone forms a core polymer, interpolymer cross-linking by calcium is favored compared to the mixed polymer. In the latter, acidic peptides may interact in such a way as to convert potential interpolymer cross-linking sites into intrapolymer sites. At $I = 0.04$ and calcium concentrations near $0.017 \, M$ (conditions for micelle formation) the monomer net charge is near -2 for α_s- and β-caseins and zero for the mixture.

Slattery and Waugh (to be published) have examined $\bar{\nu}_{\text{Ca,s}}$ and proton release, as just described, from the standpoint of monomer structure. They excluded models in which net charge is uniformly distributed, either over the molecular surface as in the model used by Scatchard et al. (1957) and Scatchard and Yap (1964) or in a selected volume as in the models of Hermans and Overbeek (1948) and Tanford (1955). First, monomer net charge is essentially concentrated on short acidic peptide segments. Second, to account for hydrodynamic properties the acidic peptides must dominate large solvent volumes (Waugh et al., 1970). Finally, distributed-charge models cannot account for calcium binding near $\bar{Z} = 0$ or for charge reversal. Slattery and Waugh have used the Harris-Rice discrete-charge approach to calculate the effects of electrostatic interaction on binding (Harris and Rice, 1954, 1956; Rice and Harris, 1954). Models based on Structure I for α_s- and β-casein acidic peptide composition were developed using customary bond angles and distances. The acidic peptide of of α_s-casein is considered to be a torus (Fig. 6) and that of β-casein an extended rod, and on these all fully charged groups are placed so as to minimize electrostatic free energy at $I = 0.04$.

The fundamental equations used in calculations are as follows. For monomer electrostatic free energy

$$W_{el} = \frac{\epsilon^2}{D} \sum_{j=1}^{n} Z_j \sum_{k>j}^{n} Z_k \frac{\exp(-\kappa r_{jk})}{r_{jk}} \tag{5}$$

Here, ϵ is the protonic charge, D is the dielectric constant (taken as 45), Z_j or Z_k is the charge on the jth or kth site, κ is the Debye parameter using the solution ionic strength, and r_{jk} is the distance between the jth and kth sites.

For electrostatic interaction in ion binding,

$$\bar{\nu}_i = \sum_{j=1}^{n} \frac{k_{ij}^{\circ} a_i \exp[-\phi_{ij}(\bar{Z})]}{1 + \sum_{q=1}^{m} k_{q,j}^{\circ} a_q \exp[-\phi_{qj}(\bar{Z})]} \tag{6}$$

In this equation, n is the total number of binding sites, m is the number of different ion species competing for these sites, a_q is the activity of the qth ion species, and $\bar{\nu}_i$ is the total binding of the ith species, in moles ion per mole protein. The $k^\circ_{q,j}$ are the intrinsic binding constants of the qth ion species to the jth site. The intrinsic binding constants of interest are those for the binding of hydrogen, calcium and sodium to phosphate groups ($k^\circ_{H,P}$, $k^\circ_{Ca,P}$ and $k^\circ_{Na,P}$) and carboxylate groups ($k^\circ_{H,C}$, $k^\circ_{Ca,C}$ and $k^\circ_{Na,C}$). Each term of the sum over j represents the average binding of the ith ion species to the jth site. In (6), $\phi_{ij}(\bar{Z})$ is equal to $(\partial W_{el}/\partial Z_j)\, NZ_i/RT$ (Tanford, 1961), which is the change in molar electrostatic free energy as binding alters the charge on the jth site, keeping other fixed charges constant. Here N is Avogadro's number, R is the molar gas constant, and T is the temperature. Differentiation of (5) yields, per mole,

$$\phi_{ij}(\bar{Z}) = \frac{NZ_i}{RT}\frac{\partial W_{el}}{\partial Z_j} = \frac{\epsilon^2 NZ_i}{RTD}\sum_{k=1,\,k\neq j}^{n} Z_k \frac{\exp(-\kappa r_{jk})}{r_{jk}} \quad (7)$$

When a doubly charged calcium ion is bound at a particular site, $\phi_{ij}(\bar{Z})$ is twice that for a singly charged ion. The electrostatic influence at the jth site, due to other sites, for a divalent ion will then be the square of that for a monovalent ion. Since the charge on the body is low, it is assumed that ions bind to body groups (phosphate and carboxylate) at their k° values.

Given a peptide structure which is fully charged, (7) readily yields the $\phi_{ij}(\bar{Z})$ for each site. However, binding alters the average charge at a site, and this will affect the electrostatic contribution of that site to all other sites. What is required is the bound-ion distribution at equilibrium. To obtain this, an iterative process, using a set of k° and (6) and (7), is applied until constant values of $\phi_{ij}(\bar{Z})$ and $\bar{\nu}_i$ appear.

A number of binding constants for simple compounds are available in the literature. For the second binding of H^+ to orthophosphate, $k^\circ_{H,P}$ is about 7×10^6. However, for alkyl phosphates, Kumler and Eiler (1943) found values as low as 3×10^6, and Alberty et al. (1951) reported values of an apparent association constant $k^\circ_{H,P}$ near 1×10^6 for AMP-5, α-AMP-3, and β-AMP-3. Use of activity coefficients would correct this upward by a factor of about two. Davies and Hoyle (1953) obtained apparent binding constants for Ca^{2+} to HPO_4^{2-} (orthophosphate) ranging from 400 to 600, while the results of Smith and Alberty (1956a) gave an apparent constant of 50 ± 2 at 25°C and $I = 0.2$. Calculation of the activity coefficients, using the data of Kielland (1937), gives a corrected value of $k^\circ_{Ca,P} = 560 \pm 22$. Measurements of pH were used by Smith and Alberty (1956b) to determine $k^\circ_{Na,P} \sim 4$. Strauss and Ross (1959), using conductance and electrophoresis methods, obtained $k°_{Na,P}$ near unity, and this value has been used uniformly in calculations. Simple organic acids give $k°_{H,C} = 7 \times 10^4$. The binding constant $k°_{Ca,C}$ has not been measured. Slattery and Waugh use ratios of known constants to make estimates. The binding constant $k°_{Na,C}$ is obviously negligible.

An examination of initial calcium binding at $I = 0.04$, using maximally expanded acidic peptides (maximum r_{jk}) for both proteins, suggests the following set of $k°$: $k°_{H,P} = 9 \times 10^5$, $k°_{Ca,P} = 72$, $k°_{H,C} = 7 \times 10^4$, $k°_{Ca,C} = 5.6$ and $k°_{Na,P} = 1$.

Calculations show that the geometries of the acidic peptides (which fix the r_{jk}) must change with ionic strength and binding; otherwise the calculated and observed calcium binding sequences always have large divergences. It is noted that changes in the conformation parameters which adjust r_{jk} must account for effects other than alterations in peptide conformation alone, such as spatial crowding of charges on peptides resulting from increases in the degree of association in core polymer and precipitate formation. For this reason conformation parameters are considered adjustable, and values are obtained such that calculated and observed $\bar{v}_{Ca,s}$ are in agreement. A calculated proton binding is then obtained using each parameter value, and calculated proton release is obtained from binding sequences. These are then compared with observed proton release.

Comparisons of the calculated and observed sequences of proton release, with increasing I and $\bar{v}_{Ca,s}$, reveal situations in which calculated proton release is smaller or larger than observed. This suggests that in the system itself, there are small proton sources in addition to those explicitly recognized in model calculations thus far, and also small proton sinks, which have not been recognized. To account for observed proton releases, electrostatic effects which alter $k°$ and cause appropriate groups to act in the required fashion are required. A single example will illustrate a phenomenon which in fact could be complex. Consider that an appropriate group is on the body and that it has a p$K°$ close to the ambient pH of 6.6. The imidazole side chain of histidine is such a group, pK 5.6–7.0, and there are six on the bodies of both α_s-casein (Ho and Waugh, 1965) and β-casein (de Konig and van Rooijen, 1965). If, in the monomer, such a group is close to a negative charge, with increasing I it will act as a proton source; if it is close to a positive charge, it will act as a proton sink. It will be expected to act as a source or a sink if, as a result of peptide collapse or increasing association, it is brought progressively into proximity to a positive or negative charge, respectively. Sources of $\sim$0.8 proton are always required for both proteins for maximum alterations in I alone, and a source of $\sim$0.4 proton is required for β-casein also for increasing $\bar{v}_{Ca,s}$, at any I employed. With increasing $\bar{v}_{Ca,S}$, α_s-casein appears not to require a source or a sink at $I = 0.04$ or 0.08, but it requires a sink of ~ 0.5 proton at $I = 0.16$.

The values $k^\circ_{H,P} = 9 \times 10^5$ and $k^\circ_{Ca,P} = 72$, selected from calculations, are lower than literature values. Increasing $k^\circ_{Ca,P}$ significantly requires a peptide expansion at $I = 0.04$ (increase in r_{jk}) which appears to be physically unattainable. Increasing $k^\circ_{H,P}$ would require, for α_s-casein, a large additional proton source for increasing I and a large additional proton sink for increasing $\bar{v}_{Ca,S}$. If both k° are increased, for example, if $k^\circ_{H,P} = 3 \times 10^6$ and $k^\circ_{Ca,P} = 240$, peptide expansion is unrealistic. A small source would be required for increasing I (~ 0.5 protons) and a large sink (~ 1.5 protons) for increasing $\bar{v}_{Ca,S}$.

Folsch and Österberg (1959, 1961) have measured apparent association constants at $I = 0.15$ and 25°C of a variety of small peptides containing O-phosphorylated serine. The range of pK is 5.4–6.0. The authors noted that their association constants may be lower than those of Kumler and Eiler (1943) as a result of the fact that the compounds used carry a positive charge on an amino nitrogen atom neighboring the phosphate group. Folsch and Österberg were also careful to note that the constants they gave are apparent constants and that binding of sodium may contribute to their departure from intrinsic constants.

Österberg (1961) has obtained titration curves for the acidic peptide of α_s-casein. He calculated an intrinsic constant for the second phosphate hydrogen of $k^\circ = 2 \times 10^5$ at $I = 0.15$ and 25°C. The electrostatic factor w is found to be 0.071. Österberg assumed that over the range of binding for the second hydrogen of phosphate ($\bar{Z}$ from -13 to -22) a plot of pH $+ \log \bar{v}/(n-\bar{v})$ against $\bar{Z}$ should give a straight line whose intercept is $\log K^\circ$ and slope $0.868\, w$. The constants given were thus obtained. By assuming w constant, the further assumption was made that the peptide has a fixed geometry. Österberg, in his dicussion, noted that the geometry is probably not fixed. If not, as is expected, w becomes a function of $\bar{Z}$, a common observation for flexible macromolecules. To these must be added the observation that w, at $I = 0.15$, would not be expected to be 0.07 ($\phi(\bar{Z}) \simeq -3.5$) but considerably less. Österberg obtained, inexplicably, a fluctuating progression of w between 0 and 0.071 when the various acidic groups were examined as described. Finally, the binding of sodium would be expected to shift apparent k° to lower values. The problems associated with an interpretation of the titration curve of the acidic peptide are only slightly less than those involved in an examination of calcium binding and proton release, as given above. It is evident that local electrostatic effects will affect proton binding and peptide conformation, and these must be taken into account in obtaining intrinsic binding constants.

Monomer charge reversals at $0.08\, M$ calcium are calculated by Waugh et al. (1971) to be about $+7$ for α_s-casein, $+4$ for β-casein, and $+6$ for an α_s-β-casein mixture. Reversal must be due to calcium binding and cannot be accounted for unless, similar to the situation for proton release, interactions take place which are not explicitly recognized in model calculations. When charge reversal takes place, solvation is least (Fig. 8), and monomers are closely packed. There is then no problem in developing sufficiently small r_{jk} so that strong electrostatic interactions can occur. Evidently it is required that the groups be positioned by the specificity of the monomer interaction so that local cooperative electrostatic effects produce the required binding. This would obviously not be expected from average properties. Cooperative and hysteresis effects will be examined further in Section IX.

VII. The Micelle Coat

The stabilizing characteristics of the covalent polymers of native κ-casein have been dealt with in Section V and will be considered further in Section VIII. Important results have been obtained from particular studies of monomers and polymers. Stable monomers, obtained by reduction and carboxymethylation (CM-κ-caseins), have been shown by Mackinlay and Wake (1964) to stabilize α_s-casein. This has been confirmed by Woychik et al. (1966), Schmidt et al. (1966), and Talbot and Waugh (1970). The individual CM-κ-casein components are also able to stabilize α_s-casein. This has been shown by Mackinlay and Wake (1965) for the major component which contains negligible carbohydrate as well as for the other components. It is clear that the carbohydrate-free macropeptide must be capable of generating a noninteracting surface. Mackinlay and Wake have also found that micelles stabilized by carbohydrate-free κ-casein can be clotted by rennin.

Talbot and Waugh (1970) have examined surface activities of a variety of monomeric and polymeric materials: (a) SH-κ-casein-(monomer); (b) SS-κ-casein-(monomer), obtained by reoxidizing SH-κ-casein in the presence of dissociating agents; (c) standard (natural) covalent polymeric κ-caseins and covalent polymers obtained by reoxidation of aqueous solutions of κ-casein; (d) CM-κ-casein; (e) CAM-κ-casein; and (f) standard κ-caseins with one to three ϵ-amino groups per monomer coupled with each of a variety of reagents. They have also examined the molecular interactions involved in the production of standard κ-casein by the procedure of McKenzie and Wake (1961). Preparations were compared according to their capacity to stabilize α_s-casein as micelles at $R_i = 10$ and their efficiency to transform large micelles to micelles of smaller size at $R_s = 8$. Mestastable colloid of α_s-casein was used in the latter test (see the footnote in Section VI.A, Table III and Section VIII). Both tests were carried out at $I \sim 0.04$, pH 6.6, and 37°C. Talbot and Waugh came to the following conclusions:

Standard κ-casein and reoxidation products of SH-κ-casein are SS-bonded covalent polymers, most of which are larger than dimers. Their M_n and $s_{20,w}$ are stable in aqueous or dissociating solvents, which suggests chain termination—either by terminal SS-bond formation with small molecules (open chains) or by grouping in SS-bonded rings. An examination of the range of M_n and the full stabilizing capacities and size-transforming efficiencies of several preparations is considered to favor open chains. In this respect it appears that SS-κ-casein-(monomer) is a monomer bonded to two mercaptoethanol molecules.

In aqueous solution, standard κ-caseins, and preparations of covalent polymers obtained by reoxidation, yield a wide range of $s_{20,w}$ (19.6 ± 4.8 SD). Although the $s_{20,w}$ of the maximum ordinate of a centrifuge peak is reproducible for a particular preparation, heterogeneity is always indicated by a sharper trailing edge (Waugh and von Hippel, 1956; Long et al., 1958; Wake, 1959; McKenzie and Wake, 1961; Swaisgood and Brunner, 1963; Mackinlay and Wake, 1964; Garnier et al., 1964; Swaisgood et al., 1964; Noble and Waugh, 1965). There is agreement that the ultracentrifuge pattern of each preparation is relatively insensitive to pH, ionic strength, temperature and calcium concentration. Hill and Wake (1969) suggest that κ-casein association products resemble soap micelles. It seems likely that association products are closed structures in the sense that product surfaces are devoid of protein–protein reactive sites. This would result if lateral κ-κ-casein specificity of interaction positions macropeptides at the surface, as they are on casein micelles. Heterogeneity would result from a combination of polymer distribution, limitations on the geometry of packing, and the requirements for lateral interaction.

Association products are also formed by SH-κ-casein monomers in aqueous mercaptoethanol solution ($s_{20,w} = 11.0 \pm 1.25$ S). Lateral interactions appear to position SH groups so that, on reoxidation, the formation of an SS bond between κ-caseins is more probable than a bond between κ-casein and mercaptoethanol. All reoxidation products give covalent polymer gel electrophoresis patterns in which monomer bands are not evident. The same result is obtained when SH-κ-casein is reoxidized while it is on the surfaces of casein micelles.

The data of Yaguchi and Terassuk (1967) suggest that larger covalent κ-casein polymers compared to small polymers contain more κ-casein monomers of lower neuraminic acid content, possibly as a result of electrostatic interaction during polymer formation, and that during the formation of natural κ-casein, lateral κ-casein interactions again appear to make SS-bond formation of κ-κ-casein more probable than coupling to give SS-κ-casein-(monomer). Such monomers are rarely seen in standard κ-casein preparations. Most of the κ-casein in milk is in covalent polymers, the observed range of which is from trimers (Swaisgood and Brunner, 1963; Swaisgood et al., 1964) to polymers considerably larger than decamers, to give M_n of 200,000 (Talbot and Waugh, 1970). The chains appear to be open and terminated by SS-bond formation with small mercaptan molecules (possibly glutathione or cysteine).

The results of Talbot and Waugh also show that (a) since SH-κ-casein has full stabilizing capacity and transforming efficiency, SS bonds are not required to establish a conformation for interaction specificity; (b) monomeric κ-caseins, compared to covalent polymers, have increased efficiency to transform large micelles into micelles of smaller size, normal $α_s$-casein stabilizing capacity, and decreased ability to stabilize micelles against an increase in environmental calcium (descending limb of the peak); at the micelle surface, it is evidently more difficult either to insert or to remove covalent polymers compared to monomers; (c) a contribution to the association of κ-casein with $α_s$- or β-caseins comes from an electrostatic interaction between a positive site on κ-casein and the negatively charged acidic peptides on the other proteins. This is indicated by the effects of chemical modification of ε-amino groups of standard κ-casein on stabilization and size transformation and by the effects of ions on κ-casein fractionation and on κ-casein micelle surface activity. They interpret their results to suggest the following interesting mechanism for micelle formation *in vivo*. In the mammary gland, particles of $α_s$-β-caseinates are formed independently of SH-κ-casein. When introduced, SH-κ-casein stabilizes and transforms these particles to give the natural micelle distribution. Subsequently, oxidation on the micelle surface takes place in the presence of sufficient chain-terminating species to give the covalent polymer distribution in a particular milk. In this way superior properties would be utilized: those of monomers to produce a maximum micelle population surface area and those of polymers to produce population stability. In the next section it will be shown that κ-casein exists at the micelle surface as a monolayer. It has a thickness of approximately 2.8 nm.

Photooxidation of κ-casein and whole casein in the presence of methylene blue affects histidine most rapidly, tryptophan less rapidly and tyrosine least rapidly, according to the results of Hill and Laing (1965). On rennin treatment a reduced amount of nonprotein nitrogen (NPN) release is observed. Photooxidation of whole casein gives the following results: At pH 6.25 the addition of 0.025 M calcium chloride apparently leads to micelle formation but in samples where 65–79 % of the histidine has been modified, samples fail to clot or clot feebly. Zittle (1965) reported that when 100% of the histidine and 70% of the tryptophan are oxidized, κ-casein completely loses its ability to stabilize $α_s$-casein and is not acted on by rennin. Photooxidized κ-casein can also stabilize κ-casein against the effects of rennin action. Photooxidized systems have not yet been analyzed to the point where differentiation can be made with respect to alterations in association reactions and, for rennin action, nonavailability or nonreactivity of rennin-sensitive sites. Photooxidation of $α_s$-casein has been shown by Zittle (1963) to lead to an increased electrophoretic mobility and to a loss of sensitivity (precipitation) to calcium. It is interesting that Nakai et al. (1966) have oxidized the tryptophan of κ-casein with N-bromosuccinimide to the extent of 2 mole/28,000 daltons with only slight degradation of the molecule. The resulting product stabilizes $α_s$-casein.

Pepper and Thompson (1963) have dephosphorylated κ- and $α_s$-caseins, using a spleen phosphoprotein phosphatase, each to 90% or greater in extent. As expected, dephosphorylated $α_s$-casein has a considerably reduced electrophoretic mobility. The dephosphorylated κ-casein band remains diffuse. Proteolysis does not occur. Stability tests were carried out according to the method of Zittle (1961) at 0.02 M calcium at pH 7. The ionic strength was not specified. The ability of κ-casein to stabilize is not significantly impaired by dephosphorylation; 90% stabilization occurs for natural and treated protein at R_i ~8. When dephosphorylated $α_s$-casein is mixed with native κ-casein, stabilization is drastically impaired. The authors concluded that the phosphate groups of κ-casein are not an important factor leading to the formation of stable micelles but that the micelle-forming reactions are dependent on the interaction of calcium ions with ester phosphate groups of the $α_s$-casein molecules. As indicated in Section VI, the core polymer has a particular geometry, and as indicated in this section, interaction involves positively charged loci on κ-casein and the acidic peptides on $α_s$- or β-caseins. Dephosphorylation would be expected to reduce (possibly eliminate) the specificity of interaction involved in forming a coat unit, thus to reduce or eliminate stabilization.

VIII. Equilibrium Casein Micelle Systems

Noble and Waugh (1965), using mixtures of $α_s$- and κ-caseins, observe that at pH 7.1, $I = 0.075$ and 37°C, single-aliquot addition of calcium to progressively increasing concentrations produces precipitates of Ca $α_s$-caseinate at calcium concentrations below those required for complete micelle formation. This result and path dependency of micelle formation have been discussed in Section V.

The clear-cut path dependency for micelle formation has not always been observed. An example is given by Waugh (1961). Calcium was added to first-cycle casein at 2°C to give a concentration of about 0.03 M, the exact level being chosen to give a copious precipitate. After the precipitate had settled, the temperature was raised to 37°C. With time, a suspension of stable micelles appeared. A similar situation was encountered, on occasion, if the addition of a single aliquot of concentrated calcium chloride (usually tenfold final concentration) to an aliquot of casein mixture was not sufficiently rapid: A precipitate appeared first, and this transformed slowly into a suspension of stable micelles. This was observed for solubilized skim milk, first-cycle casein, and, of particular interest, mixtures of α_s- and κ-caseins. Such transformations were sufficiently frequent to suggest that the micelle state might be the equilibrium state under appropriate conditions. In considering systems in which precipitate transformed to micelles it was noted that an exact control of pH and ionic strength had not been exercised—because calcium binding liberates protons and therefore affects pH, and because differences in the salt content of first-cycle casein alters ionic strength.

Zittle and Jasewicz (1962) studied the effects of ionic strength on the stability of micelles obtained from α_s- and κ-caseins. They did not investigate calcium concentrations below 0.01 M and thus did not obtain the dip. Stabilization ratios up to $R_i = 10$ were obtained, and it was observed that results were the same at pH 6 as at pH 7. At 30°C, between $I \simeq 0$ and $I = 0.17$, assay centrifugate progressively increased. Above $I = 0.17$, apparent stabilization increased as a result of protein remaining in solution. These experiments were carried out on whole casein. Using mixtures of α_s- and κ-caseins at $R_i \sim 10$, the same behavior pattern appeared. With α_s- and κ-casein mixtures at 7°C, destabilization was in agreement with the observation of calcium splitting. Noble and Waugh (1965) also found that variation in pH from 6 to 7.5 had little effect on the stabilization behavior of α_s-κ-casein mixture but that a decrease in ionic strength below 0.075 narrowed the dip. Above $I = 0.075$ an increase in ionic strength increased the width and depth of the dip. Ionic strength appeared to be an important variable.

Path dependencies have been further examined using solubilized skim milk, first-cycle casein, and mixtures of α_s- or β-casein with κ-casein, at several values of pH and ionic strength, I (KCl plus 0.01 M imidazole). Figure 9 refers to first-cycle casein at 9 absorbance units (au) per ml, and illustrates the general result using single-aliquot addition of calcium and 75 min incubation at 37°C and pH 6.9. Systems were assay centrifuged and the

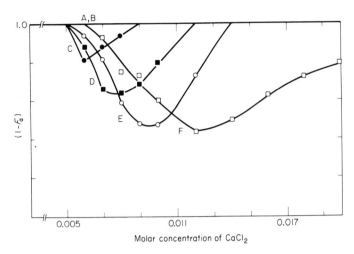

FIGURE 9. Effects of ionic strength on micelle formation at pH 6.6. Data were obtained using single-aliquot addition of calcium chloride to first-cycle casein at 9 au per milliliter. F_a is the fraction of total absorbance removed by assay centrifugation. Conditions are 0.01 M imidazole and KCl concentrations as follows: A, 0.03 M; B, 0.05 M; C, 0.07 M (●); D, 0.10 M (■); E, 0.15 M (○) and F, 0.20 M (□). A and B remain at $F_a \sim 0$.

fraction of the initial absorbance appearing in the pellet, F_a, was determined; thus $(1 - F_a)$ of Fig. 9 is the fraction remaining in the supernatant. For I below 0.07 a precipitate of Ca α_s-caseinate (the dip) does not appear; above 0.005 M calcium these systems progressively increase in turbidity to full micelle formation. In agreement with previous results, at $I = 0.07$ coalescent pellets form between 0.005 and 0.007 M calcium, after which they become progressively more particulate. The dip obviously deepens and broadens as I is increased. Other experiments show that the maximum I at which the dip is absent is near 0.055. As the pH is decreased from 7.1 to 6.3, at appropriate I in the range examined, the dip decreases slightly. These results indicate that the dip is more sensitive to an alteration in ionic strength than it is to pH, possibly as a result of the fact that, as shown by titration curves (Ho and Waugh, 1965), the intrinsic charge on α_s-casein, also presumably on β-casein, does not change appreciably between the pH limits examined. Under conditions where the dip is absent it would be expected that equilibrium dialysis could be used to establish micelle systems. This has been found to be the case. At $I = 0.05$ and pH 6.6, α_s- and κ-caseins (at $R_i = 4, 8,$ and 16; 10 mg/ml α_s-casein), on equilibrium dialysis against 0.015 M calcium, produce stable micelle systems without intermediate precipitate formation. At $I = 0.075$ and pH 6.1, Ca α_s-

caseinate precipitate appears during dialysis, and supernatants are relatively clear and contain only 9, 6, and 4% of the initial protein, respectively.

Using $I = 0.05$, pH 6.6, 37°C and final calcium concentrations near 0.02 M, Waugh and Talbot (to be published) have examined the properties of a variety of micelle systems. They used differential ultracentrifugation (see p. 24) to characterize systems containing α_s- and κ-caseins with respect to (a) the weight fractions of total protein in micelle classes of different size, (b) the total micelle population surface area, and (c) equilibrium solution protein (coat). Micelle classes were assigned radii, r_i, of 1169, 425, 143, 48.5, and 24.8 nm; supernatant protein contains particles having radii less than 15.5 nm. The major results and conclusions are now summarized.

Using equilibrium dialysis and R_i between 25 and 3 the following are observed. At $R_i = 25$, some particulate sediment (large micelles) appears at 1 g, and assay centrifugation produces a large particulate pellet. For systems at $R_i = 20, 7, 5, 4$ and 3, sediment does not appear. Differential centrifugation reveals that as R_i decreases, the average micelle size decreases, from $r_i \sim 425$ nm at $R_i = 20$ to $r_i \sim 36$ nm at $R_i = 3.0$. For each R_i, however, over 80% of the micellar protein is distributed in two or three neighboring micelle classes. The concentration of protein (including coat) in equilibrium with micelles remains low but increases slowly, from 0.15 mg/ml at $R_i = 20$ to 0.4 mg/ml at $R_i = 3.0$. At $R_i = 7$ equilibrium protein increases from only 0.17 to 0.22 mg/ml as the α_s-casein concentration is increased from 7 to 21 mg/ml. Equilibrium coat, then, increases as r_i decreases. For skim milk, these R_i systems, and other systems which are near equilibrium (to be described) the micelle population surface area and κ-casein content yield a nearly constant value of 3400 Å^2 per κ-casein monomer of $M = 20{,}000$. κ-Casein is evidently present as a monolayer at the micelle surface.

Micelle systems at $R_i = 20$ are mixed with κ-casein and systems at $R_i = 1, 2, 3, 4,$ and 5 to give systems at $R_s = 7$. Three distributions are compared: (a) the distribution predicted on the assumption that size rearrangement does not occur after mixing, (b) the distribution observed after 4 hr incubation of the mixture at 37°C and (c) the distribution observed for $R_i = 7$. In all cases, the largest and smallest micelles of the predicted distribution disappear, and there appear micelles near the size corresponding to the maximum weight fraction for $R_i = 7$. Although equilibrium is not reached in 4 hr (observed weight fractions on either side of the maximum for $R_i = 7$ are still too large), a highly significant rearrangement of micelle sizes (size transformation) has occurred. Given sufficient time it is expected that the equilibrium distribution for $R_i = 7$ will be attained.

TABLE III

Stabilization and Transformation of Metastable Colloid of Calcium α_s-Caseinate[a]

Sample	$R_s{}^b$	$F_a{}^c$	Sample	$R_s{}^b$	$F_a{}^c$
Control		0.68	5	19	0.74
1		0.98	6	14	0.62
2	191	0.93	7	9.6	0.55
3	95	0.92	8	4.8	0.30
4	48	0.90	9	1.9	0.02

[a] $I = 0.03$, 0.022 M calcium chloride, pH 6.6 and 37°C.
[b] R_s is the system weight ratio of α_s-/κ-caseins after addition of κ-casein.
[c] F_a is the fraction of the total protein in a pellet after assay centrifugation.

That there is no intrinsic upper limit to the size of a micelle, and that very large micelles can be size transformed, is shown as follows. A metastable colloid containing 7.1 mg/ml Ca α_s-caseinate was formed at 37°C, pH 7.0 and $I = 5 \times 10^{-4}$ by single-aliquot addition of calcium chloride to give 0.0045 M. No protein sedimented at 1 g, at least during 1 min. By electron microscopy, the colloid particles were observed to be spherical and to have an average size near 1 μ. Assay centrifugation of the colloid produced a coalesced pellet and $F_a \sim 0.68$ (control in Table III). An alteration of the ionic environment to that of peak micelle formation ($I = 0.03$ and 0.022 M calcium chloride at pH 6.6 and 37°C) was obtained by mixing 1.0 ml of colloid and 0.2 ml of a solution containing 0.108 M calcium chloride and 0.18 M sodium chloride. Sample 1 of Table III refers to an alteration without added κ-casein: $F_a = 0.98$ (coalesced pellet) and the protein remaining in the supernatant is consistent with Ca α_s-caseinate solubility. Additions of κ-casein, in the added salt solution, were made to give system weight ratios, R_s, of 191–1.9. One hour after addition, F_a were determined, and these are recorded for samples 2–9 in Table III. In all cases the pellets were found to be particulate.

When κ-casein accompanies a change in ionic environment it is expected, from relative diffusion constants, that the fastest reactions are ion–protein interactions and, thus, that colloid particles are rapidly altered to particles of Ca α_s-caseinate (cores) which will spontaneously coalesce. Intermediate in rate are interactions of κ-casein with cores, and least rapid are interactions between the core particles themselves. For R_s of 191–18 some protein sediments during 1 min at 1 g. For $R_s \geq 38$, the appearance of sediment at 1 g and F_a greater than 0.68 (control) shows that there is still significant coalescence of colloid particles before coating is sufficient to prevent its occurrence. At $R_s = 19$, $F_a = 0.74$, nearly that of the control. This suggests that coating is sufficiently rapid to preserve colloid particles essentially at original size. Since precipitation in the absence of κ-casein takes seconds while size transformation, to be considered next, takes minutes, immediate stabilization can hardly be attributed to a source other than a coating of the colloid particles by κ-casein. The colloid particles become the cores of large micelles.

As R_s is decreased below 19, F_a progressively decreases. There is also a progressive decrease in turbidity (apparent absorbance at 360 nm). Evidently, κ-casein sufficiently in excess of that required for size stabilization produces a spontaneous decrease in average size (size transformation). In order for a micelle to be stable to assay centrifugation it must have $r_i < 150$ nm (p. 24). If a large micelle has $r_i = 600$ nm, in order to yield stable products it must than be subdivided into a minimum of 64 micelles.

When colloid particles are established at smaller average original size, by decreasing the calcium concentration for colloid formation, the κ-casein requirement for stabilization at original size is found to be increased. Studies of time dependency at $R_s = 8$ show that the change in F_a during the first 10 min of incubation is more than 2.5 times as great as the change in the next 5 hr.

It is shown further that an appropriate micelle system can stabilize large amounts of added α_s-casein. A micelle system was established at 37°C by single-aliquot addition of calcium chloride at $R_i = 4$ to give 18 mg/ml α_s-casein, 0.04 M KCl, 0.018 M CaCl$_2$ and 0.005 M sodium cacodylate at pH 6.6. A 2.6 ml aliquot was taken and to it were added, alternately at 37°C with gentle mixing, 0.1 ml aliquots either of α_s casein at 40 mg/ml, $I = 0.003$ and pH 7, or of a buffer containing 0.04 M CaCl$_2$, 0.08 M KCl and 0.01 M sodium cacodylate at pH 6.6. The ionic environment and pH were thus maintained. A total of 1.3 ml each of buffer and α_s-casein solution (52 mg) were added to give final $R_s = 8.4$. All of the added α_s-casein was stabilized as micelles, half in micelles sufficiently large to be removed by assay centrifugation and half in micelles having r_i from 50 to 425 nm. Original micelle systems were fractionated by ultracentrifugation and supernatants tested for stabilization of α_s-casein, as described. The results suggest that, for the whole micelle system, the κ-casein required to stabilize half of the added α_s-casein comes from micelles having r_i of 24.8–48.5 nm. Most of the κ-casein required for stabilization comes from preexisting micelles and not from free κ-casein or free coat.

Having shown that micelle systems are equilibrium systems under the environmental conditions used, an examination is made of system free energy. From the Gibbs and Langmuir equations are obtained

$$\gamma_0 - \gamma_i = \frac{kT}{X_s} \ln\left(\frac{X_i}{X_i - X_s}\right) \tag{8}$$

and

$$\gamma_0 - \gamma_i = \frac{kT}{X_s} \ln(1 + K_i a_c) \simeq \frac{kT}{X_s} \ln(K_i a_c) \tag{9}$$

where γ_0 and γ_i are the surface free energies per unit area of an uncoated core, taken to be independent of size, and of a micelle at radius r_i; X_i is the micelle surface area per κ-casein molecule and X_s is this area at surface saturation ($\theta = X_s/X_i$); K_i is the surface association constant for a micelle at r_i; k is Boltmann's constant; T is temperature; and a_c is the solution coat activity with which micelles are in contact. Taking $X_s = 3.4 \times 10^{-13}$ cm² per molecule and $\gamma_0 - \gamma_i \geq 0.6$ erg/cm² in (8) yields $\theta > 0.99$. Since θ is expected to be near unity, full lateral interactions between κ-casein molecules are expected to contribute to K_i. In the following, θ is taken as unity and in (9) the assumption is made that $K_i a_c \gg 1$.

In a closed system it is recognized that subdivision of an initial single large micelle to give a uniform micelle population will produce a change in system for energy, ΔG, given by

$$\Delta G = \gamma_p \Omega_p - \gamma_q \Omega_q + CRT \ln\left(\frac{a_p}{a_q}\right) \tag{10}$$

where the subscripts q and p refer, respectively, to the initial system containing the single micelle and the final system after subdivision to give a *uniform* micelle population; Ω is the initial micelle or micelle population surface area; γ is the initial micelle or population surface free energy; C is the total moles of coat in the system; and a is the solution coat activity. Using $d(\Delta G)/dr = 0$ from (10) for a minimum in system free energy, (9), and conserving mass yields, for a uniform equilibrium population,

$$\gamma_p = \frac{kT}{X_s}\left(1 - \frac{r_p}{K_p}\left(\frac{dK}{dr}\right)_p\right) \tag{11}$$

At 37°C, $kT/X_s = 0.13$ erg/cm².

With respect to these equations, the important characteristics of micelle systems are the following. As R_i is progressively decreased, the maximum in the distribution of micelle weight fractions shifts to micelles of smaller

9. FORMATION AND STRUCTURE OF CASEIN MICELLES

size, the average micelle size decreases, and equilibrium coat activity increases. For an equilibrium system to have a stable micelle distribution it is considered that γ_i cannot be zero or negative. For γ_i positive, it is concluded that there is a dependency of K_i, and therefore γ_i, on r_i and that dK/dr is positive. In a micelle size distribution in equilibrium with coat at a_e, it appears that as r_i decreases, γ_i increases. Clearly, there is a balance between micelle population surface free energy and free energy of coat in solution and that the relation between γ_i and r_i is an important factor in determining equilibrium size distribution.

When equilibrium coat is altered, it is expected that γ_i will still increase as r_i decreases. The reactions of size transformation (subdivision and coalescence) are examined with respect to this expectation. The mechanism of subdivision is particularly interesting and may be examined by the events which take place following the addition of κ-casein (or coat) to an equilibrium micelle system. After addition, the equilibrium solution coat activity, a_e, is increased to a_c, and R is decreased. At a_c, coat must enter a micelle surface to increase surface coat activity. The core responds to this introduction: The negative free-energy change in the coat is compensated to some extent by a positive increase in free energy in the core, the latter caused by a separation of core polymers. When interactions between a core polymer (or group of core polymers) and the remainder of the core have been sufficiently reduced, subdivision takes place. In Fig. 10 a free

FIGURE 10. Penetration of κ-casein into the coat layer of a casein micelle. Core polymers of mixed α_s- and β-caseins are shown internally. The acidic peptides of β-casein and α_s-casein are bonded differently to their respective bodies (see Section VI). When κ-casein combines with the nearest free α_s- or β-casein, an expansion of the micelle surface occurs, and core polymers are separated at their junctions.

κ-casein is shown entering the surface of a micelle. Subdivision is evidently accompanied by an increase in population surface area and a decrease in a_c. It seems likely that as micelle size increases, the products of subdivision must be progressively more unequal in size. The energy barriers to a subdivision reaction are evidently those attending the insertion of surface coat and the separation of core polymers at their junctions. If a_c is increased sufficiently it is possible that large micelles develop a transient, negative surface free energy.

Studies of equilibrium systems reveal a possible source for the path dependencies observed under other environmental conditions. Obviously, *if* micelle systems are potentially equilibrium systems under the environmental conditions discussed in Section V, path dependency and failure to show a size rearrangement toward a path-independent state is due to the fact that subdivision reactions have insignificant rates. The source of a low rate must then be an unfavorable combination of energy barriers—for the insertion of surface coat and the separation of core polymers.

IX. Natural Micelles, Other Colloid Particles, and the Aqueous Phase of Milk

The ionic composition of the aqueous phase of milk, the environment, is of primary importance. At 37°C the colloidal particles in milk are, to some extent, in ionic exchange with the environment and thus in exchange with each other. As will be evident, there is the possibility that skim milk contains colloids which are structurally different from the casein micelle; for example, an apatite-like matrix of calcium–phosphate–citrate stabilized by adsorbed protein, or metastable colloid of β-caseinate or α_s-β-caseinate. The general related problems are then (a) the equilibrium levels of ionic interactants, (b) the types of structurally different colloidal particles and the contributions which small ions make to the structure of each of these, (c) the conditions of formation of each type of colloid and (d) the stability of each type of colloid in the final environment. Stability will also be determined by the extent to which cooperative and hysteresis effects permit the structure and composition of a particular colloid type to respond to alterations in the environment. The natural casein micelle will emerge as a highly organized particle of proteins, calcium, phosphate and citrate.

A. The Ionic Environment and Inorganic Colloid

The ionic environment has been examined by analyzing (a) the fluids obtained after dialysis to equilibrium of small volumes of water against a

9. FORMATION AND STRUCTURE OF CASEIN MICELLES 59

large volume of milk, (b) ultrafiltrates of skim milk, (c) the whey after removal of micelles by centrifugation and (d) the whey after rennin coagulation. Some time ago, Van Slyke and Bosworth (1915), using rennet whey, obtained values of ionic constituents which are comparable to those obtained by more elegant procedures. The definitive work is that of Davies and White (1960), who made a careful and extensive study of ultrafiltrates and equilibrium dialyzates of skim milk at 20° and 3°C. They gave their results in terms of mg constituent/100 g milk using, as a correction for measured serum values, a factor equal to the weight of water in 100 g skim milk divided by the weight of water in 100 g serum. Published values should then be increased by about 3% to obtain serum values. The concentrations of pertinent constituents are given in Table IV. There are several small dialyzable ions and molecules not accounted for in Table IV. So far the relationships of none of these to micelle structure have been considered seriously.

The averages of dialyzable and filterable components at 20°C are given in column 4 of Table IV, and in column 5 these values are converted to molarities. Jenness and Koops (1962) have used these values to develop a buffer which duplicates the ionic environment. Phosphate may present a special problem since it is present, as inorganic and organic phosphate, in a variety of forms.

Comparing environmental and total constituents, as given in Table IV, permits estimates of the amounts of constituents which are not free to filter or dialyze (column 6). Constituents clearly occurring in excess in the milk compartment are calcium, magnesium, citrate and phosphate. Potassium appears to occur as a slight positive excess in the milk compartment and chloride as a slight negative excess. Sodium and lactose appear to be the same in the two compartments, although Davies and White show, by comparing filtrate or diffusate values with total milk values for some individual samples, that sodium is also present in slight positive excess. The fact that lactose is uniformly distributed must mean that this constituent is adsorbed to, or can largely penetrate, the aqueous portions of micelles; otherwise, a significant negative excess would be observed after compensating for the volume excluded by the protein. The data of Davies and White (1960) permit comparisons for individual milk samples. Eventually, it may be necessary to make many such comparisons to gain insight into the ionic (activity) levels which are most important with respect to homeostatic control.

The differences between total and environmental levels (column 6), according to Graham's original definition, are in some form of colloid. Complete removal of all colloid ions would require the extraction of 1.4 calcium ions and 0.05 citrate ions per inorganic phosphate ion. Such calcu-

TABLE IV

SKIM MILK AND SKIM MILK ENVIRONMENT (DIALYZATE OR FILTRATE)[a]

Component	Total		Environment, 20°C		Total minus dialyzable, Δ	Environment, 3°C	
	mg/100 ml	$M \times 10^3$	mg/100 ml	$M \times 10^3$	$M \times 10^3$	mg/100 ml	$M \times 10^3$
Ca	118	29	37.2	9	20	41.9	10.4
Ca^{2+}			11.5	3			
Mg	11.7	5	7.8	3	2	8.1	3.3
Na	49	21	49.4	21	0	51	22
K	150	39	138	35	4	140	36
P_{total}	101.6	33	40.1	13		43.4	14
P_i		(25)[b]	33.7	11	14	34.6	11.2
Citrate	171.1	9	160	8	1	164.3	8.6
Cl	103	29	109	31	2	110	31
Lactose	4790	140	4750	139		4834	141
SO_4	10[c]	1					
CO_2	20[c]	4					

[a] From data of Davies and White (1960).
[b] Value in parentheses is calculated
[c] Jenness and Patton (1959).

lated ion ratios, combined with observations of mixed calcium–phosphate–citrate precipitates (see below), have, I believe, contributed substantially to the assumption that in skim milk, there are ion combinations having, for example, an apatite-like structure (Rose, 1965, 1968).

Comparison of environmental levels at 20° and 3°C (Table IV) reveals small increases at 3°C in the levels of all constituents except chloride. That some of these are significant with respect to the ion balance in the environment will be shown below.

Certain of the environmental constituents listed in Table IV interact to form soluble complexes. In addition to species such as HPO_4^{2-}, $H_2PO_4^-$, $HCit^{2-}$, and H_2Cit^-, there will be species such as $CaH_2PO_4^+$, $CaHPO_4$, $CaPO_4^-$, CaH_2Cit^+, $CaHCit$, and corresponding interactants for magnesium. When the environment is such that the solubility of any electrically neutral compound is exceeded, it may precipitate. Dissociation constants at low ionic strength for a variety of soluble calcium salts of phosphate and citrate at 25°C are given by Davies and Hoyle (1953), with corrected values for calcium citrate by Davies and Hoyle (1955).

Boulet and Marier (1960) used these and other sources of dissociation constants to examine the solubility of tricalcium citrate in synthetic mixtures at temperatures from 20° to 95°C, from pH 4.4 to 8.8 and at several ionic strengths. The solubility product relation $pk_s = 17.63 - 10.84\ (I)^{1/2}$ was found to apply over these wide limits. They then examined milk ultrafiltrate compositions at 21°C, including those of White and Davies (1958; see Table IV), those of Smeets (1955) and those of Rose and Tessier (1959) and three ultrafiltrates prepared in their laboratory. Two of the latter ultrafiltrates were found *not* to dissolve added crystals of tricalcium citrate and, therefore, appeared to be saturated or poised in this respect. The environmental levels of Ca^{2+} and Cit^{3-} were then assumed to be those which would be in equilibrium with a solid phase. Boulet and Marier calculated ionic levels from environmental levels, the sources of which are given above, and found that calculated pk_s was surprisingly uniform and averaged 14.47. If this value is inserted into the equation just given, the milk environment is predicted to have a total ionic strength of $I = 0.08$. This is in reasonable agreement with an ionic strength calculated from the environmental levels recorded in Table IV.

Tessier and Rose (1958) added calcium and citrate to skim milk at pH 6.6 and analyzed the calcium, phosphorus and citrate contents of ultrafiltrates after different levels of addition. Addition of calcium decreased levels of phosphate and citrate, addition of phosphate decreased calcium, and addition of citrate increased both filterable calcium and phosphate but decreased ionic calcium. The results were interpreted in terms of apparent solubility products, that for $CaHPO_4$ being 1.5×10^{-5}. Since calcium, phosphate and citrate were added to skim milk, this constant includes interactions with micelles and soluble proteins as well as interactions between these ionic constituents. It is, therefore, an assumption that the addition of calcium "precipitated both phosphate and citrate," if precipitate refers to a nonmicellar phase. The inference, however, is that the environment is poised.

That the ionic environment may be poised for precipitation can be demonstrated directly in another way. If an equilibrium diffusate is made against skim milk at 4°C and the temperature of the diffusate is raised to 37°C, a colloid forms first, and this eventually develops into a precipitate. Inorganic colloids can be stabilized by protein, for example, silver halide colloid. It is possible that if inorganic colloids were to occur or form in milk they would also be stabilized by protein.

Kuyper (1938) has observed coprecipitation on neutralizing acidic solutions containing calcium, phosphate and citrate. Pyne and McGann (1960) have examined the compositions of precipitates formed on neutralizing acidic solutions containing calcium, magnesium and phosphate in concentrations corresponding closely to the *totals* found in skim milk but containing different amounts of citrate; thus, 0.03 M calcium plus magnesium and 0.02 M phosphate. At the citrate level of skim milk, 0.0093 M, the precipitate composition proposed by the authors with respect to these interactants is $Ca_{10}\cdot(PO_4)_6\cdot Cit$. To these ion interactants must be added hydrogen and hydroxyl. Crystals or precipitates which form in the absence of casein micelles, or other protein, probably have structures which resemble that of apatite (Rogers, 1937; van Wazer, 1958). There is a wide variety of structures, and some of these are known to be important in determining the structure of the mineral phase of bone. The latter is reviewed by Neuman and Neuman (1953). The apatite-like minerals are revealed through X-ray diffraction patterns to have considerable regularity, in three dimensions, of ionic interactants.

Are inorganic colloid particles present in skim milk, and will the addition of calcium or phosphate to skim milk lead to initiation if they are not present or to augmentation if they are present? The possibility of coexisting casein micelles and a nonmicellar colloid of calcium–phosphate–citrate is suggested by the fact that the environment appears to be poised. Assume coexisting colloids; if the system is altered by changing temperature, pH or constituent composition, which colloid is most important in establishing environmental ion levels? It has been noted that an equilibrium diffusate of milk at 4°C produces a precipitate on warming to 37°C. What events are expected to occur when skim milk is raised from 4° to 37°C? As a result of the temperature dependence of ion–protein interactions, environmental ion concentrations will decrease: Are decreases due to ion interaction with casein micelles sufficient either to prevent the formation of a separate nonmicellar inorganic colloid if it is not already present, or to prevent augmentation if it is present? Information which bears on these questions, but is particularly relevant to micelle structure, is discussed in the remainder of this section.

The ionic calcium concentration of skim milk has been estimated in several ways. Equilibrium methods are (a) the murexide method of Smeets (1955), (b) the ion-exchange method of Christianson et al. (1954) and van Kreveld and van Minnen (1955), and (c) the ion-exchange electrode method of Affsprung and Gehrke (1956). A nonequilibrium method has been proposed by Baker et al. (1954). The values found using the equilibrium murexide method are $0.0027 \pm 0.0004\ M$ (Smeets) and $0.0025 \pm 0.0003\ M$ (Tessier and Rose, 1958). The values found using the equilibrium ion-exchange method are $0.0022\ M$ for calcium and $0.0008\ M$ for magnesium (Christianson et al. and van Kreveld and van Minnen). The ion-exchange electrode method of Affsprung and Gehrke gives $0.005\ M$ for calcium. This last method may not give activity or ion concentration directly in a complex ion mixture. It seems most likely that ionic calcium is close to $0.002\ M$, which is considered here to be a maximum. Boulet and Marier (1960) calculate levels of ionic calcium which are slightly lower than those measured.

B. Cooperative Effects

The milk environment or the buffers intended to duplicate this environment (Table IV) have the disadvantage of being sufficiently complex so that concentration or activity levels of constituent species are difficult either to measure or to calculate. The situation becomes particularly complex when alterations in milk environment or buffer composition are contemplated. This is evident from the work of Jenness and Koops (1962), who studied the effects of heat treatment on a duplicating buffer. A study has been made in this author's laboratory of simpler buffers which simulate the essential features of the milk environment. These are shown as type IV buffers in Table V. The main differences between simulating buffers and the buffer of Table IV are the use of calcium and sodium as the sole cation species and chloride, phosphate and citrate as the anion species. The chloride concentration is fixed by the concentrations of the other constituents at pH 6.6 and 37°C. Given also in Table V are representations of three partial buffers. Type I buffers contain calcium and sodium and, not listed, $0.001\ M$ sodium cacodylate to maintain pH. Type II buffers contain calcium, phosphate and sodium, and type III buffers contain

TABLE V

MICELLE FORMATION WITH SEVERAL BUFFERS[a]

Buffer type	Calcium$_o$[b] (M)	Phosphate$_o$[b] (M)	Citrate$_o$[b] (M)	Sodium$_o$[b] (M)	Ionic calcium, calculated (M)	System distribution[c]
I	0.003			0.039	0.0030	C
II	0.003	0.003		0.034	0.0024	Clear
III	0.012			0.043	0.021	C
IVa	0.013	0.006	0.010	0.051	0.0020	A or B
IVb	0.012	0.006	0.010	0.052	0.0014	A or B
IVc	0.0115	0.006	0.010	0.053	0.0011	A or B

[a] In solubilized skim milk or first-cycle casein at 37°C, pH 6.6.
[b] The subscript designates total amount.
[c] System distributions are illustrated in Table VI.

calcium, citrate and sodium. It should be noted that Table V buffers are used as equilibrium buffers. In Section VIII it is shown that the dip is absent at pH 6.6 and at the sodium chloride concentrations given in Table V (see Fig. 9). In preceding sections total calcium concentrations are used, except for those related to core formation where equilibrium calcium activity is calculated.

Column 6 of Table V gives approximations for the levels of ionic calcium using the constants of Davies and Hoyle (1953, 1955). They refer to a temperature of 25°C and to low ionic strength. No attempt has been made to correct for temperature and ionic strength; the levels of ionic calcium are given for comparison purposes only. They are, however, comparable to the ionic calcium levels calculated by Boulet and Marier (1960). As will be shown, the effects obtained by altering buffer composition are far beyond any which can be associated with errors in the calculation of ionic calcium.

Before examining results obtained with these buffers, attention should be given to the criteria on which a comparison of results is to be based. The following is a list of observations which can be used to characterize a casein micelle system or other colloid system:

(1) Formation of a coalescent caseinate precipitate.
(2) Formation of sediment (large micelles) at 1 g.
(3) Removal of the protein fraction by assay centrifugation and determination of the coalescent or particulate nature of pellet.
(4) The micelle distribution as characterized by differential ultracentrifugation.
(5) Stability of micelles to close approach, determined by the extent of spontaneous resuspension of pellets obtained by ultracentrifugation.
6) Stability of micelle systems at 0°C after their formation at 37°C.
(7) Micelle solvations.

For convenience in classification, three differential centrifugation distributions, frequently encountered in studies of micelle systems, are given in Table VI. Distribution A, typical of skim milk and a variety of reconstituted systems, has a single peak at micelle sizes which centrifuge out at 17,500 g, 20 min. Distribution B is bimodal and is given by some types of reconstituted systems. Distribution C clearly contains large micelles, but most of the protein is in the final supernatant. It should be noted that these large micelles remain suspended during assay centrifugation. This is the distribution typically obtained on dialysis or single-aliquot addition of calcium when the calcium concentration is low.

First-cycle casein has been dialyzed against the buffer types of Table

TABLE VI

Differential Centrifugation of Systems at pH 6.6 and 37°C[a]

	Distribution		
	A (normal)	B (bimodal)	C (large micelles)[b]
Centrifugation			
1900 g, 5 min	0.02	0.14	0.18
7800 g, 10 min	0.17	0.25	0.02
7800 g, 20 min	0.24	0.05	0.02
17,500 g, 20 min	0.37	0.11	0.01
70,300 g, 20 min	0.17	0.30	0.05
Supernatant	0.03	0.15	0.72

[a] Centrifugation was carried out in a swinging bucket rotor, SW 39. The numbers are fractions of the initial protein absorbance. Each specifies the fraction removed at a centrifugation, *not* including amounts removed by preceding centrifugations. Numbers in the last row are fractions remaining in the final supernatant.

[b] Protein is mainly in the supernatant.

V using volume ratios of 200 and first-cycle casein protein concentrations of 10–30 mg/ml. Table V gives representative data. In passing it should be noted that dialysis constitutes a rigorous test of incremental addition. In none of the systems studied was significant coalescent precipitate observed.

Type I buffer contains 0.003 M calcium and gives a micelle distribution of type C. The protein compartment has a grayish cast instead of the typical white opacity of skim milk. The presence of a few large micelles with protein mainly in the supernatant is indicated on differential centrifugation. If the calcium concentration is progressively increased, the micelle distribution will progressively shift to give equilibrium micelle distributions of type A in the expected concentration range of 0.01–0.015 M calcium.

Type II buffer contains phosphate. The concentration of 0.003 M calcium is near the upper limit which can exist in the presence of 0.003 M phosphate without giving a precipitate at pH 6.6 and 37°C. The protein compartment is clear. If the calcium concentration is increased over the level given, calcium phosphate precipitates in the diffusate, and the protein will go through the same sequence as for type I buffers. That is, to a first approximation, the system behaves as though it were a calcium system without phosphate. If phosphate is increased over the level given, calcium phosphate precipitates in the diffusate, and the protein solution remains clear.

Type III buffer contains citrate. At the given levels of calcium and citrate there is a sufficient excess of calcium to produce a micelle distribution of type C. If calcium concentration is increased, the protein will go through the same sequence as for type I buffers; the system again behaves as though it were a calcium system alone.

Comparing type II and type III buffers accurately requires more information than is currently available. However, it is unlikely that the calcium activity, at the particular compositions given, is greater in the type III buffer than in the type II buffer. Yet the type II buffer gives a clear protein solution. This suggests that the inorganic phosphate is exerting an inhibiting effect, most likely on core formation. This is an interesting situation, for at higher calcium concentrations there is an enhancing effect of small amounts of phosphate on core formation.

Type IV buffers contain both phosphate and citrate. Note that type IVb buffer is essentially type III buffer to which phosphate has been added. The calcium activity in type IVb buffer is certainly less than the corresponding level in type III, if both buffers have compositions as listed. In spite of this lower calcium activity, type IVb buffer gives normal or bimodal micelle distributions. Of the particular buffers listed in Table V, in fact only those of type IV will give normal micelle distributions. If the calculations of ionic calcium concentration are accepted, a normal distribution can result at an ionic calcium concentration as low as 0.0011 M (type IV). If a calcium concentration of 0.0011 M is used in a type I buffer, it is found not to be able to accomplish core formation; the protein solution remains clear. There is, therefore, a cooperative interaction of polyvalent cations, polyvalent anions and proteins in micelle formation in systems employing type IV buffers.

It should be noted that in the systems of Table V it is unlikely that the cooperative effects are due, in the protein, to the formation of a nonmicellar inorganic colloid phase independent of casein micelles. This is particularly the case for buffers of types IVb and IVc.

What might be the basis for a cooperative interaction? Many possibilities exist. The first general class would be stabilization through an ion chain involving calcium and appropriate forms of phosphate and citrate. Rose (1965) has proposed an extended chain of calcium and phosphate ions which has side links through calcium to the ester phosphates of caseins, with an occasional intercalation of a citrate–calcium unit. This chain appears to be unstable since it must frequently utilize trivalent phosphate. To this author it seems more likely that the protein binding sites will be found to be paucidisperse in type and to include, in addition to ester phos-

TABLE VII

Possible Cooperative Interactions

Protein⁻ – Ca²⁺ – H Citrate²⁻ – Ca²⁺ – H PO₄²⁻ – Ca²⁺ – ⁻Protein
Protein⁻ – Ca²⁺ – H PO₄²⁻ – ⁺Protein
Protein⁻ – Ca²⁺ – H Citrate²⁻ – ⁺Protein
Protein⁻ Ca²⁺ ⁻Protein
Protein⁺ H PO₄²⁻ ⁺Protein
Protein⁻ Ca²⁺ H PO₄²⁻ Ca²⁺ ⁻Protein

phates, protein carboxylate groups and the positively charged groups of the basic amino acids. Horst (1963) has proposed an involvement of the ε-amino groups of lysine.

An effective mechanism to account for cooperative effects is one which requires that the protein interactants have appropriate conformations and a positioning of more than one type of specific protein binding site, the specificity to be established, for example, by neighboring groups and local electrostatic environment (see Section VI). Some possible types of ion interactions are given in Table VII. These can account for the ion-binding composition as given in Table IV, column 6.

Two views of complex ion interactions have been expressed by Pyne and McGann (1960) and McGann and Pyne (1960) and reviewed by Pyne (1962). These are (a) that there are colloid domains having a structure which approaches that of a hypothetical calcium–phosphate–citrate apatite and (b) that calcium, phosphate and citrate are in some way responsible for developing the internal structure of the micelle. McGann and Pyne (1960) favor the latter view; they interpret their observations as favoring the concept of chemical links between calcium–phosphate–citrate and casein in milk. As previously discussed, similar views have been expressed by Rose (1965).

Is it possible that in skim milk there are no inorganic colloidal particles independent of casein micelles, and, thus, that the casein micelle system and the other proteins present interact in such a way as to keep constituent concentrations in the natural environment always just below levels that will lead to independent colloid formation? Only a direct experimental approach can answer this question. From what has already been stated, the presence of nonmicellar inorganic colloid may be considered ancillary, for a complete micelle system can be generated under conditions where calcium–phosphate–citrate colloid is absent, as well as under conditions where the environment might be in equilibrium with such colloid. The studies of Jenness and Koops (1962) suggest that the situation may be otherwise in processed systems.

9. FORMATION AND STRUCTURE OF CASEIN MICELLES 69

C. HYSTERESIS EFFECTS

These effects allow a micelle system, once formed, to be stable under altered environmental conditions where an equivalent micelle system does not form from protein components in solution.

Attention is first directed to the results of Pyne and McGann (1960) and McGann and Pyne (1960), which strongly suggest a temperature hysteresis; and to the results of Davies and White (Table IV) for the environmental characteristics of skim milk at 3° and 20°C. References to approaches taken by others can be found in a more comprehensive discussion by Pyne (1962). Pyne and McGann described the preparation of a material, which they term colloidal phosphate-free milk, as follows: Skim milk is acidified at 0°C to a pH of 4.8–5.0 using 1 M HCl. The acidified milk is stirred for 1 hr, after which it is dialyzed for 96 hr at 0°–5°C against extensive changes of skim milk. During this time the pH values of the two compartments equalize, and equilibrium between environments can be taken to be established.

Colloidal phosphate-free milk differs from skim milk in important respects (McGann and Pyne, 1960). Compared to skim milk, it is strikingly translucent rather than white opaque; it has a higher viscosity; it is more unstable to small amounts of added calcium (addition of as little as 0.025 M at 0°–5°C will produce a precipitate); the total inorganic phosphate is reduced to 0.012 M, which is close to the environmental level in milk (Table IV); and the total calcium is reduced to $\sim$0.020 M. Corrected for diffusible calcium, the latter leaves a binding of 0.011 M. Colloidal phosphate-free milk is similar to skim milk in its formaldehyde titration, thus giving a formaldehyde titration which is lower than that of equivalent sodium caseinate. The calcium binding in colloidal phosphate-free milk is 1.8 calcium ions per organic phosphate group compared to 3.3 for skim milk. Pyne and McGann calculated that the conversion of skim milk to colloidal phosphate-free milk requires the removal of 1.41 calcium ions per phosphate ion.

Hysteresis is indicated by the fact that colloidal phosphate-free milk does not develop and contain a micelle system equivalent to that of skim milk. The fact that colloidal phosphate-free milk remains translucent shows that if it contains micelles, they appear either to be fewer, smaller or more highly solvated than those of skim milk.

Concomitant with the experiments at 37°C, which are summarized in Table V, others were carried out at 0°–4°C using skim milk, simulating or duplicating buffers, or calcium chloride solutions to establish environmental levels by dialysis. First-cycle casein, solubilized skim milk, or mixtures of α_s-, β- and κ-caseins were used as test solutions. The existence of striking hysteresis effects was confirmed, and some additional considerations of the nature of the environment of skim milk are possible on the basis of the results summarized as follows:

As noted in Section IX.B, a type I buffer (Table V) containing 0.015 M calcium at 37°C can produce by dialysis a normal distribution of micelles with first-cycle casein. The same result is obtained with an α_s-κ-casein mixture at $R_i = 7$. After formation at 37°C, if either micelle system is cooled to 0°C, micelles persist at least over a period of several days. If the dialysis is initiated at 4°C, within 16 hr first-cycle casein becomes opaque, but the α_s-κ-casein mixture contains precipitate, and over 30% of the initial protein is removed by assay centrifugation. Hysteresis is evident in the α_s-κ-casein system. The absence of a precipitate in first-cycle casein is probably due to a stabilization of α_s-casein by β-casein. That such stabilization takes place by metastable colloid formation has been observed frequently. It will be examined below.

Hysteresis effects are quite evident in micelle systems obtained by dialysis, either against type IV buffers of Table V or against fresh whole milk. Either environment will produce typical micelle systems using first-cycle casein at 37°C, and these are found to retain much of their opacity on cooling to 0°C and to retain this opacity for at least six days. α_s-Casein is precipitated at 37°C in these environments. If dialysis is carried out at 4°C the contrast is striking: With type IVa buffer, first-cycle casein becomes only translucent and with whole milk it remains clear. At 4°C, α_s-casein will remain in clear solution using a type IVa buffer and a level of 8 mg/ml, as will an equal-weight mixture of α_s- and β-caseins. Many other examples of hysteresis could be cited.

Consider the nature of hysteresis. At 37°C cooperative interactions and the ion environment establish a micelle structure which is found to be stable at 0°C. Assuming the micelle system to be intrinsically unstable at 0°C, which it might be, failure to exhibit instability would be attributed to the rate constants associated with reversal of cooperative and conformational effects and with rearrangement of the three-dimensional structure of the system. Of course, the micelle system might be intrinsically stable in the environment at 0°C. Failure to establish micelles then would be associated with the conformations of the monomers, the rate constants for the production of cooperative effects and the rate constants for progressive rearrangement of three-dimensional structure.

If first-cycle casein is dialyzed exhaustively vs. type IVa buffer at 0°C and the first-cycle casein is then brought to 37°C, the white opacity of a typical micelle system does not develop, although the change in hue suggests some increase in particle size. The ionic environment evidently does not contain critical species at sufficiently high concentrations. On raising the temperature, protein–ion interactions (binding) increase and environmental levels are reduced to the point where complete micellization cannot be effected. The deficits are contributed by continued dialysis at 37°C.

If skim milks from individual animals or type IV buffers are used to establish environments, and first-cycle casein or solubilized skim milk are dialyzed against these, an interesting situation emerges. At 4°C, as expected, both systems remain clear or exhibit a yellow hue. At 37°C, however, the two systems differ: First-cycle casein develops an opacity comparable to that of skim milk, while solubilized skim milk develops only translucence with an orange-red hue. That micelles are present in the solubilized skim milk sample has been demonstrated by differential centrifugation. These observations raise the possibility that natural micelles are not established in the animal when all protein components are present and at environmental levels of skim milk. Natural micelles could be made using the near equivalent of first-cycle casein and whey protein added later. Possibly, all protein components are present and natural micelles are made using an ionic environment quite different from that of skim milk. This is suggested by the fact that calcium alone, at 0.015 M in a type I buffer, can cause both solubilized skim milk and first-cycle casein to develop the typical opacity of skim milk (see Waugh and Noble, 1965). After micelle manufacture, the ducts of the mammary gland might alter an initial ionic environment to that of skim milk.

Cooperative and hysteresis effects might account for the small differences in equilibrium levels of ionic constituents in skim milk, as found by White and Davies (Table IV), between 3° and 20°C. For example, assume that a true equilibrium is established between all interactants at 37°C. At any lower temperature the true equilibrium would favor the dissociation of bound ions, but as a result of hysteresis, the rates of dissociation are sufficiently slow so as to make the observation of dissociation impractical. Thus, the ionic environment appears to vary only slightly with temperature. One would expect that some binding sites would not be involved in hysteresis effects and that these would follow their equilibrium binding characteristics.

Cooperative and hysteresis effects and a requirement for an appropriate ionic environment are indicated by the results of other studies that McGann and Pyne (1960) carried out on skim milk and colloidal phosphate-free milk. The addition of calcium chloride to skim milk to give 0.025 M at 0°–40°C does not lead to instability. The same addition to colloidal phosphate-free milk readily produces precipitation. These authors attribute the stability of skim milk to the structure of the natural casein micelle and point out that this structure is absent in colloidal phosphate-free milk. Further evidence for cooperative and hysteresis effects is provided by the following: Skim milk was dialyzed for 24 hr at 0°C against a buffer 40 times its volume at pH 6.7–6.8. The buffer contained 0.004 M calcium chloride, 0.001 M magnesium chloride, 0.075 M sodium chloride, 0.005 M sodium maleate and 5% lactose. After this dialysis the skim milk contained about 90% of its original content of phosphate.

In another experiment McGann and Pyne dialyzed colloidal phosphate-free milk and skim milk against the buffer just given, after which calcium chloride was added to give 0.025 M. The two were treated with rennet and both produced a precipitate which was removed by filtration in the cold. The "β-caseinate" contents of the filtrates were then estimated by determining the protein which precipitated on warming to 30°C but which redissolved on cooling. It is unlikely that this protein fraction is pure β-casein. As will be shown, β-casein can stabilize large amounts of α_s-casein against precipitation at low temperature. The result of the experiment, however, was that about three times as much protein appeared in the temperature-dependent fraction from colloidal phosphate-free milk (0.84% protein) as appeared in the fraction from skim milk. If dialyzed colloidal phosphate-free milk and skim milk were clotted at 36°C, then cooled to 0°–2°C and filtered 12 hr later, the protein which precipitated on warming but redissolved on cooling was negligible for skim milk and 0.34% for colloidal phosphate-free milk. Evidently, much of the temperature-sensitive protein was incorporated in the clot of colloidal phosphate-free milk at the higher temperature, and was not released on cooling. From the results given earlier, cooperative and hysteresis effects can be generated by calcium alone. This might account for a part of the difference between dialyzed colloidal phosphate-free milk and skim milk clotted at 36°C and treated as described.

We now turn briefly to the observations of Choate et al. (1959). Milk was differentially centrifuged at 20°–25°C using a constant field but increasing centrifugation times. The series of centrifugates were (a) not washed, (b) washed with milk serum at 30°C or (c) washed with distilled water at 20°C. They were then dialyzed against barbiturate buffer at pH 8.4, $I = 0.1$, after which aliquots of each protein solution were either examined in the ultracentrifuge or dialyzed against skim milk at 4°C. After dialysis against barbiturate buffer, all samples were found to contain a similar distribution of smaller aggregates, $s_{20} < 10$ S, independent of the method of washing. After long dialysis against skim milk the samples were examined in the ultracentrifuge at 20°–25°C. It was observed, for all samples, that dialysis against skim milk had increased particle size over that in barbiturate buffer. In addition, a positive correlation was found between the sizes of the milk micelles in the aliquots obtained by differential centrifugation and the average sizes of the particles in each aliquot after treatment and dialysis against skim milk. This was termed "size memory." However, as the authors point out, the original micelles always had s_{20} larger than the final particles. Washing introduced differences at this level of observation. For unwashed aliquots, the micelle s_{20} values and the ratios of micelle s_{20} to final particle s_{20} were, respectively, 980 and 2.5; 870 and 3.1; 640 and 4.9; and 230 and 1.5 S. The results of Choate et al. are consistent with the results described above; namely, that dialysis against skim milk at 4°C will not restore the original colloid. They show, however, in agreement with the results of Sullivan et al. (1959), that an increased relative amount of κ-casein decreases particle size.

D. NATURAL MICELLES

An attempt is made in Sections IX.B and C to reveal and partially characterize the strong cooperative and hysteresis effects associated with the natural micelle. In agreement with this, a comparison of the properties of natural micelles with the properties of micelles obtained under conditions where systems are in equilibrium (Section VIII) suggests that natural micelle systems are not in as rapid equilibration through sets of subdivision and coalescence reactions. They may have been at some stage during their development, but it appears that they are not in the final product, milk. On the other hand, natural micelles are not entirely stable; for example, cooling milk progressively releases small amounts of ionic constituents and significant amounts of protein, particularly β-casein, into solution. Milk micelles, then, appear to have characteristics determined by a combination of effects: some fixed by hysteresis and some by equilibria which rapidly respond to changes in the environment. A number of attempts, based on a choice between complete equilibrium and nonequilibrium models, have been made to account for the properties of milk. Several examples are given by the analyses of Rose (1968) and Garnier (1969). What is evidently required is a greater understanding of those portions of micelle structure which are fixed by hysteresis effects and those which can be rapidly altered by changing environmental conditions. Important questions are the following. What is the detail of micelle structure when milk is cooled to 0°C? What portions are in rapid exchange at 37°C? What is the detail of structure at that temperature?

E. METASTABLE COLLOID

Evidence has been given in Section V to show that α_s-casein or β-casein may individually form micelle cores, and in Section VI that when they are present together they commingle at the molecular level, and that this is so for natural micelles. In the following the pertinent observation is that the skim milk environment is just sufficient to induce the formation of certain types of metastable colloid.

If α_s-casein is dialyzed against skim milk or a type IVa buffer at 37°C, a coalescent precipitate forms and the supernatant is clear. If β-casein or an equal-weight mixture of α_s- and β-caseins (8 mg/ml each) is dialyzed, precipitate does not form. Using several individual skim milks, the results indicate a range of association for both proteins: from translucent systems containing small association products to the formation of milky metastable colloid systems. A similar result is obtained by dialysis at 4°C against a type I buffer containing 0.015 M calcium. A precipitate forms with α_s-casein alone, while a mixture of α_s- and β-caseins becomes only translucent.

It is clear from these results that metastable colloid containing β-casein or an α_s-β-casein mixture could be present in milk. In fact, much of the κ-casein in milk remains in the supernatant after ultracentrifugation, and one suspects that in the final environment the coat of natural micelles may be incomplete. This will be considered again in Section X.

From the standpoint of experimental approaches it should be noted that metastable colloid systems visually resemble milk. However, they are unstable: Colloid particles will precipitate on the addition of calcium, for example, and will coalesce on centrifugation into a pellet. Milk micelles do not coalesce.

F. Minor Casein Components

Comparisons of first-cycle casein with mixtures of pure components would be expected to reveal whether the minor casein components modify coat- and core-forming reactions. Before making comparisons it should be noted that the abundances given in Section IV are uncertain, and thus the weight ratio of κ-casein to non-κ-casein in first-cycle casein is not accurately known. This makes comparison difficult, and the tentative conclusions which follow may require subsequent reexamination.

Fundamental similarities between solubilized skim milk, first-cycle casein, and α_s-κ-casein mixtures at $R_i = 7$ or 6 have been reported by Noble and Waugh (1965) and Waugh and Noble (1965). At 37°C, pH 7 and $I = 0.075$, single-aliquot addition of calcium gives similar assay supernatant protein curves, and incremental addition of calcium gives similar extensions of the descending limb of the dip. In these nonequilibrium systems there appear to be no marked differences among materials which could be ascribed to the presence of the minor casein components.

That the minor casein components might modify core- and coat-forming reactions is indicated by comparing first-cycle casein with mixtures of α_s-, β-, and κ-caseins at $R_i = 5.7$, using equal weights of α_s- and β-caseins. At equivalent concentrations, on dialysis against type IV buffers, first-cycle casein develops opacity to a greater extent than the casein mixture. It was considered that the α_s-casein fractions (Waugh et al., 1962) might be contributing to this opacity difference, since they contain more phosphorus and are less soluble than α_s- or β-casein. A preparation of first-cycle casein was fractionated according to the procedure of Waugh et al. A crude α_s-casein was applied to a DEAE-cellulose column and α_s-casein was removed. The α_s-casein fractions remained adsorbed but could be recovered by elution at a higher ionic strength. After recovery, the following solutions were compared: (a) first-cycle casein at 16 mg/ml, (b) α_s-, β-, and κ-caseins at either 8.8, 7.2, and 2.8 mg/ml or at 10.7, 5.3, and 2.8 mg/ml, and (c) α_s-casein, α_s-casein fractions, β-casein and κ-casein, either at 5.0, 3.0, 8.0, and 2.8 mg/ml or at 7.7, 3.0, 5.8, and 2.8 mg/ml, respectively. Differences between these solutions were obtained by dialysis against a type III buffer containing 0.0125 M calcium and 0.01 M citrate at pH 6.6 and 37°C. First-cycle casein and the solutions containing α_s-casein fractions became translucent with an orange or yellow hue, while the solutions containing α_s-, β-, and κ-caseins remained clear. From this comparison it is evident that the α_s-casein fractions may be interacting with α_s- and β-caseins to promote core formation under conditions where α_s- and β-caseins alone do not form cores. The mixtures given above, however, contain greater relative amounts of α_s-casein fractions than are present in first-cycle casein. At the levels of α_s-casein fractions roughly estimated for first-cycle casein, about 15% of α_s-casein, differences in opacity between mixtures of components have not been detected. It is possible that there are present in first-cycle casein as yet unspecified components which modify core- and coat-forming reactions.

X. Rennin (Chymosin) Coagulation

Extensive studies have been made of the coagulation of milk by rennin. The importance of these is obvious, and the reader is referred, for many interesting results including the preparation of rennin by Berridge, to reviews by Berridge (1951), Pyne (1962), Lindqvist (1963) and Jollès (1966) and to Chapter 12 by Mackinlay and Wake.

Since the work of Hammersten, research on rennin coagulation of milk has been based on the assumption that there is present a stabilizing substance and that this stabilizing substance is destroyed by rennin action. Largely on the basis of work by Nitschmann and his colleagues, rennin coagulation is considered to be the result of two temporal phases. In the first, the enzymatic action of rennin destroys a protective colloid, and in the second, destabilized particles aggregate. There is also a third, longer phase in which rennin and/or enzyme impurities slowly hydrolyze all casein components. Much early work focused on α-casein as the target of immediate rennin action. It was known that small amounts of nonprotein nitrogen were released during immediate rennin action and that this nonprotein nitrogen was distributed among several components.

The objective now is to present a view of rennet coagulation from the standpoint of micelle structure and alterations in micelle structure. It is convenient to discuss more recent work first. Shortly after the discovery of κ-casein, Waugh (1958) reported that it is the only protein which can be altered in the time required to clot milk and that it loses about 20% of its molecular weight in the process. The clotted milk could be fractionated to give a mixture of α_s- and β-caseins (second-cycle casein-R), and the addition of κ-casein to this mixture provided the necessary stabilization to produce micelles when calcium was added. This system could again be clotted by rennin. The addition of rennin to pure κ-casein produced para-κ-casein, which is insoluble in the presence or absence of calcium. The central role attributed to κ-casein was essentially confirmed by Jollès and Alais (1959) and Nitschmann and Henzi (1959), who found that the glycomacropeptides obtained from κ-casein are the same as those obtained from α-casein and whole sodium caseinate. Wake (1957) demonstrated that α_s-casein does not engage in this reaction. Additional studies were soon added by Wake (1959) and Nitschmann and Beeby (1960).

The kinetics of the release of peptides from sodium caseinate are first order, according to Nitschmann and Bohren (1955). The results are not without complications: Below a minimal amount of rennin the expected nonprotein nitrogen is not fully obtained. The authors suggest irreversible inhibition. As mentioned above, the structure of the sodium caseinate system is now known to be complex, and the formation of an interaction product involving para-κ-casein might protect some unaltered κ-casein. Temperature studies indicate that the activation energy is about 10 kcal/mole.

Alais et al. (1953) examined the release at 25°C of nonprotein nitrogen from milk by the action of rennin. They found the reaction to be limited and to be first order. Scott–Blair and Oosthuizen (1961, 1962) examined viscosity changes during rennin action on fat-free milk reconstituted from milk powder and on sodium caseinate. The viscosity changes in both systems were extraordinarily complex. In the milk system, the viscosity

9. FORMATION AND STRUCTURE OF CASEIN MICELLES

first decreased and then increased over the time of the study, which was in excess of 60 min. The changes in polymer structure which produce these viscosity changes are not known.

A number of investigations via electron microscopy have been carried out to obtain information concerning the sequences of physical changes which occur during the rennin coagulation of milk (Hostettler and Imhof, 1951; Baud et al., 1951, 1953; Peters and Dietrich, 1954; Hostettler and Stein, 1954; Hostettler et al., 1955; and Imhof and Hostettler, 1956.) During initial stages micelles were found to be joined by short, fine threads, and there were also present nodular fibrils. Later, these formed larger fibrils which adhered to form a three-dimensional network. The presence of threads joining micelles in early stages is particularly interesting.

Several aspects of rennin coagulation of micelles are clarified by the coat-core model. Waugh and Noble (1965) point out that the occurrence of κ-casein at the micelle surface, with the macropeptide substituents external, permits κ-casein to be attacked easily and quantitatively without micelle dissociation or other alteration in structure. Certainly, from stability characteristics of the core, including cooperative effects, and the fact that rennin does not immediately affect $α_s$- or $β$-caseins, one would not expect a dissociation or other significant rearrangement of core polymers. The fundamental consideration is then of coat. The mechanism of Garnier (1963), in which stoichiometric complexes are redistributed, seems unlikely. The sequence of events that occurs during clotting would then appear to depend, at least, on (a) whether rennin can affect the coat on micelles and (b) the relative affinities of coat for core, para-coat for core, para-coat for itself and coat for itself. The data given in preceding sections show that coat has a much greater affinity for core than for itself. In unusually stable micelle systems, using cobalt, manganese, etc. instead of calcium (Waugh, 1961), a significant dissociation of coat or para-coat from core appears to be unlikely during the time it takes for the system to clot. From this it is concluded that rennin can directly affect the coat of micelles. In such systems it is expected that clotting occurs as a result of the interaction of affected surfaces. Are micelles present in sufficient amount to form a clot? In skim milk there are present about 7×10^{13} micelles per milliliter. If the average micelle diameter is taken as 1.2×10^{-5} cm and the micelles are arranged in a simple cubic network, there will be about 2.5 micelles between network junctions. This suggests that micelles alone can easily form a clot. The clotting of dilute solutions of κ-casein is well known: There is no question that the quantity of κ-casein present in skim milk could, in itself, readily produce a firm clot. Consider a situation in which micelles alone are responsible for network formation. The situation then may be as follows. The enzymatic modification of a single coat unit on the surface of a micelle may not make the micelle reactive to the surface of an unmodified or a similarly modified micelle. In fact, it is likely that before two micelles can adhere, a certain fraction of the coat units within small surface areas (on the two micelles) must be modified. As a preliminary approach, assume that in order for micelles to interact with each other to form a three-dimensional network there must occur an average number $\bar{y}$ of areas each having $\bar{x}$ neighboring coat units converted to para-coat. Assuming that this situation can be approximated by a linear polymerization model, as developed by Flory (1953), and that the enzymatic reaction is pseudo-first order, we write

$$\bar{y} = N_0(\exp - k_1R_1t)^2(1 - \exp - k_1R_1t)^{(\bar{x}-1)} \qquad (12)$$

where k_1 is the rate constant of the enzymatic step, R_1 is the rennin concentration, t is time and N_0 is the number of κ-casein molecules in the micelle surface. Equation (12) permits a first-order release of peptide and predicts that the requisite number of areas $\bar{y}$ per micelle will appear suddenly during a short time interval just before a particular time, τ, the clotting time. An equation of the form of (12) predicts a strictly inverse relation ship between R_1 and τ. The reason for this is that the diffusion time of affected micelles has not been taken into account. If this is included it is clear that the relationship between clotting time and rennin concentration will be skewed. At infinite rennin concentration, which would correspond to instantaneous conversion of coat to para-coat, time must elapse before affected micelles can make contact and develop a network.

Under conditions where free para-coat, or para-κ-casein, contribute to gel formation the situation is more complex. The kinetics may be similar to that of the clotting of fibrinogen by thrombin, where the enzymatic step is pseudo-first order, but the polymerization of monomer fibrin to form a network involves the formation of a polymer distribution, the cross-linking of some of the larger polymers to form interpolymer nuclei and the growth and interaction of these to form a network.

Both mechanisms may contribute to the coagulation of skim milk. There is evidence for direct micelle association, and the thin strands between micelles may be fibers of para-coat. In fact, one cannot exclude the possibility that considerable rearrangement of coat and para-coat, depending on relative affinities, takes place during coagulation. According to either model, the division of clotting into temporal phases is artificial: The first two phases appear to be separated because the onset of micelle adherence and/or para-coat network formation requires the conversion of a significant fraction of total coat to para-coat.

Another purpose in introducing these simplified models is to permit a comment on complexity. A change in temperature, for example, may change more than the rate constant associated with the enzymatic part of the overall process: The micelle system may change in structure, the values of $\bar{y}$ and of $\bar{x}$ may change, the diffusion time for contact of affected micelles may change and the polymerization reactions of para-coat may change.

ACKNOWLEDGMENTS

In writing this chapter the author has drawn freely on unpublished work carried out in collaboration with several colleagues—in particular, in Sections VI, VII and VIII, with Dr. Lawrence K. Creamer, Dr. Charles W. Slattery and Dr. Bernard Talbot. In these studies we were ably assisted by Miss Carla McLaughlin, Mrs. Bette Reynolds and Mrs. Margaret Foley.

Deep appreciation is expressed to Dr. Charles Slattery and Dr. Bernard Talbot for their critical reading and for many helpful discussions.

REFERENCES

Adachi, S. (1963). *J. Dairy Sci.* **46,** 743.
Affsprung, H. E., and Gehrke, C. W. (1956). *J. Dairy Sci.* **39,** 345.
Alais, C., and Jollès, P. (1961). *Biochim. Biophys. Acta* **51,** 315.
Alais, C., Mocquot, G., Nitschmann, Hs., and Zahler, P. (1953). *Helv. Chim. Acta* **36,** 1955.
Alberty, R. A., Smith, R. M., and Bock, R. M. (1951). *J. Biol. Chem.* **193,** 425.
Anderson, L., and Kelley, J. J. (1959). *J. Amer. Chem. Soc.* **81,** 2275.
Ashworth, U. S. (1964). *J. Dairy Sci.* **47,** 351.
Baker, J. M., Gehrke, C. W., and Affsprung, H. E. (1954). *J. Dairy Sci.* **37,** 1409.
Baud, C. A., Morard, J. C., and Pernoux, E. (1951). *Compt. Rend.* **233,** 276.
Baud, C. A., Morard, J. C., and Pernoux, E. (1953). *Z. Wiss. Mikrosk.* **61,** 290.
Beeby, R. (1963). *J. Dairy Res.* **30,** 77.
Beeby, R. (1964). *Biochim. Biophys. Acta* **82,** 418.
Beeby, R. (1965). *J. Dairy Res.* **32,** 57.
Beeby, R., and Nitschmann, Hs. (1963). *J. Dairy Res.* **30,** 7.
Bennich, J., Johansson, B., and Österberg, R. (1959). *Acta Chem. Scand.* **13,** 1171.
Berridge, N. J. (1951). *In* "The Enzymes" (J. D. Sumner and K. Myrbäck, eds.), Vol. 1, p. 1079. Academic Press, New York.
Borgstrom, B. (1965). *Biochim. Biophys. Acta* **106,** 171.
Boulet, M., and Marier, J. R. (1960). *J. Dairy Sci.* **43,** 155.
Burk, N. F., and Greenberg, D. M. (1930). *J. Biol. Chem.* **87,** 197.
Burnett, G., and Kennedy, E. P. (1954). *J. Biol. Chem.* **211,** 969.
Carr, C. W. (1955). *In* "Electrochemistry in Biology and Medicine" (T. Shedlovsky, ed.), Chap. 14. Wiley, New York.
Carr, C. W., and Engelstadt, W. P. (1958). *Arch. Biochem. Biophys.* **77,** 158.
Carr, C. W., and Topol, L. (1950). *J. Phys. Colloid Chem.* **54,** 176.
Caspar, D. L. D., and Klug, A. (1962). *Cold Spring Harbor Symp. Quant. Biol.* **27,** 1.
Chanutin, A., Ludwig, S., and Masket, N. (1942). *J. Biol. Chem.* **143,** 737.
Cheeseman, G. C. (1962). *J. Dairy Res.* **29,** 163.
Cherbuliez, E., and Baudet, P. (1950a). *Helv. Chim. Acta* **33,** 398.

Cherbuliez, E., and Baudet, P. (1950b). *Helv. Chim. Acta* **33**, 1673.
Choate, W. L., Heckman, F. A., and Ford, T. F. (1959). *J. Dairy Sci.* **42**, 761.
Christianson, G., Jenness, R., and Coulter, S. T. (1954). *Anal. Chem.* **26**, 1923.
Cohen, C., and Szent-Györgyi, A. G. (1957). *J. Amer. Chem. Soc.* **79**, 248.
Cohn, E. J., and Edsall, J. T. (1943). "Proteins, Amino Acids, and Peptides." Reinhold, New York.
Cullis, A. F., Muirhead, H., Perutz, M. F., and Rossmann, M. G. (1962). *Proc. Roy. Soc., (London)* **A265**, 161.
Damodaran, G., and Ramachandran, B. V. (1940). *Biochem. J.* **35**, 122.
Davies, C. W., and Hoyle, B. E. (1953). *J. Chem. Soc.* 4134.
Davies, C. W., and Hoyle, B. E. (1955). *J. Chem. Soc.* 1038.
Davies, D. T., and White, J. C. D. (1960). *J. Dairy Res.* **27**, 171.
Debye, P., and Anacker, E. W. (1951). *J. Phys. Chem.* **55**, 644.
de Koning, P. J., and van Rooijen, P. J. (1965). *Biochem. Biophys. Res. Commun.* **20**, 241.
de Koning, P. J., van Rooijen, P. J., and Kok, A. (1966). *Biochem. Biophys. Res. Commun.* **24**, 616.
Deschamps, O. O. (1840). *J. Pharm.* **26**, 412.
Dickson, I. R., and Perkins, D. J. (1969). *Biochem. J.* **113**, 7P.
Dreizen, P., Noble, R. W., and Waugh, D. F. (1962). *J. Amer. Chem. Soc.* **84**, 4938.
Dresdner, G. W. (1965). Ph.D. thesis. Massachusetts Institute of Technology, Cambridge, Massachusetts.
Eilers, H., Saal, R. N. J., and van der Waarden, M. (1947). "Chemical and Physical Investigations on Dairy Products." Elsevier, New York.
Flory, P. J. (1953). "Principles of Polymer Chemistry." Cornell University Press, Ithaca, New York.
Fölsch, G., and Österberg, R. (1959). *J. Biol. Chem.* **234**, 2298.
Fölsch, G., and Österberg, R. (1961). *Acta Chem. Scand.* **15**, 1963.
Foltmann, B. (1969). *Biochem. J.* **115**, 3P.
Garnier, J. (1963). *Ann. Biol. Animale, Biochim. Biophys.* **3**, 71.
Garnier, J. (1966). *J. Mol. Biol.* **19**, 586.
Garnier, J. (1969). *Compt. Rend.* **268**, 2504.
Garnier, J., and Ribadeau-Dumas, B. (1969). *Compt. Rend.* **268**, 2749.
Garnier, J., Mocquot, G., and Brignon, G. (1962). *Compt. Rend.* **254**, 372.
Garnier, J., Yon, J., and Mocquot, G. (1964). *Biochim. Biophys. Acta* **82**, 481.
Gordon, W. G., Semmett, W. F., Cable, R. S., and Morris, M. (1949). *J. Amer. Chem Soc.* **71**, 3293.
Gordon, W. G., Basch, J. J., and Thompson, M. P. (1965). *J. Dairy Sci.* **48**, 1010.
Grosclaude, F., Mercier, J. C., and Ribadeau-Dumas, B. (1970). *Eur. J. Biochem.* **14**, 98.
Groves, M. L., Hipp, N. J., and McMeekin, T. L. (1958). *J. Amer. Chem. Soc.* **80**, 716.
Halwer, M. (1954). *Arch. Biochem. Biophys.* **51**, 79.
Harris, F. E., and Rice, S. A. (1954). *J. Phys. Chem.* **58**, 725.
Harris, F. E., and Rice, S. A. (1956). *J. Chem. Phys.* **25**, 955.
Hermans, J. J., and Overbeek, J. T. G. (1948). *Rec. Trav. Chim. Pays-Bas* **67**, 761.
Herskovits, T. T. (1966). *Biochemistry* **5**, 1018.
Herskovits, T. T., and Mescanti, L. (1965). *J. Biol. Chem.* **240**, 639.
Hill, R. D., and Hansen, R. R. (1963). *J. Dairy Res.* **30**, 375.
Hill, R. D., and Laing, R. R. (1965). *J. Dairy Res.* **32**, 193.
Hill, R. J. and Wake, R. G. (1969a). *Nature* **221**, 635.
Hill, R. J. and Wake, R. G. (1969b). *Biochim. Biophys. Acta* **175**, 419.

Hipp, N. J., Groves, M. L., Custer, J. H., and McMeekin, T. L. (1950). *J. Amer. Chem. Soc.* **72**, 4928.
Hipp, N. J., Groves, M. L., Custer, J. H., and McMeekin, T. L. (1952). *J. Dairy Sci.* **35**, 272.
Hipp, N. J., Basch, J. J., and Gordon, W. G. (1961a). *Arch. Biochem. Biophys.* **94**, 35.
Hipp, N. J., Groves, M. L., and McMeekin, T. L. (1961b). *Arch. Biochem. Biophys.* **93**, 245.
Ho, C., and Chen, A. H. (1967). *J. Biol. Chem.* **242**, 551.
Ho, C., and Kurland, R. (1966). *J. Biol. Chem.* **241**, 3002.
Ho, C., and Waugh, D. F. (1965a). *J. Amer. Chem. Soc.* **87**, 110.
Ho, C., and Waugh, D. F. (1965b). *J. Amer. Chem. Soc.* **87**, 889.
Ho, C., Magnusson, J. A., Wilson, J. B., Magnusson, N. S., and Kurland, R. J. (1969). *Biochemistry* **8**, 2074.
Hofman, T. (1958). *Biochem. J.* **69**, 139.
Horst, M. G. (1963). *Neth. Milk Dairy J.* **17**, 185.
Hostettler, H., and Imhof, K. (1951). *Milchwissenschaft* **6**, 351, 400.
Hostettler, H., and Stein, J. (1954). *Landwirt. Jrahrb. Schweiz* **3**, 291.
Hostettler, H., Stein, J., and Imhof, K. (1955). *Milchwissenschaft* **10**, 196.
Hutton, T. J., and Patton, S. (1952). *J. Dairy Sci.* **35**, 699.
Imhof, K., and Hostettler, H. (1956). *Schweiz. Milchztg.* **82**, 289.
Jenness, R., and Koops, J. (1962). *Neth. Milk Dairy J.* **16**, 153.
Jenness, R., and Patton, S. (1959). "Principles of Dairy Chemistry." Wiley, New York
Jollès, J. (1969). *Hoppe-Seyler's Z. Physiol. Chem.* **350**, 665.
Jollès, J., Jollès, P., and Alais, C. (1969). *Nature* **222**, 668.
Jollès, P. (1966). *Angew. Chem., Int. Ed. Engl.* **5**, 558.
Jollès, P., and Alais, C. (1959). *Biochim. Biophys. Acta* **34**, 565.
Jollès, P., Alais, C., and Jollès, J. (1961). *Biochim. Biophys. Acta* **51**, 309.
Jollès, P., Alais, C., and Jollès, J. (1962). *Arch. Biochem. Biophys.* **98**, 56.
Kalan, E. B., and Telka, M. (1959). *Arch. Biochem. Biophys.* **79**, 275; *Ibid* **85**, 273.
Kalan, E. B., Thompson, M. P., and Greenberg, R. (1964). *Arch. Biochem. Biophys.* **107**, 521.
Kalan, E. B., Thompson, M. P., Greenberg, R., and Pepper, L. (1965). *J. Dairy Sci.* **48**, 884.
Kamiyama, S., and Schmid, K. (1962). *Biochim. Biophys. Acta* **63**, 266.
Kielland, J. (1937). *J. Amer. Chem. Soc.* **59**, 1675.
Kiermeier, F., and Petz, E. (1967). *Z. Lebensm.-Unters.-Forsch.* **132**, 342; **134**, 97; and **134**, 149.
Kim, Y. K., Yaguchi, M., and Rose, D. (1969). *J. Dairy Sci.* **52**, 316.
Koshland, D. E., Nemethy, G., and Filmer, D. (1966). *Biochemistry* **5**, 365.
Krecji, L. E., Jennings, R. K., and de Spain Smith, L. (1941). *J. Franklin Inst.* **232**, 592.
Krecji, L. E., Jennings, R. K., and de Spain Smith, L. (1942). *J. Franklin Inst.* **234**, 197.
Kresheck, G. C. (1965). *Acta Chem. Scand.* **19**, 375.
Kresheck, G. C., van Winkle, Q., and Gould, I. A. (1964). *J. Dairy Sci.* **47**, 117.
Kumler, W. D., and Eiler, J. J. (1943). *J. Amer. Chem. Soc.* **65**, 2355.
Kuyper, A. C. (1938). *J. Biol. Chem.* **122**, 405.
Laurent, T. C., and Persson, H. (1965). *Biochim. Biophys. Acta* **106**, 616.
Lea, A. S., and Dickinson, W. L. (1890). *J. Physiol.* **11**, 307.
Linderstrøm-Lang, K. (1925). *C. R. Trav. Lab. Carlsberg* **16**, 48.
Linderstrøm-Lang, K. (1929). *C. R. Trav. Lab. Carlsberg* **17**, No. 9, 1.
Linderstrøm-Lang, K., and Kodama, S. (1925). *C. R. Trav. Lab. Carlsberg* **16**, No. 1, 1.

Lindqvist, B. (1963). *Dairy Sci. Abstr.* **25**, 257.
Long, J., van Winkle, Q., and Gould, I. A. (1958). *J. Dairy Sci.*, **41**, 317.
Long, J., van Winkle, Q., and Gould, I. A. (1963). *J. Dairy Sci.* **46**, 1329.
Lowndes, J., Macara, T. J. R., and Plimmer, R. H. A. (1941). *Biochem. J.* **35**, 315.
McBain, J. W., and Salmon, C. S. (1920). *Proc. Roy. Soc., (London)* **A97**, 44.
McGann, T. C. A., and Pyne, G. T. (1960). *J. Dairy Res.* **27**, 403.
McKenzie, H. A. (1967). *Advan. Protein Chem.* **22**, 56.
McKenzie, H. A., and Wake, R. G. (1959a). *Aust. J. Chem.* **12**, 712, 723.
McKenzie, H. A., and Wake, R. G. (1959b). *Aust. J. Chem.* **12**, 734.
McKenzie, H. A., and Wake, R. G. (1961). *Biochim. Biophys. Acta* **47**, 240.
Mackinlay, A. G., and Wake, R. G. (1964). *Biochim. Biophys. Acta* **93**, 378.
Mackinlay, A. G., and Wake, R. G. (1965). *Biochim. Biophys. Acta* **104**, 167.
Mackinlay, A. G., Hill, R. J., and Wake, R. G. (1966). *Biochim. Biophys. Acta* **115**, 103.
Manson, W. (1961). *Arch. Biochem. Biophys.* **95**, 336.
Manson, W. (1962). *Biochim. Biophys. Acta* **63**, 515.
Manson, W., and Annan, W. D. (1970). *Proc. Int. Dairy Congr., 18th,* **IE**, 33.
Marier, J. R., Tessier, H., and Rose, D. (1963). *J. Dairy Sci.* **46**, 373.
Mellander, O. (1939). *Biochem. Z.* **300**, 240.
Mellon, E. F., Korn, A. H., and Hoover, S. R. (1953). *J. Amer. Chem. Soc.* **75**, 1675.
Mercier, J. C., Grosclaude, F., and Ribadeau-Dumas, B. (1970). *Eur. J. Biochem.* **14**, 108.
Monod, J., Wyman, J., and Changeux, J. (1965). *J. Mol. Biol.* **12**, 88.
Nakai, S., Wilson, H. K., and Herreid, E. O. (1965). *J. Dairy Sci.* **48**, 431.
Nakai, S., Wilson, H. K., and Herreid, E. O. (1966). *J. Dairy Sci.* **49**, 469.
Neelin, J. M. (1964). *J. Dairy Sci.* **47**, 506.
Neelin, J. M., Rose, D., and Tessier, H. (1962). *J. Dairy Sci.* **45**, 153.
Neuman, W. F., and Neuman, M. W. (1953). *Chem. Rev.* **53**, 1.
Nielsen, H. C., and Lillevik, H. A. (1957). *J. Dairy Sci.* **40**, 598.
Nitschmann, Hs. (1949). *Helv. Chim. Acta* **32**, 1258.
Nitschmann, Hs., and Beeby, R. (1960). *Chimia* **14**, 318.
Nitschmann, Hs., and Bohren, H. U. (1955). *Helv. Chim. Acta* **38**, 1953.
Nitschmann, Hs., and Henzi, R. (1959). *Helv. Chim. Acta* **42**, 1985.
Nitschmann, Hs., and Zurcher, H. (1950). *Helv. Chim. Acta* **33**, 1698.
Nitschmann, Hs., Wissmann, H., and Henzi, R. (1957). *Chimia* **11**, 76.
Noble, R. W., and Waugh, D. F. (1965). *J. Amer. Chem. Soc.* **87**, 2236.
Noelken, M. (1967). *Biochim. Biophys. Acta* **140**, 537.
Noelken, M., and Reibstein, M. (1968). *Arch. Biochem. Biophys.* **123**, 397.
Österberg, R. (1961). *Biochim. Biophys. Acta* **54**, 424.
Österberg, R. (1964). *Acta Chem. Scand.* **18**, 795.
Österberg, R. (1966). "Phosphorylated Peptides: Study of Primary Structure and Metal Complexity." Almqvist and Wiksell, Uppsala, Sweden.
Overbeek, J. Th.G. (1956). *Prog. Biophys. Biophys. Chem.* **6**, 58.
Pantlitschko, M., and Grundig, E. (1958). *Monatsh. Chem.* **89**, 274.
Parry, R. M., and Carroll, R. J. (1969). *Biochim. Biophys. Acta* **194**, 138.
Payens, T. A. J. (1961). *Biochim. Biophys. Acta* **46**, 441.
Payens, T. A. J. (1966). *J. Dairy Sci.* **49**, 1317.
Payens, T. A. J., and Schmidt, D. G. (1965). *Biochim. Biophys. Acta* **109**, 214.
Payens, T. A. J., and Schmidt, D. G. (1966). *Arch. Biochem. Biophys.* **115**, 136.
Payens, T. A. J., and van Markwijk, B. W. (1963). *Biochim. Biophys. Acta* **71**, 517.
Pedersen, K. O. (1936). *Biochem. J.* **30**, 948.

Pepper, L., and Thompson, M. P. (1963). *J. Dairy Sci.* **46,** 764.
Perlmann, G. E. (1955). *Advan. Protein Chem.* **10,** 1.
Peters, I. E., and Dietrich, J. W. (1954). *Tex. J. Sci.* **6,** 442.
Peterson, R. F. (1969). Private communication.
Peterson, R. F., Nauman, L. W., and McMeekin, T. L. (1958). *J. Amer. Chem. Soc.* **80,** 95.
Pion, R., Garnier, J., Ribadeau-Dumas, B., de Koning, P., and van Rooijen, P. (1965). *Biochem. Biophys. Res. Commun.* **20,** 246.
Pujolle, J., Ribadeau-Dumas, B., Garnier, J., and Pion, R. (1966). *Biochem. Biophys. Res. Commun.* **25,** 285.
Pyne, G. T. (1962). *J. Dairy Res.* **29,** 101.
Pyne, G. T., and McGann, T. C. A. (1960). *J. Dairy Res.* **27,** 9.
Reeves, R. E., and Latour, N. J. (1958). *Science* **128,** 472.
Rice, S. A., and Harris, F. E. (1954). *J. Phys. Chem.* **58,** 733.
Rimington, C. (1941). *Biochem. J.* **35,** 321.
Rogers, H. M. (1937). "Introduction to the Study of Minerals." McGraw-Hill, New York.
Rose, D. (1965). *J. Dairy Sci.* **48,** 139.
Rose, D. (1968). *J. Dairy Sci.* **51,** 1897.
Rose, D. (1969). *Dairy Sci. Abstr.* **31,** 171.
Rose, D., and Colvin, J. R. (1966a). *J. Dairy Sci.* **49,** 351.
Rose, D., and Colvin, J. R. (1966b). *J. Dairy Sci.* **49,** 1091.
Rose, D., and Tessier, H. (1959). *J. Dairy Sci.* **42,** 969.
Scatchard, G., and Yap, W. T. (1964). *J. Amer. Chem. Soc.* **86,** 3434.
Scatchard, G., Coleman, J. C., and Shen, A. L. (1957). *J. Amer. Chem. Soc.* **79,** 12.
Schmidt, D. G. (1964). *Biochim. Biophys. Acta* **90,** 411.
Schmidt, D. G., and Payens, T. A. J. (1963). *Biochim. Biophys. Acta* **78,** 492.
Schmidt, D. G., Both, P., and de Koning, P. J. (1966). *J. Dairy Sci.* **49,** 776.
Scott-Blair, G. W., and Oosthuizen, J. C. (1961). *J. Dairy Res.* **28,** 165.
Scott-Blair, G. W., and Oosthuizen, J. C. (1962). *J. Dairy Res.* **29,** 37.
Shimmin, P. D., and Hill, R. D. (1964). *J. Dairy Res.* **31,** 121.
Shimmin, P. D., and Hill, R. D. (1965.) *Aust. J. Dairy Technol.* **20,** 119.
Shinoda, K., and Hutchinson, E. (1962). *J. Phys. Chem.* **66,** 577.
Shinoda, K., Nakagawa, T., Tamamushi, B., and Isemura, T. (1963). "Colloidal Surfactants: Some Physicochemical Properties." Academic Press, New York.
Shulman, J. H., ed. (1957). *Proc. Int. Congr. Surface Activ., 2nd.*
Slatter, W. L., and van Winkle, Q. (1952). *J. Dairy Sci.* **35,** 1083.
Slattery, C. W., and Waugh, D. F. To be published.
Smeets, G. M. (1955). *Neth. Milk Dairy J.* **9,** 249.
Smith, R. M., and Alberty, R. A. (1956a). *J. Amer. Chem. Soc.* **78,** 2376.
Smith, R. M., and Alberty, R. A. (1956b). *J. Phys. Chem.* **60,** 180.
Stark, G. R., Stein, W. H., and Moore, S. (1960). *J. Biol. Chem.* **235,** 3177.
Strauss, U. P., and Ross, P. D. (1959). *J. Amer. Chem. Soc.* **81,** 5295.
Sullivan, R. A., Fitzpatrick, M. M., Stanton, E. K., Annino, R., Kissel, G., and Palermiti, F. (1955). *Arch. Biochem. Biophys.* **55,** 455.
Sullivan, R. A., Fitzpatrick, M. M., and Stanton, E. K. (1959). *Nature* **183,** 616.
Swaisgood, H. E., and Brunner, J. R. (1962). *J. Dairy Sci.* **45,** 1.
Swaisgood, H. E., and Brunner, J. R. (1963). *Biochem. Biophys. Res. Commun.* **12,** 148.
Swaisgood, H. E., Brunner, J. R., and Lillevik, H. A. (1964). *Biochemistry* **3,** 1616.

Talbot, B., and Waugh, D. F. (1967). *Fed. Proc., Fed. Amer. Soc. Exp. Biol.* **26**, 826.
Talbot, B., and Waugh, D. F. (1970). *Biochemistry* **9**, 1807.
Tanford, C. (1955). *J. Phys. Chem.* **59**, 788.
Tanford, C. (1961). "Physical Chemistry of Macromolecules." Wiley, New York.
Tessier, H., and Rose, D. (1958). *J. Dairy Sci.* **41**, 351.
Thompson, M. P., and Kiddy, C. A. (1964). *J. Dairy Sci.* **47**, 626.
Thompson, M. P., and Pepper, L. (1962). *J. Dairy Sci.* **45**, 794.
Thompson, M. P., and Pepper, L. (1964a). *J. Dairy Sci.* **47**, 293.
Thompson, M. P., and Pepper, L. (1964b). *J. Dairy Sci.* **47**, 633.
Thompson, M. P., Tarassuk, N. P., Jenness, R., Lillevik, H. A., Ashworth, U. S., and Rose, D. (1965). *J. Dairy Sci.* **48**, 159.
Tuckey, S., Roche, H., and Clark, G. (1938). *J. Dairy Sci.* **21**, 767.
Turkington, R. W., and Topper, Y. J. (1966). *Biochim. Biophys. Acta* **127**, 366.
van Bruggen, E. F. J., Wiebenga, E. H., and Gruber, M. (1962). *J. Mol. Biol.* **4**, 1.
van Kreveld, A., and van Minnen, G. (1955). *Neth. Milk Dairy J.* **9**, 1.
van Slyke, L., and Bosworth, A. W. (1915). *J. Biol. Chem.* **20**, 135.
van Wazer, J. R. (1958). "Phosphorus and its Compounds," Vol. I. Interscience (Wiley), New York.
Verwey, E. J. W., and Overbeek, J. T. (1948). "Theory of the Stability of Hypophobic Colloids." Elsevier, Amsterdam.
von Hippel, P. H., and Waugh, D. F. (1955). *J. Amer. Chem. Soc.* **77**, 4311.
von Nägeli, C., and Schwendener, S. (1877). "Das Mikruskop." Engelmann, Leipzig; see also Frey-Wyssling, A. (1953). "Submicroscopic Morphology of Protoplasm." Elsevier, New York.
Wake, R. G. (1957). *Aust. J. Sci.* **20**, 147.
Wake, R. G. (1959). *Aust. J. Biol. Sci.* **12**, 479.
Wake, R. G., and Baldwin, R. L. (1961). *Biophys. Biochim. Acta* **47**, 225.
Warner, R. C. (1944). *J. Amer. Chem. Soc.* **66**, 1725.
Waugh, D. F. (1958). *Discuss. Faraday Soc.* **25**, 186.
Waugh, D. F. (1961). *J. Phys. Chem.* **65**, 1793.
Waugh, D. F. (1962). *Abstr. Amer. Chem. Soc. Meeting, 142nd* 62C.
Waugh, D. F., and Noble, R. W. (1965). *J. Amer. Chem. Soc.* **87**, 2246.
Waugh, D. F., and Talbot, B. To be published.
Waugh, D. F., and von Hippel, P. H. (1956). *J. Amer. Chem. Soc.* **78**, 4576.
Waugh, D. F., Ludwig, M., Gillespie, J. M., Garnier, J., Kleiner, E. S., and Noble, R. W. (1960). *Fed. Proc., Fed. Amer. Soc. Exp. Biol.* **19**, 337.
Waugh, D. F., Ludwig, M. L., Gillespie, J. M., Metton, B., Foley, M., and Kleiner, E. S. (1962). *J. Amer. Chem. Soc.* **84**, 4929.
Waugh, D. F., Creamer, L. K., Slattery, C. W., and Dresdner, G. W. (1970). *Biochemistry* **9**, 786.
Waugh, D. F., Slattery, C. W., and Creamer, L. K. (1971). *Biochemistry* **10**, 817.
Whitaker, R., Sherman, J. M., and Sharp, P. F. (1927). *J. Dairy Sci.* **10**, 361.
White, J. C. D., and Davies, D. T. (1958). *J. Dairy Res.* **25**, 236.
Whitnah, C. H., and Rutz, W. D. (1959). *J. Dairy Sci.* **42**, 227.
Whitney, R. M. (1961). Abstract IIA. *Amer. Chem. Soc. Meetings*, March.
Woychik, J. H. (1964). *Biochem. Biophys. Res. Commun.* **16**, 267.
Woychik, J. H., Kalan, E. B., and Noelken, M. E. (1966). *Biochemistry* **5**, 2276.
Yaguchi, M., and Terassuk, M. P. (1967). *J. Dairy Sci.* **50**, 1985.
Zittle, C. A. (1961). *J. Dairy Sci.* **44**, 2101.
Zittle, C. A. (1963). *J. Dairy Sci.* **46**, 607.

Zittle, C. A. (1965). *J. Dairy Sci.* **48,** 1149.
Zittle, C. A., and Custer, J. H. (1963). *J. Dairy Sci.* **46,** 1183.
Zittle, C. A., and Jasewicz, L. B. (1962). *J. Dairy Sci.* **45,** 703.
Zittle, C. A., and Walter, M. (1963). *J. Dairy Sci.* **46,** 1189.
Zittle, C. A., DellaMonica, E. S., Rudd, R. K., and Custer, J. H. (1958). *Arch. Biochem. Biophys.* **76,** 342.
Zittle, C. A., Thompson, M. P., Custer, J. H., and Cerbulis, J. (1962). *J. Dairy Sci.* **45,** 807.
Zweig, G., and Block, R. J. (1953). *J. Dairy Sci.* **36,** 427.

10 □ Whole Casein: Isolation, Properties, and Zone Electrophoresis

H. A. McKENZIE

I.	Introduction	87
II.	Methods of Isolation	90
	A. Acid Precipitation at 20°C	90
	B. Acid Precipitation at 30°C	91
	C. Acid Precipitation at 2°C	91
	D. Centrifugation in the Presence of Added Calcium(II) at 3°C	91
	E. Centrifugation in the Presence of Added Calcium(II) at 37°C	92
	F. Centrifugation in the Absence of Added Calcium(II) at 3°C	93
	G. Centrifugation in the Absence of Added Calcium(II) at 20°C	93
	H. Ammonium Sulfate Precipitation at 2°C	93
III.	Properties	94
	A. Physical Appearance and Solubility	94
	B. Yield	95
	C. Chemical Composition	96
	D. Electrophoretic Patterns	98
	E. Sedimentation	98
IV.	Controversial Components	101
	A. γ-Caseins	101
	B. Minor Glycoproteins	104
V.	Zone Electrophoresis in Casein Typing	109
VI.	Summary	113
	References	114

I. Introduction

We have seen in the preceding chapter that a fundamental necessity in the study of milk proteins is to gain an insight into the structure and composition of the casein micelles and their interactions with the proteins and ions in their environment. This is not only a problem in colloid science, because it involves an understanding of the chemistry and immunology of the individual proteins, the genetic control of their synthesis, and the recent clinical history of the animal from which the milk is derived. Most studies of the individual proteins involve the initial separation of "whole" casein from "noncasein" proteins and its subsequent fractionation to obtain the individual casein proteins. We shall be concerned in this chapter with the isolation and properties of whole casein.

Attention will be given to the zone electrophoretic examination of milk fractions for the identification of caseins. It will be seen that while much progress has been made, many real problems remain to be solved; even the definition of casein involves difficulty.

It has been pointed out in Chapters 1 and 2, Volume I, that Mülder showed in 1838 how an abundant protein could be precipitated from bovine milk by the addition of acid. A method for the "purification" of this protein, after precipitation with acetic acid, was published by Hammersten in 1900. This "acid precipitation" method has been the basis of the overwhelming majority of methods for the isolation of "whole" casein since Hammersten's period. It is amazing that a century passed after Mülder's work before there was a concerted attempt to make use of two important properties of casein in isolating it without adjustment of the pH value from neutrality: its interaction with calcium(II) and its precipitation in the presence of high concentrations of salts, for example, ammonium sulfate.

Despite the development of different methods of preparation, an operational definition of casein in terms of acid precipitation is usually employed —whole casein is the protein precipitated from skim milk at pH 4.6. The American Dairy Science Association's Committee on Nomenclature (Thompson et al., 1965) gives a definition that is more precisely expressed: "a heterogeneous group of phosphoproteins precipitated from skim milk at pH 4.6 and 20°." It is fortunate that the Committee has specified the temperature, for different precipitates are obtained at various temperatures in the range 0°–37°C, which is commonly used for acid precipitation of casein. Some workers have precipitated their casein at pH 4.6 and 2°C. We shall see that precipitation is far from complete under the latter conditions. However the Committee's definition is still not without difficulties. Even at 20°C the precipitation of phosphoprotein is probably not complete, and the casein carries with it small amounts of proteins that are not phosphoproteins, for example, transferrin, protease, and immunoglobulins. Furthermore the casein appears to be altered physically by acid precipitation at higher temperatures. During the alkali dissolution of acid casein all or part of the γ-casein may be lost (El-Negoumy, 1963; Lahav, 1965).

There are difficulties in proposing other definitions. The calcium(II) sedimentation method of von Hippel and Waugh (1955) results in a preparation that is obtained under very mild conditions, but all the casein does not seem to be isolated by this method. It is possible, as will be discussed later, to vary this method to isolate nearly all the casein. Hill and Hansen (1964) separated skim milk into four fractions on Sephadex G-100 (for a general discussion of the properties of Sephadex see Chapter 4, Volume I). They found that one of them is predominantly casein. However this material has not been subjected to exhaustive examination.

At intervals over a period of six years the author and his colleagues examined the isolation of whole casein from a variety of types of milk by several methods, including ammonium sulfate fractionation. There are many advantages of using this gentle reagent in protein fractionation (Dixon and Webb, 1961) and these apply equally well to casein. The question immediately arises as to what one is separating out by this salt fractionation. It does seem that virtually all the casein is separated, but in addition some of the immunoglobulins, the "proteose peptone," and glycoproteins containing phosphorus may be included. The amount of the immunoglobulins in normal mature milk is small (see below). It has been proposed by McKenzie (1967) that the "proteose peptones," are caseins (i.e., phosphoproteins) and that the undesirable term of "proteose peptone" should be dropped. [A revision of the nomenclature of Thompson et al. (1965) for cow milk proteins has been published by Rose et al. (1970). These authors agree with the suggestion of McKenzie (1967) but retain the term "for the present" on the flimsy ground that present knowledge of these components is incomplete.] Along with these there may be one or more additional glycoproteins containing phosphorus. At present noone knows exactly what role these proteins play in milk and whether they contribute to the stability of the micelle, although there is some recent evidence (vide infra), that they do. Thus careful consideration must be given as to whether these proteins are caseins. There certainly seems to be as much justification at present for including these proteins in the class of caseins as for including γ-casein.

We consider it preferable to isolate a "whole" casein preparation that contains all the "caseins" prepared in a gentle fashion, possibly containing a small amount of noncasein proteins which are removable in subsequent fractionation, than to start with a preparation that does *not* contain all the "casein," that may contain partly altered casein, and that contains some "noncasein" proteins.

Whatever method is used, careful attention must be given to the genetic and clinical history of the animal supplying the milk. Care must be taken not to assume that methods developed for bovine caseins will be directly applicable to the isolation of casein from other species.

A number of isolation procedures will now be described briefly. In assessing them an attempt will be made, where the relevant data are available, to compare the products with respect to the following:

(1) Physical characteristics
   appearance
   ease of separation (centrifugation)
   ease of washing
   solubility
   reprecipitation and re-solution
(2) Yield (dry weight) in terms of the percentage of total milk protein
(3) Chemical composition
   total nitrogen content
   total phosphorus content
   absorbancy ($A_{1\,cm}^{1g/dl}$) at 278 nm
   residual calcium(II)
(4) Carbohydrate moiety
(5) Zone electrophoretic patterns
(6) Sedimentation-velocity properties: s values and relative peak areas
(7) Loss of β-casein and κ-casein, especially in low-temperature methods
(8) Loss of γ-casein, especially in dissolution and washing
(9) Presence of proteolytic activity
(10) Presence of nonphosphoproteins
(11) Effect of stage of lactation
(12) Contamination with immunoglobulins due to infection

II. Methods of Isolation

The methods described are based on experience in the author's laboratory (Graham et al., 1970). The amounts of reagent needed are given per 100 ml skim milk, although this may not be the actual volume used.

A. Acid Precipitation at 20°C

Hydrochloric acid (1 M) is added slowly from a burette with the tip below the surface of the skim milk (100 ml) at 20° ± 2°C and with mechanical stirring over a period of 40–45 min until a pH value of 4.5–4.6 is attained (~4.5 ml acid). Stirring is continued for a further 30 min. The precipitate is sedimented at ~1300 g (15 min, 2000 rpm, Sorvall RC-2 centrifuge, GSA rotor). It is washed twice by suspending it in water (20 ml) and centrifuging as previously. It is then dissolved by suspending it in a volume of water (45 ml), approximately half that of the skim milk used, and titrating it with sodium hydroxide (1 M) to pH 7.0–7.2. Considerable pulverization of lumps with a glass rod is necessary at intervals to assist solution. After a period of 2–3 hr most of the casein will dissolve to give an opalescent solution. The precipitation, washing, and dissolution procedures are repeated. Some workers extract the casein with alcohol and ether to remove residual lipid. This procedure is to be deprecated in the author's opinion since it may cause irreversible changes in the protein. Most residual lipid may be removed by high-speed centrifugation.

B. Acid Precipitation at 30°C

The procedure is similar to that described in Section A, except that all operations are carried out at 30°C. Some workers prefer this temperature to 20°C, finding the precipitate more granular and easily separated. However other problems, for example, irreversible changes, may be greater.

C. Acid Precipitation at 2°C

This procedure is similar to that described in Section A, except that all operations are carried out at 2°–3°C (cold room) and the acid is added until a pH value of 4.3 is attained. This pH value is preferred, as precipitation is by no means complete at pH 4.6 and 2°C. The centrifugation is carried out at 14,600 g (35 min, 9500 rpm, Sorvall RC-2, GSA rotor). During centrifugation it is essential to set the refrigeration plate temperature to such a value that the temperature of the rotor *contents* does not rise above 3°C. Ice-cold distilled water is used in the washing. It is important in the low-temperature method to check the actual temperature in the reaction vessel at all times.

D. Centrifugation in the Presence of Added Calcium(II) at 3°C

The procedure used is similar to the original one of von Hippel and Waugh (1955) for the preparation of first-cycle casein, except that a lower calcium(II) level is used and citrate or ion-exchange resins or Sephadex may be used instead of oxalate to remove calcium(II). Skim milk (100 ml) is cooled to 3°C, and a 2 M $CaCl_2$ solution (3.5 ml) is added with constant stirring and control of pH to 6.6–6.8 (final added calcium concentration = $\sim$0.07 M instead of 0.13 M used in the original method). The mixture is centrifuged at 105,000 g (90 min, 30,000 rpm, 4°C, Spinco L, Rotor 30, Beckmann Instruments, Palo Alto, California). Three layers result in the centrifuge tube. At the bottom there is a pale yellow opalescent gel; on top of this there is a small amount of an opalescent viscous liquid layer and the main supernatant. Each of these layers is removed. The bottom two layers are retained for the casein preparation and are suspended in a 0.08 M sodium chloride plus 0.07 M calcium chloride solution (25 ml). Any gel is broken up by hand with a glass rod and dispersed carefully. The resulting opalescent solution may be taken directly to the next stage, involving calcium removal, or it may be recentrifuged first and the resultant bottom layers suspended in 0.08 M sodium chloride (25 ml).

In the oxalate procedure for calcium removal, 1.5 M potassium oxalate (10 ml) is added while keeping the pH in the range 6.6–7.0 with 0.1 M oxalic acid. The calcium oxalate precipitate is removed by centrifugation (for problems see von Hippel and Waugh, 1955). The supernatant is dialyzed against 0.05 M sodium chloride or water.

In the citrate procedure the bottom layers are brought into solution by suspension in 0.1 M potassium citrate (60 ml), keeping the pH constant at 6.5 ± 0.1. The gel is broken up carefully and the mixture is gently stirred for a few hours. It is then placed in a dialysis sac and dialyzed vs. 0.1 M potassium citrate. This is followed by dialysis against 0.05 M sodium chloride or water.

In the ion-exchange resin procedure (Bohren and Wenner, 1961) the bottom layers are suspended in water (~30 ml). The calcium(II) is exchanged against sodium(I) at pH 6.2–6.7 with a sodium–hydrogen cation-exchange resin (16 ml sodium saturated Amberlite IR–120, Rohm and Haas, Co., Philadelphia, Pennsylvania, plus 2.4 ml 1 M HCl). (The resin has been previously stirred for 5 min, allowed to stand for 2 hr, rinsed four times with 50 ml distilled water by decantation and filtered on a Büchner funnel.) After stirring the resin–casein suspension for 40–50 min about 95% of the calcium should be replaced by sodium. The resin is removed by centrifugation (20 min, 1500 rpm, Sorvall RC-2 centrifuge, SS34 rotor).

A Sephadex procedure has been described by Talbot and Waugh (1967) for the removal of calcium(II) from second-cycle casein, and presumably it can be adapted to first-cycle casein.

E. CENTRIFUGATION IN THE PRESENCE OF ADDED CALCIUM(II) AT 37°C

This method has been used by Waugh et al. (1962) and Noble and Waugh (1965). Skim milk (100 ml) is made 0.07 M in added calcium chloride (3.5 ml 2 M CaCl$_2$) at 37°C, the pH being maintained at 7 by the addition of

2 M Tris. Micelles are removed by centrifugation at 37°C (18,000 rpm, 90 min, Spinco L rotor). They are then dispersed into 0.01 M calcium chloride and recentrifuged. This is followed by dispersal into 0.07 M potassium citrate (60 ml) at 8°C and pH 6.5 (addition of 2 M sodium acetate). After a few hours, further potassium citrate (3.5 ml 1 M) is added, and the preparation is dialyzed. (Waugh et al. dialyze against 0.08 M sodium acetate, but other media may be used.) Calcium(II) may also be removed by any of the methods described in Section D.

F. CENTRIFUGATION IN THE ABSENCE OF ADDED CALCIUM(II) AT 3°C

The procedure is similar to that described in Section D, except that no calcium(II) is added to the milk.

G. CENTRIFUGATION IN THE ABSENCE OF ADDED CALCIUM(II) AT 20°C

The procedure is similar to that described in Section D except that no calcium(II) is added to the milk and the centrifugation is carried out at 20°C. Bohren and Wenner (1961) have described a procedure for obtaining residual casein from the supernatant of the centrifugation. This method is discussed further below.

H. AMMONIUM SULFATE PRECIPITATION AT 2°C

Ammonium sulfate (26.4 g) is added to skim milk (100 ml) with mechanical stirring over a period of $\sim$40 min at 2°C (not less than 0°C or greater than 3°C) for a further 75–90 min, and the precipitate is collected by centrifugation (Sorvall RC-2, GSA rotor, 9500 rpm, 35 min, 14,600 g). During the centrifugation it is important to set the refrigerant coil temperature at such a value that the temperature of the casein mixture in the rotor does not rise above 3°C. A supernatant volume of $\sim$90 ml is obtained from 100 ml skim milk. The precipitate is dissolved in water (50 ml water added initially, i.e., about half the volume of skim milk, making sure the total volume does not exceed 90 ml) using mechanical stirring and occasional maceration of the precipitate with a glass rod having a flattened end to help disperse any large lumps. Solution of the casein is complete in about 50 min. The volume is made up to 100 ml with water. In all these operations careful attention is paid not to allow the temperature to rise above 3°C. For a period, this procedure was varied to include washing the precipitate with ammonium sulfate, but it was felt eventually that little was gained and there was a tendency for the precipitate to become more difficult to dissolve. A further precipitation is carried out with ammonium sulfate (24 g), the solution being stirred for 90 min after addition of all the salt. The precipitated casein is separated by centrifugation, as before, and then dissolved in water so that the final volume is ~60 ml, solution being complete in 1 hr. The final solution is dialyzed exhaustively against glass-distilled water or 0.1 M NaCl.

III. Properties

A. Physical Appearance and Solubility

The most striking thing about whole caseins prepared by various procedures is their difference in appearance. Acid casein at temperatures above ~10°C has a characteristic white granular appearance, reminiscent of an inorganic precipitate rather than a protein. As the temperature of precipitation is lowered, the precipitate becomes more hydrated and more gel-like in appearance so that near 2°C it is an opalescent pale yellow gel. However precipitation is far from complete at pH 4.6–4.7, and it is necessary to lower the pH to 4.3 to get a reasonable yield. Both the low-temperature and high-temperature preparations are difficult to redissolve. Considerable care must be exercised in the re-solution with alkali to minimize irreversible changes: pH excesses must be avoided and the pH of the mixture not allowed to rise above 7.0–7.2 during titration. The low-temperature preparation becomes somewhat less gel-like and more granular on reprecipitation.

Washing the acid precipitate with water is a controversial procedure. According to El-Negoumy (1963) repeated washing results in loss of γ-casein from the precipitate. Thus exhaustive washing is not recommended. Lahav (1965) has claimed that purification of acid casein by repeated alkaline dissolution and acid precipitation may cause loss of γ-casein. He considers that divergences in opinion concerning the amount of γ-casein in whole casein can be reconciled when it is realized how readily γ-casein may be lost in the repeated alkaline treatment of casein. Lahav advances no detailed theory of the nature of any reactions involved.

The granular appearance of acid casein as normally prepared may be contrasted with the appearance of casein obtained by high-speed centrifugation of skim milk, in either the presence or absence of added calcium(II). The casein centrifugate consists of an opalescent gel plug with a very viscous layer of liquid on top of it. The amount of the liquid layer decreases with increasing added calcium(II) or increasing temperature. The calcium caseinate obtained can be redispersed in dilute calcium chloride solution and converted to a casein solution on removal of calcium(II) by precipitation or dialysis.

The appearance of the centrifuged casein gel from skim milk is somewhat similar to the one obtained by ammonium sulfate precipitation; on the other hand the latter gel is soft and readily soluble in water and hence cannot be washed with water. There is slight loss of ease of dissolution on reprecipitation.

B. YIELD

In normal mature bovine skim milk, about 80% of the total protein is casein. This estimate is based on area analysis of moving-boundary electrophoresis patterns (see Rolleri et al., 1956; Chapter 2, Volume I). There are breed differences but these are probably not great for Western dairy cattle. Comparisons of the yield from various preparative procedures have been made by Graham et al. (1970). All preparations have been made from milk from individual cows. In some cases, comparisons have been made in milk from normal and subclinical mastitic quarters of the same animal.

1. *Normal Milk*

The yield of whole casein from a given sample of normal milk depends on the method of preparation. That from acid casein prepared at 20° or 30°C is ~80% of the total milk protein. The yield of acid casein (pH 4.3) prepared at 2°C is somewhat lower. In ammonium sulfate casein preparations from normal milk, a slightly higher yield of protein is usually obtained (~3% more). The yield obtained from centrifuged casein preparations is dependent on the amount of added calcium(II). At a level of 0.06 M added calcium(II), the yield is usually the same as for the ammonium sulfate preparation and higher than that for acid casein.

It is of interest to compare the results for the calcium procedure with those of Bohren and Wenner (1961). They found that in the absence of added calcium chloride, the yield at 4°C was 85–90% of the yield of acid casein (precipitation at 35°C) by the method of Rowland (1938); at 20°C the yield was 93–97%. They found that the yield could be increased to the acid casein yield by second centrifugation after the addition of 6 mmole of calcium chloride to 100 ml of supernatant from the first centrifugation. At a level of 6 mmole calcium chloride/100 ml milk, maintaining the pH at 6.7, centrifugation at 50,000 g for 3 hr at 20°C, gave a yield similar to that for acid casein. Higher levels of calcium chloride did not result in greater yields.

2. *Subclinical Mastitic Milk*

It is known that there is an increase in immunoglobulin content (mainly IgG fast) of the milk from the infected quarters in subclinical mastitis. The level of some other blood proteins in the milk is also increased, for example, serum albumin (see Chapter 3, Volume I; Lascelles et al., 1967, 1969). The absolute yield of whole casein from the infected quarters in the experiments of Graham et al. was less than that from the normal quarters. (This is to be expected since the total protein content is less.) The yields from the acid precipitation and calcium-centrifugal methods for the mastitic quarters were similar to one another but these were both less than the low-temperature ammonium sulfate procedure. However there were considerable differences in chemical composition in the casein for a given preparation between normal and abnormal quarters and between preparations from given abnormal quarters. These results are discussed below.

C. CHEMICAL COMPOSITION

The chemical composition of whole casein prepared from the milk of normal quarters and subclinical mastitic quarters of a given animal by various preparative procedures may be compared in the typical results of Graham et al. given in Table I. There is no appreciable difference in nitrogen and phosphorus contents among the various preparations from normal quarters. However the nitrogen content in some ammonium sulfate preparations and the phosphorus content in some calcium(II) preparations have sometimes been a little higher than for preparations by the other methods. This is due to the difficulty in removing ammonium sulfate and calcium phosphate from the respective preparations. The hexosamine, hexose, and *N*-acetylneuraminic acid (NANA) contents are appreciably lower in the acid caseins than in the other preparations. There are small variations in the ratio of the sugar components for the different types of preparation on a given sample of normal milk and for the same type of preparation on different samples of normal milk. There is no difference in the analyses between acid casein prepared at 20° and 30°C, or between calcium preparations at 2° and 37°C.

In comparing analyses for the different types of preparation from subclinical mastitic quarters of the same animal, one striking difference is the low phosphorus content of the ammonium sulfate preparations. There are also appreciable increases in hexosamine, hexose, and NANA contents for a

TABLE I

Chemical Composition of Some Whole Casein Preparations (g/100 g)[a]

Property	Source of milk	Method of preparation				
		30°C, acid	20°C, acid	2°C, $(NH_2)SO_4$	2°C, calcium(II)	37°C, calcium(II)
N Content	Normal quarter	15.3	15.4	15.5	15.3	15.2
	S.C.M.[b]		15.2	15.3	15.2	
P Content	Normal	0.79	0.78	0.79	0.78	0.84
	S.C.M.		0.78	0.65	0.84	
Hexosamine	Normal	0.25	0.26	0.30	0.31	0.29
	S.C.M.		0.31	0.42	0.36	
Hexose	Normal	0.23	0.23	0.27	0.31	0.30
	S.C.M.		0.28	0.40	0.41	
NANA	Normal	0.45	0.45	0.49	0.50	0.50
	S.C.M.		0.52	0.56	0.51	

[a] After Graham et al. (1970).
[b] Subclinical mastitic quarter.

given type of casein preparation from the mastitic milk compared with the equivalent type of preparation from the normal milk. The proportional increase is different for the various sugars. The increase in a given type of carbohydrate is not the same for the different types of preparation. Thus the use of subclinical mastitic milk does not result in preparations of whole casein whose composition is changed similarly or simply, as compared with casein from normal milk.

D. Electrophoretic Patterns

There have been few comparisons made of electrophoretic patterns of whole casein prepared by different procedures. Reference has already been made to the loss of γ-casein in acid casein preparations. This is sufficiently great in certain procedures for the preparation of acid casein that it can be seen both in moving-boundary and zone electrophoretic patterns. Bohren and Wenner (1961) found that the supernatant from casein prepared by centrifugation in the absence of added calcium chloride had a relatively much higher percentage of β- and γ-casein compared with the centrifugate, the latter representing only 85–95% of the total casein. Wake and Baldwin (1961) found no apparent difference in zone electrophoretic patterns between first-cycle casein (prepared according to the method of von Hippel and Waugh, 1955) and acid casein.

Graham *et al.* (1970) compared the electrophoretic patterns of a number of preparations using a starch-gel–urea method similar to that of Wake and Baldwin (1961), and the starch-gel–urea–mercaptoethanol discontinuous tris-borate system and continuous tris-glycinate systems developed by McKenzie and Treacy (1970). They found that there were no *qualitative* differences in the different types of preparations from normal milk. However in certain comparative runs at equal concentrations of total protein, the κ-casein bands in the acid casein preparations appeared to stain less intensely than the κ-casein bands in the preparations by other methods. Some types of milk had very low (or no) γ-casein contents. Preparations of whole casein by the various methods using mastitic milk gave electrophoretic patterns similar to one another. However, at moderate subclinical levels of infection, patterns differed from those of casein from normal milk in that they had an intense band in the position of immunoglobulin (IgG).

E. Sedimentation

At present it is not possible to make precise interpretations of sedimentation-velocity patterns for whole casein. This is due to the problems involved in interpreting transport patterns of interacting systems as discussed generally in Chapter 7, Volume I, by McKenzie and Nichol. Furthermore there are difficulties in obtaining reproducibility in patterns not only for whole casein but for synthetic mixtures of individual caseins, such as α_s-, β-, and κ-caseins. It will be recalled that Waugh has written in Chapter 9 (p.33): "the many interactions into which the caseins can enter, with themselves and with each other in the absence of calcium, make a particular interaction pattern which is temporarily reproducible but strongly dependent on unknown experimental conditions, including the past history of the sample."

Halwer (1954) showed that both α- and β-caseins undergo increasing self-association with increasing electrolyte concentration at pH 7. He found that the polymerization was rapid for α-casein but slow for β-casein. Sullivan et al. (1955) showed that β-casein was largely dissociated to the monomeric form at low temperature near pH 7. More extensive studies have been made of the association of α_s- and of β-caseins (see Chapter 7, Volume I). Payens and Schmidt (1966) found that $\alpha_{s,1}$-casein B at 2°C, pH 6.6 (barbiturate), $I = 0.2$, has a $s_{20,w}^{\circ}$ of 6.75 S, $s_{20,w}$ being 8.2 S at a protein concentration of 10 g/liter. Later Schmidt et al. (1967) obtained a somewhat lower value of 6.0 S for $\alpha_{s,1}$-casein BC at a concentration of 7.0 g/liter. Payens and van Markwijk (1963) studied the association of β-casein and concluded that at 4°C β-casein behaves as a monomer, but at 8.5°C polymers are formed. A value of $s_{20,w}$ of 1.5 S was found at 4°C and pH 7.5 (barbiturate) for a protein concentration of 10 g/liter.

The situation for κ-casein is more complex (see Chapter 8, Volume I). Most of the current preparations of κ-casein are highly associated in solution at neutral pH. The chains of κ-casein are linked by intermolecular disulfide bridges, and the polymers so formed are themselves able to undergo considerable polymerization through noncovalent forces. Beeby (1964) has suggested that the SS linkages are formed during the isolation procedures currently used for κ-casein, that the residues concerned occur in the reduced form in milk and that the SH groups are masked by calcium(II). Most other workers are of the opinion that the κ-casein chains are synthesized originally with SH groups but that the cysteine is converted to intermolecular cystine bridges in the mammary gland. The author, on the other hand, is inclined to agree with Beeby (see below).

Mackinlay and Wake (1964) obtained $s_{20,w}$ values of 13.2 and 9.1 S for κ-casein and S-sulfo-κ-casein, respectively, at pH 7.0 ($I = 0.1$, phosphate) at 20°C. However in 40% dimethylformamide the respective values were 4.7 and 1.5 S. Thus even the S-sulfo derivative is polymerized in the absence of a dissociating agent such as urea or dimethylformamide (see also Swaisgood and Brunner, 1963).

In their early studies of first-cycle (whole) casein von Hippel and Waugh (1955) obtained a bimodal sedimentation pattern for this casein at 4°C and pH 7.0 (phosphate, $I = 0.19$). The slow-moving material had an s_{20} of 1.3 S and the fast-moving material had an s_{20} of 4.5 S. Later Waugh and von Hippel (1956) reported 1.3 and 7.5 S for first-cycle casein, and 1.3 and 6.0 S for acid casein. They interpreted the patterns in terms of slow-moving material arising from β-casein monomers and the fast-moving material from α_s-κ complexes. It was considered that the preferred ratio for such complexes was 4:1 (see also Waugh, 1958). This view was later abandoned by Noble and Waugh (1965), as has been discussed by Waugh in Chapter 9. Noble and Waugh found, on sedimentation of synthetic mixtures of α_s- and κ-caseins, that α–κ interaction occurs readily at 37°C but requires pretreatment with urea or alkali at 20°C, and does not occur at 5°C. When interaction occurs at 37°C the complex has an α–κ weight ratio of ~ 1 and an s_{20} of ~ 7 S. Patterns for first-cycle casein at 2°C had peaks with s_{20} values of 3 and 5 S. Mixtures of his second-cycle casein fraction P and fraction S gave patterns at 2°C with peaks corresponding to those of fractions P and S individually, i.e., peaks of 3, 5 and 15 S. These peaks were interpreted as being characteristic of β-, α_s-, and κ-caseins, respectively. Thus it was concluded that when fractions S and P are mixed at 2°C there is no appreciable α_s–κ interaction.

Graham et al. (1970) compared sedimentation patterns for whole casein prepared by various methods from normal and subclinical mastitic milk. Their patterns were obtained at low temperature in order to minimize association of β-casein. In some fifty sedimentation patterns of preparations obtained from normal mature milk a slow-moving, slow-spreading peak and a fast-moving, rapidly spreading peak were observed. The former peak had an $s_{20,w}$ value, for a total protein concentration of 10 g/liter, usually in the range 1.2–1.4 S, although some preparations had values of as low as 1.05 and as high as 1.55 S. No specific reason for this variation can be advanced. It did not seem to be related simply to the solvent used (0.1 M NaCl or phosphate buffer), the time of standing on dilution of the concentrated stock protein solution, the age of the preparation, the type of preparation or the stage of lactation. Part of the variation arises from the fact that the s values of the slow peak were calculated without the peak being completely separated from the meniscus (see Chapter 7, Volume I). The fast-moving peak exhibited considerable polydispersity. This was variable and the rate of movement of the fast-moving peak gives no idea of its extent. The $s_{20,w}$ of the peak was usually in the range 4.5–6.0 S; however values as low as 3 and as high as 12 S have been observed. It was noted that there appears to be more heterogeneity in polymer size in this fast peak for preparations involving addition of calcium chloride (even when using phosphate buffer as solvent). The greater polymerization may be due to incomplete removal of calcium(II). The ratio of the area of the fast peak to the slow peak is of the order of 1.4:1, although values as low as 1.0:1 and 1.7:1 have been observed.

IV. Controversial Components

A. γ-CASEINS

Following the resolution by Mellander (1939) of moving-boundary electrophoretic patterns of whole casein into three peaks, designated the α-, β-, and γ-peaks in order of decreasing mobility at pH 7, attempts were made to separate casein into three components with electrophoretic properties comparable to these peaks. It is now realized of course that this was a gross oversimplification: No single protein component corresponds to each respective peak. While material corresponding crudely to the α_1- and β-peaks has been much investigated, the γ-peak has been comparatively neglected until the last few years.

Cherbuliez and Baudet (1950) separated a γ-fraction by making use of differences in solubility of various casein fractions at pH 4.6 and 4°C and at pH 4.9 and 40°C. Later Hipp et al. (1952) took advantage of differences in solubility of α-, β- and γ-fractions in urea and in alcohol. These procedures result in products that appear to be heterogeneous on urea–starch-gel electrophoresis (see Wake and Baldwin, 1961). Subsequently Groves et al. (1962) prepared acid casein at 25°C, dissolved it, and adjusted the temperature to 2°C and the pH to 4.0. They rejected the precipitate formed and warmed the supernatant to 26°C. A precipitate then formed, and it was dissolved and subjected to chromatography on a DEAE-cellulose column at pH 8.3 (0.005 M phosphate). The protein fraction eluted with the front was designated "temperature sensitive" since it is soluble at 2°C but precipitates at 25°C. The γ-fraction is eluted at 0.02 M phosphate concentration.

Thompson et al. (1965) have defined γ-casein, in terms of urea solubility, as being soluble in 3.3 M urea but insoluble in 1.7 M urea, at pH 4.7, with ammonium sulfate present. In view of recent work on γ-casein this definition is probably inadequate (see Rose et al., 1970). The question must be considered as to whether γ-casein is in fact a true casein. It appears to contain phosphorus, although the content is low (∼0.10%). However it is considered by some workers not to be synthesized in the mammary gland but to be transported from the blood. If this is so it may not be desirable to class it as a casein. However the recent work of Groves and his collaborators indicates that it is genetically linked to β-casein (see Groves, 1969). It is of interest to note that Murthy and Whitney (1958) compared the properties of IgG immunoglobulin fractions, isolated from colostral and mature milk, with γ-casein. They had reasonably similar electrophoretic properties but differed in a number of other properties. No phosphorus was detected in the immunoglobulin fractions but the γ-fraction had 0.107%. Irrespective of whether γ-casein is a true casein or not, it is this author's opinion, in the light of recent work, that the γ-casein fraction is a more interesting and important fraction than has been realized. Some of this work will now be described briefly.

El-Negoumy (1963) noted (as referred to in Section III) that γ-caseins are lost from whole acid casein on repeated washing at pH 4.6. Despite being aware of this problem El-Negoumy (1967) was unable to detect γ-casein bands in urea–2-mercaptoethanol–starch gel electrophoretic patterns of 219 samples of acid casein. He prepared alcohol fractions from 24 samples of acid casein from Holstein cows and from 20 samples of Jersey cows. Bands, designated as γ-bands, were observed in electrophoretic patterns of all these alcohol fractions. The bands observed have been designated γ_1, γ_2, γ_3, γ_4, and γ_5. El-Negoumy also ascribed genetic polymorphism to the γ_2- and γ_4-bands. It seems to this author that all of these bands have been arbitrarily assigned to γ-casein. He is also puzzled as to why El-Negoumy could not detect any of them in acid casein. This is discussed further below.

Groves and Kiddy (1968) and Groves and Gordon (1969) have studied the polymorphism of bovine γ-casein and have examined the relationship of slow-moving bands in acrylamide zone electrophoresis to β- and γ-casein variants. They defined γ-casein as that fraction eluted in 0.02 M phosphate buffer of pH 8.3 from their DEAE-cellulose columns. Typical disc electrophoretic patterns obtained for the γ- region are shown in Fig. 1.

It was concluded, on the basis of acrylamide electrophoresis in 4 M urea at pH 9.5, that there are two γ-casein variants, A and B, and that their occurrence is genetically related to the β-casein variants A and B. No γ-casein was detected in milk samples homozygous in β-casein C. All samples examined had a band, designated TS (temperature sensitive). Samples typed γ-casein A or AB had a band, designated R, of mobility intermediate between that of A and B. The AB type also had a slow band, designated S. The B type had band S but not band R.

In 8 M urea at pH 4.3, samples typed β-casein A^1, A^2, and A^3 showed corresponding γ-casein bands A^1, A^2, and A^3, respectively. In the case of γ-casein A, B, and AB, the TS band was replaced at low pH by one fast, one slow, or two bands, respectively.

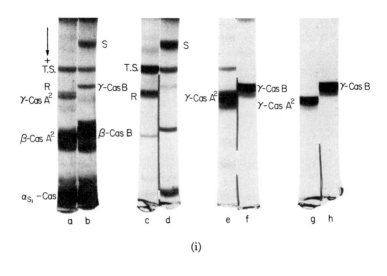

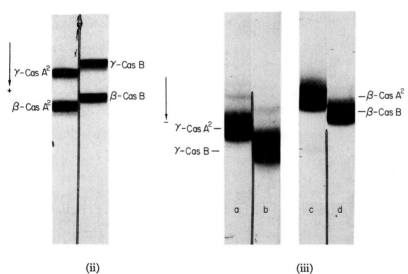

FIGURE 1. (i) Disc-gel electrophoretic patterns (pH 9.6, 4 M urea) of (a) whole casein, containing β-casein A² and γ-casein A²; (b) whole casein, containing β-casein B and γ-casein B; (c) TS fraction, A² type; (d) TS fraction, B type; (e) γ-casein fraction, A² type; (f) γ-casein fraction, B type; (g) purified γ-casein A²; (h) purified γ-casein B. (ii) Disc-gel electrophoretic patterns (pH 9.6, 4 M urea) of mixtures of (a) γ-casein A², β-casein A² and (b) γ-casein B, β-casein B. (iii) Disc-gel electrophoretic patterns (pH 4.3, 8 M urea) of (a) γ-casein A², (b) γ-casein B, (c) β-casein A², (d) β-casein B (Groves and Gordon, 1969).

TABLE II

Comparison of Composition of γ- and β-Caseins: Whole Number of Residues per Molecule Containing Five Gly[a]

Component[b]	γ-A² (25,020 daltons)[c]	γ-B (25,130 daltons)[c]	β-A¹ (23,550 daltons)[c]	β-A² (23,590 daltons)[c]	β-B (23,700 daltons)[c]	β-C (23,260 daltons)[c]
Lys	12	12	11	11	11	12
His	6	7	6	5	6	6
Arg	3	4	4	4	5	4
Asp	9	9	9	9	9	9
Thr	10	10	9	9	9	9
Ser	13	12	15	15	14	15
Glu	39	39	39	39	39	37
Pro	41	40	33	34	33	33
Gly	5	5	5	5	5	5
Ala	6	6	5	5	5	5
Val	20	20	18	18	18	18
Met	7	7	6	6	6	6
Ile	8	8	10	10	10	10
Leu	23	23	21	21	21	21
Tyr	5	5	4	4	4	3
Phe	11	11	9	9	9	9
Trp	1	1	1	1	1	1
P	1	1	5	5	5	4
Hexose	1	1				

[a] Data of Groves and Gordon (1969).
[b] Amide NH_3 determinations have been omitted because significant differences could not be demonstrated.
[c] Molecular weight calculated from the compositions shown.

Groves and his colleagues have suggested, on the basis of amino acid analysis, that the two TS bands represent one pair of polymorphs, and the R, and S bands represent another pair, and that the γ- and β-caseins are more closely related than the TS-, R-, and S-caseins.

Groves and Gordon (1969) were able to purify the genetic variants of γ- and β-casein from acid casein. A comparison of the amino acid compositions of γ- and β-caseins, based on this work, is given in Table II.

B. Minor Glycoproteins

There is associated with whole casein a group of glycoproteins, at least some of which are phosphoproteins, and may eventually be best considered as caseins. The first of these have been loosely called "proteose peptones."

Osborne and Wakeman (1918) appear to have been the first workers to describe such a fraction. They found that variable amounts of material remained in solution following acid precipitation of the casein and precipitation of the "albumins" and "globulins" by boiling the acid whey. They concluded that this fraction did not arise from hydrolytic degradation of the main milk proteins and that it was present to the extent of ∼0.2 g/liter. Later Rowland (1938) found that a portion of the milk serum proteins was not rendered acid precipitable at pH 4.6, following heat treatment of skim milk at 95°–100°C for 30 min. This fraction could be divided into two fractions on the basis of solubility in 0.5 saturated ammonium sulfate solution. Aschaffenburg (1946) prepared a fraction from both acid whey and rennin whey by salting it out of the heat-cleared whey with ammonium sulfate. His fraction was completely insoluble in 0.5 saturated ammonium sulfate and hence does not conform to the original definition of a peptone. He called this heterogeneous fraction σ-proteose on account of its marked surface activity.

Subsequently Larson and Rolleri (1955) designated certain peaks in moving-boundary electrophoresis of solutions of acid whey at pH 8.6 as *possibly* being "proteose peptones." There was a small peak, "8," with a mobility a little greater than that of serum albumin at pH 8.6, a larger peak, "5," with a mobility intermediate between that of α-lactalbumin and β-lactoglobulin at pH 8.6, and a small peak "3," with a mobility a little less than that of α-lactalbumin but of the order of that of faster immunoglobulins.

Aschaffenburg and Drewry (1959) observed that six bands were present in filter-paper electropherograms of preparations made by the Rowland method. The major band corresponded to Larson and Rolleri's peak "5" and the five small bands to their peak "3." The same six bands were found in concentrates prepared from ultrafiltrates of unheated milk. Thus the material causing these bands occurs in mature milk and does not require heating and acidification to produce it. Aschaffenburg found that he could prepare this heterogeneous fraction by salting out from acid whey with sodium sulfate to a concentration of 120 g/liter. It would appear from the yellow stain given with bromphenol blue on the filter-paper strips that all bands contained carbohydrate.

Jenness (1959) found that he could precipitate a protein mixture containing material corresponding to peak "5" by saturating skim milk with sodium chloride. The precipitate was redispersed, the casein removed at pH 4.6, the supernatant concentrated, the lactoperoxidase absorbed on an ion-exchange resin, and the material corresponding to peak "5" obtained by selective precipitation at pH 4.5. It had the characteristics summarized in Table III.

TABLE III

SOME MINOR GLYCOPROTEINS FROM BOVINE MILK

Protein	Mobility, pH 8.6 ($cm^2/V/sec$)	Zone electrophoresis	N (g/100 g)	P (g/100 g)	Solubility, rennin action
"Component 8"	$-8 \times 10^{-5\,f,h}$; F, -9.3×10^{-5}; S, $-9.2 \times 10^{-5\,j}$	2 Bands (pH 8.6)[g]	14[f]; 12.3–13.8[g]; F, 13.3; S, 12.3[j]:	1.2[f]; F, 2.4; S, 2.0[j]:	
"Component 5"	$-4.5 \times 10^{-5\,e,h}$; $-4.8 \times 10^{-5\,j}$		12.3–13.8[g]; 13.8[j]:	1.2[e]; 1.0[j]	Sol.; No[e]
"Component 3"	$-3.0 \times 10^{-5\,h}$; $-3.8 \times 10^{-5\,j}$		13.1[i,j]	0.5[i,j]	
Phosphoglycoprotein (M1 type)	$-5.8 \times 10^{-5\,c,b}$	Multiple bands[c]	13.4[c,b]	0.44[c]	
Glycoprotein-a	$-1.26 \times 10^{-5\,d}$	Multiple bands[d]			

Protein	Carbohydrate (g/100 g)	Amino acid residue	Mol wt (daltons)	$s_{20,w}$ (S)
"Component 8"	Hexose: F, 1.4; S, 4.5; hexosamine: F, 0.3; S, 2.5; NANA: F, 0.4; S, 3.3[g,j]		F, 4100; S, 9900[g,j]	1.0 (pH 7)[f]; 1.2; 1.41[g]; F, 0.8; S, 1.4[j]:

"Component 5"	Hexose: 0.9; hexosamine: 0.2; NANA: 0.3^{a,i}	High: Pro; absent: Cys^a	14,000^{a,i} 0.8^a
"Component 3"	Galactose, mannose: 7.2; glucosamine, galactosamine: 6.0; fucose: 1.0; NANA: 3.0^i	High: Glu, Asp, Ser, Thr; low: aromatic, Met; absent: Cys^i	200,000 (pH 8.6)^i; 40,000 (5 M GHCl)^j
Phosphoglycoprotein (M1 type)	Hexose: 3.1; N-acetylhexosamine: 24; NANA: 4^c (ex-colostrum)	High: Glu, Thr, Ile; low: basic; N.A.: Cys^c	10,000 (milk)^c 0.8 (pH 7, 12)^b
		N-Terminal variable^c	10,000–13,000 (colostrum)^c 1.0–1.2^c
Glycoprotein-a	Hexose 3.1^d	High: Val, Lys; low: Met, His, Ile, Phe; Present: Cys^d	48,000^d 4.0^d

^{a} Legend: Sol., soluble in the presence of calcium(II); No, not clotted by rennin; F, fast component; S, slow component; GHCl, guanidine hydrochloride.
^{b} Bezkorovainy (1965).
^{c} Bezkorovainy (1967).
^{d} Groves and Gordon (1967).
^{e} Jenness (1959).
^{f} Kolar and Brunner (1965).
^{g} Kolar and Brunner (1968).
^{h} Larson and Rolleri (1955).
^{i} Ng and Brunner (1967).
^{j} Rose et al. (1970).

Brunner and Thompson (1961) compared the Rowland (1938), Aschaffenburg (1946) (σ-proteose), Jenness (1959), and Weinstein et al. (1951) fractions and the soluble membrane protein fraction of Herald and Brunner (1957). All of these fractions were heterogeneous in moving-boundary electrophoresis at pH 8.6 and 2.4 and showed considerable boundary spreading in their sedimentation-velocity patterns at pH 8.6 ($I=0.1$). All had nitrogen contents in the range 10–14 g/100 g protein, ash contents of 3–7 g/100 g and phosphorus contents of 0.6–1.5 g/100 g and contained hexose sugars. Thompson and Brunner (1959) found that their Rowland type of preparation had a hexose content of 2.9 g/100 g; hexosamine, 1.2 g/100 g; fucose, 0.7 g/100 g; and sialic acid, 2.0 g/100 g.

Kolar and Brunner (1965) released material, similar to peak "8" from casein micelles by heating, by the addition of acid to pH 4.6, by the action of rennin, by freezing or by dehydration. They also isolated it from acid casein and from acid whey. Kolar and Brunner consider this material to be the tenacious "contaminant" in κ-casein preparations and that it may be identical to the fraction called λ-casein by Long et al. (1958). They found that it migrates as a single band in urea–starch and polyacrylamide gels. Its characteristics are summarized in Table III.

Ng and Brunner (1967) isolated "component 3" from skim milk heated to 95°C for 20 min, using a combination of ammonium sulfate and gel electrophoretic fractionation. Subsequently Kolar and Brunner (1968) isolated enriched preparations of components "3," "5," and "8" from unheated as well as heated milk. Their characteristics are also summarized in Table III.

Bezkorovainy (1965) found that he could retain acid glycoproteins from bovine blood serum, bovine colostral acid whey and bovine mature milk acid whey on a DEAE-cellulose column at pH 4.5 with a low ionic strength buffer. Subsequent elution resulted in predominantly M–2 glycoprotein-like material in the first peak and M–1 glycoprotein-like material in the second peak. Chromatography on CM-cellulose columns was used to purify these fractions further. In the case of blood serum he was able to isolate orosomucoid from the M–1 fraction, and M–2 acid glycoprotein from the M–2 fraction. Colostrum acid whey contained appreciable amounts of orosomucoid and the M–2 acid glycoprotein. The M–1 fraction containing orosomucoid also contained additional glycoproteins, at least one of which appears to be similar to one found in mature milk. Mature milk acid whey contained only traces of the M–2 blood serum acid glycoprotein in the first peak and no detectable orosomucoid in the second peak, but there was a phosphoglycoprotein specific to milk in the second peak. This protein has a mobility of -3.0×10^{-5} cm^2/V/sec at pH 4.5, i.e. slower than human orosomucoid but faster than that of M–2 glycoprotein. It has a considerably lower carbohydrate fraction than orosomucoid.

10. WHOLE CASEIN 109

Later Bezkorovainy (1967) examined the physical and chemical properties of bovine M-1 acid glycoproteins from colostrum and mature milk whey in greater detail. These properties are summarized in Table III. Although Bezkorovainy obtained single peaks in moving-boundary electrophoresis (pH 4.5) and sedimentation studies (pH 5, 7, 12) his fractions contained several N-terminal amino acids and some gave rise to several bands in gel electrophoresis. An examination of the amino acid composition of this phosphoglycoprotein and that of bovine casein glycomacropeptide leads one to wonder if the two are not the same in varying states of homogeneity.

Groves and Gordon (1967) have isolated a glycoprotein from bovine milk and have named it tentatively, glycoprotein-a. They fractionated acid bovine whey on DEAE-cellulose columns according to the method of Groves (1965). Fraction IF was chromatographed on phosphocellulose. The IF-2 fraction eluted with 0.1 M sodium phosphate at pH 6.0 gave a single peak in moving-boundary electrophoresis at pH 8.5 (I = 0.1, Veronal) having a mobility of -1.26×10^{-5} cm^2/V/sec. Two bands were observed on acrylamide disc electrophoresis at low pH and two peaks (4 and 7 S) were observed in sedimentation-velocity experiments at pH 7.0. These two peaks were separated on Sephadex G-200 after three passes. The 4 S peak is glycoprotein-a (yield, 50 mg/liter milk) and the 7 S peak IgG immunoglobulin (yield, 30 mg/liter milk). The properties of glucoprotein-a are summarized in Table III.

It is apparent that despite discrepancies in the above work, there are several minor phosphoglycoproteins present in mature milk. However when milk is treated with acid and heated, some of the phosphoglycoprotein and peptide material detected is probably the result of the rather rigorous treatment. It is important to establish unequivocally which of the minor phosphoglycoproteins do occur naturally in milk. This will be possible only if the gentlest possible treatments are used.

V. Zone Electrophoresis in Casein Typing

The detection of "true" heterogeneity of casein fractions in electrophoresis is difficult owing to the individual caseins interacting with themselves, with one another and with ions or molecules present in the electrophoresis buffer systems. Some of these general problems are discussed in Chapter 7, Volume I.

It seemed to Wake and Baldwin (1961) that if reliable information on casein heterogeneity were to be obtained a zone electrophoretic method would have to be used in which the caseins were dissociated into their monomeric forms. They aimed to (a) disperse all of the casein components into their monomeric form by carrying out the electrophoresis in 7 M urea solution and (b) take advantage of the high resolution offered by the Smithies (1959) starch-gel method of electrophoresis. The most suitable buffer system they tried was the discontinuous tris-citrate buffer system (pH 8.6) of Poulik (1957). They found, on electrophoresis of whole acid casein, over twenty bands, although only a few were major bands. Wake and Baldwin numbered the different bands according to their relative positions in the gel, the distance from the starting slot to an especially well-defined band being set at 1.00.

The reproducibility of the band positions was found to be very satisfactory. After elution of a number of zones from the gel they were rerun. The patterns obtained in each case were reproducible with respect to the number of bands observed and their mobilities. One unsatisfactory feature was that a smeared zone was observed for κ-casein with some sharp bands superimposed on the smeared band.

Neelin et al. (1962) examined the effect of urea concentration in causing dissociation and of pH and buffer type on resolution. Mobilities increased with increasing pH, but above pH 7.0 the separation of major zones diminished. They preferred pH 7.0–7.2 (cacodylate) for comparison of more mobile bands but pH 8.2–8.4 (Veronal) for slower moving bands. In the former pH range, 5.5 M urea was used, but in the latter range, only 4.8 M urea was used in order to retain gel consistency. The problem of "smearing" and lack of complete reproducibility of κ-casein behavior still remained. Subsequently Neelin (1964) suggested the use of small amounts of 2-mercaptoethanol in the urea–gel buffer in the hope of dispersing the κ-casein, since small amounts of SS had been reported by others in κ-casein (Waugh, 1958; Jollès et al., 1962). Addition of mercaptoethanol overcame the smearing effect, improved the resolution in this region and enabled genetic variants of κ-casein to be detected (see also Waugh, 1962; Schmidt, 1964).

Mackinlay and Wake (1964) prepared S-sulfo-κ-casein and S-carboxymethyl-κ-casein and showed that these derivatives of reduced κ-casein gave rise to well-defined bands in urea–starch-gel electrophoresis.

Aschaffenburg and Thymann (1965) developed a method of 7 M urea–starch-gel electrophoresis, using a tris-EDTA–boric acid buffer (pH 8.9) and milk samples that had prior treatment with mercaptoethanol but had not been subjected to fractionation. It is possible with this method to get good resolution for the caseins and some of the whey proteins. [β-Lactoglobulins B and C are *not* resolved in the original method but are resolved in a subsequent modification (Aschaffenburg and Michalak, 1968).]

Peterson (1963) found that he obtained good resolution of α_s- and β-caseins by use of acrylamide-gel electrophoresis with a 4.5 M urea, tris-EDTA–boric acid (pH 9.0) mixture. This method has been applied very successfully by Thompson and his collaborators for the detection of genetic polymorphism (see Chapter 7, Volume I) in $\alpha_{s,1}$- and β-caseins. Aschaffenburg (1964) has shown how the acrylamide method may be used directly for typing by electrophoresis of whole milk samples.

Peterson and Kopfler (1966) then followed up the earlier work of Nauman et al. (1960) that had indicated variation in the histidine content of a peptide isolated from a tryptic digest of β-casein prepared from pooled milk. The histidine variation was found by Peterson and Nauman (1963) to be associated with the A type of β-casein, but they could not differentiate these A variants by electrophoresis at alkaline pH values. Peterson and Kopfler found that on acrylamide electrophoresis in urea–formic acid–acetic acid at pH 3.0, they were able to resolve the A type of β-casein into three bands and that these could be resolved from the bands due to types B and C. They suggested a new, but confusing, nomenclature for the variants. Fortunately this nomenclature was withdrawn in a subsequent communication (Kiddy et al., 1966). The original nomenclature for A, B, and C (order of decreasing mobility at alkaline pH) was restored. However they called the three A variants detected at pH 3.0, A^1, A^2, A^3 in decreasing order of mobility *at pH 3.0*. On the basis of the work of Peterson et al. (1966) it was concluded that there are three A^2 variants. Thus it can be seen that the complexity of the β-caseins means that special methods of zone electrophoresis are needed for the detection of β-casein variants. The resolution of these methods and the nomenclatures employed are shown in the schematic diagram of Fig. 2. The isolation and properties of the β-casein variants are discussed in detail in Chapter 11 by Thompson.

Methods for the zone electrophoretic examination of caseins have been developed empirically. An amazing amount has been achieved with such empirical methods. Nevertheless it is the author's opinion that the time is overdue for a development of systematic procedures based on the general principles enunciated in Chapter 7, Volume I. Some such methods are being developed in the author's laboratory. They draw attention to the special need of using methods that are adequate to reduce the κ-casein completely and maintain it in the reduced state, and to enable good resolution of the κ-bands from the bands due to γ-casein and other components.

Even if close attention is paid to the underlying electrophoretic principles and the interactions of caseins, the development of adequate procedures for zone electrophoresis of caseins is a formidable problem. The recent discoveries of new genetic variants of the important components, $\alpha_{s,1}$-, β- and κ-caseins, and of considerable numbers of minor components have

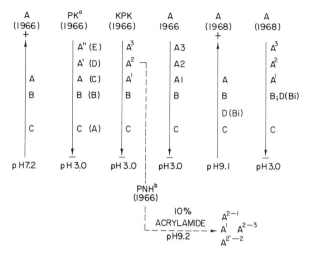

FIGURE 2. β-Caseins determined by zone electrophoresis. Explanatory notes: (a) Peterson and Nauman (1963) detected heterogeneity in the A variant by peptide analysis. Peterson and Kopfler (1966) resolved A into three variants at pH 3.0 and introduced the confusing nomenclature in brackets. This nomenclature was replaced by Kiddy et al. (1966) as shown. (b) Peterson et al. (1966) claim the resolution of A^2 as shown. A^1 and A^{2-3} were distinguished by their different mobilities at pH 3.0. Key to workers: A (1961), Aschaffenburg (1961); PK (1966), Peterson and Kopfler (1966); KPK (1966), Kiddy et al. (1966); A (1966), Aschaffenburg (1966); A (1968), Aschaffenburg (1968); PNH (1966), Peterson et al. (1966).

increased the complexity. An example of this is the characterization on the minor α_s- components, $\alpha_{s,0^-}$, $\alpha_{s,2^-}$, $\alpha_{s,3^-}$, $\alpha_{s,4^-}$, $\alpha_{s,5^-}$, $\alpha_{s,6^-}$, discussed in Chapters 9 and 18. The present position for whole casein is summarized in Fig. 3. It must be emphasized in considering this schematic that the positions of the minor components relative to particular genetic variants are not known at present in exact detail for most, if not all, gel and buffer systems.

In the course of the preparation of this manuscript the author was asked by many protein chemists and university teachers to include notes on the various electrophoretic methods currently available for examination of milk proteins. Such notes have been included in the Appendix. No particular type of vertical or horizontal apparatus or gel (for example, starch or acrylamide) is considered superior for all milk proteins. Each method has its proponents; each has its advantages and disadvantages. The worker must make his own assessment of the methods for his own particular problem.

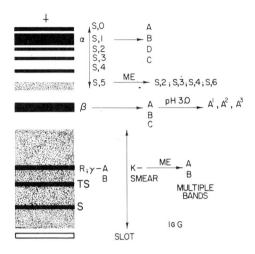

FIGURE 3. Schematic representation of zone electrophoresis at alkaline pH of whole casein from milk of Bos taurus. The relative positions shown for the zones are only approximate and may depend appreciably on the compositions of the gel and buffer. Diffuse zones for κ-casein and $\alpha_{s,5}$-casein that are altered by 2-mercaptoethanol are indicated by stippling. Minor components, such as the phosphoglycoproteins, are not shown. The detailed resolution of β-caseins is shown in Fig. 2.

VI. Summary

It has been the object of this chapter to show that considerable care must be taken in the isolation of whole casein if we are to obtain a product that is worthy of careful fractionation. There is no point in applying sophisticated chemical, physical and biological techniques to the study of the properties of caseins that have been irreversibly altered in the process of separation. Likewise, considerable care must be taken to avoid rejection of important components, even if they are minor ones, during fractionation procedures. Considerable care must be taken in the choice of milk samples for fractionation. The effect of subclinical mastitis on the isolation of whole casein has been demonstrated in the work of Graham et al. (1970). It is important to realize that this level of mastitis was subclinical, that it was only detectable by California (CMT) or rapid (RMT) mastitis tests. (The effects of such levels on heat stability, and micelle formation will be discussed elsewhere by L. Bailey, J. Feagan, H. A. McKenzie, and M. C. Taylor.) Problems in the fractionation of casein from milk prepared from infected animals or from colostral milk mean that it is difficult to determine if the amount or even the type of carbohydrate moiety of κ-casein changes during lactation or during infection.

It is recommended that whole casein be isolated from normal mature milk under conditions leading to minimum alteration of components, especially oxidation of SH groups. Ammonium sulfate fractionation and centrifugation in the presence of calcium(II) offer the best possibilities of success.

γ-Casein and possibly other components moving slowly in electrophoresis at alkaline pH value occur in some milk samples to a greater extent than most workers seem to have realized. The presence of these components and various phosphoglycoproteins and glycoproteins must be assessed carefully in the future. There has been inadequate assessment of the role of proteases in casein fractions. Their role must be carefully assessed in the future (for a start in this direction see Zittle, 1965). The methods described in this chapter are suitable for bovine caseins. Methods will need to be developed individually for the milk of other species.

REFERENCES

Aschaffenburg, R. (1946). *J. Dairy Res.* **14,** 316.
Aschaffenburg, R. (1961). *Nature* **192,** 431.
Aschaffenburg, R. (1964). *Biochim. Biophys. Acta* **82,** 188.
Aschaffenburg, R. (1966). *J. Dairy Sci.* **49,** 1284.
Aschaffenburg, R., and Drewry, J. (1959). *Proc. Int. Dairy Congr., 15th, 1959* **3,** 1631.
Aschaffenburg, R., and Michalak, W. (1968). *J. Dairy Sci.* **51,** 1849.
Aschaffenburg, R., and Thymann, M. (1965). *J. Dairy Sci.* **48,** 1524.
Aschaffenburg, R., Sen, A., and Thompson, M. P. (1968). *Comp. Biochem. Physiol.* **25,** 177.
Beeby, R. (1964). *Biochim. Biophys. Acta* **82,** 418.
Bezkorovainy, A. (1965). *Arch. Biochem. Biophys.* **110,** 558.
Bezkorovainy, A. (1967). *J. Dairy Sci.* **50,** 1368.
Bohren, H. U., and Wenner, V. R. (1961). *J. Dairy Sci.* **44,** 1213.
Brunner, J. R., and Thompson, M. P. (1961). *J. Dairy Sci.* **44,** 1224.
Cherbuliez, E., and Baudet, P. (1950). *Helv. Chim. Acta* **33,** 398.
Dixon, M., and Webb, E. C. (1961). *Advan. Protein Chem.* **16,** 197.
El-Negoumy, A. M. (1963). *J. Dairy Sci.* **46,** 768.
El-Negoumy, A. M. (1967). *Biochim. Biophys. Acta* **140,** 503.
Graham, E. R. B., McKenzie, H. A., Murphy, W. H., and Symons, L. (1970). In preparation. Preliminary accounts have also been presented by H. A. McKenzie (1967) and Research Seminar, Australian Dairy Board, Melbourne (1969).
Groves, M. L. (1965). *Biochim. Biophys. Acta* **100,** 154.
Groves, M. L. (1969). *J. Dairy Sci.* **52,** 1155.
Groves, M. L., and Gordon, W. G. (1967). *Biochemistry* **8,** 2388.
Groves, M. L., and Gordon, W. G. (1969). *Biochim. Biophys. Acta* **194,** 421.
Groves, M. L., and Kiddy, C. A. (1968). *Arch. Biochem. Biophys.* **126,** 188.
Groves, M. L., McMeekin, T. L., Hipp, N. J., and Gordon, W. G. (1962). *Biochim. Biophys. Acta* **57,** 197.

Halwer, M. (1954). *Arch. Biochem. Biophys.* **51,** 79.
Hammarsten. O. (1900). "Textbook of Physiological Chemistry." Wiley, New York.
Herald, C. T., and Brunner, J. R. (1957). *J. Dairy Sci.* **40,** 948.
Hill, R. D., and Hansen, R. R. (1964). *J. Dairy Res.* **31,** 291.
Hipp, N. J., Groves, M. L., Custer, J. H., and McMeekin, T. L. (1952). *J. Dairy Sci.* **35,** 272.
Jenness, R. (1959). *J. Dairy Sci.* **42,** 895.
Jollès, P., Alais, C., and Jollès, J. (1962). *Arch. Biochem. Biophys.* **98,** 56.
Kiddy, C. A., Peterson, R. F., and Kopfler, F. C. (1966). *J. Dairy Sci.* **49,** 742.
Kolar, C. W., Jr., and Brunner, J. R. (1965). *J. Dairy Sci.* **48,** 772.
Kolar, C. W., Jr., and Brunner, J. R. (1968). *J. Dairy Sci.* **51,** 945.
Lahav, E. (1965). Private communication.
Larson, B. L., and Rolleri, G. D. (1955). *J. Dairy Sci.* **38,** 351.
Lascelles, A. K., Mackenzie, D. D. S., and Outteridge, P. M. (1967). Paper read a Research Seminar, Australian Dairy Board, Sydney.
Lascelles, A. K., Mackenzie, D. D. S. and Outteridge, P. M. (1969). In "Protides of the Biological Fluids: Proceedings of the 16th Colloquium, Brugges, 1968" (H. Peeters, ed.), p. 639. Pergamon, Oxford.
Long, J., van Winkle, Q., and Gould, I. A. (1958). *J. Dairy Sci.* **41,** 317.
McKenzie, H. A. (1967). *Advan. Protein Chem.* **22,** 55.
McKenzie, H. A., and Treacy, G. B. (1970). In preparation.
Mackinlay, A. G., and Wake, R. G. (1964). *Biochim. Biophys. Acta* **93,** 378.
Mellander, O. (1939). *Biochem. Z.* **300,** 240.
Mülder, G. J. (1838). *Ann. Pharm.* **28,** 73.
Murthy, G. K., and Whitney, R. McL. (1958). *J. Dairy Sci.* **45,** 1.
Nauman, L. W., Peterson, R. F., and McMeekin, T. L. (1960). *Abstr. Amer. Chem. Soc. Meeting, 138th,* P70C.
Neelin, J. M. (1964). *J. Dairy Sci.* **47,** 506.
Neelin, J. M., Rose, D., Tessier, H. (1962). *J. Dairy Sci.* **45,** 153.
Ng, W. S., and Brunner, J. R. (1967). *J. Dairy Sci.* **50,** 950.
Noble, R., and Waugh, D. F. (1965). *J. Amer. Chem. Soc.* **87,** 2236.
Osborne, T. B., and Wakeman, A. J. (1918). *J. Biol. Chem.* **33,** 243.
Payens, T. A. J., and Schmidt, D. G. (1966). *Arch. Biochem. Biophys.* **115,** 136.
Payens, T. A. J., and van Markwijk, B. W. (1963). *Biochim. Biophys. Acta* **71,** 517.
Peterson, R. F. (1963). *J. Dairy Sci.* **46,** 1136.
Peterson, R. F., and Kopfler, F. C. (1966). *Biochem. Biophys. Res. Commun.* **22,** 388.
Peterson, R. F., and Nauman, L. W. (1963). *Abstr. Amer. Chem. Soc. Meeting, 152nd* P76C.
Peterson, R. F., Nauman, L. W., and Hamilton, D. F. (1966). *J. Dairy Sci.* **49,** 601.
Poulik, M. D. (1957). *Nature* **180,** 1477.
Rolleri, G. D., Larson, B. L., and Touchberry, R. W. (1956). *J. Dairy Sci.* **39,** 1683.
Rose, D., Brunner, J. R., Kalan, E. B., Larson, B. L., Melnychyn, P., Swaisgood, H.E., and Waugh, D. F. (1970). *J. Dairy Sci.* **53,** 1.
Rowland, S. J. (1938). *J. Dairy Res.* **9,** 40.
Schmidt, D. G. (1964). *Biochim. Biophys. Acta* **90,** 411.
Schmidt, D. G., Payens, T. A. J., van Markwijk, B. W., and Brinkhuis, J. A. (1967). *Biochem. Biophys. Res. Commun.* **27,** 448.
Smithies, O. (1959). *Advan. Protein Chem.* **14,** 65.
Sullivan, R. A., Fitzpatrick, M. M., Stanton, E. K., Annino, R., Kissel, G., and Palermiti, F. (1955). *Arch. Biochem. Biophys.* **55,** 455.

Swaisgood, H. E., and Brunner, J. R. (1963). *Biochem. Biophys. Res. Commun.* **12**, 148.
Talbot, B., and Waugh, D. F. (1967). *J. Dairy Sci.* **50**, 950.
Thompson, M. P., and Brunner, J. R. (1959). *J. Dairy Sci.* **42**, 369.
Thompson, M. P., Tarassuk, N. P., Jenness, R., Lillevik, H. A., Ashworth, U. S., and Rose, D. (1965). *J. Dairy Sci.* **48**, 159.
von Hippel, P. H., and Waugh, D. F. (1955). *J. Amer. Chem. Soc.* **77**, 4311.
Wake, R. G., and Baldwin, R. L. (1961). *Biochim. Biophys. Acta* **47**, 225.
Waugh, D. F. (1958). *Discuss. Faraday Soc.* **25**, 186.
Waugh, D. F. (1962). *Abstr. Amer. Chem. Soc. Meeting, 142nd* **P62C**.
Waugh, D. F. and von Hippel, P. H. (1956). *J. Amer. Chem. Soc.* **78**, 4576.
Waugh, D. F., Ludwig, M. L., Gillespie, J. M., Melton, B., Foley, M., and Kleiner, E. S. (1962). *J. Amer. Chem. Soc.* **64**, 4929.
Weinstein, B. R., Duncan, C. W., and Trout, G. M. (1951). *J. Dairy Sci.* **34**, 570.
Zittle, C. A. (1965). *J. Dairy Sci.* **48**, 771.

11 □ α_s- and β-Caseins*

M. P. THOMPSON

I.	Introduction	117
II.	α_s-Caseins	120
	A. Nomenclature of α_s-Caseins	120
	B. Genetic Polymorphism of $\alpha_{s,1}$- and β-Caseins	121
	C. Preparation of α_s-Caseins	133
	D. Composition and Structure	139
	E. Physical Properties of $\alpha_{s,1}$-Caseins	148
	F. Nature of Organic Phosphates in Caseins	153
III.	β-Caseins	155
	A. Nomenclature	155
	B. Preparation	156
	C. Composition and Structure	160
	D. Physical Properties	166
IV.	General Considerations	168
	References	169

I. Introduction

It is significant that the bulk of our current knowledge regarding the complexity of the milk protein system has been accumulated during the past 15 years. Nevertheless, our current concepts of the chemistry of casein must be viewed in light of nearly a century of accumulated data. The history of the chemistry of casein has been reviewed in Chapter 1, Volume I. It suffices to recall here that casein was long regarded as homogeneous until the work of Linderstrøm-Lang and Kodama (1925). Mellander (1939) concluded on the basis of moving-boundary electrophoresis that whole casein consists of three discernible components, termed α-,

*The commas in the subscripts of the α_s-caseins have been used by the Editor throughout this treatise to designate that $\alpha_{s,1}$-, $\alpha_{s,2}$-, etc., are subclasses of α_s-. No comma is used in the A. D. S. A. nomenclature or by *Chemical Abstracts*. The author of this chapter prefers the A. D. S. A. nomenclature to that used in this treatise and recommends its adoption.

β-, and γ-casein. This nomenclature has been retained to this day. Warner (1944) was the first to attempt the fractionation of α-, β-, and γ-caseins from whole casein, while Hipp et al. (1952a,b) further refined the isolation of these fractions by using urea or alcohol methods to achieve separations.

Despite the efforts of the above research workers, a paucity of information existed as to the factors responsible for the nature and/or stability of casein micelles. The studies of Waugh and von Hippel (1956) were the most important of the decade in disclosing that casein contains a protective colloid, a concept first advanced by Linderstrøm-Lang and Kodama (1925). They clearly demonstrated that a calcium-soluble casein (termed κ-casein) is capable of stabilizing calcium-insoluble casein (termed $α_s$-casein) against precipitation with calcium ions. Additionally, κ-casein is the protein attacked by the enzyme rennin during the primary phase of the reaction leading to destabilization of the colloidal system. Waugh and von Hippel (1956) employed a normal constituent of milk, calcium(II), to achieve the fractionation of the components of whole casein.

Most workers of the present decade have, apparently, been more concerned with the composition of those macromolecules comprising the casein micelle than with the construction of the micelle itself, and with good reason. To understand the construction of the total unit, one must first concern oneself with the nature of those components (and the forces which lead to interaction) comprising the total unit. A significant development of the 1960s which has led to a more thorough examination of casein components is the use of starch-gel electrophoresis (Wake and Baldwin, 1961) in the presence of urea to dissociate casein. Dissociation of the proteins by urea and subsequent zonal electrophoresis is sufficient to disclose the presence, in whole casein, of at least 20 components, all of which appear to be true constituents of the system. Starch-gel and polyacrylamide-gel electrophoresis have been valuable tools to the protein chemist in general.

In the author's opinion, the discovery of genetic polymorphism in the caseins has been the single most important contribution of the 1960s. In this regard the plaudits must go to Aschaffenburg (1961), who anticipated that the caseins, like the β-lactoglobulins (Aschaffenburg and Drewry, 1955), would be genetically variable; his anticipation came to fruition with the discovery of β-casein polymorphism (Aschaffenburg, 1961). Using simple paper electrophoresis he showed that β-casein exists in three forms, A, B, and C, an observation which opened up a vast new area of research to scientists in the area of casein chemistry. Additional proof of Aschaffenburg's observations was supplied by Thompson et al. (1964), who showed that the major component of casein, $α_{s,1}$-, is also genetically variable (Thompson et al., 1962). Neelin (1964) suggested that κ-casein, too, is genetically variable, a suggestion substantiated by several others (Woychik, 1964; Schmidt, 1964; Aschaffenburg and Thymann, 1965). Definite proof of κ-casein polymorphism, determined from family studies, has been supplied only by Thymann and Larsen (1965) and Aschaffenburg et al. (1968a).

Casein has been generally defined as that group of phosphoproteins from milk which are precipitated at pH 4.6 and 20°C. Accordingly, any proteins remaining in solution are considered whey proteins, an observation which may not be correct due to the greater solubility of some casein components at the above conditions. For example, milk component "5" (Brunner et al., 1960), a phosphoprotein, is not precipitated at pH 4.6 from skim milk but upon concentration will precipitate at this pH. The α-casein complex, which is insoluble in 4.5 M urea at pH 4.6–4.8, has been subdivided to include $α_{s,1}$-, $α_s$-like caseins, κ- and "m" or λ-caseins, while the fraction soluble in 3.3 M urea, but insoluble in 1.7 M urea is the β-casein fraction. Indeed, the urea fractionation of casein is not clearcut—some κ-casein remains in urea solution with β-casein, for example (Thompson, 1966)—but the method is certainly valid in producing gross enrichment of certain casein components, and casein components can be defined operationally on their solubility in urea solutions.

An alternate, and perhaps more desirable, definition of whole casein, based upon the insolubility of these proteins in dilute $CaCl_2$ after centrifugation of skim milk at designated gravities and temperatures, seems warranted to this author. Certainly, casein components can be defined (in part, at least) on the basis of their solubility in different concentrations of calcium(II); κ-casein, for example, is typified by its solubility over a wide range of calcium concentrations and temperatures. First-cycle soluble casein (Waugh and von Hippel, 1956) is prepared by high-speed centrifugation of skim milk following the addition of 0.07 M $CaCl_2$. Most of the micellar casein is centrifuged down at 25°C and above.

This chapter on $α_s$- and β-caseins is not intended to be an historical document on all that is known about these proteins; references have been selected which contribute to a clearer understanding of the areas under discussion, and while some are deleted it is not to be construed that the quality of work is not excellent. The author has liberally referred to the abundance of research in the area of casein polymorphisms, its significance in establishing a flexible nomenclature scheme for the caseins, and its importance in delineating differences in chemically related components. He has further included sections on the current nomenclature of caseins, their methods of preparation and some of their physical properties. More detailed physical chemistry, protein–protein interactions, and factors leading to the formation of casein micelles have been discussed in Chapter 9 by Professor David F. Waugh.

Finally, it is of importance to note that little is known regarding the primary structure of casein molecules, and a paucity of information exists as to the formation and composition of micellar casein. Methodology is available to explore these two areas, and one would anticipate development of these areas so that a clearer picture of this colloidal system will emerge. Several review articles concerning caseins are available; they include those of Lindqvist (1963), Jollès (1966), and four separate reports of the Committee on Nomenclature and Methodology of the American Dairy Science Association (Jenness et al., 1956; Brunner et al., 1960; Thompson et al., 1965; Rose et al., 1970).

II. α_s-Caseins

A. Nomenclature of α_s-Caseins

The Committee on Nomenclature and Methodology of Milk Proteins of the American Dairy Science Association has been actively engaged in establishing a nomenclature scheme for milk proteins. Their reports have been considered in Chapters 2 and 6, Volume I. Only the detailed nomenclature for α_s- and β-caseins will be considered in this chapter.

Nomenclature for calcium-sensitive α-casein is a difficult problem. Waugh (1958) termed the fraction α_s- because it denoted the calcium-sensitive fraction of α-casein. Waugh et al. (1962) termed it $\alpha_{s,1,2}$-casein, a term still descriptive, but one which was considered confusing by others (Lindqvist, 1963; Thompson and Kiddy, 1964; Thompson et al., 1965) who felt that $\alpha_{s,1,2}$- are the genetic variants $\alpha_{s,1}$-B and $\alpha_{s,1}$-C. Thompson and Kiddy (1964) termed the A, B, and C' variants of α_s-casein as $\alpha_{s,1}$-caseins A, B, and C, a suggestion subsequently adopted in the nomenclature of milk proteins. They further proposed that since starch-gel electrophoresis zones 1.04 and 1.00 qualified by definition as α_s-casein (i.e., calcium sensitivity and stabilization by κ-casein), they should be tentatively designated $\alpha_{s,2}$- and $\alpha_{s,3}$-caseins, respectively, to distinguish them from the genetically variable $\alpha_{s,1}$-casein.

A zone numbering system, patterned after the one suggested by Wake and Baldwin (1961), was adopted to further identify the genetic polymorphs. This system, applied to starch-gel electrophoresis and polyacrylamide-gel electrophoresis, is shown in Table I. The nomenclature of $\alpha_{s,1}$-B, according to this system, is $\alpha_{s,1}$-B(1.13) by polyacrylamide-gel electrophoresis and $\alpha_{s,1}$-B(1.10) by starch-gel electrophoresis. The scheme also allowed for the naming of new variants yet to be discovered in the $\alpha_{s,1}$-casein series. The usefulness of this scheme has already been realized

TABLE I

RELATIVE MOBILITIES OF $\alpha_{s,1}$- AND β-CASEIN VARIANTS BY STARCH-GEL AND POLYACRYLAMIDE-GEL ELECTROPHORESIS[a]

Variant	Starch	Acrylamide
$\alpha_{s,1}$-A	1.18	1.22
$\alpha_{s,1}$-D	1.13[c]	1.15[d]
$\alpha_{s,1}$-B	1.10	1.13
$\alpha_{s,1}$-C	1.07	1.10
$\alpha_{s,2}$-(zone 1.04)	1.04	1.03
$\alpha_{s,3}$-(reference zone)	1.00	1.00
β-A[b]	0.80	0.65
β-B	0.76	0.61
β-D		0.58
β-C	0.70	0.54

[a] Using the zone reference system of Wake and Baldwin (1961). All mobilities except those designated otherwise are from Thompson et al. (1965).
[b] Alkaline gel electrophoresis only.
[c] de Koning and van Rooijen (1967).
[d] Grosclaude et al. (1966).

with the discovery of $\alpha_{s,1}$-D(1.15) by Grosclaude et al. (1966) by polyacrylamide-gel electrophoresis and the reference to it by de Koning and van Rooijen (1967) as $\alpha_{s,1}$-D(1.13) by starch-gel electrophoresis. It is interesting that the Committee on Milk Protein Nomenclature used the example $\alpha_{s,1}$-D(1.14) by polyacrylamide-gel electrophoresis to illustrate how the scheme would be applied to the naming of newly discovered variants which differ in net negative charge at alkaline pH values.

B. GENETIC POLYMORPHISM OF $\alpha_{s,1}$- AND β-CASEINS

1. Mode of Inheritance

The serendipitous discovery of genetic polymorphism in $\alpha_{s,1}$-casein by Thompson et al. (1962) has led to a fruitful study of the genetics and chemistry of the polymorphs. To date, four variants of the $\alpha_{s,1}$- series (locus symbol $\alpha_{s,1}$-Cn) are known to exist. They are termed A, B, D, and C (in order of decreasing electrophoretic mobility) and their synthesis is controlled by four allelic autosomal genes with no dominance. The phenotypes correspond to the genotypes; i.e., A(A/A), AB(A/B), AC(A/C), etc.

Several interesting aspects have emerged from the study of casein polymorphism. The synthesis of specific $\alpha_{s,1}$-casein polymorphs is breed

TABLE II

GENE FREQUENCIES AND BREED SPECIFICITY OF $\alpha_{s,1}$-CASEIN VARIANTS

Breed	$\alpha_{s,1}$-Casein variant		
	A	B	C
Holstein	0.08[a]	0.87	0.05
Guernsey		0.77	0.23
Jersey		0.72	0.28
Brown Swiss		0.94	0.06
Ayrshire		1.00	0
Shorthorn		1.00	0
Zebu (Indian)		0.16	0.84
Boran (African)		0.33	0.67

[a] Gene frequency of $\alpha_{s,1}$-A is not representative of the breed since it seems to be restricted to one given blood line (Kiddy et al., 1964).

specific (Table II). This aspect will be discussed in considerable detail. Polymorphism of $\alpha_{s,1}$-casein is not universal in Western dairy cattle breeds (*Bos taurus*); Ayrshire and Shorthorn (Aschaffenburg et al., 1968a), for example, possess genes for control of synthesis of $\alpha_{s,1}$-Cn^B only. Guernsey and Jersey, on the one hand, possess genes for the control of both $\alpha_{s,1}$-Cn^B and $\alpha_{s,1}$-Cn^C; $\alpha_{s,1}$-Cn^B predominates in both breeds, but Jersey show a higher gene frequency of the C variant. Holstein, on the other hand, also synthesize B and C (as well as A) but show a low gene frequency (0.05) for the C variant.

Thymann and Larsen (1965) reported an extensive survey (over 2000 cattle) of milk protein polymorphism in Danish cattle—RDM, SDM, and Jersey. (Incidentally, the Danish workers were the first to demonstrate by family studies that κ-casein variation is genetically controlled, although others have made unsupported suggestions that it is.) RDM and SDM synthesize essentially $\alpha_{s,1}$-Cn^B; Thymann and Larsen's data (Table III) on Jersey cattle, $\alpha_{s,1}$-Cn^C (0.05), differ from those reported by Kiddy et al. (1964) for Jersey cattle bred in the United States. While the gene frequencies for β-casein agree between both groups, it is not unexpected that the gene frequency of a particular polymorph (in this case, $\alpha_{s,1}$-Cn) will vary from one herd or location to another as the author and his colleagues have often observed. Cuperlovic et al. (1964) observed in 59 Yugoslavian and Hungarian cattle that $\alpha_{s,1}$-Cn^B predominates (0.88) when they examined the caseins of Simmentaler, Frisisk × Simmentaler, RDM × Bŭsa, and Bŭsa. Sandberg (1967) examined the milk

TABLE III
Gene Frequencies in $\alpha_{s,1}$-Casein from Danish Cattle

Breed	$\alpha_{s,1}$-Casein variant		
	A	B	C
RDM	0.005	0.98	0.01
SDM	0	1.00	0.004
Jersey	0	0.95	0.05

of 193 Swedish red and white (SRB), 320 Swedish Friesian (SLB) and 85 Swedish polled (SKB) and found little variation in gene frequencies of $\alpha_{s,1}$- and β-caseins among the three breeds.

Aschaffenburg et al. (1968a) examined the caseins of Indian and African Zebu cattle (Bos indicus) for casein polymorphism. Interestingly (Table IV), $\alpha_{s,1}$-Cn^C predominates in these cattle; contrast this with Western cattle, and a significant difference is evident. The gene frequencies of six breeds in Indian Zebu cattle are also shown in Table IV, the difference in gene frequency of $\alpha_{s,1}$-Cn^C in Bos indicus as compared with Bos taurus being clearly demonstrated. The good agreement of observed phenotypes with expected values from Hardy-Weinberg calculations are shown in Table V. κ-Casein polymorphism is reviewed in Chapter 12. However it is necessary to consider certain aspects of it in this chapter. κ-Casein A predominates in Indian Zebu (as well as African Zebu) and doubtless it

TABLE IV
Gene Frequencies of $\alpha_{s,1}$-, β-, and κ-Caseins in Breeds of Zebu Cattle[a]

Breed[b]	$\alpha_{s,1}$-		β-			κ-	
	B	C	A	B	D	A	B
Hariana (72)	0.21	0.79	0.79	0.21	0	0.79	0.21
Sahiwal (21)	0.05	0.95	0.98	0.02	0	0.90	0.10
Tharparker (7)	0.14	0.86	0.86	0.14	0	0.86	0.14
Deshi (56)	0.06	0.94	0.96	0.05	0.03	0.82	0.14
Gir and Red Sindhi (5)	0	1.00	1.00	0	0	1.00	0

[a] Using 161 samples.
[b] Number of samples for each indicated in parentheses.

is inherited in a straightforward Mendelian manner as shown by daughter-dam comparisons and Hardy-Weinberg expectation (Table V). To date, Jersey cattle are the only breed in which κ-casein B predominates.

Grosclaude et al. (1966) reported the occurrence of a new $\alpha_{s,1}$-casein variant, $\alpha_{s,1}$-Cn^D, in French Flemande cattle. In this breed they found gene frequencies for B, C, and D of 0.87, 0.09, and 0.04, respectively. The D variant has also been observed by Dr. W. Michalak in Polish cattle and may yet be observed in other breeds of cattle.

Kiddy et al. (1964) reported that $\alpha_{s,1}$-Cn^A is restricted to the Holstein breed and, in fact, considered that the mutation may be a relatively new one. Certainly, in the United States at least, this variant has become reasonably widespread due to (a) artifical breeding and (b) its association with high-producing cattle. The Danish workers Thymann and Larsen (1965) suggested, however, that the $\alpha_{s,1}$-Cn^A variant is not restricted to American Holsteins and does, in fact, occur in RDM cattle (gene frequency, 0.005). However, since the purported A was not compared with authentic A, further proof of identity was needed. Recent studies by Farrell et al. (1970) on chymotryptic "fingerprints" of authentic A compared with Danish A show the two to be identical. The frequency of the A variant in American Holsteins is 0.08, but this value is not derived from a random sample of the breed. The author has examined a single casein sample from a New Zealand Friesian cow, supplied by Dr. L.

TABLE V

A REPRESENTATIVE PATTERN OF INHERITANCE OF $\alpha_{s,1}$-, β-, AND κ-CASEINS[a]

Caseins	Observed	Expected
$\alpha_{s,1}$-Caseins		
B	3	2.4
BC	28	29.3
C	91	90.3
β-Caseins[b]		
A	92	90.3
AB	25	29.3
B	3	2.4
κ-Caseins[b]		
A	95	95.7
AB	33	31.7
B	2	2.6

[a] Hardy-Weinburg expectations; all Zebu breeds.
[b] Includes eight samples in addition to those tabulated for $\alpha_{s,1}$-caseins.

11. α_s- AND β-CASEINS

TABLE VI
Gene Frequencies and Breed Specificity of β-Casein Variants

Breed	By alkaline gel electrophoresis				By acid gel electrophoresis[a]
	β-A	β-B	β-C	β-D	
Holstein	0.98	0.02	0	0	A^1, A^{2*}, A^3
Guernsey	0.98	0.004	0.16	0	A^1, A^{2*}
Jersey	0.62	0.38	0	0	A^1, A^{2*}
Brown Swiss	0.79	0.19	0.02	0	A^1, A^{2*}
Ayrshire	1.00	0	0	0	A^{1*}, A^2
Shorthorn	1.00	0	0	0	
Zebu (India)	0.85	0.13	0	0.02	A^1, A^{2*}
Boran (Africa)	0.93	0.05	0	0.02	

[a] Asterisk denotes form of β-casein predominating by acid gel electrophoresis.

Creamer, which was phenotyped $\alpha_{s,1}$-AB. Doubtless, the A variant will be observed elsewhere and new variants of the $\alpha_{s,1}$- series will be reported.

The brilliant studies of Aschaffenburg (1961) were the first reported on genetic variation in any of the caseins. He clearly demonstrated breed specificity of the occurrence of β-caseins A, B, and C (locus symbol, β-Cn), of which A is common to all breeds of dairy cattle. β-Casein C (Table VI) has been observed in Guernsey (Aschaffenburg, 1961), Brown Swiss (Thompson et al., 1964) and Hungarian and Yugoslavian dairy cattle (Cuperlovic et al., 1964). In Yugoslavian cattle (Simmentaler, Frisisk × Simmentaler, and Bŭsa) the gene frequency is low (0.005), whereas in Hungarian cattle it is 0.10 or close to that of Guernsey (0.16). β-Casein B is fairly common in Jersey, Brown Swiss and Indian Zebu as is shown in Table VI, but its frequency is low in Hungarian and Yugoslavian cattle.

β-Casein D, observed in Indian and African Zebu (Aschaffenburg et al., 1968a), has not been observed in Western breeds of cattle. It has, however, been observed only in low frequency (0.02) in a few of the many breeds of Zebu cattle.

2. Linkage of Genes

King et al. (1965) and Grosclaude et al. (1964) concurrently but independently reported a close correlation between the loci controlling $\alpha_{s,1}$-Cn and β-Cn polymorphism. Certain combinations of $\alpha_{s,1}$-Cn and

TABLE VII
PATTERN OF OCCURRENCE OF $\alpha_{s,1}$- AND β-CASEIN PHENOTYPES IN JERSEY COWS

$\alpha_{s,1}$-Casein	β-Casein		
	A	AB	B
B	60	73	56
BC	45	82	
C	35		

β-Cn are common (i.e. $\alpha_{s,1}$-Cn^B, β-Cn^A; $\alpha_{s,1}$-Cn^{BC}, β-Cn^A), while others ($\alpha_{s,1}$-Cn^{BC}, β-Cn^B; $\alpha_{s,1}$-Cn^C, β-Cn^{AB}; and $\alpha_{s,1}$-Cn^C, β-Cn^B) occur infrequently, if at all, and are considered forbidden combinations. The notable absence of these last three classes is observed with Jersey cattle as shown in the contingency table (Table VII).

Patterns similar to the Jersey breed emerge on comparing Guernsey and Brown Swiss cows; however, the expectation of observing certain combinations is admittedly small (Tables VIII and IX). Close linkage of the $\alpha_{s,1}$-Cn and β-Cn loci has also been observed in Indian and African Zebu cattle; no crossing-over of loci has been observed (Aschaffenburg et al., 1968a).

King et al. (1965) remarked

> The present finding that the two kinds of variants ($\alpha_{s,1}$- and β-) do not occur independently suggests that the two loci might be linked, and in the Jersey herd, for example, the following chromosomes would be postulated: $\alpha_{s,1}$-Cn^B, β-Cn^A; $\alpha_{s,1}$-Cn^B, β-Cn^B; and $\alpha_{s,1}$-Cn^C, β-Cn^A. Since the $\alpha_{s,1}$-Cn^B, β-Cn^A combination prevails in all breeds, the other combinations are assumed to be the result of mutations in either $\alpha_{s,1}$-Cn or β-Cn. The combination of $\alpha_{s,1}$-Cn^C, β-Cn^B apparently does not occur, so the linkage would have to be very close unless recombinant types were at a selective advantage.

Grosclaude et al. (1964) further elaborated on the linkage of $\alpha_{s,1}$-Cn and β-Cn. They concluded that the loci are either identical (pleotropic) or closely correlated. If the latter hypothesis is true then the upper limit of the distance between the two loci can be estimated by the following reasoning: In effect, if x is the percentage of recombinations expressing the distance between the two loci, and n is the number of useful pairs, x being small, the number of recombinants follows a law of Poisson of parameter nx; the probability, $P_{(0)}$, that recombinants will not be observed is then $P_{(0)} = e^{-nx}$. By fixing the threshold of probability at 0.05, Gros-

TABLE VIII

PATTERN OF OCCURRENCE OF $\alpha_{s,1}$- AND β-CASEIN PHENOTYPES IN GUERNSEY COWS

$\alpha_{s,1}$-Casein	β-Casein			
	A	AB	AC	C
B	85	5	11	1
BC	52	2	5	
C	9			

Claude et al. (1965) found x to be 0.038. The distance between the two loci is then $<$ 3.8 units of recombination.

More recently, Grosclaude et al. (1965) have proposed a close correlation of the κ-Cn locus with the $\alpha_{s,1}$- and β-Cn loci. They conclude that there is a 0.95 probability that the distance between the κ-Cn and the pair of $\alpha_{s,1}$-Cn, β-Cn loci is less than 2.8 units of recombination. Therefore the linkage between the $\alpha_{s,1}$-Cn and β-Cn loci appears to be closer than with these two and κ-Cn. Thymann and Larsen (1965) concluded that the genes involved in the synthesis of β-, $\alpha_{s,1}$-, and κ-casein types are not transmitted independently. This could be explained by postulating that the genes responsible for the variations in the three caseins belong to the same locus, although close linkage of 2–3 loci could not be excluded.

Reasonable assurance of the linkage of genes controlling the synthesis of the major milk proteins ($\alpha_{s,1}$-, β-, and κ-caseins) adds to the growing number of examples of genetically linked synthesis of chemically related proteins; i.e., β- and δ-chain hemoglobin variants (Boyer et al., 1963) and egg-white proteins (Buvanendran, 1964).

TABLE IX

PATTERN OF OCCURRENCE OF $\alpha_{s,1}$- AND β-CASEIN PHENOTYPES IN BROWN SWISS COWS

$\alpha_{s,1}$-Casein	β-Casein			
	A	AB	B	AC
B	146	69	4	2
BC	22	5		
C	2			

3. Additional Variants and Species Differentiation

The number of variants, four for $\alpha_{s,1}$- and six for β-casein, is not fixed and one expects that additional polymorphism will be observed in bovine caseins. Perhaps it is advisable to point out that electrophoretic methods, for the most part, disclose differences in charged amino acids; it is unlikely that neutral substitutions would be observed by conventional zonal electrophoresis systems.

Aschaffenburg and Thompson (1967) have observed that polymorphism exists in the "α_s-" fraction of caprid casein; the mode of inheritance is not as simple as that for the bovine, and its nature is still obscure. King (1966) has observed an equally complex situation in the "α_s-" fraction of ovine casein where three variants are observed, but he has stated that their inheritance is straightforward. He also observed variation in β-casein from the Dorset Horn breed and believed this was due to a simple alternative allele. Variation in the β-casein-like fraction has also been reported for sow milk (Glasnak, 1968). Finally, Aschaffenburg et al. (1968b) have observed limited polymorphism in the β-casein of Indian buffalo (*Bubalus bubalis*) which appears to be inherited in a straightforward Mendelian manner. $\alpha_{s,1}$-Casein of this species is not identical to that of *Bos*, but polymorphism has not been observed in this fraction of the casein to date.

4. Methods of Phenotyping $\alpha_{s,1}$-Casein Polymorphs

An important consideration in the study of casein polymorphism is the method used in phenotyping. Methods of phenotyping $\alpha_{s,1}$-, β-, and κ-caseins are included in this section. (For a more thorough review, see Thompson, 1970.)

Aschaffenburg (1961) first applied paper electrophoresis to a study of β-casein polymorphism. The method involved horizontal zonal electrophoresis in a 6.0 M urea, pH 7.15, citrate–phosphate buffer which closely simulated the buffer conditions in milk. While β-caseins A, B, and C were clearly distinguished (Fig. 1) the method was of little value in phenotyping $\alpha_{s,1}$- and κ-caseins. Concurrent with the studies of Aschaffenburg on β-caseins polymorphism, Thompson et al. (1962), Kiddy et al. (1964), and Thompson et al. (1964), applied the more sophisticated method of starch-gel electrophoresis to the phenotyping of $\alpha_{s,1}$- and β-caseins. Starch-gel electrophoresis, adapted by Wake and Baldwin (1961) to a study of the heterogeneity of caseins, is carried out in pH 8.6, 7.0 M urea in a discontinuous buffer system. The results of this method are clearly seen in Fig. 2 where β-caseins A, B, and C are resolved as discrete zones. Phenotyping of κ-casein is accomplished using essentially the method of Wake

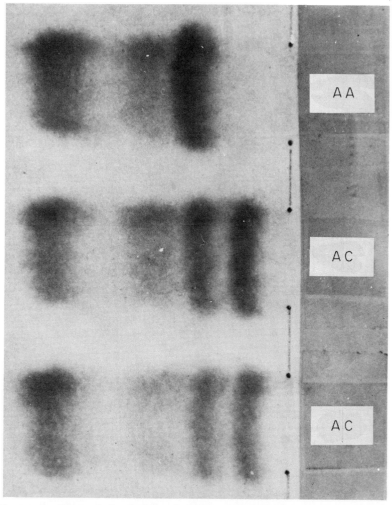

FIGURE 1. Paper electrophoresis of whole caseins at pH 7.15, citrate–phosphate buffer in 6.0 M urea showing the separation of β-caseins A and C. (Original electrophoretogram courtesy of Dr. R. Aschaffenburg, Oxford, England.)

and Baldwin by adding 2-mercaptoethanol to the warm gel solution prior to pouring.

Aschaffenburg and Thymann (1965) have employed thin starch-gel electrophoresis for the phenotyping of $\alpha_{s,1}$-, β-, and κ-caseins. The method has the added advantage of disclosing some β-lactoglobulin phenotypes, thereby yielding the complete complement of milk protein polymorphs (with the exception of α-lactalbumins) from a single milk sample in a single gel electrophoresis analysis. Thymann and Larsen (1965) have used this method to phenotype a large number of milk samples in Denmark. The method of thin starch-gel electrophoresis has been perfected by Michalak (1967); the excellent results of his technique are shown in Fig. 3. Most of the major proteins of cow milk are clearly resolved, including β-lactoglobulin D. α-Lactalbumin migrates in the vicinity of $\alpha_{s,1}$-casein.

In addition to starch-gel electrophoresis, polyacrylamide-gel electrophoresis by the procedure of Peterson (1963) has been used extensively in the phenotyping of $\alpha_{s,1}$-, β-, and κ-caseins. Electrophoresis is conducted at pH 9.1, 6–7% (w/v) cyanogum in 4.5 M urea. Electrophoresis is per-

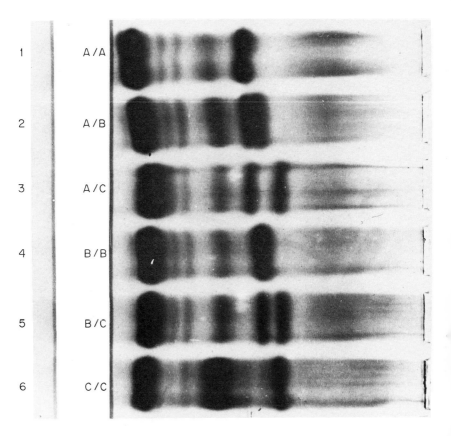

FIGURE 2. Starch-gel electrophoresis patterns of whole casein at pH 8.6, 7.0 M urea (Thompson et al., 1964) were of the six phenotypes of β-casein. Patterns 1–6 are β-caseins A, AB, AC, B, BC, and C, respectively.

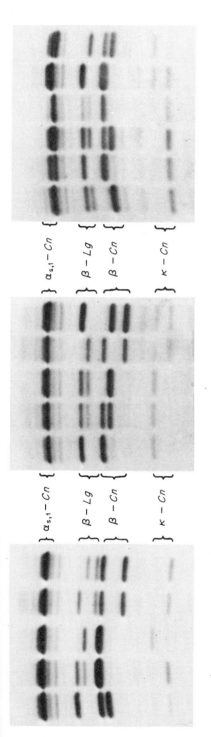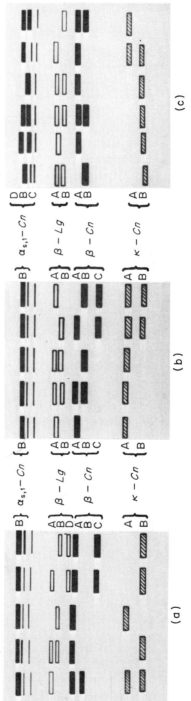

FIGURE 3. Simultaneous phenotyping of caseins and β-lactoglobulins by thin starch-gel electrophoresis at pH 8.4, borate buffer, in the presence of mercaptoethanol (Michalak, 1967). (Gel patterns courtesy of Dr. Wieslaw Michalak, Warsaw, Poland.)

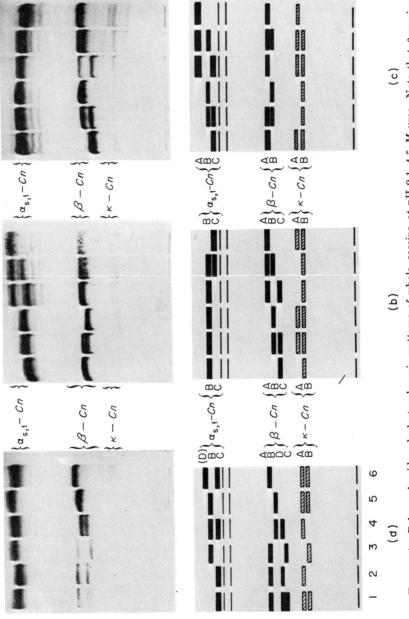

FIGURE 4. Polyacrylamide-gel electrophoresis patterns of whole caseins at pH 9.1, 4.5 M urea. Note that β-caseins A, D and C are shown in pattern 1 of gel A, the β-casein C having been added to whole β-casein AD.

formed in the vertical position at a temperature of 14°C. Aschaffenburg (1964) has used horizontal zonal electrophoresis in polyacrylamide to achieve the separation of the casein polymorphs. Polyacrylamide-gel electrophoresis is capable of phenotyping κ-casein upon the addition of the reductant 2-mercaptoethanol to either the sample or the gel solution.

It is the opinion of this author that polyacrylamide is a more desirable supporting medium than starch because of (a) ease of operation, (b) brevity of electrophoresis, and (c) resolution of casein components. The method has the disadvantage of not being adaptable to the simultaneous typing of β-lactoglobulin in the presence of β-casein (Thompson, 1966). Figure 4 shows the separation by polyacrylamide-gel electrophoresis of three of the four known types of $\alpha_{s,1}$- and all of the four known β-casein variants disclosed by electrophoresis at pH 9.1.

Additional polymorphism in the caseins, that is the splitting of β-A into A^1, A^2, and A^3 (Peterson and Kopfler, 1966), is not disclosed by gel electrophoresis at alkaline pH values but is clearly distinguished at pH 2.8 in a formic–acetic acid buffer system. The gel concentration is increased to 10% (w/v), and the use of N,N,N',N'-tetramethylethylenediamine as a catalyst is necessary. Figure 5 shows the separation of β-casein C, B, A^1A^2, and A^1A^3. β-Casein D (*Bos indicus*) migrates at the same rate of β-B, while neither $\alpha_{s,1}$- nor κ-casein polymorphs can be distinguished adequately at pH 2.8.

C. Preparation of α_s-Caseins

Because many modifications of preparations of calcium-sensitive α-caseins have been reported, the author has limited the presentation of methods to those which he feels have been most widely used. However, Table X shows the chemical composition of some calcium-sensitive α-caseins whose preparations are not reported here, in addition to those described by the following methods.

Waugh et al. (1962) described in detail the unique preparation of $\alpha_{s,1,2}$-casein (earlier termed α_s-casein) from first-cycle soluble casein, which was obtained by high-speed centrifugation of skim milk at 37°C and pH 7.0 following the addition of sufficient $CaCl_2$ to make the milk 0.07 M in added calcium(II). A solution of first-cycle casein was cooled to 0°–1°C, and 5 M $CaCl_2$ was added to give a 0.17 M solution at pH 7.0. After standing overnight, the mucilaginous precipitate was collected and dissolved in 0.05 M potassium citrate at pH 6.5 and 2°C. The solution was diluted with distilled water at 5°–7°C and 5 M $CaCl_2$ was added to give a final concentration of 0.07 M. The resulting precipitate was dis-

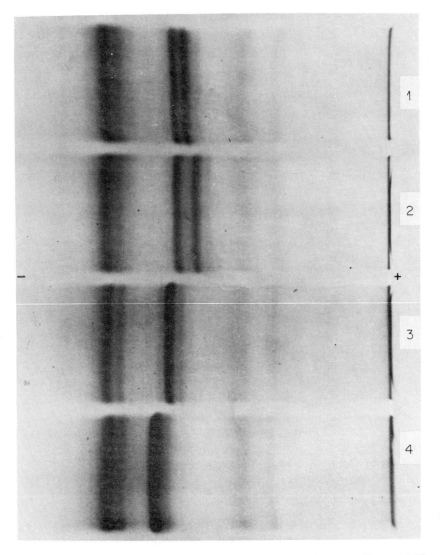

FIGURE 5. Polyacrylamide-gel electrophoresis of β-casein variants at pH 2.8, 4.5 M urea, in formic–acetic acid buffer (Peterson and Kopfler, 1966). Phenotypes from patterns 1–4 are β-A^1A^2, A^1A^3, B, and C. (Courtesy of Dr. Fred C. Kopfler, Philadelphia, Pennsylvania.)

solved in 0.05 M potassium citrate and stored for further use. The $\alpha_{s,1,2}$ was separated from other α_s-casein fractions by step-wise elution from DEAE-cellulose columns, 4.5 M urea in an 0.01 M imidazole–HCl buffer system.

TABLE X
Some Properties of Reported Components of the α-Casein Complex

Calcium-sensitive component	Nitrogen (g/100 g)	Phosphorus (g/100 g)	s_{20} (S)	$A_{1\,cm}^{1g/dl}$	P^I	End group C	End group N	Reference
$\alpha_{s,1}$-Casein A	15.10	1.01		10.10	4.95^a	Trp	Arg	Thompson and Kiddy (1964)
$\alpha_{s,1}$-Casein B	15.34	1.01		10.05	5.15^a	Trp	Arg	Thompson and Kiddy (1964)
$\alpha_{s,1}$-Casein C	15.40	1.01		10.03	5.15^a	Trp	Arg	Thompson and Kiddy (1964)
α_s-Casein	15.10	1.01		10.20	4.4			Zittle and Custer (1963)
α-Caseins (1.07 and 1.10)	14.00	1.12	1.64^b				Arg	Schmidt and Payens (1963)
$\alpha_{s,1,2}$-Casein	14.70	1.03	4.55^d	10.1	5.16^c	Trp–Leu–Try		Waugh et al. (1962)
α_R-Casein		1.18	3.00^d					Long et al. (1958)
α_1-Casein	14.10	0.85			4.3–4.7			McMeekin et al. (1959)

^{a} Isoionic point at 25°C (Thompson and Gordon, 1967).
^{b} Denotes characteristics of monomer form.
^{c} Isoionic point at 20°C (Ho and Waugh, 1965).
^{d} Denotes characteristics of associated form.

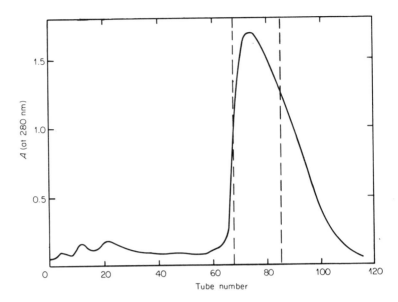

FIGURE 6. Elution diagram (absorbancy, A, at 280 nm, vs. tube number) of $\alpha_{s,1}$-casein B from DEAE-cellulose, pH 7.0, imidazole–HCl, 3.3 M urea buffer; 540 mg protein charge (Thompson and Kiddy, 1964).

Schmidt and Payens (1963) prepared their calcium-sensitive α-casein (termed starch-gel components 1.07 and 1.10) from Waugh and von Hippel's second-cycle soluble casein followed by further purification using the method of Zittle et al. (1959). The material was purified by zone electrophoresis in a column of packed cellulose powder at pH 6.5, 5 M urea. According to these authors, zones 1.07 and 1.10 probably corresponded to α_s-($\alpha_{s,1}$-) B and C. Calcium-sensitive α-casein prepared by zone electrophoresis appears to be quite homogeneous by starch-gel electrophoresis.

While the previous two methods used calcium precipitation of whole casein from skim milk, Zittle and Custer (1963) dissolved acid casein in 6.6 M urea and acidified with 3.5 M H_2SO_4. Upon dilution of the above mixture with water, a precipitate slowly forms which is rich in $\alpha_{s,1}$- and β-caseins while the supernatant is a good source of κ-casein. The precipitate is purified by the earlier method of Zittle et al. (1959) in which NaCl is added in the urea fractionation procedure to dissociate the α-casein complex. The crude $\alpha_{s,1}$-casein can be purified by precipitation of contaminants from an ethanol–water solution at pH 7.2 with ammonium acetate. In addition to yielding a reasonably pure $\alpha_{s,1}$-casein, the method is basically simple, and large scale preparations can be made in a relatively short time.

Thompson and Kiddy (1964) described the preparation of $\alpha_{s,1}$-A, B, and C using a number of different principles. Whole α-casein was prepared by a slight modification of the urea fractionation method of Hipp et al. (1952b) in which 4 M CaCl$_2$ was added to give a 0.40 M solution at 2°–4°C to precipitate crude $\alpha_{s,1}$-casein at pH 7.2. After removal of calcium, the crude $\alpha_{s,1}$- was dissolved in 50% ethanol–water and α_s-like caseins were precipitated by the addition of ammonium acetate. Further purification was accomplished by gradient elution from DEAE-cellulose columns at pH 7.0, 3.3 M urea (Fig. 6). The chromatographed proteins were demonstrated to be of high purity by starch- and polyacrylamide-gel electrophoresis. Other procedures involving the fractionation of $\alpha_{s,1}$-casein using urea have been described by El-Negoumy (1966) and Gehrke et al. (1966).

$\alpha_{s,1}$-Caseins can also be prepared by column chromatography of whole casein. The $\alpha_{s,1}$-, purified by rechromatography, is suitable for further study. While the quantity of a desired component obtained by this method

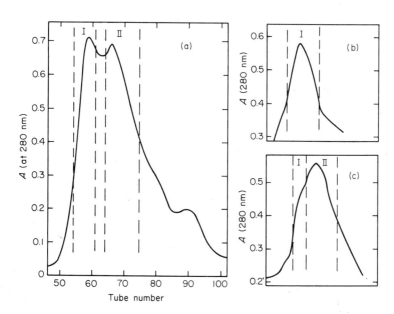

FIGURE 7. Elution diagram (absorbancy, A, at 280 nm, vs. tube number) of zones 1.00 ($\alpha_{s,3}$; fraction I) and 1.04 ($\alpha_{s,2}$; fraction II) from DEAE-cellulose, pH 7.0, imidazole–HCl, 3.3 M urea buffer in mercaptoethanol; 500 mg protein charge; (b) and (c) represent rechromatography of peaks I and II in (a), respectively (Hoagland et al., 1970).

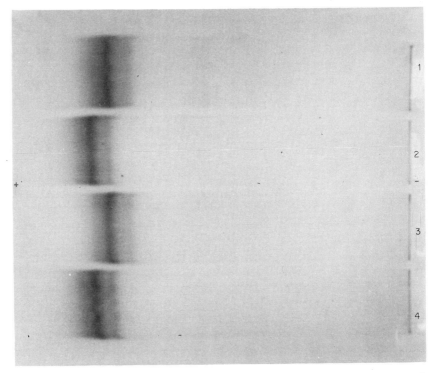

FIGURE 8. Polyacrylamide-gel electrophoresis patterns of fraction I (zone 1.00) and fraction II (zone 1.04) of Fig. 7 prior to rechromatography. Patterns 1–4: zone 1.00, zone 1.04, zone 1.00, and zone 1.04, respectively. Buffer conditions are identical to Fig. 4 (Hoagland et al., 1970).

is generally small, no prior fractionations are necessary and the fraction is not unduly exposed to possible harmful reagents. Thompson (1966) has demonstrated, as many others have, that $\alpha_{s,1}$-casein can be purified by direct chromatography of whole casein.

$\alpha_{s,2}$-Casein (zone 1.04) and $\alpha_{s,3}$-casein (zone 1.00) have been prepared by obtaining the fraction soluble in 3.3 M urea by the method of Aschaffenburg (1963b). After dilution to 1.0 M urea, the precipitate is washed free of urea, dissolved at pH 7.2, and an equal volume of absolute ethanol is added. Ammonium acetate serves to precipitate most of the $\alpha_{s,2}$- and $\alpha_{s,3}$-caseins which, unlike $\alpha_{s,1}$-caseins, do not remain in solution. They are further purified by DEAE-cellulose chromatography in 3.3 M urea, as described by Thompson and Kiddy (1964) for $\alpha_{s,1}$-casein variants, but 2-mercaptoethanol is added, as described by Thompson (1966). The partial separation of these components on DEAE-cellulose is shown in Fig. 7 and their purity after a single chromatographic separation in Fig. 8.

D. COMPOSITION AND STRUCTURE

1. *Amino Acid Analysis and Peptide Mapping of $\alpha_{s,1}$-Casein Variants*

The availability of cows homozygous for $\alpha_{s,1}$-caseins has been valuable for obtaining these variants for chemical and physical analyses. Heretofore it was tacitly assumed that calcium-sensitive α-casein was the same from one cow to another. Two laboratories have independently reported the amino acid composition of $\alpha_{s,1}$-B and C, the most common of the known variants of the $\alpha_{s,1}$-series (Table XI). The general agreement on the analyses of the two variants is remarkable, and both groups agree that $\alpha_{s,1}$-B differs from C by one more glutamic acid and one less residue per

TABLE XI

AMINO ACID COMPOSITION OF THE GENETIC VARIANTS OF $\alpha_{s,1}$-CASEIN[a]

Acid	$\alpha_{s,1}$-Casein variant					
	A[b]	B[b]	B[c]	C[b]	C[c]	D[d]
Asp	16.8	18.1	17.9	18.2	17.6	18.0
Thr	6.7	6.0	5.7	6.1	5.8	6.3
Ser	17.8	17.3	16.8	17.6	16.6	15.8
Glu	46.6	46.4	47.0	45.5	45.9	47.0
Pro	20.6	20.3	19.5	20.4	20.0	20.5
Gly	10.7	10.7	10.7	11.8	11.8	10.9
Ala	9.9	10.8	11.0	10.8	11.0	10.4
Val	11.9	13.4	13.2	13.6	13.4	13.1
Met	5.9	5.7	5.7	5.7	5.6	5.3
Ile	13.6	13.1	12.9	13.3	13.0	13.0
Leu	17.3	20.3	20.1	20.5	20.1	20.1
Tyr	12.1	11.6	11.4	11.7	11.4	11.2
Phe	7.5	9.6	9.4	9.7	9.4	9.6
Trp	2.8[e]	2.7[e]	3.4	2.8[e]	3.5	3.2
Lys	18.1	17.0	16.3	17.0	16.2	16.6
His	6.2	6.1	6.0	6.1	6.0	5.8
Arg	6.2	7.2	7.0	7.2	7.0	7.2
NH_3	27.1	31.1	32.3	29.7	33.1	33.0
PO_3H	11.3	11.3	11.5	11.3	11.7	11.4

[a] Residues of amino acid per 28,600 mol wt, except for $\alpha_{s,1}$-A, which is calculated on the basis of 28,000 mol wt.
[b] Gordon et al. (1965).
[c] de Koning and van Rooijen (1965).
[d] de Koning and van Rooijen (1967).
[e] More recent analyses indicate this value to be 2.0 residues (Rose et al., 1970).

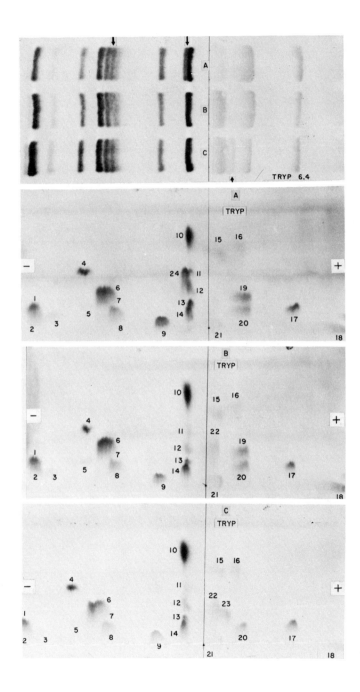

28,600 mol wt. This substitution, involving a single amino acid (Glu/ Gly), involves the coding triplets $GA^{A/G}/GG^{A/G}$. The only significant discrepancy in analyses between the Dutch and American workers involves the percentage of tryptophan in these proteins. Recently Spies (1967) has presented values of 1.62, 1.59, and 1.57 ± 0.02% per $\alpha_{s,1}$-A, B, and C, respectively, or about two residues for a molecular weight of 28,600 daltons.

The rarer of the few known variants of the $\alpha_{s,1}$-series, A and D, have also been analyzed. Reportedly (Table XI), $\alpha_{s,1}$-D varies from B in a Pro/Ser substitution which would involve the coding triplets CCA/UCA. Admittedly, serine is lower in D than B, but proline, differing by only 2.5%, is a questionable difference. Further, the explanation that D is differentiated from B on the basis of a conformational change is doubtful, and this author would expect the tryptic fingerprints to disclose the difference peptide as was shown for the A, B, and C variants by Kalan et al. (1966).

$\alpha_{s,1}$-Casein A is different from the B, C, and D variants in amino acid composition. Notably, A is devoid of three residues of leucine, two residues of phenylalanine, and one residue each of valine, alanine and arginine. It is evident that A is not a mutation of B in the sense that the term is generally applied (see below).

While the amino acid analysis (Gordon et al., 1965) of a protein is suggestive of the difference between it and other chemically related mutants, the final proof becomes evident when different peptides are isolated and analyzed. Kalan et al. (1966) hydrolyzed $\alpha_{s,1}$-A, B, and C by trypsin, pepsin and chymotrypsin, followed by high-voltage electrophoresis and second-dimension ascending paper chromatography, in an attempt to locate the difference peptides (Figs. 9, 10, 11). In comparing the B and C variants, the authors concluded that the amino acid substitution (Glu/Gly) is probably located in a pair of difference peptides from the above digests. With all three digests the B variant gave rise to an acidic peptide which was absent in C. A less acidic peptide appeared in C, and this mobility difference is explainable on the basis of an excess of glutamic acid in B. Similar electrophoretic behavior has been demonstrated for the aspartic acid/glycine difference pair in β-lactoglobulins A and B (Kalan et al., 1962). Furthermore, specific staining reactions

---

FIGURE 9. Peptide patterns from tryptic digests of $\alpha_{s,1}$-caseins A, B, and C. The top panel shows one-dimensional high-voltage paper electrophoresis at pH 6.4, 40 V/cm, 2 hr, with arrows marking differences. The lower three panels are "fingerprints" produced by two-dimensional electrophoresis and chromatography. Peptides located in equivalent positions are given the same number (Kalan et al., 1966).

142 M. P. THOMPSON

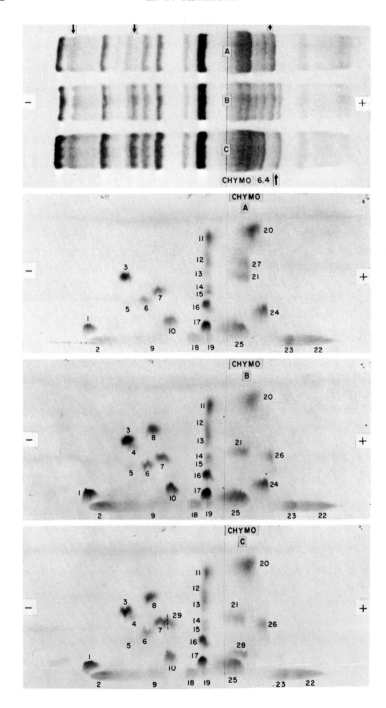

of the different peptides of B and C indicate the presence of tyrosine, methionine, histidine, and tryptophan in the vicinity of the amino acid substitution. Thompson et al. (1967a) have isolated and analyzed chymotryptic difference peptides from B and C and have shown conclusively an excess of one glutamic acid in B (-1 glycine) as compared with C. The glutamic acid/glycine substitution is a familiar one known to occur in hemoglobin and tobacco mosaic virus coat protein, and the reverse substitution in tryptophan synthetase.

Recent developments concerning the primary structure of the $\alpha_{s,1}$-caseins are deserving of much attention. Grosclaude et al. (1969) isolated the chymotryptic difference peptides from $\alpha_{s,1}$-B and C (Fig. 2). The primary structure of the peptides was determined as

$$\text{Asn-Ser-} \begin{matrix} \text{Glu} \\ \\ \text{Gly} \end{matrix} \text{-Lys-Thr-Thr-Pro-Met-Leu-Trp-OH}$$

This peptide was determined to be positioned in the C-terminal portion of the $\alpha_{s,1}$- molecule. We have verified the positioning of this peptide by first reacting native $\alpha_{s,1}$-casein B with carboxypeptidase A to remove C-terminal tryptophan and penultimate leucine, followed by chromatography of the reaction mixture to remove residual carboxypeptidase. The protein was subsequently digested with chymotrypsin. The resulting two-dimensional fingerprint showed (a) the loss of tryptophan from the $\alpha_{s,1}$-B/C difference peptide and (b) the repositioning of the B/C difference peptide. Subsequently Mercier et al. (1970) determined the sequence of 48 residues at the COOH terminal end of the B variant.

The second development concerns the N-terminal amino acid sequence of the $\alpha_{s,1}$-caseins. From the amino acid and end-group analyses of chymotryptic peptide 25 (Fig. 10), combined with our knowledge of the specificity of chymotrypsin, an empirical formula of the N-terminal sequence follows:

$$\text{H}_2\text{N-Arg-(Asp}_2\text{Glu}_8(\text{NH}_2)_5\text{Ser}_2\text{Pro}_3\text{Ala}_2\text{Val}_4\text{Ile}_2\text{Leu}_3\text{Lys}_2\text{His})\text{Pro-Arg-His-Leu}$$

This large peptide (34 residues) carries a net charge of -1 and may be positioned adjacent to the phosphorus-rich region of the molecule.

Earlier, Gordon et al. (1965) had concluded that $\alpha_{s,1}$-A was not, in the sense that the term is ordinarily applied, a mutation of B because of

---

FIGURE 10. Peptide patterns from chymotryptic digestion of $\alpha_{s,1}$-caseins A, B, and C. Conditions of electrophoresis, chromatography, and peptide numbering are as in Fig. 9.

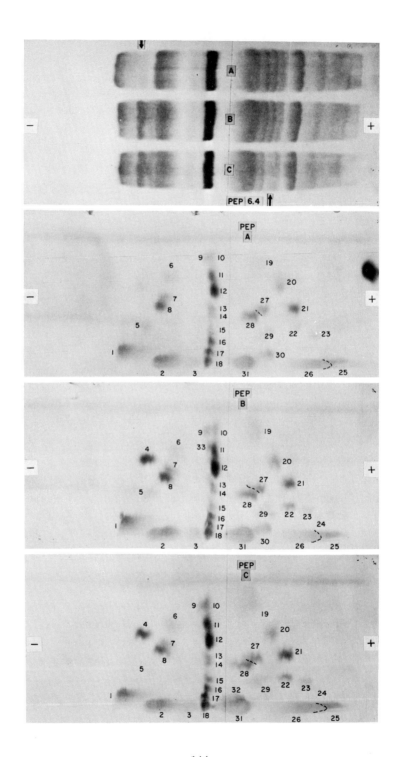

gross differences in amino acid substitution. However, if the amino acids deficient in A were distributed throughout the molecule one would expect the peptide maps to differ greatly from B or C; they do not (Figs. 9, 10, 11). The outstanding characteristics of the A maps are the total absence of one or two major spots, the observations of which suggest the hypothesis that A is missing an intact segment (or several segments) of the molecule, the remainder of which is similar to the B variant. Identical end groups (see below) tend to rule out the possibility of a missing N- or C-terminal portion.

Thompson *et al.* (1969a) have explained the difference between the $\alpha_{s,1}$-A and B variants as involving two mutational events which probably did not occur simultaneously. The first event suggests a straightforward amino acid substitution, Lys/Gln, (codons $AA^{A/G}/AA^{U/C}$). The second event involved the deletion of a largely nonpolar segment of the molecule. A portion of this segment, a tripeptide, peptide 8, isolated from chymotryptic digests of $\alpha_{s,1}$-B, (Fig. 10) has been sequenced as Arg-Phe-Phe. Valine, alanine and three leucine residues are yet to be located and sequenced. Multiple deletions of the type exhibited by $\alpha_{s,1}$-casein A, while not common, have been demonstrated to occur (Bradley *et al.*, 1967) and may have arisen by unequal crossing-over between the closely linked $\alpha_{s,1}$-Cn and β-Cn genetic loci (King *et al.*, 1965; Thompson *et al.*, 1969a).

Clarification of the chemistry of the α_s-like caseins has been made in several independent studies. Annan and Manson (1969) isolated these minor caseins by chromatography on SE-Sephadex in 8 M urea at pH 4.0. These researchers retained the terminology of $\alpha_{s,1}$-casein but contributed materially to the nomenclature of the other α_s-like caseins. The component migrating immediately ahead of $\alpha_{s,1}$-casein was termed $\alpha_{s,0}$-; the component immediately behind they termed $\alpha_{s,2}$-. Zones 1.04 and 1.00 which were formerly termed $\alpha_{s,2}$- and $\alpha_{s,3}$- (Thompson and Kiddy, 1964; Thompson *et al.*, 1965) were termed (appropriately) $\alpha_{s,3}$- and $\alpha_{s,4}$-. $\alpha_{s,5}$-Casein migrated ahead of β-casein but could be converted to $\alpha_{s,3}$- and $\alpha_{s,4}$- in the presence of reducing agents (Hoagland *et al.*, 1970). $\alpha_{s,0}$-Casein and $\alpha_{s,1}$-casein have an identical C-terminal sequence of Leu-Trp-OH whereas $\alpha_{s,2}$-, $\alpha_{s,3}$-, $\alpha_{s,4}$- and $\alpha_{s,5}$- probably have C-termini of Leu-Tyr-OH.

According to the definition of α_s-casein (calcium-precipitated and stabilized by κ-casein against precipitation by Ca^{2+}), $\alpha_{s,2}$- and $\alpha_{s,5}$- qualify as α_s-caseins. The amino acid analyses of $\alpha_{s,3}$- and $\alpha_{s,4}$-caseins

---

FIGURE 11. Peptide patterns from peptic digestion of $\alpha_{s,1}$-caseins A, B, and C. Conditions of electrophoresis, chromatography, and peptide numbering are as in Fig. 9.

TABLE XII

TENTATIVE AMINO ACID COMPOSITION OF $\alpha_{s,3}$- AND $\alpha_{s,4}$-CASEINS

	Residue number[a]					
	$\alpha_{s,3}$-			$\alpha_{s,4}$-		
Amino acid	2×	4×	Residue[b]	2×	4×	Residue[b]
Asp	6.6	13.2	13	6.2	12.4	12
Thr	5.2	10.4	10	4.9	9.8	10
Ser	5.8	11.6	11–12	5.2	10.4	10–11
Glu	14.4	28.8	29	13.9	27.8	28
Pro	4.0	8.0	8	3.7	7.4	7
Gly	1.0	2.0	2	0.95	1.9	2
Ala	2.9	5.8	6	2.8	5.6	6
½ Cys[c]	0.37	0.74	1	0.36	0.72	1
Val	4.8	9.6	10	4.8	9.6	10
Met	1.4	2.8	3	1.2	2.4	2–3
Ile	3.8	7.6	8	3.6	7.2	7
Leu	4.5	9.0	9	4.4	8.8	9
Trp	3.6	7.2	7	3.4	6.8	7
Phe	2.1	4.2	4	2.1	4.2	4
Lys	7.5	15.0	15	8.0	16.0	16
His	1.4	2.8	3	1.4	2.8	3
Arg	1.9	3.8	4	2.0	4.0	4

[a] Calculated by molar ratios: arginine = 1.
[b] Calculated on the basis of four residues of arginine.
[c] ½ Cys not corrected for destruction.

are presented in Table XII. Clearly the amino acid analyses of these components differ markedly from the $\alpha_{s,1}$-casein variants (Table XI). The most striking difference is the presence of half cystine which suggests the presence of one disulfide bond per molecule although two SH groups cannot be ruled out. The ratio of basic amino acids also differs from $\alpha_{s,1}$-caseins. Although $\alpha_{s,4}$-casein contains one more residue of lysine than $\alpha_{s,3}$-, the charge difference is partially offset by $\alpha_{s,3}$- possessing one more residue each of aspartic and glutamic acids. Therefore, the faster mobility of $\alpha_{s,3}$-casein can be explained on that basis.

Additional studies on $\alpha_{s,3}$- and $\alpha_{s,4}$-caseins are needed, including peptide fingerprinting, the determination of molecular weights by physical methods, and elemental analyses.

While no polymorphism of the α_s-like caseins has been observed in Western breeds of cattle, it is suggestive in Eastern breeds (Zebu) of cattle (Aschaffenburg et al., 1968a). Michalak (1967) observed the total absence of zones 1.00 and 1.04 in a few RDM cattle (Poland). Why this should be true is not altogether clear.

2. *End-Group Analysis of $\alpha_{s,1}$-Caseins*

Mellon et al. (1953) and Wissmann and Nitschmann (1957) carried out amino-terminal end-group analysis on α-casein by dinitrophenylation. Both investigation groups found arginine and lysine to be the major amino-terminal residues of α-casein, a protein now known to consist of two major fractions, $\alpha_{s,1}$- and κ-caseins. Manson (1961), aware of the heterogeneity of whole α-casein, undertook an investigation of the N-terminal structure of calcium-sensitive casein (α_s-) which appeared to be pure. His results, like those of the above researchers, disclosed the presence of arginine, but he did not detect the presence of lysine. Calculated on the basis of N-terminal arginine, the molecular weight of α_s- was found to be 31,000 daltons. Similarly, Kalan et al. (1964) observed only N-terminal arginine in $\alpha_{s,1}$-caseins A, B, and C by the 2,4-dinitrofluorobenzene (FDNB) method of Sanger, an observation in accord with Schmidt and Payens (1963) as well as with Manson (1961). Analysis of the phenylthiohydantoin derivatives of the amino acids by the Edman procedure revealed only phenylthiohydantoin arginine, which confirmed the results of the FDNB procedure (Kalan et al., 1964). Ganguli et al. (1964), also using this procedure, observed that buffalo α-, β-, and γ-caseins contained the same N-terminal amino acids, arginine and lysine. In addition, no quantitative difference in the N-terminal residues of caseins from buffalo and cow milk was observed.

The situation with regard to the C-terminal sequence of $\alpha_{s,1}$-caseins has not been interpreted as definitely as has the N-terminal studies. Waugh et al. (1962) presented results on $\alpha_{s,1,2}$-caseins which indicated that tryptophan, leucine and tyrosine were released from these proteins by the action of carboxypeptidase A. These workers suggested that two nonidentical chains (presumably $\alpha_{s,1}$- and $\alpha_{s,2}$-, which are probably $\alpha_{s,1}$-B and $\alpha_{s,1}$-C) exist, one terminating in tryptophan, the other in leucine. While Kalan et al. (1964) did not rule out this possibility (Fig. 12), they suggested that the genetic variants (A, B, and C) of $\alpha_{s,1}$-caseins, which appeared as single zones on starch- or polyacrylamide-gel electrophoresis, terminated in tryptophan with penultimate leucine followed by a second leucine in the sequence. Their data, then, inferred a Leu-Leu-Trp sequence for all three variants since tryptophan was quantitatively released within 15 min from the start of enzyme reaction, followed closely by the quantitative release of one residue of leucine within 5 hr. Kalan et al. (1964) did

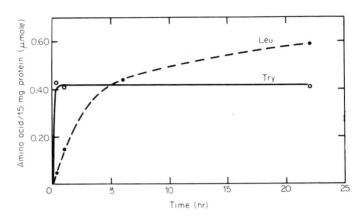

FIGURE 12. Action of carboxypeptidase A on $\alpha_{s,1}$-casein B. Weight ratio of enzyme to substrate, 1:100; pH 7.8 (unbuffered system); T = 37°C. Aliquot of digest equivalent to 13.0 mg of protein (moisture-free basis) analyzed at indicated times (Kalan et al., 1964).

not consider a third chain terminating in tyrosine likely as suggested by Waugh et al. (1962). The molecular weight values for $\alpha_{s,1,2}$-casein, calculated from C-terminal tryptophan released, lie in the region of 27,300 daltons, while the values of Kalan et al. (1964), calculated on the same basis, were 26,100, 31,300 and 31,300 daltons for $\alpha_{s,1}$-A, B, and C, respectively. The latter values appear too high compared with physical determinations on the molecular weights of the $\alpha_{s,1}$-casein variants (see below) and the more recent chemical value of 23,600 given by Grosclaude et al. (1970).

E. PHYSICAL PROPERTIES OF $\alpha_{s,1}$-CASEINS

1. Association of $\alpha_{s,1}$-Caseins

Thompson (1964) and Schmidt and Payens (1963) have reported that the tendency of $\alpha_{s,1}$-B and C to associate is different. This situation is reminiscent of β-lactoglobulins A and B from cow milk which, like $\alpha_{s,1}$-caseins, differ only slightly in amino acid composition. With both protein classes, carboxyl-containing amino acids are involved in the substitution.

Payens and his associates in The Netherlands have been pioneers in the study of the association behavior of the $\alpha_{s,1}$-caseins. Some of the more well-defined association characteristics of $\alpha_{s,1}$-casein C can be seen in the concentration-dependence of the weight-average molecular weight ($\bar{M}_w$)

shown in Fig. 13 (see also Chapter 7, Volume I). Figure 13 may be interpreted as follows:

(1) The tendency to associate is enhanced by increasing concentration; the particle weight increases rapidly with increasing concentration.

(2) With increasing temperature the particle weight also increases, indicating that the association is endothermic.

(3) At zero concentration the particle weight of the subunit approaches 113,000 daltons.

(4) The plots of $\bar{M}_w$ vs. concentration are essentially linear. Payens and Schmidt (1966) have interpreted these results as follows: (a) The subunits of $\alpha_{s,1}$-C undergo a number of rapidly consecutive association steps at least up to hexamerization. (b) The consecutive association constants are all equal.

Table XIII, representing the thermodynamic parameters of the association, suggests that the subunits are held together by hydrophobic

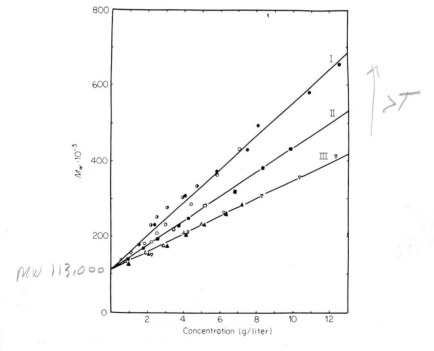

FIGURE 13. Concentration dependence of the weight-average molecular weight of $\alpha_{s,1}$-casein C at pH 6.5 and different concentrations (Payens, 1966).

TABLE XIII

THERMODYNAMIC PARAMETERS OF THE CONSECUTIVE ASSOCIATION STEPS OF $\alpha_{s,1}$-CASEIN C[a]

Equilibrium	ΔH (kcal/mole)	ΔG (kcal/mole)	ΔS (eu)
Dimerization	+6.5	−3.2	+34
Trimerization	+13.2	−3.1	+58
Tetramerization	−0.6	−3.2	+9
Pentamerization	+3.3	−3.1	+23

[a] At pH 6.4; Payens (1966).

bonding. Interestingly, $\alpha_{s,1}$-B possesses similar behavior to C except that the molecular weight of the polymerizing unit of B approaches 87,500 daltons instead of 113,000 daltons or a difference of the monomeric weight of a single subunit, 27,000 daltons.

2. Molecular Weights

Neilsen and Lillevik (1957) reported the molecular weight of whole α-casein to be in the range of 27,000 daltons by osmotic pressure measurements in the urea, whereas Burk and Greenberg (1930) reported a value of 33,600 daltons for whole casein. Melnychyn and Wolcott (1965), also employing osmotic pressure measurements in urea, observed a value of 24,000 daltons for $\alpha_{s,1}$-casein.

Dreizen et al. (1962) reported a value of 27,000 ± 1000 daltons for $\alpha_{s,1,2}$-caseins. $\alpha_{s,1,2}$-Caseins were dissolved directly in pH 12.0 buffer; at ionic strengths above 0.3 M the light-scattering data were generally in agreement. Schmidt and Payens (1963) also studied the molecular weight of calcium-sensitive α-casein at elevated pH (12.2 in glycine buffer, 0.5 M) and arrived at an unusually low molecular weight of 16,500 daltons by sedimentation analyses. A more recent value of 23,000 daltons ± 2000 has been reported by these authors (Schmidt et al., 1967). Noelken (1967) confirmed the results of Schmidt and Payens but concluded that at pH 12.2, glycine buffer has the peculiar ability of slowly and irreversibly degrading $\alpha_{s,1}$-casein B. No such degradation occurred in sodium phosphate buffer at pH 12.3, the pH at which Dreizen et al. (1962) analyzed $\alpha_{s,1,2}$-caseins. Noelken, by sedimentation equilibrium, pH 7.0, 3 M guanidine hydrochloride, observed a molecular weight of 24,600 daltons for $\alpha_{s,1}$-B. Garnier (1967) reported 24,300 daltons for this variant. By light-scattering studies in anhydrous formic acid, chloroethanol, and 26%

aqueous methanol, Swaisgood and Timasheff (1968) established a molecular weight of 26,900 daltons for $\alpha_{s,1}$-C. This value coincides with that of Dreizen et al. (1962) for $\alpha_{s,1,2}$- but is about 10% higher than the value of 24,600 daltons for $\alpha_{s,1}$-B by Noelken (1967).

Most of the data to date on the molecular weight of $\alpha_{s,1}$-caseins, including physical (light-scattering, sedimentation equilibrium, and osmotic pressure measurements) and chemical data (end-group analysis and molecular weights calculated from amino acid analyses), is consistent with molecular weights of the $\alpha_{s,1}$-variants (excluding A, which appears to be about 1000 daltons less than B) in the range of 24,000 to 29,000 daltons. Molecular weights calculated from end groups range from 27,000 to 30,000 daltons. The general agreement using a variety of techniques in a wide range of aqueous and nonaqueous solvent systems is gratifying.

3. *Stabilization of $\alpha_{s,1}$-Caseins*

Waugh and von Hippel (1956) and Waugh (1958) first demonstrated (a) that α_s-casein and κ-casein form a stoichiometric complex (weight ratio of 4) as demonstrated by sedimentation analyses and (b) that in the presence of calcium(II), α_s-casein is protected from precipitation by κ-casein. These observations are fundamental to a concept of the structure of micellar casein although some have suggested that the desired α_s-κ weight ratio should be unity. Perhaps the reason for this observed difference is that current methods of preparation yield purer preparations than the initial preparations used by Waugh and his co-workers. Noble and Waugh (1965) have suggested that the $\alpha_{s,1}/\kappa$ interaction "occurs spontaneously at 37°, requires pretreatment with urea or high pH to occur at 20°, and occurs neither spontaneously nor after such pretreatment at temperatures near 5°!" The interaction occurs at unity (1:1 weight ratio of $\alpha_{s,1}/\kappa$) with an s_{20} near 7.5 S (Garnier et al., 1964b). For a complete discussion see Chapter 9.

The characteristic solubilization of $\alpha_{s,1}$- by κ-casein in the presence of calcium(II) is shown in Fig. 14, where the three $\alpha_{s,1}$-variants (A, B, and C) are shown. Similar stabilization data have been reported by Zittle (1961), Thompson and Pepper (1962), Pepper and Thompson (1963), Schmidt et al. (1966), and Noble and Waugh (1965). The essential feature is that all of the stabilization ratios approach a value of 10. $\alpha_{s,1}$-Casein A (Fig. 14) is stabilized more fully at lower ratios of $\alpha_{s,1}/\kappa$, a feature characteristic of this variant.

Several factors could materially affect the weight ratio of 10. One might expect that $\alpha_{s,1}$-casein polymorphic forms would have an effect on this ratio; they do not, at least in model systems, as shown by Schmidt et al.

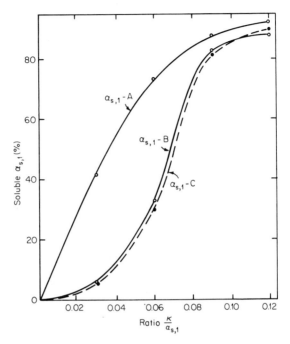

FIGURE 14. Stabilization curves for $\alpha_{s,1}$-caseins A, B, and C obtained by the method of Zittle (1961). From Thompson and Kiddy (1964).

(1966). Further, Thompson and Pepper (1962) showed that the removal of sialic acid from κ-casein did not materially affect its ability to stabilize $\alpha_{s,1}$-casein, nor did the removal of κ-casein phosphate affect this property (Pepper and Thompson, 1963). However, when phosphate residues were removed from $\alpha_{s,1}$-casein, Pepper and Thompson (1963) showed that κ-casein could no longer stabilize this component at the usual weight ratios of $\alpha_{s,1}/\kappa$. These authors interpreted this observation to mean that calcium bound to organic phosphate contributed materially to the strength of the complex (Yamauchi et al., 1967). Assuming that all phosphate groups in $\alpha_{s,1}$-caseins are available and in the form of monoesters, approximately 11 atoms of calcium would be bound per molecule of 28,000 daltons molecular weight. Certainly other functional groups, particularly COO$^-$, could bind additional Ca^{2+}.

4. Characteristics of $\alpha_{s,1}$-Casein A

General agreement exists as to the insolubility of $\alpha_{s,1}$ caseins B and C in 0.40 M calcium chloride at temperatures of 1°–35°C and probably above. β-Caseins remain soluble to about 18°C, abruptly precipitate and can be redissolved by cooling. κ-Casein, however, is readily soluble in calcium chloride, which certainly differentiates it from either an $\alpha_{s,1}$- or β-casein. $\alpha_{s,1}$-Casein A is a peculiar protein, as has already been reported; in addition to its unusual amino acid composition it is soluble in 0.40 M CaCl$_2$ at 1°C, a condition which persists to 33°C, at which point the protein precipitates. It can be readily dissolved at 28°C. Of further interest is the fact that equal mixtures of $\alpha_{s,1}$-A and B remain soluble at 1°C, B precipitates at 18°C and the whole mixture is insoluble at 33°C. Examination of the supernatant from a mixture of A and B at 25°C, following centrifugation, shows that a portion of the A variant is soluble, an important observation because of the infrequent occurrence of homozygous $\alpha_{s,1}$-A cows. Additionally, we have noted that $\alpha_{s,1}$-A and κ-casein (3:1 weight ratio) do not form micelles characteristic of $\alpha_{s,1}$-C/κ-mixtures when examined by electron microscopy.

F. Nature of Organic Phosphates in Caseins

Much speculation and controversy exists as to the nature of phosphorus bonds, which are characteristic of the caseins. Perlmann (1954a, 1955) dephosphorylated α-casein and reported that its phosphorus was present as 40% phosphomonoester (R-O-P), 40% as phosphodiester (R-N-P-O-R) and 20% as pyrophosphate. She also reported (Perlmann, 1954b) that at least 70% of the phosphorus of β-casein was present as phosphodiester (R-O-P-O-R). On the basis of the phosphate liberated from whole casein, α-casein, and three different casein fractions obtained by proteolytic digestion by two different enzyme preparations, Thoai et al. (1954) concluded that phosphomonoesters and phosphodiesters were present in whole casein.

Considerable doubt has subsequently been cast on the validity of these suggestions. Sampath Kumar et al. (1957) treated α-, β-, and unfractionated casein with a spleen phosphatase and found identical pH maxima and Michaelis constants for the three fractions; 80% of the phosphorus was released as orthophosphate after 3 hr of incubation. They concluded that the phosphorus bond was the same for all fractions studied. Additionally, Sundararajan and Sarma (1957) found that two different vegetable phosphatases from ground nut and soya beans liberated phosphorus from the caseins in an identical fashion, although the rate of phosphorus liberated depended on the enzyme employed. Nevertheless, these findings are in accord with the hypothesis of a single type of linkage in whole α- and β-caseins.

Evidence as to the exact nature of the phosphorus linkage has come from several workers. Among the first to propose the hypothesis of a monoester type of linkage were Hipp et al. (1952a), who measured titration curves of α- and β-casein. Based on the perfect agreement of the number of ionizable groups liberated at two different pH values with the number of ionizable groups calculated from amino acid composition and phosphorus content, it was concluded that such agreement between two independent sets of data is possible only if the phosphorus exists in one form, namely monoester. Isolation of a phosphopeptide by Peterson et al. (1958) from β-casein containing phosphorus and hydroxyamino acids in a molar ratio of 1:1, and containing 75% of the phosphorus originally present in the fraction, was strong evidence that all phosphorus exists as monoester (see also Schormuller et al., 1966).

Hofman (1955) provided further evidence in favor of monoesters from enzymatic and chemical studies. An extensive study by Kalan and Telka (1959) provided suggestive evidence of the presence of similar phosphate linkages in α- and β-caseins and whole caseins. Orthophosphate was liberated from these caseins by three different phosphatases; the rate and amount of phosphorus released was not influenced by prior incubation with phosphodiesterase or pyrophosphatase. These authors appear to favor the existence of phosphomonoesters in casein. While Zittle and Bingham (1959) and Bingham and Zittle (1963) have shown that alkaline and acid phosphatases act on caseins to release inorganic phosphorus, the nature of the phosphate linkage was not considered by these workers. Likewise, although Pepper and Thompson (1963) showed that 96% of the phosphorus of α_s- ($\alpha_{s,1}$-) casein could be liberated by spleen phosphoprotein phosphatase, they made no mention of the type of phosphate bond involved. It is of interest to note from the studies of Revel and Racker (1960) that phosphoprotein phosphatase releases phosphorus from whole casein at a rapid rate, that phosphoserine and phosphothreonine are attacked at 1% of the rate of whole casein, but that the rate increases on phosphoserine-containing peptides and appears to be influenced by the chain length of the peptide.

Ho and Kurland (1966) have reported that P nuclear magnetic resonance data on α_s- ($\alpha_{s,1}$-) casein is consistent with disubstituted pyrophosphate or phosphodiester compounds rather than with phosphomonoester compound. These data agree with the earlier suggestions of Perlmann (1955), which were seemingly in doubt. Some of the interpretations of Ho and Kurland (1966), however, are difficult to reconcile in view of our current knowledge of the composition of $\alpha_{s,1}$-casein. First, they refer to the earlier work of Ho and Waugh (1965), who reported good agreement between the number of charged groups of $\alpha_{s,1}$-casein as determined by titration and amino acid data. Side-chain carboxyl groups were 40 ± 3 per molecule of 27,300 daltons from titration data, but 46 ± 3 by amino acid analyses. They further point out that this difference is minimized or eliminated if one considers that the phosphorus is in the form of diester or pyrophosphate. A more logical explanation is that a discrepancy in amino acid analyses has occurred, especially in side-chain carboxyl groups (aspartic acid), as shown by the amino acid analyses reported by Gordon et al. (1965) and de Koning et al. (1965). Second, the authors refer to the studies of Kalan et al. (1964) and Waugh et al. (1962) on the C-terminal amino acids of $\alpha_{s,1}$-caseins. Kalan et al. (1964) favored the conclusion that $\alpha_{s,1}$-A, B, and C were single-chained polypeptides with a probable sequence of Leu-Leu-Trp. Although they did not rule out the possibility of two chains, one terminating in tryptophan, the other in leucine, a logical interpretation of the data seemed to be a sequential one. Ho et al. (1969) have concluded in a later paper that the phosphate groups are in monoester (serine) linkages and Grosclaude et al. (1970) have shown that 8 of the 9 phosphate groups are in an acidic peptide (see also Chapters 9, 18).

III. β-Caseins

A. NOMENCLATURE

Aschaffenburg (1961) termed the three β-casein variants he discovered as A, B, and C. This discovery, confirmed by Thompson et al. (1964) in American dairy cattle, prompted them as well as the Nomenclature Committee of the American Dairy Science Association to adopt Aschaffenburg's nomenclature and to ascribe relative mobility values for the variants on starch- and polyacrylamide-gel electrophoresis (see Table I, Section II.A). However, Peterson and Kopfler (1966) clearly demonstrated by acid gel electrophoresis that β-casein A existed in three forms which they termed C, D, and E. The difficulty with this proposal was that C, by alkaline gel electrophoresis, became A, while B remained B. Kiddy et al. (1966) reversed the earlier nomenclature proposal and termed the caseins C, B, A^1, A^2, and A^3 by acid gel electrophoresis. Clearly, then, C and B by acid gel electrophoresis are identical to C and B by alkaline gel electrophoresis. Aschaffenburg (1966) prefers to term the new β-caseins A1, A2, and A3. Which nomenclature will gain acceptance will depend, in part at least, on the usage, but either seems satisfactory.

The phenotyping of β-casein D (*Bos indicus*) is best accomplished by alkaline gel electrophoresis (Aschaffenburg et al., 1968) instead of acid gel electrophoresis (Peterson and Kopfler, 1966). This variant migrates between B and C at pH 9.1, while at pH 2.8 it migrates with the same mobility as β-B.

B. PREPARATION

Warner (1944) reported the first successful chemical fractionation of β-casein as shown by free-boundary electrophoresis. β-Casein was prepared by warming the filtrate, obtained from α-casein fractionation, to room temperature and adjusting the pH to 4.9, at which point crude β-casein precipitated. This precipitate was dissolved at pH 6.0, chilled to 2°C, and cold 0.01 N HCl was added. The precipitate formed at this pH was filtered off and dissolved with NaOH. The solution was then brought to pH 4.5 at 2°C where chiefly α-casein was precipitated. After warming to room temperature, β-casein precipitated and was further purified by repeated low pH and temperature fractionation. β-Casein prepared by this method showed an analysis of 15.53% (w/w) nitrogen and 0.61% (w/w) phosphorus.

While the method of Warner is acceptable for obtaining β-casein, large amounts of this protein are obtained with difficulty. Two methods, devised by Hipp et al. (1952a), have been reported. The first method, alcohol fractionation, is based on the solubility of casein fractions in 50% (v/v) alcohol in the presence of ammonium acetate at various pH values. Three fractions of casein (termed A, B, and C) are obtained; fraction C, which contains 44% (w/w) α-casein, 50% β-casein, and 6% γ-casein, is used as the starting source for β-casein. This fraction is dissolved with dilute NaOH, cooled to 2°C, and HCl is added to pH 4.5, at which point most of the α-casein precipitates. When the clear filtrate is warmed to 32°C, β-casein precipitates while γ- remains in solution. The β-casein may be further purified by extraction at low pH and temperature. It is finally purified by reprecipitation five times at pH 4.3 and 2°C. β-Casein prepared by this method contains 15.33% nitrogen and 0.55% phosphorus and has a relative mobility of -3.05×10^{-5} cm^2/V/sec, at pH 8.4, $I = 0.10$ M in Veronal buffer.

The second method, urea fractionation, is widely used today as an excellent procedure for obtaining crude β-casein. Whole casein is dissolved in 6.6 M urea; after removing α-casein by diluting to 4.63 M urea, the solution is again diluted to 3.3 M, where β-, γ-, and other casein components remain in solution. Upon dilution to 1.7 M urea, crude β-casein, which may be further purified by recycling from 3.3 M urea, is obtained. The β-casein is higher in nitrogen (15.47%) and phosphorus (0.64%) than β-casein obtained by the aqueous alcohol method and is compositionally very similar to the β-casein of Warner (1944). Urea-fractionated β-casein has a mobility of -3.15×10^{-5} cm^2/V/sec. Aschaffenburg (1963b) has simplified the urea method of Hipp et al. (1952a,b). Crude whole casein is dissolved at pH 7.5 in 3.3 M urea, and the pH adjusted to pH 4.6, whereupon the bulk of the protein insoluble ($α_{s,1}$- and κ-) in 3.3 M urea precipitates. After adjusting the pH of the 3.3 M urea supernatant to 4.9, followed by dilution to about 1 M urea and warming to 30°C, β-casein flocculates. The crude β-casein is dispersed in 3.3 M urea at pH 7.5, adjusted to pH 4.6 at 37°C and centrifuged. Finally, the supernatant is adjusted to pH 4.9 diluted to 1 M urea at 30°C and the β-casein is centrifuged. All of the β-casein variants can be prepared by the urea-fractionation methods.

Although the above three methods yield β-caseins of good purity as judged by moving-boundary electrophoresis, Wake and Baldwin (1961) have demonstrated that they are contaminated to varying degrees with impurities, as shown by starch-gel electrophoresis. Realizing this fact, Garnier et al. (1964a) purified β-casein variants by DEAE-cellulose–urea chromatography employing the method of Ribadeau-Dumas (1961). Crude β-casein was obtained by a linear NaCl elution of components from DEAE-cellulose, in 3.3 M urea–imidazole buffer at pH 7.0. The crude β-casein was further purified by rechromatography at pH 8.2 in an 0.01 M tris buffer. β-Casein A obtained by this method contained 0.52% phosphorus but a decidedly low nitrogen content of 14.4% as compared with the values of other researchers (Table XIV). The French workers reported their fractions to be immunologically pure. Groves et al. (1962)

TABLE XIV

SOME PROPERTIES OF β-CASEIN FRACTIONS

Fraction	Nitrogen (g/100 g)	Phosphorus (g/100 g)	P/N ratio	$A_{1\,cm}^{1g/d\,l}$	Reference to preparation
β-Casein	15.53	0.61	0.0397		Warner (1944)
	15.33	0.55	0.0359		Hipp et al. (1952); alcohol
	15.47	0.64	0.4140		Hipp et al. (1952); urea
	15.35	0.48	0.0313		Groves et al. (1962)
β-Casein A	14.40	0.52	0.0360	4.6[a]	Garnier et al. (1964a)
	15.18	0.59	0.0389	4.6[b]	Thompson and Pepper (1964a,b)
β-Casein B	15.33	0.57	0.0372	4.7[b]	Thompson and Pepper (1964a,b)
β-Casein C	15.35	0.50	0.0324	4.5[b]	Thompson and Pepper (1964a,b)

[a] Wavelength is 278 nm.
[b] Wavelength is 280 nm.

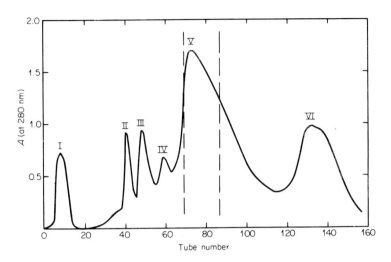

FIGURE 15. Elution diagram of crude β-casein A from DEAE-cellulose, pH 7.0, 3.3 M urea in the presence of mercaptoethanol (Thompson, 1966).

had earlier prepared β- and γ-caseins by column chromatography at pH 8.3, sodium phosphate buffer, in the absence of urea, and obtained β-casein of excellent purity as shown by both acid and alkaline starch-gel electrophoresis.

Thompson and Pepper (1964b) reported the purification of β-caseins A, B, and C by DEAE-cellulose–urea chromatography of fractions obtained by the method of Hipp et al. (1952a,b). While β-caseins of apparent high purity were obtained as disclosed by zonal electrophoresis, the incorporation of mercaptoethanol to reduce κ-casein disulfide revealed that the β-casein preparations were not as pure as originally suspected; they contained $\sim$5% κ-casein. Thompson (1966), therefore, modified the column purification of β-casein variants by adding 2-mercaptoethanol to the protein solution applied to the DEAE-cellulose column, thereby insuring the removal of κ-casein. The elution profile of β-casein A is shown in Fig. 15, while the purity of the eluted β-casein is demonstrated in the polyacrylamide-gel electrophoresis patterns shown in Fig. 16. The usefulness of DEAE-cellulose–urea column chromatography in separating components from whole casein when mercaptoethanol is incorporated into the system is shown in Figs. 17 and 18. The elution profile and gel electrophoresis analyses of a whole casein phenotyped $\alpha_{s,1}$-B, β-AC, and κ-B is also shown. Obviously the separation of κ-casein from other casein components is complete, while the β-casein AC separation is excellent.

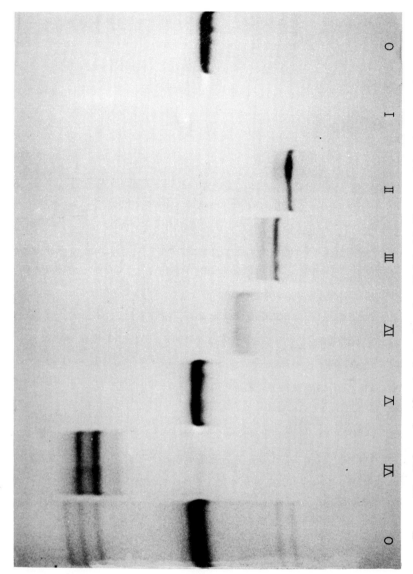

FIGURE 16. Polyacrylamide-gel electrophoresis, pH 9.1, 4.5 M urea, of fractions shown in Fig. 15.

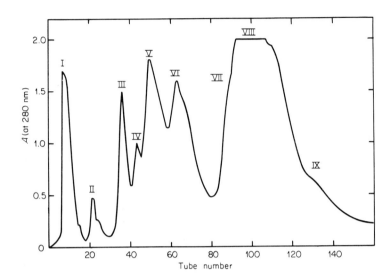

FIGURE 17. Elution diagram of whole casein typed $\alpha_{s,1}$-B, β-AC, κ-B from DEAE-cellulose, pH 7.0, 3.3 M urea in the presence of mercaptoethanol (Thompson, 1966).

The separation of two very closely related β-casein variants, B and D (*Bos indicus*) is shown in Figs. 19 and 20. Clearly, neither β-B nor β-D are contaminated with each other. β-Caseins prepared by column chromatography in the presence of 2-mercaptoethanol appear to be pure by immunoelectrophoresis (Thompson, 1966).

For the preparation of crude β-caseins, the methods of Warner (1944) and Hipp et al. (1952a,b), both aqueous alcohol and urea, and modification of Hipp's urea fractionation method by Aschaffenburg (1963a,b,c) are acceptable. Since both of the chromatographic purification procedures of Garnier et al. (1964a) and Thompson (1966) yield immunologically pure β-caseins, both methods may be deemed satisfactory and will yield essentially the same product.

C. COMPOSITION AND STRUCTURE

1. *End-Group Analysis*

In 1953, Mellon et al. (1953) determined that β-casein from a pooled milk source contained N-terminal arginine. In addition, they also reported the presence of 2.4 mole of lysine per 100,000 g of protein as an N-terminal amino acid. Kalan et al. (1965), using the 2,4-dinitrofluoro-

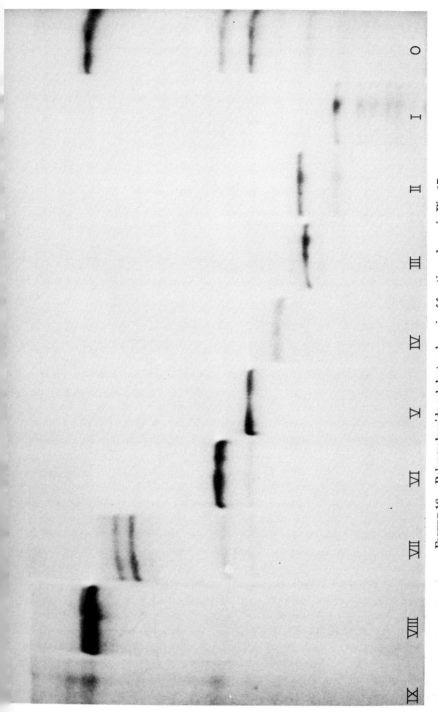

FIGURE 18. Polyacrylamide-gel electrophoresis of fractions shown in Fig. 17.

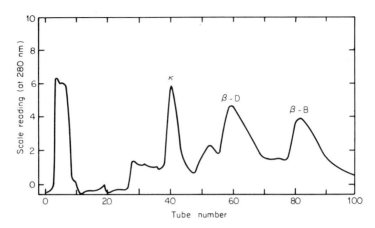

FIGURE 19. Elution diagram of whole casein ($\alpha_{s,1}$-region not shown) typed $\alpha_{s,1}$-BC, β-BD, κ-B from DEAE-cellulose, pH 7.0, 3.3 M urea in the presence of mercaptoethanol (Thompson et al., 1969b).

benzene method of Sanger, confirmed the presence of N-terminal arginine in β-caseins A, B, and C but could not detect the presence of N-terminal lysine. The presence of N-terminal arginine was confirmed by the Edman degradation procedure; the second step of this procedure strongly suggested that glutamic acid was adjacent to arginine in the N-terminal sequence of all three β-casein variants. More recently Peterson et al. (1966) confirmed that arginine was indeed N-terminal in the β-caseins studied, which included A, B, and C.

Excellent agreement has also been observed in the C-terminal sequence of β-caseins. Dresdner and Waugh (1964) reported the presence of one mole of valine and two moles of isoleucine per mole (24,000 daltons) of β-casein(s) but did not indicate their sequence in the molecule. Apparently they were working with β-caseins A, AB, and AC because genetic typing on starch gel is comparable with paper electrophoresis. Kalan et al. (1965) in an extensive examination of the C-terminal sequence of β-caseins A, B, and C concluded that the most probable sequence was Ile-Ile-Val after treatment of the protein with carboxypeptides A (Fig. 21). Valine was most rapidly released followed very closely by isoleucine; thus, a possible sequence of Ile-Val-Ile was not ruled out, although the former sequence was more probable. On the basis of C-terminal valine released, Kalan et al. (1965) determined the molecular weights of β-caseins A, B, and C to be in the expected range of 24,000–25,000 daltons. Similarly Peterson et al. (1966) suggested molecular weight values of β-caseins to lie in the region 23,471 to 24,113 daltons based on residue weights of amino acids, and they concluded that valine was the C-terminal amino acid of the β-casein variants.

Two interesting features emerge from the study of end groups of β-caseins. First, a C-terminal sequence of valine and isoleucine accounts for a very hydrophobic region in the β-casein molecule, a feature considered later. Second, and perhaps equally important, is the possibility that the phosphopeptide of Peterson *et al.* (1958) may actually exist in the

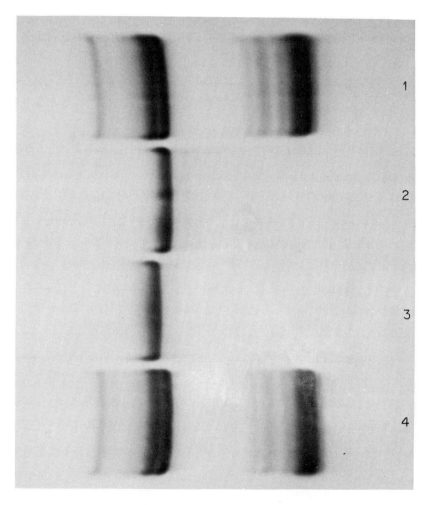

FIGURE 20. Polyacrylamide-gel electrophoresis of β-caseins shown in Fig. 19. Patterns 1 and 4 represent initial casein; patterns 2 and 3 are β-B and β-D, respectively.

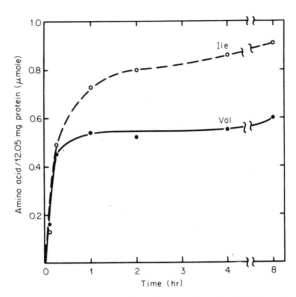

FIGURE 21. Action of carboxypeptidase A on β-casein A. Weight ratio of enzyme to substrate, 1:100; pH 8.2 (unbuffered system); T = 37°C. Aliquot of digest equivalent to 12.05 mg protein (moisture-free basis) analyzed at indicated times (Kalan et al., 1965).

N-terminal region of the molecule since both its N-terminal and C-terminal residues are arginine. All of the phosphorus of β-casein exists in the phosphopeptide. If confirmed that the phosphopeptide is in the N-terminal sequence, then β-casein would possess a strongly hydrophilic N-terminal end, while the C-terminal sequence would contribute a hydrophobic portion.

2. Amino Acid Composition

The availability of the genetic variants of β-caseins, from cows homozygous for a particular mutant, has prompted investigations into the amino acid substitutions responsible for the observed differences in electrophoretic mobility of these proteins. A careful inspection of Table XV shows a compilation of amino acid analyses of the β-casein variants to date. Unfortunately, following the original analyses of β-casein A it was found by acid gel electrophoresis to exist in at least three forms, A^1, A^2, and A^3 (Peterson and Kopfler, 1966; Kiddy et al., 1966). Therefore, the results of the analyses of β-A may be spurious. For example (Table XV), the group at Jouy-en-Josas found 5.5 residues of histidine while their associates in Ede found 6.0. Obviously the two laboratories were not analyzing the same variant; the European group analyzed a mixture of β-A^1 (6.0 histidine) and β-A^2 (5.0 histidine). β-Casein A^3 (not shown)

TABLE XV

Amino Acid Composition of β-Casein Variants

Amino acid	β-Casein A					β-Casein B				β-Casein C			β-Casein D (Zebu)
	Jouy[a]	Ede[a]	A¹ [b]	A¹ [c,d]	A²⁻² [b]	Jouy[a]	Ede[a]	Peterson[b]	Gordon[c,d] Zebu[c,d]	Jouy[a]	Ede[a]	Peterson[b]	Gordon[c,d]	Gordon[c,d]
Lys	11.3	10.9	10.6	11	11.0	11.2	10.9	11.1	11	11	12.6	11.8	11.6	12	12
His	5.5	6.0	5.9	6	5.0	6.0	5.9	5.9	6	6	6.2	6.0	6.2	6	5
Arg	4.0	4.0	3.9	4	4.2	4.8	5.1	4.6	5	5	3.8	3.9	4.1	4	4
Asp	9.1	9.7	9.2	10	9.8	9.0	9.2	9.4	9	10	9.0	9.4	9.2	10	10
Thr	8.7	9.2	9.1	10	9.2	8.9	9.8	9.1	9	10	9.3	9.0	9.1	10	11
Ser	14.9	15.3	15.0	15	14.7	14.0	14.9	13.9	14	15	16.2	13.5	14.7	15	16
Glu	39.5	40.2	39.7	39	38.0	39.0	38.7	39.6	40	37	38.5	40.1	37.9	38	37
Pro	33.4	33.3	33.7	35	33.3	33.7	32.4	33.1	36	33	33.1	34.1	33.7	35	34
Gly	5.4	5.2	5.0	5	5.0	5.1	5.3	5.0	5	5	5.4	5.2	5.0	5	5
Ala	5.2	5.2	5.0	6	5.4	5.2	6.1	5.4	5	6	5.1	5.3	5.2	7	7
Val	20.0	19.1	18.7	19	18.3	19.8	18.7	18.9	20	17–18	19.5	19.4	17.9	19	18
Met	6.2	6.0	5.9	6	5.8	6.3	6.2	5.9	6	5	5.8	5.9	5.4	6	5
Ile	9.9	9.7	9.4	10	9.5	9.8	9.9	9.9	10	10	9.6	10.0	9.5	10	10
Leu	22.3	22.1	21.3	21	21.3	22.1	21.4	21.3	23	20	22.0	21.8	21.3	21	20
Typ	3.9	3.9	3.9	5	4.0	4.3	4.5	3.9	4	4	3.9	3.9	3.8	5	4
Phe	9.0	9.1	8.8	9	8.8	9.1	8.9	8.9	9	8	8.8	8.0	8.7	9	8
Try	1.3	1.0	1.0	1	1.0	1.1	1.1	1.0	1	1	1.0	1.1	1.0	1	1

[a] Pion et al. (1965).
[b] Peterson et al. (1966).
[c] Thompson et al. (1969b).
[d] Rounded off to nearest integer (tentative data).

possesses but 4 histidine residues per mole (24,000 daltons); thus the A^1, A^2, and A^3 variants contain 6, 5, and 4 residues of histidine, respectively. In general, the analyses of β-casein A(s) are in good agreement, but assignment of specific triplet coding to these variants as well as to β-B, C, and D seems difficult at this time.

Agreement among values reported for β-casein B (*Bos taurus*) are, for the part, better than for β-A, which should be expected. In doubt, however, are the exact residue numbers for proline, glutamic acid, leucine and valine. But it is certain that β-casein B possesses 11 lysine, 6 histidine, and 5 arginine. β-Casein C possesses 12 lysine, 6 histidine, and 4 arginine. Thompson and Gordon (1968) observed 7.0 alanine residues in β-C, a value considerably higher than the 5.0 reported by others. They also observed a higher number (6) in β-A^1. Additionally they observed 5, instead of 4, residues of tyrosine in β-C.

Interesting comparisons can be made between β-casein B of Jersey (*Bos taurus*) with β-casein B of Zebu (*Bos indicus*) (Table XV). While the content of basic amino acids is identical from these two breeds, noteworthy differences in methionine, phenylalanine, glutamic acid, threonine, serine, and alanine are apparent. These differences do not manifest themselves in changes in electrophoretic mobility in either alkaline or acid gel electrophoresis. This is significant when comparisons of supposedly identical proteins are being made, especially when tracing the origin of breeds of cattle. Fingerprinting of chymotrypsin digests of β-B from *Bos taurus* and *Bos indicus* reveals some differences in the peptide maps. Apparently, β-B of Western breeds is the result of multiple substitutions of β-B from Eastern breeds with no apparent effect on electrophoretic behavior, an observation deserving much caution in assigning identity on the basis of the criterion of R_m. Interesting, however, is the fact that chymotrypsin fingerprinting of Zebu $\alpha_{s,1}$-C and β-A^2 are identical with their counterparts in Western breeds!

Zebu β-casein D, observed on the subcontinent of India and also in East Africa, is more similar to β-C than to either the B or A variants. The only significant differences between the D and C variants are the following: one less histidine in D, one less methionine and possibly one less glutamic acid. It seems more reasonable to assume that β-C is a mutation of β-D involving multiple substitutions than to assume that it is a mutation of β-B involving an even more complex multiple substitution pattern.

D. PHYSICAL PROPERTIES

β-Casein must be regarded as the most interesting of all the caseins from a thermodynamic point of view. Sullivan *et al.* (1955) were the first to consider seriously the physical behavior of β-casein. They showed this protein to form aggregates at room temperature while sedimentation below 15°C gives an s_{20} of 1.57 S and a molecular weight of 24,100 daltons. Von Hippel and Waugh (1955) have also observed the tendency of β-casein to associate with rising temperature. The molecular weight given by sedimentation is in agreement with that obtained with viscosity measurements (D_{20} = 6.05 × 10^{-7} cm²/sec) at low temperature. Sullivan et al. (1955) reported $\bar{v}_{20}$ of 0.739 ml/g and f/f_o of 1.83 and suggested that β-casein is a rodlike molecule.

An extensive investigation on the temperature-dependent association behavior of β-casein has been reported by Payens and van Markwijk (1963). Measurements of molecular weight at 4°C show the minimum molecular weight of the monomer to be 25,000 daltons. At 8.5° and 13.5°C, following prior equilibration at these temperatures, threadlike polymers are formed which are influenced to a large extent by concentration (Fig. 22). The ultracentrifugal patterns for β-casein at 8.5°C are shown in Fig. 23; the effect of concentration is evident. The rate of association appears

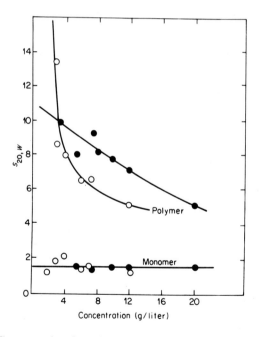

FIGURE 22. Concentration dependence of the sedimentation coefficient of β-casein monomers and polymers; ●—experiment at 8.5°C; ○—experiments at 13.5°C (Payens and van Markwijk, 1963).

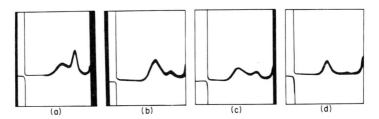

FIGURE 23. Typical sedimentation patterns of β-casein solutions at 8.5°C. Experimental conditions: barbituate buffer (pH 7.50; $I = 0.20$); Phywe air-driven ultracentrifuge; synthetic boundary cell at 45,000 rpm. (a) 2%, after 100 min; (b) 1%, after 65 min; (c) 0.8%, after 80 min; (d) 0.33%, after 25 min (Payens and van Markwijk, 1963).

low; at 8.5°C the degree of polymerization amounts to about 22 while at 13.5°C it is appreciably higher. This author (1968) has observed similar behavior of the three genetic variants of β-casein A, B, and C; however, β-casein C tends to associate more than A or B. Each variant, however, possesses an s_{20} of about 1.50 S at 4°C. Thompson et al. (1967b) have further shown that the removal of C-terminal valine and penultimate isoleucine residues results in the virtual inability of β-caseins to associate at 8.5°C. Apparently these hydrophobic amino acids are essential in the polymerization reaction.

Interesting observations on the aggregation behavior of β-casein A has been reported by Hoagland (1966). Under normal conditions, i.e., pH 6.86, $I = 0.20$ in phosphate buffer, β-casein sediments with a major component of 9.3 S. However, succinylated β-casein did not form this fast-sedimenting peak, which is generally associated with this protein. Acetylated β-casein also had a lower S value than did untreated β-casein. Hoagland interpreted these findings to mean that an increase in net negative charge of the modified protein affected the aggregation behavior of β-casein. He further observed that succinylated and acylated β-caseins were not as sensitive to calcium ions as was untreated β-casein.

IV. General Considerations

In the foregoing pages the author has considered those areas of research in which considerable knowledge has been accumulated. The reader may well challenge the depth of knowledge, especially where protein–protein interactions are concerned. He may justly propose, for example, that to know more about the bonding sites involved in interactions, more must be known regarding the primary structure of the caseins. While certain segments of casein molecules are being examined for primary structure it is doubtful, at the present pace of research on caseins, that the full sequence will be worked out in the forseeable future.

Additionally, some of the current methods of preparation are fraught with problems, particularly in the use of reagents whose action on secondary structure is unknown. The early methods of fractionation of casein using calcium(II) used by von Hippel and Waugh (1955) and Waugh and von Hippel (1956) deserve more attention than is given by many researchers. Along with simplicity, they utilize a normal constituent of milk, calcium(II), which one would doubt affects any physical properties of these proteins.

We are admonished to reevaluate our definitions of the many casein components and to base our definitions on a clear understanding of the solubility of given components in the presence of calcium at specific temperatures and pH conditions. Evidently we have tacitly assumed, by qualitative observations, that caseins behave in a given way—they often do not on a quantitative basis. Furthermore, a paucity of information exists concerning the physical chemistry of casein molecules, needed data which would contribute to more precise definitions of the caseins. The author feels that the research group of Professor David Waugh and his colleagues at Massachusetts Institute of Technology has taken the most deliberate and valuable approach to those factors which are necessary in the formation of casein micelles. Other laboratories may well pattern their programs along these lines of endeavor.

ACKNOWLEDGMENTS

The author thanks his wife, Virginia, for her efforts in preparing this manuscript. He also thanks those who have so generously contributed in supplying figures and helpful comments.

REFERENCES

Annan, W. D., and Manson, W. (1969). *J. Dairy Res.* **36**, 259.
Aschaffenburg, R. (1961). *Nature* **192**, 431.
Aschaffenburg, R. (1963a). *J. Dairy Res.* **30**, 251.
Aschaffenburg, R. (1963b). *J. Dairy Res.* **30**, 259.
Aschaffenburg, R. (1963c). *In* "Man and Cattle" (A. E. Mourant and F. E. Zeuner, eds.), pp. 50–54. Royal Anthropological Institute Occasional Paper No. 18, London.
Aschaffenburg, R. (1964). *Biochim. Biophys. Acta* **82**, 188.
Aschaffenburg, R. (1965). *J. Diary Sci.* **48**, 128.
Aschaffenburg, R. (1966). *J. Dairy Sci.* **49**, 1284.
Aschaffenburg, R., and Drewry, J. (1955). *Nature* **176**, 218.
Aschaffenburg, R., and Thompson, M.P. (1967). Unpublished data.

Aschaffenburg, R., and Thymann, M. (1965). *J. Dairy Sci.* **48**, 1524.
Aschaffenburg, R., Sen, A., and Thompson, M. P. (1968a). *Comp. Biochem. Physiol.* **25**, 177.
Aschaffenburg, R., Sen, A., and Thompson, M. P. (1968b). *Comp. Biochem. Physiol.* **27**, 621.
Bingham, E. W., and Zittle, C. A. (1963). *Arch. Biochem. Biophys.* **101**, 471.
Boyer, S. H., Rucknagel, D. L., Weatherall, D. J., and Watson-Williams, E. J. (1963). *Amer. J. Hum. Genet.* **15**, 438.
Bradley, T. B., Wohl, R. C., and Rieder, R. F. (1967). *Science* **157**, 1581.
Brunner, J. R., Ernstrom, C. A., Hollis, R. A., Larson, B. L., Whitney, R. McL., and Zittle, C. A. (1960). *J. Dairy Sci.* **43**, 901.
Burk, N. F., and Greenberg, D. M. (1930). *J. Biol. Chem.* **87**, 197.
Buvanendran, V. (1964). *Genet. Res.* **5**, 330.
Cuperlovic, M., Kovacs, G., and Thymann, M. (1964). "Annual Report, Royal Veterinarian College," pp. 107–111. Sterility Research Institute, Copenhagen.
de Koning, P. J., and van Rooijen, P. J. (1965). *Biochem. Biophys. Res. Commun.* **20**, 241.
de Koning, P. J., and van Rooijen, P. J. (1967) *Nature* **213**, 1028.
Dreizen, P., Noble, R. W., and Waugh, D. F. (1962). *J. Amer. Chem. Soc.* **84**, 4938.
Dresdner, G. W., and Waugh, D. F. (1964). *Fed. Proc., Fed. Amer. Soc. Exp. Biol.* **23**, 474.
El-Negoumy, A. M. (1966). *J. Dairy Sci.* **49**, 1461.
Farrell, H. M., Jr., Thompson, M. P., and Larsen, B. (1971). *J. Dairy Sci.* **54**, 423.
Ganguli, N. C., Sud, S. K., and Bhalerao, V. R. (1964). *Indian J. Diary Sci.* **19**, 119.
Garnier, J. (1967). *Biopolymers* **5**, 473.
Garnier, J., Ribadeau-Dumas, B., and Mocquot, G. (1964a). *J. Dairy Res.* **31**, 131.
Garnier, J., Yon, J., and Mocquot, G. (1964b). *Biochim. Biophys. Acta* **82**, 481.
Gehrke, C. W., Chun, P., and Oh, Y. H. (1966). *Separ. Sci.* **1**, 431.
Glasnak, V. (1968). *Comp. Biochem. Physiol.* **25**, 355.
Gordon, W. G., Basch, J. J., and Thompson, M. P. (1965). *J. Dairy Sci.* **48**, 1010.
Grosclaude, F., Garnier, J., Ribadeau-Dumas, B., and Jeunet, R. (1964). *Compt. Rend.* **259**, 1569.
Grosclaude, F., Pujolle, J., Garnier, J., and Ribadeau-Dumas, B. (1965). *Compt. Rend.* **261**, 5229.
Grosclaude, F., Pujolle, J., Garnier, J. and Ribadeau-Dumas, B. (1966) *Ann. Biol. Anim., Biochim. Biophys.* **6**, 215.
Grosclaude, F., Mercier, J.-C., and Ribadeau-Dumas, B. (1969). *Compt. Rend.* **268**, 3133.
Grosclaude, F., Mercier, J.-C., and Ribadeau-Dumas, B. (1970). *Eur. J. Biochem.* **14**, 98.
Groves, M. L., McMeekin, T. L., Hipp, N. J., and Gordon, W. G. (1962). *Biochim. Biophys. Acta* **57**, 197.
Hines, H. C., Kiddy, C. A., Brum, E. W., and Arave, C. W. (1968). *Genetics* **62**, 401.
Hipp, N. J., Groves, M. L., and McMeekin, T. L. (1952a). *J. Amer. Chem. Soc.* **74**, 4822.
Hipp, N. J., Groves, M. L., Custer, J. H., and McMeekin, T. L. (1952b). *J. Dairy Sci.* **35**, 272.
Ho, C., and Kurland, R. J. (1966). *J. Biol. Chem.* **241**, 3002.
Ho, C., and Waugh, D. F. (1965). *J. Amer. Chem. Soc.* **87**, 110.
Ho, C., Magnusson, J. A., Wilson, J. B., Magnusson, N. S., and Kurland, R. J. (1969). *Biochemistry* **8**, 2074.
Hoagland, P. D. (1966) *J. Dairy Sci.* **49**, 783.

Hoagland, P. D., Thompson, M. P., and Kalan, E. B. (1970). *J. Dairy Sci.* (in press).
Hofman, T. (1955). *Proc. 3rd Int. Biochem. Cong., Brussels, 1955* p. 26.
Hoogendoorn, M. P., Moxley, J. E., Hawes, R. O., and MacRae, H. F. (1969). *Can. J. Anim. Sci.* **49,** 331.
Jenness, R., Larson, B. L., McMeekin, T. L., Swanson, A. M., Witnah, C. H., and Whitney, R. McL. (1956) *J. Dairy Sci.* **39,** 536.
Jollès, P. (1966). *Angew Chem., Int. Ed. Engl.* **5,** 558.
Kalan, E. B., and Telka, M. (1959). *Arch. Biochem. Biophys.* **79,** 275.
Kalan, E. B., Gordon, W. G., Basch, J. J., and Townend, R. (1962). *Arch. Biochem. Biophys.* **96,** 376.
Kalan, E. B., Thompson, M. P., and Greenberg, R. (1964). *Arch. Biochem. Biophys.* **107,** 521.
Kalan, E. B., Thompson, M. P., Greenberg, R., and Pepper, L. (1965). *J. Dairy Sci.* **48,** 884.
Kalan, E. B., Greenberg, R., and Thompson, M. P. (1966). *Arch. Biochem. Biophys.* **115,** 468.
Kiddy, C. A., Johnston, J. O., and Thompson, M. P. (1964). *J. Dairy Sci.* **47,** 147.
Kiddy, C. A., Peterson, R. F., and Kopfler, F. C. (1966). *J. Dairy Sci.* **49,** 742.
King, J. W. B. (1966). Personal communication.
King, J. W. B., Aschaffenburg, R., Kiddy, C. A., and Thompson, M. P. (1965). *Nature* **206,** 324.
Linderstrøm-Lang, K., and Kodama, S. (1925). *C. R. Trav. Lab. Carlsberg* **16,** 48.
Lindqvist, B. (1963). *Dairy Sci. Abstr.* **25,** 257.
Long, J., van Winkle, Q., and Gould, I. A. (1958). *J. Dairy Sci.* **41,** 317.
McKenzie, H. A., and Wake, R. G. (1961). *Biochim. Biophys. Acta* **47,** 240.
McMeekin, T. L., Hipp, N. J., and Groves, M. L. (1959). *Arch. Biochem. Biophys.* **83,** 35.
Manson, W. (1961). *Arch. Biochem. Biophys.* **95,** 336.
Mellander, O. (1939). *Biochem. Z.* **300,** 240.
Mellon, E. F., Korn, A. H., and Hoover, S. R. (1953). *J. Amer. Chem. Soc.* **75,** 1675.
Melnychyn, P., and Wolcott, J. M. (1965). *J. Dairy Sci.* **48,** 780.
Mercier, J.-C., Grosclaude, F., and Ribadeau-Dumas, B. (1970). *Eur. J. Biochem.* **14,** 108.
Michalak, W. (1967). *J. Dairy Sci.* **50,** 1319.
Neelin, J. M. (1964). *J. Dairy Sci.* **47,** 506.
Nielsen, H. C., and Lillevik, H. A. (1957). *J. Dairy Sci.* **40,** 598.
Noble, R. W., and Waugh, D. F. (1965). *J. Amer. Chem. Soc.* **37,** 2236.
Noelken, M. (1967). *Biochim. Biophys. Acta* **140,** 537.
Payens, T. A. J. (1966). *J. Dairy Sci.* **49,** 1317.
Payens, T. A. J., and Schmidt, D. G. (1966). *Arch. Biochem. Biophys.* **115,** 136.
Payens, T. A. J., and van Markwijk, B. W. (1963). *Biochim. Biophys Acta* **71,** 517.
Pepper, L., and Thompson, M. P. (1963). *J. Dairy Sci.* **46,** 764.
Perlmann, G. E. (1954a). *Nature* **174,** 273.
Perlmann, G. E. (1954b). *Biochim. Biophys. Acta* **13,** 452.
Perlmann, G. E. (1955). *Advan. Protein Chem.* **10,** 1.
Peterson, R. F. (1963). *J. Dairy Sci.* **46,** 1136.
Peterson, R. F., and Kopfler, F. C. (1966). *Biochem. Biophys. Res. Commun.* **22,** 388.
Peterson, R. F., Nauman, L. W., and McMeekin, T. L. (1958). *J. Amer. Chem. Soc.* **80,** 95.
Peterson, R. F., Nauman, L. W., and Hamilton, D. F. (1966). *J. Dairy Sci.* **49,** 601.

Pion, R., Garnier, J., Ribadeau-Dumas, B., de Koning, P. J., and van Rooijen, P. J. (1965). *Biochem. Biophys. Res. Commun.* **20,** 246.
Revel, H. R., and Racker, E. (1960). *Biochim. Biophys. Acta* **43,** 465.
Ribadeau-Dumas, B. (1961). *Biochim. Biophys. Acta* **54,** 400.
Ribadeau-Dumas, B., Maubois, J. L., Mocquot, G., and Garnier, J. (1964). *Biochim. Biophys. Acta* **82,** 494.
Rose, D., Brunner, J. R., Kalan, E. B., Larson, B. L., Melnychyn, P., Swaisgood, H. E., and Waugh, D. F. (1970). *J. Dairy Sci.* **53,** 1.
Sampath Kumar, K. S. V., Sundararajan, T. A., and Sarma, P. S. (1957) *Enzymologia* **18,** 228.
Sandberg, K. (1967). *Acta Agr. Scand.* **17,** 127.
Schmidt, D. G. (1964). *Biochim. Biophys. Acta* **90,** 411.
Schmidt, D. G., and Koops, J. (1965). *Neth. Milk Dairy J.* **19,** 63.
Schmidt, D. G., and Payens, T. A. J. (1963). *Biochim. Biophys. Acta* **78,** 492.
Schmidt, D. G., Both, P., and de Koning, P. J. (1966). *J. Dairy Sci.* **49,** 776.
Schmidt, D. G., Payens, T. A. J., van Markwijk, B. W., and Brinkhuis, J.A. (1967). *Biochem. Biophys. Res. Commun.* **27,** 448.
Schormuller, J., Hans, R., and Belitz, H. D. (1966). *Z. Lebensm.-Unters.-Forsch.* **131,** 65.
Spies, J. R. (1967). *Anal. Chem.* **39,** 1412.
Sullivan, R. A., Fitzpatrick, M. M., Stanton, E. K., Annino, R., Kissel, G., and Palermiti, F. (1955). *Arch. Biochem. Biophys.* **55,** 455.
Sundararajan, T. A., and Sarma, P. S. (1957). *Enzymologia* **18,** 234.
Swaisgood, H. E., and Timasheff, S. N. (1968). *Arch. Biochem. Biophys.* **125,** 344.
Thoai, N., Roche, J., and Pin, P. (1954). *Bull. Soc. Chim. Biol.* **36,** 483.
Thompson, M. P. (1964). *J. Diary Sci.* **47,** 1261.
Thompson, M. P. (1966). *J. Dairy Sci.* **49,** 792.
Thompson, M. P. (1968). Unpublished data.
Thompson, M. P. (1970). *J. Dairy Sci.* **53,** 1341.
Thompson, M. P., and Gordon, W. G. (1967). *J. Dairy Sci.* **50,** 941.
Thompson M. P., and Gordon, W. G. (1968) Unpublished data.
Thompson, M. P., and Kiddy, C. A. (1964). *J. Dairy Sci.* **47,** 626.
Thompson, M. P., and Pepper, L. (1962). *J. Dairy Sci.* **45,** 794.
Thompson, M. P., and Pepper, L. (1964a). *J. Dairy Sci.* **47,** 293.
Thompson, M. P., and Pepper, L. (1964b). *J. Dairy Sci.* **47,** 633.
Thompson, M. P., Kiddy, C. A., Pepper, L., and Zittle, C. A. (1962). *Nature* **195,** 1001.
Thompson, M. P., Kiddy, C. A., Johnston, J. O., and Weinberg, R. M. (1964). *J. Dairy Sci.* **47,** 378.
Thompson, M. P., Tarassuk, N. P., Jenness, R., Lillevik, H. A., Ashworth, U.S., and Rose, D. (1965). *J. Dairy Sci.* **48,** 159.
Thompson, M. P., Greenberg, R., and Kalan, E. B. (1967a). Personal communication.
Thompson, M. P., Kalan, E. B., and Greenberg, R. (1967b). *J. Dairy Sci.* **50,** 767.
Thompson, M. P., Farrell, H. M., and Greenberg, R. (1969). Unpublished data.
Thompson, M. P., Farrell, H. M., Jr., and Greenberg, R. (1969a). *Comp. Biochem. Physiol.* **28,** 471.
Thompson, M. P., Gordon, W. G., Pepper, L., and Greenberg, R. (1969b). *Comp. Biochem. Physiol.* **30,** 91.
Thymann, M., and Larsen, B. (1965). "Annual Report, Royal Veterinarian College," pp. 225–250. Sterility Research Institute, Copenhagen.

Thymann, M., and Moustgaard, J. (1964). "Annual Report, Royal Veterinarian College," pp. 99–106. Sterility Research Institute, Copenhagen.
von Hippel, P. H., and Waugh, D. F. (1955). *J. Amer. Chem. Soc.* **77,** 4311.
Wake, R. G., and Baldwin, R. L. (1961). *Biochim. Biophys. Acta* **47,** 225.
Warner, R. C. (1944). *J. Amer. Chem. Soc.* **66,** 1725.
Waugh, D. F. (1958). *Discuss. Faraday Soc.* **25,** 186.
Waugh, D. F., and von Hippel, P. H. (1956). *J. Amer. Chem. Soc.* **78,** 4576.
Waugh, D. F., Ludwig, M. L., Gillespie, J. M., Melton, B., Foley, M., and Kleiner, E. S. (1962). *J. Amer. Chem. Soc.* **84,** 4929.
Wissmann, H., and Nitschmann, Hs. (1957). *Helv. Chim. Acta* **40,** 356.
Woychik, J. H. (1964). *Biochem. Biophys. Res. Commun.* **16,** 267.
Yamauchi, K., Takemoto, S., and Tsugo, T. (1967). *Agr. Biol. Chem.* **31,** 54.
Zittle, C. A. (1961). *J. Dairy Sci.* **44,** 2101.
Zittle, C. A., and Bingham, E. W. (1959). *J. Dairy Sci.* **42,** 1772.
Zittle, C. A., and Custer, J. H. (1963). *J. Dairy Sci.* **46,** 1183.
Zittle, C. A., and Walter, M. (1963). *J. Dairy Sci.* **46,** 1189.
Zittle, C. A., Cerbulis, J., Pepper, L., and DellaMonica, E. S. (1959). *J. Dairy Sci.* **42,** 1897.

12 ☐ κ-Casein and Its Attack by Rennin (Chymosin)

A. G. MACKINLAY AND R. G. WAKE

I. The General Properties of κ-Casein 175
   A. Introduction 175
   B. Procedures for the Isolation of κ-Casein 178
   C. Chemical Composition of κ-Casein 185
   D. Intermolecular Disulfide Bonding in κ-Casein and the Effects of Disulfide Cleavage 188
   E. Heterogeneity of κ-Casein 189
   F. Conclusion 197
II. The Action of Rennin (Chymosin) on κ-Casein 198
   A. Introduction 198
   B. κ-Casein as the Substrate for Rennin 200
   C. Nature of the Rennin-Sensitive Linkage 201
   D. Heterogeneity, Origin, and Chemical Composition of the Products of Rennin Action 204
   E. Sizes of the Reaction Products 209
   F. Conclusion 210
   References 212

I. The General Properties of κ-Casein

A. INTRODUCTION

Although solutions of whole casein free of calcium(II) are clear, the addition of low concentrations of calcium(II) causes the solution to turn opalescent; this is due to the formation of a stable suspension of large colloidal aggregates or micelles.* In 1929 Linderstrøm-Lang, having established that casein was not a single protein, suggested that it consisted of

---

* The definition of the term *micelle* and the structure of the *casein micelle* are considered in Chapter 9 by Waugh (see also McKenzie, 1967). The early history of the casein *protective colloid* is considered in Chapter 1, Volume I, by McMeekin.

at least two components, one insoluble and the other soluble in the presence of calcium(II). This latter component, or "protective colloid", interacted with the insoluble component and thereby prevented its precipitation. He proposed that the protective colloid is destroyed by rennin* during the clotting of milk when casein is converted in the presence of calcium(II) into insoluble paracasein. The notion of the protective colloid was generally accepted by subsequent workers, and many attempts were made to identify and isolate it. Finally, in 1956, Waugh and von Hippel identified a casein fraction, amounting to about 15% of the total, which fitted the description of the protective colloid and which they called κ-casein.

Prior to the identification of κ-casein the available evidence indicated that α-casein was the component specifically altered by rennin action. α-Casein has the highest mobility of the three components observed in moving-boundary electrophoresis of whole casein at neutral pH. Nitschmann and Lehmann (1947) observed that α-casein, under conditions of moving-boundary electrophoresis where it moved as a single component, could be resolved into two peaks after treatment with rennin. Cherbuliez and Baudet (1950) prepared paracasein, fractionated it by procedures already developed for casein and showed that β- and γ-caseins were unaltered. Similarly, it was shown by Alais et al. (1953) that the nonprotein nitrogen (NPN), released from whole casein by rennin and soluble in 12% trichloracetic acid (TCA), originated from α-casein and not from β-casein.

The essential advance made by Waugh and von Hippel was their discovery of a method which gave a separation of whole casein into fractions which were soluble or insoluble in the presence of calcium(II) (see Fig. 1). They were then able to demonstrate that a component, κ-casein of the soluble fraction, is the micelle stabilizer and the substrate for rennin. As starting material they used whole casein which had previously been characterized electrophoretically and in the ultracentrifuge (von Hippel and Waugh, 1955). To a solution of whole casein held at 2°C and from which calcium(II) had previously been removed, calcium(II) was added to a final concentration of 0.25 M. The temperature was raised to 37°C, and the mixture centrifuged. The precipitate (*fraction P*) accounted for 80% of the total protein. Calcium(II) was removed from the supernatant giving a fraction designated *fraction S*. Removal of all calcium(II) from fraction P once more gave a clear solution; however addition of low concentrations of calcium(II) led to the formation of a silty precipitate. This observation suggested that a stabilizing factor had been removed by the first calcium(II) treatment.

* The term *chymosin* may be used as an alternative to *rennin* and *prochymosin* an alternative to *prorennin*. See the introduction to Part D and Chapter 13.

12. κ-CASEIN AND ITS ATTACK BY RENNIN (CHYMOSIN)

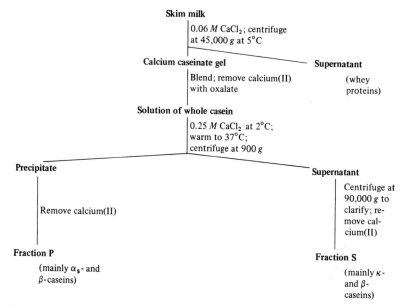

FIGURE 1. The Waugh and von Hippel procedure for obtaining fractions P and S from casein micelles.

Examination of the sedimentation rate of fraction S in the ultracentrifuge revealed that it was heterogeneous with respect to size, and a new aggregated species having a sedimentation coefficient of approximately 13 S was present. Waugh and von Hippel concluded that the 13 S peak contained κ-casein. By means of reconstitution experiments, they established that the κ-casein is responsible for micelle stabilization. In these experiments the addition of increasing quantities of fraction S to fraction P led to the progressive stabilization of the latter in the presence of calcium(II). Treatment of fraction S with rennin caused the formation of a precipitate, para-κ-casein, the disappearance of the 13 S peak from the sedimentation pattern, and abolished the ability of fraction S to stabilize fraction P.

In their original paper Waugh and von Hippel referred to the component of fraction P which is insoluble in the presence of calcium(II) as α-casein. This material is now generally referred to as α_s-casein, the term α-casein referring to preparations containing both α_s- and κ-caseins (see Section III, Chapter 2, Volume I). McKenzie and Wake (1959) have shown that most of the α-casein preparations reported up to that time had in fact contained both α_s- and κ-caseins, thus reconciling the evidence which pointed to α-casein as the micelle stabilizer with the findings of Waugh and von Hippel. The latter workers also obtained evidence that in aqueous solution, at low temperature and in the absence of calcium(II), α_s- and κ-caseins form a complex having a weight ratio of 4:1 and a sedimentation coefficient of 7.5 S. This seemed to indicate that the failure of previous methods of fractionation to separate α_s- and κ-caseins could have been due, in part at least, to the difficulty of dissociating this complex. In a more recent investigation, however, Noble and Waugh (1965) were not able to observe the complex under these conditions (see also Chapter 9 by Waugh).

κ-Casein has been identified in milk obtained from a number of species, including the cow, goat, sheep and human. We will be concerned in this chapter almost exclusively with bovine κ-casein, but κ-casein variants from other species are dealt with briefly in Chapter 18. It has been suggested that as well as its primary role as micelle stabilizer, κ-casein may function as a lipase (Yaguchi et al., 1964; Rout et al., 1966). Fox et al. (1967) considered that lipase of milk is usually associated with κ-casein and is a minor component of the casein system. However Gaffney et al. (1966) concluded that lipase is associated with other proteins besides κ-casein in milk. Downey and Murphy (1970) have studied the binding of pancreatic lipase to casein micelles and casein complexes. They consider that pancreatic lipase interacts similarly to milk lipase and conclude that micelle structure is not necessary for the binding of pancreatic lipase to casein and that its activity is impaired on binding. It is of interest, however, that the activity of purified milk lipase (Fox and Tarassuk, 1968) is not inhibited by skim milk (Patel et al., 1968). It will not be considered in this chapter but is discussed in Chapter 16 by Groves.

B. PROCEDURES FOR THE ISOLATION OF κ-CASEIN

The identification of κ-casein and of its functions as micelle stabilizer and substrate for rennin was proposed without actually isolating it from the slowly sedimenting (1.3 S) material also present in fraction S. Many subsequent reports have appeared giving methods for obtaining κ-casein free from the other components of whole casein. In general most methods for isolating κ-casein involve one or other of two approaches: Either (a) an initial step is employed to separate α_s- and κ-caseins, such as treatment with concentrated solutions of calcium salts, followed by further purification of material corresponding to fraction S of Waugh and von Hippel, or (b) α-casein is isolated and κ-casein then obtained by "splitting" it away from the α_s-casein. In view of the fact that treatment with con-

12. κ-CASEIN AND ITS ATTACK BY RENNIN (CHYMOSIN)

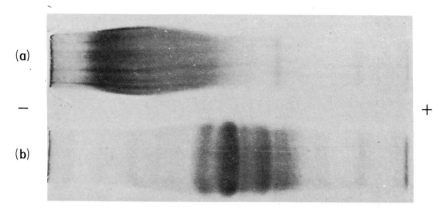

FIGURE 2. Starch-gel electrophoresis (according to Wake and Baldwin, 1961) of (a) κ-casein and (b) SCM-κ-casein.

centrated calcium(II) gives good separation of $\alpha_{s,1}$- and κ-caseins this is an extremely useful step in any preparative procedure.

In the following section, a brief description will be given of the methods which have been most commonly used to prepare κ-casein; they are summarized in Table I. A summary of characteristics which have been reported for various preparations is given in Table II.

1. *The Alcohol Fractionation Method (McKenzie and Wake, 1961)*

This procedure consists essentially of the preparation of fraction S followed by its fractionation in 50% ethanol. Two precipitations from alcohol yield a product which gives a 13 S peak on sedimentation with no evidence of the more slowly sedimenting 1.3 S (β-casein) material. Starch-gel electrophoresis in the presence of urea (Wake and Baldwin, 1961) shows a single smeared zone which extends from close to the starting slot (see Fig. 2a). Usually there are barely detectable amounts of α_s- and β-caseins present. Because of the spreading behavior of κ-casein during gel electrophoresis the presence of more slowly migrating contaminants is difficult to detect. Small amounts of such species, including some para-κ-casein, are in fact usually found in these preparations. These impurities may be detected by electrophoretic analysis of disulfide-reduced κ-casein and can be removed by column chromatography of κ-casein in dissociating solvents (Mackinlay and Wake, 1965). The yield of κ-casein by this method of preparation is about 20%, assuming 15% κ-casein in whole casein. The final product is an efficient stabilizer of micelles. Neelin et al. (1962) reported that it was less efficient in this regard than κ-casein prepared by

TABLE I

PROCEDURES FOR ISOLATING κ-CASEIN

Method	Summary of procedure	Reference
Alcohol fractionation	Fraction S prepared from acid casein; protein precipitated with Na_2SO_4, redissolved and dialyzed to give a concentrated solution in 0.005 M NaCl; made 50% in ethanol, and κ-casein precipitated by addition of CH_3COONH_4; reprecipitated from 50% ethanol to improve purity	McKenzie and Wake (1961)
Trichloracetic acid–urea	Crude α-casein dissolved in 6.6 M urea, cooled to 3°C and TCA added to 12%; precipitate of α_s-casein removed; supernatant brought to pH 7, TCA and urea removed; $CaCl_2$ added to 0.25 M at 3°C; precipitate removed, supernatant adjusted to pH 11.3 and κ-casein precipitated at pH 4.4	Swaisgood and Brunner (1962)
Urea–sulfuric acid	Acid casein dissolved in 6.6 M urea and acidified with H_2SO_4 to pH 1.3–1.5; after standing, precipitate filtered off, and κ-casein precipitated from the supernatant with $(NH_4)_2SO_4$; further purified by precipitation from aqueous ethanol	Zittle and Custer (1963)
Preparation of α_3-casein	α-Casein treated with $CaCl_2$ in the cold and precipitate of α_1-casein removed by centrifugation; calcium(II) removed from supernatant, and α_3-casein centrifuged out at high speed; purified further by recentrifugation	Hipp et al. (1961)
Morr	α_s-Casein removed from α-casein by precipitation with $CaCl_2$; κ-casein precipitated from supernatant by adjusting to pH 4.7	Outlined by Neelin et al. (1962)

Method	Description	Reference
Cheeseman	Crude α-casein dissolved at pH 6.8–7.0; all traces of calcium(II) removed by oxalate, solution cooled to 4°C and made 3.3 M with respect to urea; adjusted to pH 4.9 and diluted to 2 M urea; precipitate of α_s-casein removed and supernatant treated with 0.2 M CaCl$_2$ at 4°C; after warming, insoluble material removed and κ-casein precipitated at pH 4.6 and 4°C	Cheeseman (1962)
Craven and Gehrke	Fraction S, containing 0.25 M CaCl$_2$, prepared from acid casein; made 3.0 M in urea and 0.35 M in K$_2$C$_2$O$_4$, adjusted to pH 4.2 and warmed to 35°–40°C; α_s-, β-, and γ-caseins precipitate leaving pure κ-casein in supernatant	Craven and Gehrke (1967)
Isolation from whole casein by DEAE-cellulose chromatography in urea	Acid casein in 0.01 M imidazole–HCl, pH 7, and 4.5 M urea adsorbed onto DEAE-cellulose; the various proteins separated by elution with a gradient of NaCl (0–0.6 M) in 0.02 imidazole–HCl, pH 7, and 3.3 M urea; κ-casein fraction further purified by rechromatography	Ribadeau-Dumas et al. (1964)
DEAE-Cellulose chromatography of crude κ-casein, fraction S	Crude κ-casein, fraction S, adsorbed onto DEAE-cellulose in CH$_3$COOH–NaOH, pH 6.25 and 3°C; β-casein eluted ahead of κ-casein by a simultaneous gradient of CaCl$_2$ (0–0.25 M) and pH (6.25–4.8); κ-casein further purified by rechromatography	Hill (1963)
Sephadex gel filtration	Whole casein in 0.005 M tris-citrate, pH 8.6, 6.6 M urea passed through Sephadex G-150 or G-200 at room temperature; κ-casein separates from the other proteins by eluting in the void volume	Yaguchi et al. (1967)

other methods. Although an α_s/κ ratio of 4:1 as used by these authors usually gives complete stabilization, they found only 37 and 71% stabilization for two different preparations obtained by this method. Further instances of variation in stabilizing ability of κ-casein preparations are mentioned below. Talbot and Waugh (1967, 1970) have described a modification of the McKenzie-Wake procedure which gives higher yields of κ-casein.

2. *The Trichloracetic Acid–Urea Method (Swaisgood and Brunner, 1962)*

This procedure was developed from the observation that a portion of the soluble product of rennin action on κ-casein, the "glycomacropeptide", is soluble in 12% trichloracetic acid (Wake, 1959b). The initial step consists of the addition of trichloracetic acid (to a final concentration of 12%) to a solution of casein dissolved in 6 M urea at 3°C. The resulting supernatant, rich in κ-casein, is then further purified by treatment with contrated calcium(II) and by precipitation at pH 4.4. Yields of up to 9% of the total casein are obtained by this method. When the procedure was applied to whole casein the supernatant contained some β-casein which was difficult to remove. The best way to minimize the amount of contaminating β-casein was found to be by trichloracetic acid–urea fractionation of purified α-casein. A similar difficulty in removing small amounts of β-casein was reported by Wake (1959a).

κ-Casein prepared by this method was estimated to be 92% pure and was an efficient stabilizer of micelles. Rennin released 29% by weight in a form souble at pH 4.7. The usual spreading zone is observed on starch-gel electrophoresis, while variable quantities of contaminating components are evident in patterns for the disulfide-reduced derivative (Neelin, 1964). A further step involving centrifugation in 0.1 M NaCl may be used to remove most of these.

3. *The Urea–Sulfuric Acid Method (Zittle and Custer, 1963)*

Zittle and Custer have reported a procedure for the isolation of κ-casein which combines features of the two methods already outlined. This method involves an initial fractionation in concentrated urea under acid conditions, followed by alcohol fractionation of the supernatant as employed by McKenzie and Wake. Whole casein is dissolved in 6.6 M urea, and sulfuric acid is added until pH 1.3–1.5 is reached. On standing, a precipitate containing α_s- and β-caseins forms. The supernatant contains κ-casein with only small amounts of α_s- and β-caseins. It amounts to 7–12% of the original whole casein. This procedure is thus straightforward and gives a good yield. Further purification is achieved by one or more precipitations from aqueous alcohol to give a product which analyzes as a single spreading zone by gel electrophoresis.

An indication that artifacts may sometimes arise as a result of this preparative procedure comes from the work of Woychik (1965). In the course of typing κ-caseins from individual cows by gel electrophoresis after disulfide bond reduction (conditions under which discrete components resolve, see below), he observed that the patterns of components in some κ-caseins isolated by this method were altered compared to those obtained before isolation. Evidently a modification may occur during isolation which leads to an increase in net negative charge. The nature of this alteration is not known.

Zittle and Custer found that preparations of κ-casein purified by alcohol precipitation occasionally had less stabilizing ability than the crude material from which they were obtained. They suggested that κ-casein purified in this manner should always be checked for stabilizing ability. They also noted that the final preparations gave solutions showing slight turbidity due presumably to traces of lipid. Extraction of dried preparations with organic solvents such as acetone, ethanol and ether impaired the solubility of κ-casein and its ability to stabilize micelles. The loss of stabilizing ability did not occur when less pure preparations were extracted.

4. The Preparation of α_3-Casein

Hipp et al. (1961) have described the preparation from α-casein of a fraction, termed α_3-casein, which displays some of the properties of κ-casein. When α-casein is subjected to moving-boundary electrophoresis at pH 2.35 in HCl–NaCl buffer it is resolved into three components, α_1-, α_2- and α_3-casein in order of decreasing mobility. Upon treatment of α-casein with concentrated calcium(II), α_1-casein precipitates and α_2- and α_3-casein are found in the supernatant. The supernatant is then further fractionated by high-speed centrifugation until a pellet, containing α_3- but no α_2-casein, is obtained. α_3-Casein is shown to be related to κ-casein by the presence of sialic acid and by the release from it by rennin of nonprotein nitrogen soluble in 2% trichloracetic acid. The total nonprotein nitrogen released is low, however, compared with the values reported for κ-casein, indicating that sizable proportions of other components are present. A significant difference between α_3- and κ-caseins lies in their solubilities, α_3-casein being soluble to the extent of only 0.26% at pH 6.9. The stabilizing ability of α_3-casein is lower than that of the α-casein from which it is prepared. The sedimentation coefficient of α_3-casein in aqueous solution was observed to be 23.1 S at pH 7.0 at both 2° and 25°C.

5. Other Chemical Methods

Among other procedures used to obtain κ-casein is that of Morr, as outlined by Neelin et al. (1962). This consists of the calcium(II) fractionation of α-casein, followed by the precipitation of κ-casein at pH 4.7. Starch-gel analysis of material prepared by this method indicates the presence of significant quantities of β-casein along with a number of other minor components (Neelin et al., 1962). Cheeseman (1962) reported another method in which κ-casein is obtained from α-casein by calcium(II) and urea fractionation. Craven and Gehrke (1967) have described what they claim is an "exceptionally mild" procedure for preparing κ-casein. It is based on the use of calcium oxalate as a "carrier precipitate" to enhance the removal of other caseins from fraction S in 3 M urea at pH 4.2 and 35°–40°C. None of these preparative procedures would appear to offer great advantages over others.

6. Isolation of κ-Casein by Ion-Exchange Chromatography and Gel Filtration

Ribadeau-Dumas et al. (1964) reported on the fractionation of whole casein by DEAE-cellulose chromatography in the presence of urea. κ-Casein essentially free of contaminants is obtained by this procedure. Noble and Waugh (1965) found that the stabilizing ability of κ-casein obtained by chromatography in urea is comparable to that obtained by alcohol fractionation. The chromatographic conditions employed to separate the components of disulfide-reduced κ-casein have also been used to isolate disulfide-reduced κ-casein directly from whole casein (Mackinlay and Wake, 1965), although complete separation of β- and κ-caseins was not obtained.

Another method for the preparation of κ-casein which involves DEAE-cellulose chromatography of fraction S was reported by Hill (1963). In this case, however, chromatography was done without the inclusion of a dissociating agent in the eluting buffer. This results in the presence of a considerable amount of β-casein in the final product. The heterogeneity of the material prepared by this method is evident in the gel patterns obtained by Hill and Hansen (1963). The suggestion made by these authors and by Beeby (1963) that κ-casein consists of three quite distinct components, one containing all the cystine, another containing all the sialic acid and another containing neither, is not in accord with the bulk of the evidence obtained using other preparations. It is not possible at present to exclude the possibility that the preparation of Hill and Hansen contains, as well as the usual material which spreads during gel analysis, other components which have some of the properties of κ-casein. It would seem desirable however not to include these under the term κ-casein.

Sephadex gel filtration has also been applied to the isolation of κ-casein. Hill (1965) was able to achieve a partial purification by gel filtration of fraction S on Sephadex G-100 or G-200 at low temperature. (The properties of Sephadex G-100 and G-200 are discussed in Chapter 4, Volume I.) More recently, Yaguchi et al. (1967, 1968) have isolated κ-casein, apparently free from other casein components, by gel filtration of whole casein and crude κ-casein preparations (∼1 g/run) in 6 M urea at pH 8.6 (0.005 M tris-citrate buffer) using Sephadex G-150 columns (95 cm × 2.5 cm) at room temperature. κ-Casein, presumably cross-linked by disulfide bridges, is eluted in the void volume.

It is practically certain that none of the bulk preparative procedures gives rise to products which are *completely* free of other casein components. Thus κ-casein prepared by alcohol fractionation gives the cleanest patterns on starch-gel analysis but is found to contain easily detectable amounts of slow-moving impurities when the analysis is carried out on the disulfide-reduced derivative. A final column chromatographic step is therefore essential if completely pure preparations are required. In situations where only relatively small quantities of κ-casein of a high degree of purity are required, direct isolation by column chromatography of whole casein may be preferable.

C. Chemical Composition of κ-Casein

Like all the other caseins, κ-casein contains phosphorus. It is the only major casein having a significant quantity of bound carbohydrate, amounting to approximately 5% of the total and made up of galactosamine, galactose and sialic acid approximately in the molar ratio 1:1:1 (Jollès et al., 1962, 1964). It would appear from the work of Wheelcock and Sinkinson (1969) that the galactosamine is N-acetylated.

The overall amino acid composition of κ-casein was established originally by Jollès and his co-workers (Jollès et al., 1962). Most of the commonly occurring amino acids are present. As with the other caseins there is a relatively high proportion of proline (8.8%), and this probably accounts for the almost complete absence of any α-helical or other ordered structures (Herskovits, 1966). However, unlike the other caseins it contains a significant quantity of half cystine (1.4%). In these initial studies κ-casein obtained from pooled milk was used. Because of the possibility of the presence of small amounts of other caseins in such preparations and the now well-established genetic polymorphism in κ-casein, the amino acid content of more highly purified material obtained from one or other of the genetic variant series has considerably more significance. Such data

TABLE II

Properties of κ-Casein Obtained by Various Procedures

Preparation	Sialic acid content (%)	Phosphorus content (%)	$A^{1g/dl}_{280,1cm}$ [a]	$s_{20,w}$ [a]	Reference
Alcohol fractionation		0.14		14.0	Pepper and Thompson (1963)
	2.5	0.22		14.0	Thompson and Pepper (1962)
	0.65, 1.25				Marier et al. (1963)
	2.4	0.22			Jollès et al. (1962)
			10.5		Garnier et al. (1962)
Trichloracetic acid–urea	1.4	0.22			Swaisgood et al. (1964)
	2.14				Marier et al. (1963)
		0.35		12.9	Swaisgood and Brunner (1962)
Urea–sulfuric acid	1.94	0.30	12.2		Zittle and Custer (1963)
			11.7		Nakai et al. (1965)
	1.81				Marier et al. (1963)
Cheeseman	0.79				Cheeseman (1962)
α_3-Casein	1.21	0.35	15.6	23.1	Hipp et al. (1961)

[a] These have been determined in salt solution at room temperature near pH 7.

have been obtained by several groups of workers for bovine κ-casein A and B, para-κ-casein and the macropeptide. They are reproduced in Table IV and will be discussed below.

In comparing the chemical composition of κ-casein obtained by various preparative procedures, favorite quantities for determination have been the phosphorus and sialic acid contents (see Table II). The spread of the values indicates that neither of these quantities can as yet be used to characterize κ-casein nor do they provide much information regarding purity. Determinations of the phosphorus content of chromatographically purified fractions of S-carboxymethyl-κ-casein (SCM-κ-casein) have given values centering around 0.16% or approximately 1 atom/20,000 g. It is evident that the presence of small amounts of relatively phosphorus-rich contaminating components will lead to much higher values. The range of values for sialic acid content may also be due to varying proportions of the carbohydrate-rich and carbohydrate-free components (see below, and Chapter 6, Volume I; Chapter 10). Recent observations suggest that the sialic acid content (and presumably that of the other sugars) of κ-casein may vary with the stage of lactation (Mackinlay, 1965). Consequently variations in the sialic acid content of whole casein could be a reflection of this rather than of variation in the overall proportion of κ-casein. As will become clear, it is for the same reason, i.e., variation in carbohydrate content, that measurements of the total nonprotein nitrogen released by rennin provide a better indication of the amount of κ-casein present in a given preparation than do determinations of the nonprotein nitrogen soluble only in 12% trichloracetic acid. The best check of homogeneity of κ-casein preparations is provided by starch-gel analysis both before and after disulfide bond reduction. Small amounts of β-casein are more readily detected in patterns for the unreduced protein while slower-moving contaminants and traces of para-κ-casein can be detected only when the reduced derivative is analyzed.

The desirability of checking the stabilizing ability of κ-casein preparations has been stressed by Zittle and Custer (1963) (see also Chapter 9). Among treatments which have been reported to result, on occasions, in loss of stabilizing activity are (a) precipitation from 50% alcohol or from trichloracetic acid–urea at 25°C, (b) extraction with organic solvents, and (c) high-speed centrifugation as employed in the preparation of α_3-casein. Possibly the loss of stabilizing activity results from irreversible aggregation reactions which prevent interaction with α_s-casein in the presence of calcium(II).

The functional monomer of κ-casein has a molecular weight in the range 17,000–20,000 daltons (see below), and most likely it is made up of a single polypeptide chain. There is evidence from at least three laboratories that the C-terminal amino acid is valine (Jollès et al., 1963; de Koning et al., 1966; Pujolle et al., 1966). However, there is only partial agreement concerning the *sequence* of amino acids at this end of the molecule.

The N-terminal residue of κ-casein has not yet been identified unequivocally, but it may be glutamine. Rennin cleaves the molecule at a position approximately one-third along its length from the C-terminus to give insoluble para-κ-casein and the soluble macropeptide made up of the C-terminal portion (see Section II). The amino acid contents of these two portions have been compared by Jollès and his collaborators (Jollès et al., 1961, 1962, 1963), by Kalan and Woychik (1965) and de Koning et al. (1966). The sequence of the macropeptide has been partially resolved (Delfour et al., 1966; Jollès et al., 1968) and shows some unusual features. These will be discussed below (see also Fig. 10, in which present knowledge of the structure of bovine κ-casein is summarized). The macropeptide region contains the sites for attachment of the carbohydrate as well as the single phosphate which is linked to a serine residue (Jollès et al., 1961). Both half-cystine residues are located in the para-portion (Jollès et al., 1962; Nakai et al., 1965), and most workers have considered that all of it is present as SS, none as SH (see below and Mackinlay and Wake, 1964). Waugh (1961) has stated that κ-casein is the only component of whole casein which contains cystine, and the effect of treatment to break disulfide bonds on the starch-gel electrophoretic pattern of whole casein supports this view, although the presence of cystine in some of the minor casein components is not definitely excluded. It is clear, however, that the bulk of the cystine in casein is present in κ-casein, and it is possible that determination of the cystine content of whole casein would be useful in detecting variations in its content of κ-casein (see Chapter 5, Volume I).

D. INTERMOLECULAR DISULFIDE BONDING IN κ-CASEIN AND THE EFFECTS OF DISULFIDE CLEAVAGE

Studies of the sedimentation properties of κ-casein prepared by the trichloracetic acid–urea method and by alcohol fractionation indicate that the disulfide bonds in these preparations are intermolecular (Swaisgood et al., 1964; Mackinlay and Wake, 1964). That is to say they link together individual κ-casein molecules to form large aggregates. This was first suggested by the observation that although κ-casein analyzed by starch-gel electrophoresis gives a slowly migrating, spreading zone, treatment with reagents which break SS bonds gives rise to a number of discrete components which migrate more rapidly in the gel (Waugh, 1962) and which are all susceptible to rennin action (Mackinlay and Wake, 1964). The effect of SS cleavage on the gel pattern of κ-casein is shown in Fig. 2 where κ-casein and SCM-κ-casein are compared. The latter, prepared by reduction of SS bonds followed by alkylation of the SH groups, shows two major components and several minor ones.

It has not been possible to dissociate κ-casein completely in any solvent. On the other hand reduced κ-casein dissociates readily in a wide range of solvents such as 50% acetic acid, 4 M guanidine HCl and 20% dimethylformamide. The apparent dissociation of κ-casein at pH 12 has been shown to result from the rupture of SS bonds under these conditions (Mackinlay and Wake, 1964). Swaisgood et al. (1964) performed molecular weight measurements on κ-casein containing intact disulfide bonds in several solvents and obtained a value near 60,000 daltons for the smallest component. However, the presence of much larger species was also evident under these conditions. It seems likely that the cross-linking of κ-casein by disulfide bonds occurs randomly and that the degree of cross-linking varies widely. Woychik (1965) investigated whether all the disulfide of κ-casein was intermolecular by looking for variations in the ease of reduction but found none. Additional evidence that intramolecular SS bonds are absent comes from the consideration that κ-casein contains two half-cystine residues per peptide chain. If these were to form an intramolecular bond to any significant extent then discrete components might be observed during starch-gel electrophoresis. This is not observed either at pH 8.6 or at pH 3.2.

Molecular weight determinations carried out on disulfide-reduced κ-casein indicate a value of 17,000–20,000 daltons. Swaisgood and Brunner (1963) determined values of 18,000 and 20,000 daltons for reduced, unfractionated κ-casein in urea and guanidine solutions, respectively. Woychik et al. (1966) performed measurements in 5 M guanidine HCl on two fractions of reduced κ-casein containing no carbohydrate and obtained values of 17,800 and 18,400 daltons. Mackinlay (1965), using similar material, obtained values of 18,700–20,400 daltons in 20% dimethylformamide, 50% acetic acid and pH 12 phosphate buffer. These values are consistent with those calculated from amino acid contents (Pujolle et al., 1966).

One report has appeared to the effect that SH groups, not SS, are present in κ-casein (Beeby, 1964). The thiol groups were detectable in a crude preparation of κ-casein only when all calcium(II) had been removed by treatment with ethylenediaminetetraacetate (EDTA) or oxalate. They were no longer present after freeze-drying, indicating that they were easily oxidized. Beeby also found that sulfite caused the appearance of the same SH titer as EDTA treatment and concluded that sulfite in this case was removing calcium(II) rather than breaking disulfide bonds. The conclusion drawn from these observations is that calcium(II) normally masks the SH groups. However, it is clear that at least in the case of κ-casein prepared by alcohol fractionation, all of the thiol titrated after sulfite treatment is originally present as SS, since the half-cystine content calculated on this basis agrees with that determined by performic acid oxidation (Jollès et al., 1962; Nakai et al., 1965; Mackinlay and Wake, 1964). Beeby's observation draws attention to the possibility that disulfide-reduced κ-casein may be more nearly "native" than the unreduced material. It is possible that κ-casein is incorporated in vivo into the micelle in the SH form and that interactions with calcium(II) and with the other components of the micelle are necessary to maintain the reduced form. The need for further investigation is certainly indicated here.

E. HETEROGENEITY OF κ-CASEIN

Gel electrophoresis of κ-casein obtained from pooled milk and treated to break disulfide bonds reveals the presence of 7–8 discrete bands (Schmidt,

1964; Woychik, 1964, 1965; Mackinlay and Wake, 1964). The two slowly moving major bands of SCM-κ-casein have relative mobilities of 0.52 and 0.59, respectively (see Wake and Baldwin, 1961).

There is now evidence that the total number of bands present in such samples is considerably greater than 8 (Mackinlay et al., 1966). The fact that all of the components giving rise to these bands are degraded by rennin leads to the conclusion that they are related species and are not components of a larger basic unit. Rather, it seems that all the components share a basic polypeptide chain and that the heterogeneity arises through modification of this basic structure in various ways. At least three distinct types of modification have been recognized in this connection, and these will be considered in turn.

1. *Variations between κ-Caseins from Individual Cows*

A number of workers have observed variation in the pattern of components in samples of disulfide-reduced κ-casein obtained from individual cows (Woychik, 1964; Schmidt, 1964; Neelin, 1964). To date all samples examined have shown one of three different patterns. Examples of these may be seen in Fig. 3 where samples of whole casein are shown analyzed in a starch-gel containing mercaptoethanol. Only the major components

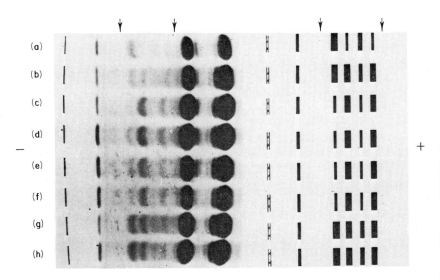

FIGURE 3. Starch-gel electrophoresis of whole casein samples in the presence of mercaptoethanol showing genetic variants of κ-casein. The κ-casein bands are enclosed by arrows. (b), (g) and (h) are κ-casein type AB; (a) is κ-casein type B; (c), (d), (e) and (f) are κ-casein type A.

TABLE III
DISTRIBUTION OF κ-CASEIN TYPES AMONG DIFFERENT BREEDS OF COWS[a]

Breed	Type		
	A	AB	B
Holstein	99	36	3
Guernsey	26	17	4
Brown Swiss	9	19	5
Jersey	0	10	38
Ayrshire	1	2	
Total	135	84	50

[a] After Woychik (1965).

of κ-casein can be seen clearly under these conditions. Samples (b), (g), and (h) contain both major components in approximately equal amounts while sample (a) contains only the more slowly migrating of the two. The remainder of the samples all contain the more quickly migrating major component. A genetic basis for the observed variation is now generally accepted (Aschaffenburg and Thymann, 1965; Larsen and Thymann, 1966). Woychik et al. (1966) designate κ-caseins containing only the more quickly migrating major component as κ-casein A and those containing the slower as κ-casein B. This designation will be used here.*

That it is possible to type κ-casein satisfactorily on the basis of patterns where only the major components are observable, such as in Fig. 3, follows from an examination of isolated κ-caseins A and B (Mackinlay et al., 1966). Both variants show similar patterns of heterogeneity, and the mobility difference observed between the major components is also found to exist between corresponding minor components in each series. Schmidt et al. (1966) have observed slight but reproducible differences between the distribution of minor components in the two series. The two purified genetic variants, κ-caseins A and B, are compared in Fig. 4.

Data on the distribution of κ-casein types among different breeds were first obtained by Woychik (1965) and are shown in Table III. Larsen and

---

* In a previous publication (Mackinlay et al., 1966) the terms κ-casein A and B were used to distinguish between the components, both occurring in each genetic variant series, which give rise to the major and minor para-derivatives. The system of nomenclature which has been generally adopted to distinguish between genetic variants of the same protein has been used here to avoid confusion: A and B refer to the two electrophoretically distinguishable genetic variants of κ-casein; κ-casein A migrates more quickly at pH 8.6.

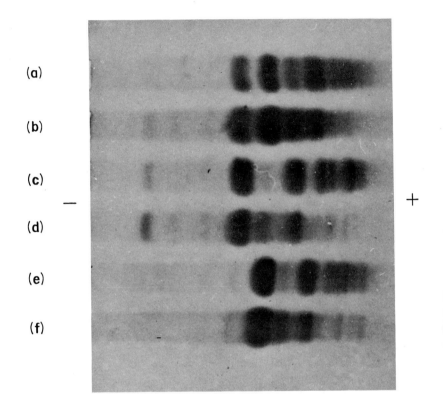

FIGURE 4. Starch-gel electrophoresis of the purified genetic variants of κ-casein in the presence of mercaptoethanol (after Schmidt et al., 1966). (a) Type AB; (b) neuraminidase-treated AB; (c) type B; (d) neuraminidase-treated B; (e) type A; (f) neuraminidase-treated A.

Thymann (1966) made a study of milk protein polymorphism in Danish cattle and concluded that $\alpha_{s,1}$-, β- and κ-casein types are controlled by genes at the same locus. Much more comprehensive data for the occurrence of κ-casein in various breeds of cattle in several countries have been obtained. Aschaffenburg (1968) has prepared a very comprehensive table of the gene frequencies. The A allele tends to be predominant in the majority of breeds.

The amino acid compositions of the major components of type A and B κ-caseins, para-κ-casein and the macropeptides have been determined by several groups. The data are compared in Table IV. It appears that κ-

TABLE IV

AMINO ACID COMPOSITION OF THE MAJOR COMPONENT OF κ-CASEINS A AND B, PARA-κ-CASEIN, AND THE MACROPEPTIDE
(NUMBER OF RESIDUES PER MOLECULE)

Amino acid	κ-Casein				Para-κ-casein		Macropeptide	
	A		B				A	B
	WKN[a]	SBK[b]	WKN[a]	SBK[b]	KYR[c]	KW[d]	KRK[e]	KRK[e]
Asp	12	12	11	11	7	8.2	4.3	3.5
Thr	14	14	13	14	3	5.2	9.8	8.8
Ser	12–13	12–13	12–13	12–13	7	7.6	4.7	4.8
Glu	27	27	27	27	18	17.9	8.8	8.7
Pro	20	18	20	19	13	12.9	6.6	6.4
Gly	3	2–3	3	2–3	1	1.7	1.1	1.1
Ala	13–14	13–14	14	14	9	9.7	4.2	4.9
Val	10–11	10–11	11	10–11	5	6.5	4.9	4.9
Met	2	2	2	2	1	1.3	0.7	0.7
Ile	11	11–12	12	12	6	7.4	4.9	5.7
Leu	8	8–9	8	8–9	7	8.0	1.0	1.0
Tyr	8	9	8	8–9	9	8.8		
Phe	4	4	4	4	4	4.1		
Lys	9	9	9	9	7	7.0	2.8	2.9
His	3	3	3	3	3	2.9		
Arg	5	5	5	5	5	5.0		
Cys/2	2	1	2	1	2	1.8		
PO₃H		1		1				
Trp[f]					1	1		

[a] Woychik et al. (1966).
[b] Schmidt et al. (1966); the data of Schmidt et al. have been recalculated on a molecular weight basis of 19,000 daltons.
[c] Kim et al. (1969).
[d] Kalan and Woychik (1965).
[e] De Koning et al. (1966).
[f] Data of Spies (1967).

casein A contains one more aspartic acid (or asparagine) residue and one more threonine residue than κ-casein B. The results would also suggest one additional alanine and isoleucine residue in κ-casein B on the basis of a molecular weight of 19,000 daltons. The major components of the two variants thus probably differ to the extent of a single charged group, and this is consistent with the observed mobility difference during electrophoresis. Since both κ-caseins give the same para-κ-caseins it can be concluded that the charge difference is located in the soluble product of rennin action, the macropeptide. This has now been verified (de Koning et al., 1966; Armstrong et al., 1967).

As far as is known the two major components of κ-casein differ only in the amino acids mentioned, and alterations in other properties as a result of these substitutions have not yet been reported.

2. Heterogeneity due to Variations in Carbohydrate Content

The heterogeneity with respect to carbohydrate content of the soluble material released from κ-casein by rennin (see Section II) suggested that κ-casein itself varies in this regard, although at the time it could not be excluded that more than one peptide might be split from each molecule. Attempts to fractionate κ-casein on the basis of carbohydrate content were unsuccessful until it was realized that the individual monomer units of κ-casein were extensively cross-linked by SS bonds. Fractionation of disulfide-reduced κ-casein has now been successfully carried out by three groups independently. In all cases fractionation was accomplished using DEAE-cellulose chromatography, and comparable separations of components were obtained. Mackinlay and Wake (1965) used SCM-κ-casein as starting material, Woychik et al. (1966) used S-carboxyamidomethyl-κ-casein (SCAM-κ-casein), while Schmidt et al. (1966) started with the reduced, nonalkylated material. Because of its complexity, it has been difficult to obtain completely homogeneous fractions of κ-casein. However, the study of fractions enriched in particular components has led to an understanding of the types of variations responsible for the observed heterogeneity. Pujolle et al. (1966) claim to have obtained more purified components by repeated chromatography.

Carbohydrate analyses carried out on fractions obtained by column chromatography of pooled SCM-κ-casein have shown that the two major components do not contain significant quantities of carbohydrate, while for the minor components, the carbohydrate content increases with increasing electrophoretic mobility at pH 8.6. A fraction containing the three most quickly migrating components contained approximately 10% carbohydrate (Mackinlay and Wake, 1965). No significant variation in phosphorus content has been found among the fractions. The analytical results obtained by the other two groups of workers who fractionated separately each of the two genetic variant series are in agreement with these observations. The mobility difference between the components of a single series is thus probably mainly due to varying contents of negatively charged sialic acid. Figure 4, taken from Schmidt et al. (1966), shows the effect of neuraminidase removal of sialic acid in preferentially reducing the mobility of the fastest migrating components.

Woychik et al. (1966) have compared the amino acid compositions of fractions enriched in various minor components with that of the corresponding major component obtained from the same genetic variant series. In each case the amino acid compositions were found to be very similar, differences amounting to only one or two residues for a few amino acids. It is not known if such differences contribute of the heterogeneity observed by electrophoresis. Confirmation of the differences in amino acid composition between the major and minor components appears desirable. Pujolle et al. (1966) have shown that carboxypeptidase A releases exactly the same eleven amino acids from the C-terminal region of each of the seven components identified by them in reduced κ-casein A and completely separated from one another by repeated chromatography.

Since all the carbohydrate of κ-casein is contained in the macropeptide released by rennin it should be possible to observe a pattern of heterogeneity in the macropeptide similar to that which is seen for κ-casein. This is found to be the case as reference to Fig. 8 shows. The similarity is even greater when comparison is made with SCM-κ-casein analyzed by starch-gel electrophoresis at pH 3.2, conditions which are closer to those used to analyze the macropeptide. A better resolution of some of the minor components occurs under these conditions than at pH 8.6.

In spite of the large variation in carbohydrate content the major and minor components of κ-casein have very similar properties, being equally effective in micelle stabilization and as a substrate for rennin. The only difference so far observed is that the major components lacking carbohydrate tend to precipitate in salt-free solution at pH 7, while the carbohydrate-rich components are quite soluble. A difference in solubility properties is also seen in 2% trichloracetic acid (Mackinlay and Wake, 1965). While no function for the carbohydrate in κ-casein has been established it is well to keep in mind that experimentally the measure of overall stabilizing ability is made under conditions quite different from those which exist in milk. The observation (Mackinlay, 1965) that the proportion of the carbohydrate-rich components is increased in κ-casein obtained from colostrum may indicate that the function of the carbohydrate is expressed during the early stages of lactation.

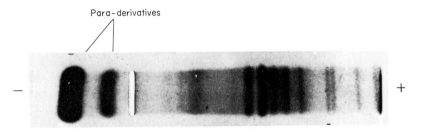

FIGURE 5. Starch-gel electrophoresis (according to Wake and Baldwin, 1961) of a rennin–SCM-κ-casein reaction mixture showing the resolution of the insoluble product, para–SCM-κ-casein.

3. *Variation in the Para-κ-casein Portion of the Molecule*

The action of rennin on disulfide-reduced κ-casein results in the appearance of two para-κ-casein derivatives, a major and a minor one, as is shown in Fig. 5 (Mackinlay and Wake, 1965). Kalan and Woychik (1965) have also observed two components in reduced para-κ-casein, but in roughly equal proportions. A close examination of fractions obtained by chromatography of κ-caseins A and B showed that minor components present in each of the two types give rise separately to the minor para-component during rennin treatment. This has been directly shown for the case of the minor component which runs immediately in front of the major component in each of κ-caseins A and B (Mackinlay et al., 1966). Since all the fractions obtained by column chromatography of either pooled or individual κ-caseins (always containing more than a single component) give rise to both para-derivatives, it follows that two distinct sets of components exist within each κ-casein variant series; their proportions are those of the two para-derivatives obtained when rennin is allowed to act on unfractionated κ-casein.

The origin of these two sets of para-components has become controversial. It is known from a comparison by starch-gel electrophoresis of SCM-, SCAM- and a mixed SCM/SCAM derivative of κ-casein that a single charge difference exists between the major and minor para-derivatives and presumably therefore between the two sets of components in each κ-casein type (Mackinlay, 1965). Hill and Wake (1969a) have found that the two para-compounds (I and II) formed form S-carboxymethyl-κ-casein have identical C-terminal amino acid sequences suggesting they both arise from cleavage at the same position in the κ-casein molecule. They found that in 7 M urea, para-compound I was slowly converted to para-compound II over several days. This was considered to be due to reaction with cyanate present in the urea. Kim et al. (1969) found that the amino acid composition of the major and minor para-κ-caseins differed only in lysine and homocitrulline content, the major fraction having more lysine and no homocitrulline. They concluded that the minor components are present only in para-κ-casein preparations that have been treated with urea and that they result from carbamylation of lysine residues.

F. CONCLUSION

There are at least two distinct sources of heterogeneity in κ-casein: amino acid substitutions which distinguish the type A and type B κ-caseins and variation in carbohydrate content. There is possibly a third source of variation situated in the para-region of the molecule. The simplest scheme which can account for the way in which the heterogeneity of κ-casein arises *in vivo* is that the major components observed in κ-caseins A and B are the first to be synthesized and that a portion of these are then modified by the addition of varying amounts of carbohydrate to their macropeptide region. At the same time, and independently, a small proportion of the molecules may be modified in the para-region. This scheme seems adequate to account for the heterogeneity of κ-casein as it is detected electrophoretically but may need modification if other sources of heterogeneity, perhaps not involving charge differences, are discovered. The present state of our knowledge concerning the primary structure of κ-casein is summarized in Fig. 10.

The picture which emerges for κ-casein is consistent with the role proposed for it by Waugh and Noble (1965). In their model for the casein micelle they assign a position on the outside of the micelle for κ-casein where it is required to interact both with α_s-casein and with the surrounding solvent (see Chapter 9). The solubility properties of para-κ-casein and of the macropeptide would lead us to expect that these portions of the molecule interact with α_s-casein and with the solvent, respectively.* A higher degree of specificity would be expected to be required in the interaction with α_s-casein. Thus most of the variation in κ-casein is found to occur in the macropeptide region where fewer restrictions operate. It

---

* If the carbohydrate, attached to different extents to only a portion of the κ-casein monomers, does not account solely for the preferential interaction of the macropeptide region with the aqueous environment, it is possible that the nature of the amino acid side-chain residues themselves is responsible. In this connection the results of Hill (1969) are interesting. His amino acid analyses suggest a polarity in the distribution of hydroxylic amino acid residues along the polypeptide chain. This is also obvious from the results of Delfour *et al.* (1966). The relative proportion of serine and threonine residues rises strikingly toward the C-terminus. Hill considers that this distribution is intimately associated with the micelle-stabilizing properties of the molecule. This work has been published in detail by Hill and Wake (1969b).

is also in this region that the phosphorus of κ-casein is located (Jollès et al., 1962), and Pepper and Thompson (1963) have shown that it is not essential to the stabilizing ability of κ-casein. Finally, as Waugh and Noble point out, this model accounts nicely for accessibility of the sensitive bond to rennin.

II. The Action of Rennin (Chymosin) on κ-Casein

A. INTRODUCTION

The ability of proteolytic enzymes to clot milk is well known (see review by Berridge, 1954). Milk clotting by rennin, a gastric enzyme which occurs predominantly in the new-born mammal, particularly the calf, has been the most intensely studied. This is because rennin is most commonly used to clot milk commercially as in the first stage (renneting) of cheesemaking. The structure of rennin (chymosin) and that of its inactive precursor, prorennin (prochymosin), are discussed by Foltmann in Chapter 13. Some other powerful clotting agents are pepsin, chymotrypsin and papain. We will be concerned here with the chemical transformation which rennin causes in milk and which is responsible for clotting.

The pH activity curve for rennin acting as a general proteolytic enzyme on denatured hemoglobin is shown in Fig. 6 (Berridge, 1945). Activity is greatest at pH 3.8 and drops to zero at pH 5.0. The enzyme, acting at

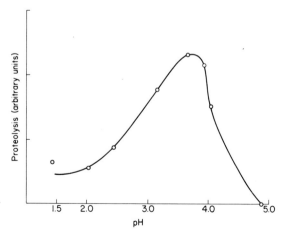

FIGURE 6. The pH activity curve for crystalline rennin acting as a general proteolytic enzyme on denatured hemoglobin (after Berridge, 1945).

12. κ-CASEIN AND ITS ATTACK BY RENNIN (CHYMOSIN) 199

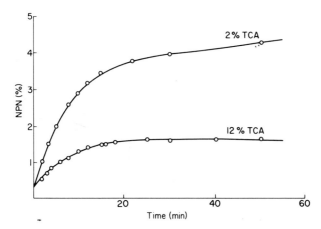

FIGURE 7. The release of NPN from whole casein by rennin at pH 6.7 (after Nitschmann and Bohren, 1955).

pH 4 on the B chain of insulin, shows a relatively broad specificity similar to that of pepsin (Fish, 1957). Undoubtedly, the main function of rennin is to digest milk in the stomach of young animals where the pH is between 2.8 and 4.5 (Berridge, 1954). Clotting of milk, however, can occur rapidly near neutral pH (6.6–6.8) and under conditions where general proteolytic activity would be extremely low. When it clots milk, rennin seems to be exerting a restricted and apparently specific action. The fact that this occurs under conditions which are unfavorable for general proteolysis has perhaps been one of the main reasons which has influenced investigators to look for an effect other than the splitting of one or more peptide bonds as the primary cause of clotting.

Rennin does not act to clot milk directly. Rather, it first brings about a chemical transformation, and the milk clots as a result of this. The two phases can be clearly distinguished (Berridge, 1954), and Nitschmann and his associates (Alais et al., 1953; Nitschmann and Keller, 1955) have shown that the first (or primary) phase can be followed by the formation of nitrogen (nonprotein nitrogen) which is soluble in trichloracetic acid. With low concentrations of rennin the amount of nonprotein nitrogen increases gradually before visible clotting occurs. The nonprotein nitrogen stems from the casein, and, more specifically, we now know that it originates from the κ-casein fraction.

The primary reaction will take place if rennin is allowed to act on whole casein near neutral pH. In the absence of calcium(II) the reaction mixture remains perfectly clear—no clotting occurs. Figure 7 (Nitschmann and Bohren, 1955) shows that the amount of nitrogen which finally becomes soluble depends upon the concentration of trichloracetic acid used to precipitate the protein. Approximately 1.5% of the total nitrogen becomes soluble in 12% trichloracetic acid and 4% of the nitrogen in 2% trichloracetic acid. The larger amount of material obtained with the lower concentration of trichloracetic acid represents the total soluble reaction product. It can also be obtained by precipitating the casein at pH 4.6. Only that portion of the soluble product which contains relatively large amounts of carbohydrate is soluble at the higher concentration of trichloracetic acid (see below).

B. κ-Casein as the Substrate for Rennin

When κ-casein was identified as the micelle stabilizer it was natural to look for it as being the site of specific attack by rennin and the source of the nonprotein nitrogen. A destruction of the stabilizing properties of κ-casein would afford a relatively simple explanation of clotting (Waugh and von Hippel, 1956). Wake (1957) and Garnier (1958a) independently showed that κ-casein is specifically the source of the nonprotein nitrogen. The rather selective degradation of κ-casein by rennin is also apparent when one examines casein–rennin reaction mixtures by starch-gel electrophoresis in concentrated urea (Wake and Baldwin, 1961).

With κ-casein at pH 6.7 and in the absence of calcium(II) rennin causes the formation of a precipitate, para-κ-casein, while 25–30% of the total nitrogen remains in solution (Wake, 1957; Beeby and Nitschmann, 1963). Approximately the same amount of material remains soluble if the para-κ-casein is removed at the end of the reaction either by centrifugation or by precipitation at pH 4.7. The soluble material is nondialyzable and has been referred to as a "macropeptide." The specific reaction which rennin catalyzes and which, in turn, causes clotting can therefore be presented as

$$\kappa\text{-casein} \xrightarrow{\text{rennin}} \text{para-}\kappa\text{-casein} + \text{macropeptide}$$
$$\text{(insoluble)} \quad \text{(soluble)}$$

Evidently when rennin acts on whole casein in the absence of calcium(II) the para-κ-casein must interact with other components so as to remain soluble.

If one follows the release of nonprotein nitrogen from κ-casein by using 12% trichloracetic acid, about two-thirds of the soluble product is precipitated along with the para-κ-casein. Only that portion which contains the carbohydrate of κ-casein remains in solution. This portion has been called the "glycomacropeptide" (Nitschmann et al., 1957; Nitschmann and

12. κ-CASEIN AND ITS ATTACK BY RENNIN (CHYMOSIN)   201

Beeby, 1960). Because of the unique occurrence among the caseins of a relatively large amount of carbohydrate in the micelle stabilizer, all of which appears in the soluble product after treatment with rennin, there has been a tendency to account for the stabilizing properties of κ-casein by virtue of its carbohydrate. The chemistry of the total soluble reaction product (or macropeptide) therefore has been largely neglected while most attention has been paid to the glycomacropeptide portion of it.

C. NATURE OF THE RENNIN-SENSITIVE LINKAGE

While there is general agreement that κ-casein is the specific substrate for rennin when it clots milk, there is considerable controversy as to the precise nature of the transformation which it undergoes to give the reaction products. Perhaps the main reason for this has been the inability to demonstrate the existence of a rennin-sensitive peptide bond in intact κ-casein. This has led to a search, with some apparent success, for an alternative type of rennin-sensitive bond. It has also been suggested that the *primary* effect of rennin is to bring about a dissociation of a "κ-casein complex" held together by secondary bonds (Beeby and Nitschmann, 1963), a covalent bond being broken subsequent to this. It will be seen later in this discussion that the most recent evidence favors the existence of a rennin-sensitive peptide bond in κ-casein.

When κ-casein is attacked by rennin, low molecular weight material sometimes appears in the soluble portion along with the macropeptide. This does not represent a primary product of enzyme action; rather it is due either to a continuing nonspecific proteolysis by rennin or to a release of adsorbed impurities in the preparation (Alais, 1956; Nitschmann and Henzi, 1959; Jollès *et al.*, 1963). A careful study of the peptide moieties obtained by the action of rennin on a single genetic variant of κ-casein gave good agreement between the summation of amino acids in para-κ-casein plus glycomacropeptide and total κ-casein (Kalan and Woychik, 1965).

Although many workers have considered that both para-κ-casein and the macropeptide are heterogeneous, it has been generally assumed that the primary action of rennin is to break only a single bond in intact κ-casein. It has been shown that the heterogeneity of the rennin reaction products is a direct result of the presence in κ-casein of a series of closely related components, most likely containing a common rennin-sensitive linkage (see Section I). The assumption of a single site of rennin attack is thus probably a valid one.

The existence of a specific rennin-sensitive peptide bond would show up by the appearance of a new N-terminal amino acid in one of the reaction products. The chemistry of the carbohydrate-rich portion of the soluble product was first investigated by Nitschmann et al. (1957). In particular, no free α-amino group could be detected in this material. Significant amounts of α-amino groups were also absent from para-κ-casein (Wake, 1959b) and it was therefore concluded that rennin could not be exerting its effect by hydrolyzing a specific peptide bond. Subsequently, both para-κ-casein and the carbohydrate-rich portion of the soluble product were studied in more detail by Jollès and his co-workers (Jollès et al., 1961, 1962). In agreement with the earlier results no N-terminal group appeared in either of the products as the result of rennin action on κ-casein. However, a new C-terminal sequence appeared in para-κ-casein.

Garnier and his co-workers (Garnier, 1958b; Garnier et al., 1962) have followed the progress of rennin action on both whole casein and κ-casein by means of the pH-stat. In both cases, at neutral pH, protons were released with kinetics similar to those for the formation of nonprotein nitrogen. During the primary reaction at pH 6.9, 1.5×10^{-6} mole H^+ per gram of whole casein and 1.8×10^{-5} mole H^+ per gram of κ-casein were produced. With κ-casein the same result was obtained over the pH range 5.4–7.4. Clearly, this is not the result which would be expected if one were observing simply the effect of the rupture of a peptide bond. No proton release was observed at pH 3.2, and it was therefore concluded that a new ionizable group with a pK between 4 and 5 was appearing. It was thought most likely to be a carboxyl arising from the hydrolysis of a rennin-sensitive ester linkage, and this fitted in with the appearance of a new C-terminal sequence in para-κ-casein. It was suggested that the rennin-sensitive linkage involved the C-terminus of para-κ-casein esterified to an alcoholic group in the macropeptide.

Jollès et al. (1963) proceeded to characterize the linkage in more detail. On reduction of κ-casein with lithium borohydride a precipitate (P) and a supernatant (S) were obtained. These two products were very similar in chemical composition to para-κ-casein and the macropeptide, respectively. More specifically, phenylalaninol was detected in the precipitate (P) and it was assumed to be in an equivalent position to the C-terminal phenylalanine which appears in para-κ-casein. The conclusion drawn from these observations was that "$LiBH_4$ reduces the rennin-sensitive linkage in κ-casein which seems to be an ester linkage involving the carboxyl group of the C-terminal phenylalanine residue of para-κ-casein."

Thus, the rejection up to 1965 by most workers of a specific peptide bond in κ-casein as being the site of attack when rennin clots milk rested largely on two types of experimental data. They were (a) the nonappearance of a new N-terminal amino acid in either of the reaction products (macropeptide and para-κ-casein) and (b) the same liberation of protons during rennin action over the pH range 5.4–7.4. In 1965 a report from Jollès' group (Delfour et al., 1965), however, indicated the need for a reappraisal of the problem. They reported that there is an N-terminal methionine residue in the glycomacropeptide which until then had escaped detection because it is rapidly destroyed during total hydrolysis. The reason for the unusual lability of this N-terminal methionine residue is not clear. It has since been detected by other workers (de Koning et al., 1966). With regard to the earlier studies by Jollès and his associates where lithium borohydride was allowed to act on κ-casein, it is clear that this reagent is splitting the same linkage as rennin. However, we now know that this in itself does not rule out the possibility that the linkage involved is a peptide bond. Crestfield et al. (1963) have shown that sodium borohydride, a similar but milder reducing agent, can cleave a peptide bond in ribonuclease. Concerning the pH-stat data of Garnier and his co-workers, the interpretation made by this group does not take into account the possible release of protons due to aggregation reactions which accompany the formation of para-κ-casein and which could mask the effect of the primary chemical reaction. This is precisely what happens during the clotting of fibrinogen by thrombin (Mihályi, 1954). Macdonald and Thomas (1970) have studied the action of rennin on SCM-κ-casein using subsequent treatment with 6 M urea to prevent aggregation reactions that could mask the proton release. They found that an ionizing group with an apparent pK of 7.6 is released by the action of rennin. It was concluded that this group is the new amino group formed by proteolytic action of the rennin on a peptide linkage between the para-κ-casein and the macropeptide.

Very convincing evidence that the primary step in milk clotting by rennin consists of hydrolysis of a labile peptide bond, linking the C-terminus of para-κ-casein to the N-terminus of the macropeptide in intact κ-casein, has been obtained by Jollès et al. (1968). They isolated a tryptic peptide containing an intact phenylalanine–methionine peptide linkage.

Because of the early finding of phosphoamidase activity in preparations of crystalline rennin (Holter and Li, 1950), and the conclusion by Perlmann (1955) that α-casein (α_s-κ-casein complex) contains NPO cross-links, D'yachenko (1959) suggested a phosphoamidase action for rennin as an explanation of its milk-clotting ability. However, the observation by Fish (1957) that phosphoamidase activity is due to a contaminant, taken along with the results of Hofman (1958) which do not support Perlmann's findings, make this suggestion unlikely (for further discussion of the phosphate esters of casein, see Chapter 9.) In addition, at least two other proteolytic enzymes, pepsin and chymotrypsin, are known to bring about a very similar, if not the same, primary reaction on κ-casein as rennin (Habermann et al., 1961; Dennis and Wake, 1965).

While Beeby and Nitschmann (1963) would agree that rennin causes the rupture of a specific covalent bond in κ-casein they have suggested that this is not the primary effect of the enzyme. Rather, they consider that the first step in rennin action is to bring about the dissociation of a "κ-casein complex" which results in the liberation of a component which is subsequently degraded to yield the glycomacropeptide. Experimentally, Beeby and Nitschmann found that precipitation of κ-casein at pH 4.7 or treatment with urea resulted in the release of material containing sialic acid and was accompanied by the formation of insoluble material resembling para-κ-casein. The acid precipitation experiments have been repeated by Alais and Jollès (1964). These workers analyzed more closely the nature of the small amount of soluble material obtained under these conditions as well as the protein which remained. In particular, they found that the soluble material did not have the expected high carbohydrate content and that para-κ-casein was not formed. Perhaps the difference in the experimental results obtained by these two groups of workers will be traced to a difference in the κ-casein preparations used. Beeby and Nitschmann used preparations obtained by two methods and it is even possible that these differed from one another, particularly with respect to the amount of other caseins which are always present in at least trace amounts.

A possible explanation of Beeby and Nitschmann's results is that the several components of κ-casein (see Section I) were not completely cross-linked by SS bridges in one or both of their preparations so that some of the components rich in sialic acid were preferentially soluble at pH 4.7. However, this would not explain the formation of para-κ-casein. Unfortunately, the nature of their insoluble material resembling para-κ-casein was not investigated in any detail. Also, it is difficult to see how the theory of rennin action put forward by these workers fits in with the observation that all of the components of κ-casein, those lacking carbohydrate as well as those rich in carbohydrate, give rise to the same para-derivatives (Mackinlay and Wake, 1965; Woychik et al., 1966).

D. HETEROGENEITY, ORIGIN, AND CHEMICAL COMPOSITION OF THE PRODUCTS OF RENNIN ACTION

The resolution of reduced and alkylated κ-casein into its individual components by starch-gel electrophoresis has greatly facilitated studies on the heterogeneity and origin of the insoluble and soluble products of rennin action.

Rennin attacks SCM-κ-casein in the same way that it acts on κ-casein. Para-SCM-κ-casein precipitates and at least 25% of the nitrogen remains

12. κ-CASEIN AND ITS ATTACK BY RENNIN (CHYMOSIN)

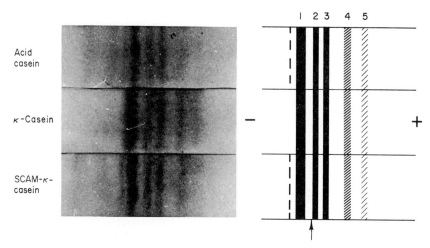

FIGURE 8. Analysis of the total soluble product, formed by the action of rennin on various casein substrates, by paper electrophoresis in acetate at pH 4 (Armstrong et al., 1967).

in solution (Mackinlay and Wake, 1964). Para-SCM-κ-casein resolves into two discrete bands, a major and a minor one, on starch-gel electrophoresis in concentrated urea (see Fig. 5). The two bands on the cathodic side of the starting slot are the para-derivatives.* The material on the anodic side comprises unreacted SCM-κ-casein components and some impurities originally present in the preparation. The soluble product does not appear under these conditions; it can be examined by paper.electrophoresis at pH 4 (Armstrong et al., 1967). The total soluble products obtained from whole casein, κ-casein and SCM-κ-casein are compared in Fig. 8. In each case it resolves into a number of components, and there is no major difference among the products obtained from these three substrates.

1. The Insoluble Product

It has been shown that the two para-derivatives are both primary products of rennin action. In other words they are formed independently of one another and are not derived from precursor products (Mackinlay et al., 1966). The same two para-derivatives are obtained from either of the genetic variants of κ-casein (and SCM-κ-casein) which analyze differently from one another under the same conditions of electrophoresis.

---

* De Koning (1967) has observed the production of three para-derivatives from reduced and nonalkylated κ-casein. See Section I.E.3.

Obviously, in the two genetic variant series the portion of each component which gives rise to a para-κ-casein fragment must be the same. At least the major para-derivative, and possibly both, would contain a C-terminal phenylalanine. The two derivatives probably differ only very slightly. That they both contain the same number of alkylated sulfhydryl groups and have an overall net charge difference of unity at pH 8.6 can be inferred from a comparison of the electrophoretic behavior of the para-SCM- and and para-SCAM-derivatives (Mackinlay, 1965). Also, there is no tendency for them to separate from one another under optimal conditions of Sephadex gel filtration (Hill, 1965). Thus they probably have similar sizes.

The two para-derivatives are formed in unaltered proportions under a wide variety of conditions of rennin treatment—the temperature has been varied from 10° to 30°C, the pH from 2 to 7.8, and the reaction has been carried out with the substrate in various forms and in organic solvent–water mixtures. This result is contrary to an earlier suggestion (Mackinlay and Wake, 1965) that they arise from an attack by rennin at two different positions in κ-casein. Hill and Wake (1969) have confirmed that the two para-κ-caseins (designated I and II) are derived from two types of precursor molecule (κ-casein I and II respectively) and this is consistent with an attack by rennin at only one position in the polypeptide chain of each SCM-κ-casein component.

Until this work, C-terminal end-group analyses had been carried out only on products that are now known to be heterogeneous. An inhibiting feature in previous work has been the difficulty in separating the two para-κ-caseins. Ion-exchange chromatography had been only partially successful (Hill, 1965). In their recent work Hill and Wake (1969) were able to fractionate the para-κ-casein into two components by preparative polyacrylamide-gel electrophoresis in concentrated urea, using two buffers with different pH values and distinct functions. The C-terminal analysis (carboxypeptidase A and B) indicated the sequence -Leu-Ser-Phe-COOH for para-components I and II. It was shown that component I could be converted to component II by prolonged action of urea, and a third component could be derived from II. The reaction was concluded to be one of carbamylation due to traces of cyanate in the urea. The para-κ-caseins used in the work had been prepared by procedures involving contact with urea. However when they were prepared by procedures not involving dissociation by urea, the two para-κ-caseins were still obtained. Thus Hill and Wake concluded that component II is not an artifact. They made the assumption that carbamylation of component I during their analytical starch-gel–urea electrophoresis is negligible. As mentioned in Section I.E. this conclusion has been disputed by Kim et al. (1969), who conclude that component II is an artifact, that there is only one para-κ-casein and that it is derived by attack by rennin at only one position in the κ-casein peptide chain. (The findings of both groups are different from that of Woychik et al., 1966, who considered that two para-derivatives arise from rennin attack on a single κ-casein component.)

The amino acid composition of para-κ-casein has been investigated by Jollès et al. (1962, 1963), Kalan and Woychik (1965), de Koning et al. (1966) and more recently by Kim et al. (1969). It contains all of the half cystine and aromatic amino acids of κ-casein as well as all of the histidine and arginine, but none or very little of the carbohydrate. Its C-terminal sequence has been determined by several groups of workers. There is some lack of agreement in the results. The C-terminal sequence was found by Jollès et al. (1962, 1963) to be -Leu-Phe-COOH, only trace amounts of N-terminal amino acids being detected. However, de Koning (1967) suggested a different C-terminal sequence, -(Val)-Leu-Ser-Phe-COOH, from studies on para-κ-casein derived from the A variant. Jollès et al. (1969) found the C-terminal sequence His-Pro-Pro-His-Leu-Ser-Phe-COOH, as shown in Fig. 10.

2. The Soluble Product

The presence of at least six discrete components in each genetic variant series of κ-casein and the attack by rennin at just one position in each to give one of two types of para-derivative would mean that the soluble portion should reflect some of the heterogeneity in the original substrate. This has now been demonstrated (Armstrong et al., 1967). An electrophoretic separation at pH 4 of the total soluble product into at least five discrete macropeptide bands is shown in Fig. 8. It is considered that the small amount of material, which is more positively charged and which varies in quantity in the three products analyzed, probably represents secondary degradation products or impurities in the original substrate preparaations. Under these conditions of electrophoresis the soluble products obtained from either of the two genetic variants of κ-casein show no difference. However, the expected difference does appear at pH 4.7 in pyridine–acetate (Armstrong et al., 1967). The resolution of the components of the soluble product is not achieved by paper electrophoresis in pyridine–acetate buffer at pH 6.5 (Blondel-Queroix and Alais, 1964).

Kinetic studies have shown that the five components of the soluble product are formed independently of one another and are not derived from precursor products. If the mixture of components is treated with 12% trichloracetic acid the major, less negatively charged ones precipitate. The carbohydrate-rich material or glycomacropeptide, represented by the faster-moving bands, remains soluble. The major component of the soluble product, which is derived from the major component of reduced and alkylated κ-casein, can be obtained free from the others by a combination of trichloracetic acid fractionation and chromatography on carboxymethyl–cellulose (Armstrong et al., 1967).

These preliminary observations are in accord with the suggestion that each component of reduced and alkylated κ-casein is split at a single position in its polypeptide chain to yield one or other of the para-derivatives along with a single soluble product component. The latter would account for approximately 30% of the original molecule. The origin of the soluble product components has been investigated by analyzing the material obtained from various purified SCM-κ-casein B components. Overall, the results were as one would have predicted. The fraction containing the major slowest-moving SCM-κ-casein component gives rise mainly to the major, slowest-moving soluble component. Fractions containing the faster-moving, carbohydrate-rich SCM-κ-casein components yield correspondingly faster-moving, carbohydrate-rich soluble products. In all cases there are additional minor components present in the soluble products. It is considered that they are not primary products of rennin action.

The amino acid and carbohydrate compositions of the soluble product, mainly the glycomacropeptide portion of it, have been investigated by many workers (for example, Nitschmann et al., 1957; Jollès and Alais, 1959; Nitschmann and Beeby, 1960; Jollès et al., 1961, 1969; Alais and Jollès, 1961; Alais et al., 1964; Kalan and Woychik, 1965; de Koning et al., 1966; Fiat et al., 1968; Kuwata et al., 1969). As mentioned previously the soluble product contains all of the carbohydrate originally present in κ-casein. Of the three sugars, galactosamine, galactose, and sialic acid, the last can be largely removed from κ-casein by neuraminidase and must therefore be terminal. Galactosamine is linked directly to the peptide chain (Jollès et al., 1964). Malpress and Seid-Akhavan (1966) concluded that the linkage is glycosidic in bovine and human casein glycopeptides. Fiat et al. (1968) also found a glycosidic linkage between the galactosamine and threonine in the bovine glycopeptide as well as in whole κ-casein. The galactosamine is probably N-acetylated (Wheelcock and Sinkinson, 1969). The carbohydrate-rich portion or glycomacropeptide fraction has been separated into at least three components which have different total carbohydrate contents (Jollès and Alais, 1959; Alais and Jollès, 1961). The total amino acid composition was determined in two of the components and found to be the same. The amino acid compositions of the total soluble product and the glycomacropeptide portion have also been compared and found to be very similar (Alais et al., 1964); this has been verified by de Koning et al. (1966). It is suggested on the basis of these findings that all of the macropeptide components in the soluble product have at least very similar peptide backbones (except for the amino acid replacements between the genetic variants), the main difference being the attachment of different total amounts of carbohydrate.

Hill et al. (1970) have determined the amino acid compositions of the highly purified major macropeptide from the A and B variant. The analyses showed the same differences as were reported earlier by de Koning et al. (1966); namely, the A macropeptide has one extra aspartic acid (or asparagine residue) and one more threonine residue, but one less alanine and isoleucine compared with the B macropeptide.

Carboxypeptidase liberates the same four amino acids, Ser, Thr, Ala and Val, from the glycomacropeptide as from intact κ-casein. It has been suggested therefore that the glycomacropeptide is situated at the C-terminal side of κ-casein and that there is only a single C-terminal sequence (Jollès et al., 1962). A start has been made on the elucidation of the amino acid sequence of the glycomacropeptide portion of the soluble product (Delfour et al., 1966) and this has already revealed some interesting features. It contains a total of 52–58 amino acids and, as mentioned previously, a methionine residue has now been shown to occupy the N-terminal position. The sequence of the first eighteen amino acids at the N-terminal end is shown in Fig. 10. Interestingly this one-third portion of the glycomacropeptide, which is attached directly to the para-part in intact κ-casein, contains all of its lysine residues, three of the four aspartic residues, half of the proline residues, no attached carbohydrate moieties and only two of the 14–17 serine plus threonine residues. This means that one in three of the 34–40 residues at the C-terminal end of κ-casein is serine or threonine. (The significance of these findings has been discussed by Hill and Wake, 1969b; see also p. 197.)

E. Sizes of the Reaction Products

The carbohydrate-free component of reduced and alkylated κ-casein has a molecular weight of 17,000–20,000 daltons (see Section I). The glycomacropeptide portion of the soluble rennin reaction product has been quoted to have a size of 6000–8000 daltons (Nitschmann et al., 1957). This is consistent with the release of approximately 30% of the original nitrogen into the soluble product. The para-κ-casein monomer should therefore have a size twice that of the macropeptide.

The realization of the nature of the heterogeneity of κ-casein and that of its rennin reaction products should allow meaningful molecular weight determinations of them. In terms of stoichiometry it seems as if initial studies in this direction would most profitably concern the products derived from the major carbohydrate-free components of κ-casein.

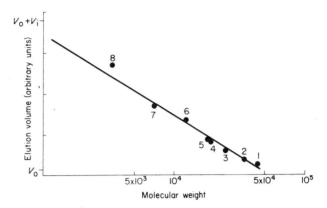

FIGURE 9. A comparison of the sizes of SCM-κ-casein and its rennin reaction products by Sephadex (G-100) gel filtration in 7 M urea at pH 8.6 (after Hill, 1965). (1) SCM-ovalbumin, (2) SCM-pepsin, (3) SCM-chymotrypsinogen, (4) SCM-κ-casein, (5) SCM-β-lactoglobulin, (6) para-SCAM-κ-casein, (7) DNP-macropeptide, (8) B chain of insulin oxidized by performic acid.

Some preliminary studies have already been made by Hill (1965). He has compared the relative sizes of the major component of SCM-κ-casein, para-SCAM-κ-casein, and the major component of the macropeptide fraction by Sephadex gel filtration in concentrated urea. A plot of elution volume vs. molecular weight is shown in Fig. 9 for these three fractions, together with a number of proteins most of which have been reduced and S-carboxymethylated. In this plot, molecular weights suggested by Mackinlay (1965), 19,500 ± 600, 12,500 ± 1000 and 7000 ± 1000 daltons, have been assumed for SCM-κ-casein, para-SCAM-κ-casein and the DNP-macropeptide, respectively. As expected, SCM-κ-casein elutes just before SCM-β-lactoglobulin (mol wt 18,000 daltons), while para-SCAM-κ-casein and the DNP-macropeptide come off in the order and in approximately the positions which would be expected if they are the result of rennin breaking a bond approximately two-thirds along the length of the polypeptide chain of κ-casein. Considering the precision of molecular weight values expected from this technique, the true significance of the result is that it gives the relative sizes of the reaction products rather than accurate molecular weight values. From its amino acid composition the minimum molecular weight of the peptide portion of the glycomacropeptide would be close to 6000 daltons (Nitschmann and Henzi, 1959; Jollès et al., 1961).

F. CONCLUSION

Chemical and physical studies by a number of workers on κ-casein and its rennin reaction products have given a considerable amount of informa-

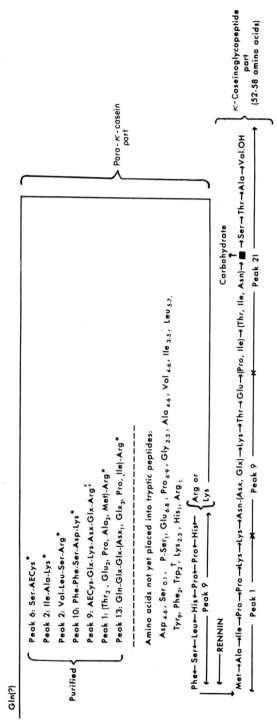

FIGURE 10. Partial amino acid sequence of bovine κ-casein based on present knowledge from the sources discussed in the text and a representation of Jollès et al. (1969). The A variant contains + 1 Asx, + 1 Thr, − 1 Ala, − 1 Ile with respect to the B variant. The location of the difference amino acids in the sequence is unknown. The several components of a given variant differ in the amount of carbohydrate present and this varies from 0–10%. The carbohydrate moiety is N-acetylgalactosamine, galactose, and N-acetylneuraminic acid in the molar ratio 1:1:1. Jollès (1969) has shown that the glycopeptide has a glycosidic linkage between threonine and galactosamine. Legend: AECys, aminoethylcystine; *, tryptic peptide; †, sequences Trp-Ser and Trp-Glu established by Jollès et al. after reaction with bromosuccinimide; ‡, tryptic peptide; no split occurred after AECys and Lys residues (Jollès et al., 1969); ■, further amino acids of the C-terminal tryptic peptide: Asp$_1$, Thr$_{6-8}$, Ser$_{4-5}$, P-Ser$_1$, Glu$_1$, Pro$_{3-4}$, Gly$_1$, Ala$_{2-3}$, Val$_{3-4}$, Ile$_2$, Leu$_1$. (For additional studies see Chapter 18.)

211

tion concerning the primary structure of the κ-casein monomer. This has been summarized in Fig. 10 (Jollès et al., 1969; see also Figs. 1 and 2 of Hill and Wake, 1969b).

At the present time the most plausible explanation for the rapid and apparently specific degradation of κ-casein by rennin near neutral pH is that a particularly labile bond exists within the molecule. That it is primarily a feature of the bond rather than a unique property of rennin is suggested by the ability of at least two other proteolytic enzymes, pepsin and chymotrypsin, to degrade κ-casein in essentially the same way as rennin does. The unusual lability of the bond also shows up by its ease of reduction by lithium borohydride.

If the bond in question is a peptide, one has to ask what renders it so susceptible to attack by these various methods. That it might not be just a feature of a Phe-Met sequence is suggested by the observation that treatment of whole casein under conditions which oxidize methionine residues has no affect on the rennin clotting of micelles prepared from such material (Hill and Laing, 1965). The observation that photooxidation of κ-casein, under conditions which affect histidine and tryptophan and not methionine residues, stabilizes the bond against rupture by rennin (Hill and Laing, 1965; Zittle, 1965) suggests that other amino acids, perhaps in the vicinity of the rennin-sensitive linkage, somehow contribute to its lability. Further support for this comes from the work of Hill (1968) and Hill and Craker (1968).

REFERENCES

Alais, C. (1956). *Proc. Int. Dairy Congr., 14th, 1956* **2**, 823.
Alais, C., and Jollès, P. (1961). *Biochim. Biophys. Acta* **51**, 315.
Alais, C., and Jollès, P. (1964). *Ann. Biol. Animale, Biochim. Biophys.* **4**, 79.
Alais, C., Mocquot, G., Nitschmann, Hs., and Zahler, P. (1953). *Helv. Chim. Acta* **36**, 1955.
Alais, C., Blondel-Queroix, J., and Jollès, P. (1964). *Bull. Soc. Chim. Biol.* **44**, 973.
Armstrong, C. E., Mackinlay, A. G., Hill, R. J., and Wake, R. G. (1967). *Biochim. Biophys. Acta* **140**, 123.
Aschaffenburg, R. (1968). *J. Dairy Res.* **35**, 447.
Aschaffenburg, R., and Thymann, M. (1965). *J. Dairy Sci.* **48**, 1524.
Beeby, R. (1963). *J. Dairy Res.* **30**, 77.
Beeby R. (1964). *Biochim. Biophys. Acta* **82**, 418.
Beeby, R. and Nitschmann, Hs. (1963). *J. Dairy Res.* **30**, 7.
Berridge, N. J. (1945). *Biochem. J.* **39**, 179.
Berridge, N. J. (1954). *Advan. Enzymol.* **15**, 423.
Blondel-Queroix, J., and Alais, C. (1964). *Bull. Soc. Chim. Biol.* **44**, 963.
Cheeseman, G. C. (1962). *J. Dairy Res.* **29**, 163.
Cherbuliez, E., and Baudet, P. (1950). *Helv. Chim. Acta* **33**, 1673.
Craven, D. A., and Gehrke, C. W. (1967). *J. Dairy Sci.* **50**, 940.
Crestfield, A. M., Moore, S., and Stein, W. H. (1963). *J. Biol. Chem.* **238**, 622.

de Koning, P. J. (1967). Thesis, University of Amsterdam, Amsterdam.
de Koning, P. J., van Rooijen, P. J., and Kok, A. (1966). *Biochem. Biophys. Res. Commun.* **24,** 616.
Delfour, A., Jollès, J., Alais, C., and Jollès, P. (1965). *Biochem. Biophys. Res. Commun.* **19,** 452.
Delfour, A., Alais, C., and Jollès, P. (1966). *Chimia* **20,** 148.
Dennis, E. S., and Wake, R. G. (1965). *Biochim. Biophys. Acta* **97,** 159.
Downey, W. K., and Murphy, R. F. (1970). *J. Dairy Res.* **37,** 47.
D'yachenko, P. F. (1959). *Proc. Int. Dairy Congr., 15th, 1959* **2,** 629.
Fiat, A. M., Alais, C., and Jollès, P. (1968). *Chimia* **22,** 137.
Fish, J. C. (1957). *Nature* **180,** 345.
Fox, P. F., and Tarassuk, N. P. (1968). *J. Dairy Sci.* **51,** 826.
Fox, P. F., Yaguchi, M., and Tarassuk, N. P. (1967). *J. Dairy Sci.* **50,** 307.
Gaffney, P. J., Harper, W. J., and Gould, I. A. (1966). *J. Dairy Sci.* **49,** 921.
Garnier, J. (1958a). *Proc. Int. Symp. Enzyme Chem. Tokyo, Kyoto,* **2,** 524.
Garnier, J. (1958b). *Compt. Rend.* **247,** 1515.
Garnier, J., Mocquot, G., and Brignon, G. (1962). *Compt. Rend.* **254,** 372.
Habermann, W., Mattenheimer, H., Sky-Peck, H., and Sinohara, H. (1961). *Chimia* **15,** 339.
Herskovits, T. T. (1966). *Biochemistry* **5,** 1018.
Hill, R. D. (1963). *J. Dairy Res.* **30,** 101.
Hill, R. D. (1968). *Biochem. Biophys. Res. Commun.* **33,** 659.
Hill, R. D. (1969). *J. Dairy Sci.* **52,** 902.
Hill, R. D., and Craker, B. A. (1968). *J. Dairy Res.* **35,** 13.
Hill, R. D., and Hansen, R. R. (1963). *J. Dairy Res.* **30,** 375.
Hill, R. D., and Laing, R. R. (1965). *J. Dairy Res.* **32,** 193.
Hill, R. J. (1965). M.Sc. thesis, University of Sydney, Sydney.
Hill, R. J. (1969). Ph.D. thesis, University of Sydney, Sydney.
Hill, R. J., and Wake, R. G. (1969a). *Biochim. Biophys. Acta* **175,** 419.
Hill, R. J., and Wake, R. G. (1969b). *Nature* **221,** 635.
Hill, R. J., Naughton, M. A., and Wake, R. G. (1970). *Biochim. Biophys. Acta* **200,** 267.
Hipp, N. J., Groves, M. L., Custer, J. H., and McMeekin, T. L. (1952). *J. Dairy Sci.* **35,** 272.
Hipp, N. J., Groves, M. L., and McMeekin, T. L. (1961). *Arch. Biochem. Biophys.* **93,** 245.
Hofman, T. (1958). *Biochem. J.* **69,** 139.
Holter, H., and Li, S.O. (1950). *Acta Chem. Scand.* **4,** 1321.
Jollès, J., Alais, C., and Jollès, P. (1968). *Biochim. Biophys. Acta* **168,** 591.
Jollès, J., Jollès, P., and Alais, C. (1969). *Nature* **222,** 668.
Jollès, P. (1969). *Hoppe-Seyler's Z. Physiol. Chem.* **350,** 665.
Jollès, P., and Alais, C. (1959). *Biochim. Biophys. Acta* **34,** 565.
Jollès, P., Alais, C., and Jollès, J. (1961).. *Biochim. Biophys. Acta* **51,** 309.
Jollès, P., Alais, C., and Jollès, J. (1962). *Arch. Biochem. Biophys.* **98,** 56.
Jollès, P., Alais, C., and Jollès, J. (1963). *Biochim. Biophys. Acta* **69,** 511.
Jollès, P., Alais, C., Adam, A., Delfour, A., and Jollès, J. (1964). *Chimia* **18,** 357.
Kalan, E. B., and Woychik, J. H. (1965). *J. Dairy Sci.* **48,** 1423.
Kim, Y. K., Yaguchi, M., and Rose, D. (1969). *J. Dairy Sci.* **52,** 316.
Kuwatu, T., Niki, R., and Arima, S. (1969). *J. Agr. Chem. Soc. Jap.* **43,** 183.
Larsen, B., and Thymann, M. (1966). *Acta Vet. Scand.* **7,** 189.

Linderstrøm-Lang, K. (1929). *C. R. Trav. Lab. Carlsberg* **17**, 1.
McCabe, E. M., Brunner, J. R., and Lillevik, H. A. (1969). *J. Dairy Sci.* **52**, 1093.
Macdonald, C. A., and Thomas, M. A. W. (1970). *Biochim. Biophys. Acta* **207**, 139.
McKenzie, H. A. (1967). *Advan. Protein Chem.* **22**, 55.
McKenzie, H. A., and Wake, R. G. (1959). *Aust. J. Chem.* **12**, 712.
McKenzie, H. A., and Wake, R. G. (1961). *Biochim. Biophys. Acta* **47**, 240.
Mackinlay, A. G. (1965). Thesis, University of Sydney, Sydney.
Mackinlay, A. G., and Wake, R. G. (1964). *Biochim. Biophys. Acta* **93**, 378.
Mackinlay, A. G., and Wake, R. G. (1965). *Biochim. Biophys. Acta* **104**, 167.
Mackinlay, A. G., Hill, R. J., and Wake, R. G. (1966). *Biochim. Biophys. Acta* **115**, 103.
Malpress, F. H., and Seid-Akhavan, M. (1966). *Biochem. J.* **101**, 764.
Marier, J. R., Tessier, H., and Rose, D. (1963). *J. Dairy Sci.* **46**, 373.
Mihályi, E. (1954). *J. Biol. Chem.* **209**, 733.
Nakai, S., Wilson, H. K., and Herreid, E. O. (1965). *J. Dairy Sci.* **48**, 431.
Neelin, J. M. (1964). *J. Dairy Sci.* **47**, 506.
Neelin, J. M., Rose, D., and Tessier, H. (1962). *J. Dairy Sci.* **45**, 153.
Nirenberg, M., Leder, P., Bernfield, M., Brimacombe, B., Trupin, J., Rottman, F., and O'Neal, C. (1965). *Proc. Natl. Acad. Sci. U.S.* **53**, 1161.
Nitschmann, Hs., and Beeby, R. (1960). *Chimia* **14**, 318.
Nitschmann, Hs., and Bohren, H. U. (1955). *Helv. Chim. Acta* **38**, 1953.
Nitschmann, Hs., and Henzi, R. (1959). *Helv. Chim. Acta* **42**, 1985.
Nitschmann, Hs., and Keller, W. (1955). *Helv. Chim. Acta* **38**, 942.
Nitschmann, Hs., and Lehmann, W. (1947). *Helv. Chim. Acta* **30**, 804.
Nitschmann, Hs., Wissmann, H., and Henzi, R. (1957). *Chimia* **3**, 76.
Noble, R. W., and Waugh, D. F. (1965). *J. Amer. Chem. Soc.* **87**, 2236.
Patel, C. V., Fox, P. F., and Tarassuk, N. P. (1968). *J. Dairy Sci.* **51**, 1879.
Pepper, L., and Thompson, M. P. (1963). *J. Dairy Sci.* **46**, 764.
Perlmann, G. E. (1955). *Advan. Protein Chem.* **10**, 1.
Pujolle, J., Ribadeau-Dumas, B., Garnier, J., and Pion, R. (1966). *Biochem. Biophys. Res. Commun.* **25**, 285.
Ribadeau-Dumas, B., Maubois, J. L., Mocquot, G., and Garnier, J. (1964). *Biochim. Biophys. Acta* **82**, 494.
Rout, T. P., Webb, E. C., and Masters, C. J. (1966). *Abstr. Aust. Biochem. Soc. Meeting, Brisbane.*
Schmidt, D. G. (1964). *Biochim. Biophys. Acta* **90**, 411.
Schmidt, D. G., Both, P., and de Koning, P. J. (1966). *J. Dairy Sci.* **49**, 776.
Spies, J. R. (1967). *Anal. Chem.* **39**, 1412.
Swaisgood, H. E., and Brunner, J. R. (1962). *J. Dairy Sci.* **45**, 1.
Swaisgood, H. E., and Brunner, J. R. (1963). *Biochem. Biophys. Res. Commun.* **12**, 148.
Swaisgood, H. E., Brunner, J. R., and Lillevik, H. A. (1964). *Biochemistry* **3**, 1616.
Talbot, B., and Waugh, D. F. (1967). *J. Dairy Sci.* **50**, 950.
Talbot, B., and Waugh, D. F. (1970). *Biochemistry* **9**, 2807.
Thompson, M. P., and Pepper, L. (1962). *J. Dairy Sci.* **45**, 794.
von Hippel, P. H., and Waugh, D. F. (1955). *J. Amer. Chem. Soc.* **77**, 4311.
Wake, R. G. (1957). *Aust. J. Sci.* **20**, 147.
Wake, R. G. (1959a). *Aust. J. Biol. Sci.* **12**, 479.
Wake, R. G. (1959b). *Aust. J. Biol. Sci.* **12**, 538.
Wake, R. G., and Baldwin, R. L. (1961). *Biochim. Biophys. Acta* **47**, 225.
Waugh, D. F. (1961). *J. Phys. Chem.* **65**, 1793.
Waugh, D. F. (1962). *Abstr. Amer. Chem. Soc. Meeting, 142nd* **P62C**.

Waugh, D. F., and Noble, R. W. (1965). *J. Amer. Chem. Soc.* **87,** 2246.
Waugh, D. F., and Von Hippel, P. H. (1956). *J. Amer. Chem. Soc.* **78,** 4576.
Wheelcock, J. V., and Sinkinson, G. (1969). *Biochim. Biophys. Acta* **194,** 597.
Woychik, J. H. (1964). *Biochem. Biophys. Res. Commun.* **16,** 267.
Woychik, J. H. (1965). *Arch. Biochem. Biophys.* **109,** 542.
Woychik, J. H., Kalan, E. B., and Noelken, M. E. (1966). *Biochemistry* **5,** 2276.
Yaguchi, M., Tarassuk, N. P., and Abe, M. (1964). *J. Dairy Sci.* **47,** 1167.
Yaguchi, M., Davies, D. T., and Kim, Y. K. (1967). *J. Dairy Sci.* **50,** 940.
Yaguchi, M., Davies, D. T., and Kim, Y. K. (1968). *J. Dairy Sci.* **51,** 473.
Zittle, C. A. (1965). *J. Dairy Sci.* **48,** 1149.
Zittle C. A., and Custer, J. H. (1963). *J. Dairy Sci.* **46,** 1183.

13 ☐ The Biochemistry of Prorennin (Prochymosin) and Rennin (Chymosin)

B. FOLTMANN

I. Introduction 217
II. Assay of Rennin 218
   A. Experimental Performance of the Milk-Clotting Test 219
   B. Definition of a Rennin Unit and Calculation of the Results 220
III. Preparation of Prorennin and Rennin. 222
   A. Prorennin 222
   B. Rennin 224
   C. Chromatographic Heterogeneity of Prorennin and Rennin . . . 227
IV. Formation of Rennin from Prorennin. 230
   A. Experimental Observation of the Activation of Prorennin . . . 231
   B. A Possible Mechanism for the Activation of Prorennin 234
V. Physical and Chemical Properties 236
   A. Molecular Weights 236
   B. Isoelectric Points and Solubility 239
   C. Chemical Composition 240
   D. Stability 244
VI. Proteolytic Activity 246
   A. General Proteolytic Activity 246
   B. Proteolytic Specificity 248
VII. A Comparison of Rennin and Pepsin 249
   References 251

I. Introduction

Rennin (EC 3.4.4.3) is the predominant milk-clotting enzyme in the fourth stomach of the calf. The use of this enzyme in cheesemaking has been known since ancient times. In the introduction to his studies on rennet and rennet-like enzymes, Peters (1894) included comprehensive quotations from the prescientific literature. The first rational attempt at isolating the active principle in cheese rennet was recorded by the French pharmacist Deschamps (1840); he suggested the name *chymosin*, derived from *chymos* (juice). The name *rennin*, originating from cheese rennet, was introduced 50 years later by Lea and Dickinson (1890). The term *chymosin* was used for many years in European literature, but it did not gain acceptance in English literature, and *rennin* was officially accepted by the Enzyme Commission (1961). It is unfortunate that the official list of enzyme names now contains two different enzymes with almost identical spelling: *rennin* (EC 3.4.4.3) from calf stomach and *renin* (EC 3.4.4.15) from the kidney. Experience has shown that because of this similarity in spelling, the two enzymes are often confused. In the time between the preparation of the final manuscript for the present chapter in November 1968 and minor revision in 1970, there has been a reintroduction of the term *chymosin*. In 1969 the editors of "Methods in Enzymology" agreed upon a change in the trivial nomenclature from rennin to chymosin (see Foltmann, 1970b). Accordingly the name chymosin has been introduced in the headings of this chapter, while the name rennin remains in the text (see also the footnote, p. 2).

Fuld (1902) reviewed the literature of the last decades of the 19th century. The name *chymase* was used by Zunz (1911) and by Oppenheimer (1926, 1939), who published reviews with numerous references. In his review on milk and milk proteins Porcher (1930) also dealt with the milk-clotting enzyme. Holter (1941) described methods used in the investigations of rennin, and more recent literature has been reviewed by Berridge (1951, 1954, 1955) and by Ernstrom (1965). In a review on casein, Lindqvist (1963) has surveyed the action of rennin on casein. Garnier *et al.* (1968) have published a comprehensive review on casein and rennin. The present chapter is partly based on an earlier review by Foltmann (1966).

II. Assay of Rennin

The assay of rennet activity on the basis of the proteolytic activity of rennin was suggested by van Dam (1912) and Mulder and Radema (1947). The most widely used method, however, has been the direct observation of milk-clotting activity. The clotting of milk by means of rennin is described in detail in Chapter 12 (see also Chapter 9). It should be pointed out that the process consists of two stages: (a) a primary enzymic transformation of κ-casein, which loses its stabilizing effect toward the remaining part of the casein complex; and (b) a secondary aggregation or coagulation stage in which a gel or a precipitate of the so-called calcium paracaseinate is formed.

A. Experimental Performance of the Milk-Clotting Test

The primary enzymic stage may be followed by measurement of the peptide liberated from κ-casein, but for routine determination of rennin activity the accomplishment of the secondary stage is always used. Visual observation of the clotting point has the disadvantage that it depends on a subjective judgment, and attempts have been made to substitute this method with more objective means. These have been concentrated mainly on viscosimetric methods. Just before visual clotting occurs there is a sharp rise in the viscosity of the milk, and this sudden increase of viscosity may be taken as the clotting point. Viscosity has been observed with torsion viscosimeters (Holter, 1932; King and Melville, 1939; Hostettler and Stein, 1954) or by measuring the flow rate of the milk through a narrow tube (Hostettler and Rüegger, 1950). Storrs (1956) has used a special apparatus he constructed himself. However, the accuracy and the reproducibility of the viscosimetric methods seem to be no better than can be obtained by visual observation of the clotting point. Consequently, the latter method is commonly used for routine determinations of rennin activity.

The clotting point, or clotting time, is generally defined as the time at which a thin film of milk breaks into visible flakes or grains. The determination may be performed by dipping a dark spatula in the milk and observing the liquid flowing down the spatula, or drops of milk may be placed at the wall of a test tube and their flow down the wall observed. A more elaborate setup is described by Sommer and Matsen (1935); in this, wide-mouthed bottles are placed on tilted rollers partly immersed in a thermostat. During the test the bottles are slowly rotated, and depending on the speed, thinner or thicker films of milk are formed over the walls of the bottles. Berridge (1952a,b) has used a similar principle. In his experiments, test tubes of milk are connected to rubber bungs fixed on tilted, slowly rotating shafts. The technique of Berridge has been adopted in the rennet test published by the British Standards Institution (1963). Yet another way of performing the milk-clotting test is by using bifurcated glass tubes. The enzyme solution is placed in one of the branches and the milk in the other. After temperature equilibration in a thermostat, the enzyme solution and the milk are rapidly mixed, and during the test the mixture is kept flowing slowly from one branch to the other and back again. In this way up to six different samples may be observed at once (Foltmann, 1962).

A modified micromethod for the quantitative estimation of rennets and other proteolytic enzymes has more recently been suggested by Lawrence and Sanderson (1969). Previously, Cheeseman (1963) found that the estimation of rennet activity by diffusion of the enzyme in a calcium caseinate agar gel was much inferior to the milk-clotting assay. However, Lawrence and Sanderson have modified the technique and recommend it as a micro-method.

B. DEFINITION OF A RENNIN UNIT AND CALCULATION OF THE RESULTS

Soxhlet (1877) suggested that the activity of rennet should be expressed by the volume of milk which one volume of rennet is able to clot in 40 min at 35°C. Although different milk samples show very great variations in their ability to clot, the "unit" of Soxhlet has been widely used in dairying. In the scientific literature of more recent years, milk-clotting activity is most often expressed as suggested by Berridge (1945). The substrate used by Berridge is reconstituted skim milk, 12 g spray-dried skim milk powder being dissolved in 100 ml 0.01 M calcium chloride. One rennin unit (RU) is then defined as the rennin activity which will clot 10 ml of the substrate in 100 sec at 30°C. Different samples of milk powder, however, also show variations in clotting ability after reconstitution. This difficulty may be overcome by testing the properties of any given batch of skim milk powder with a rennin preparation of known activity. In order to facilitate this, the British Standards Institution provides ampules of freeze-dried rennet powder of standardized activity.* It should be noted that in "Methods in Enzymology" (see Foltmann, 1970b) the following definition is suggested: A solution of chromatographically purified B-rennin (chymosin B) with an absorbancy of 1.00 at 278 nm contains 100 chymosin units (rennin units) per milliliter.

Segelcke and Storch (1870, 1874) stated that rennin activity is inversely proportional to clotting time. This simple relationship has often been used for the calculation of rennin activity, although it is only valid under special experimental conditions. Holter (1932) pointed out that the relation between clotting time and rennin activity is best described if a correction for the time lag of the secondary coagulation stage is introduced:

$$RU(T - t) = k \qquad (1)$$

where RU means the amount of rennin added, T is the observed clotting time, t is a correction for the time lag, and k is a constant which depends on the experimental conditions.

Comparison of the primary reaction, expressed by the nonprotein nitro-

* In the author's investigations the clotting abilities of different milks have been matched against a standard of crystalline rennin stored at −15°C (Foltmann, 1959a, 1962).

13. THE BIOCHEMISTRY OF PRORENNIN AND RENNIN

gen (NPN) liberated, with the visually observed clotting point shows that variations in coagulability of different milk samples are mainly due to differences in the time lag t (Foltmann, 1959c). The same series of experiments also show that t will decrease if the temperature is raised from 25° to 42°C; at the latter temperature, the clotting point will coincide with the end of the primary, enzymic stage of the clotting process (pH of the milk, 6.7). If the clotting experiments are carried out at constant temperature, t will decrease when the pH is lowered. At pH 6 and 25°C, the clotting will occur almost simultaneously with completion of the enzymic stage. This means that the rule of inverse relationship between rennin activity and clotting time holds true only under certain conditions of temperature and pH. Generally, it is advisable to calculate the milk-clotting activity of rennin from (1). It should be borne in mind, however, that since the two stages of the milk-clotting process are overlapping, (1) represents only an empirical approximation. Nevertheless, as shown by Foltmann (1959c, 1962), and as Fig. 1 illustrates, (1) is valid within relatively large variations of rennin concentrations.

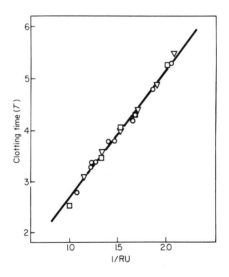

FIGURE 1. Variation of clotting time, T, with rennin concentration, RU, where $T = 2.5 \ (1/RU) + 0.3$ (Foltmann, 1962). Reconstituted skim milk (10 ml) plus rennin solution (1 ml). Each series of points, △, ○, and □, represents dilutions from one rennin stock solution. The activity of each stock solution was determined as the average of two assays with clotting times of about 4 min. The experiments were carried out at 30°C.

III. Preparation of Prorennin and Rennin

A. PRORENNIN

Like other proteinases from the digestive tract, rennin is secreted as an inactive precursor which is generally called prorennin. The proenzyme is converted into an active enzyme in the acid environments of the stomach.

Hammersten (1872) reported that he had found in calf stomachs "a substance which was not rennet, but which after reaction with acid gave rise to rennet." Only a few attempts at purifying the proenzyme have been made. Compared to rennin, prorennin is remarkably stable at neutral or weakly alkaline pH. Use has been made of this property in all published procedures for preparation of prorennin.

Kleiner and Tauber (1932) extracted minced mucosa of the fourth stomach of the calf with a 2% suspension of calcium carbonate; after filtration, the extract was precipitated twice with magnesium sulfate, and the precipitate was dried *in vacuo*. It is difficult to evaluate the purity of the preparation of Kleiner and Tauber, but it was probably rather impure.

Foltmann (1958) extracted dried, finely cut calf stomachs with 2% sodium bicarbonate. After two hours of stirring at room temperature, the tissue was removed by centrifugation. The raw extract was then clarified by a two-stage precipitation with aluminum sulfate and disodium phosphate. (This procedure is a modification of a clarification for cheese rennet suggested by van der Burg and van der Scheer, 1937.) After each step, the precipitate was spun down and the extract was finally clear and yellowish. The prorennin was precipitated twice by saturation with sodium chloride, and the precipitates were collected by centrifugation. The major part of the salt was removed from the last precipitate by dialysis before freeze-drying. The potential rennin activity, PRU, of this partially purified prorennin was 560 PRU/mg N, which is about half of the activity of crystalline rennin. These results are summarized in Table I. It may be added that six preparations of partially purified prorennin have been made according to this scheme, and although these have not been followed up by nitrogen determinations after each step, the main features have been the same as those described in Table I.

From the partially purified preparation, prorennin has been obtained by chromatography on columns of DEAE-cellulose (Foltmann, 1960b, 1962). The columns were equilibrated with 0.1 M phosphate buffer, pH 5.8; in this buffer most of the nonenzymic material passed through the column without adsorption. Prorennin was eluted with 0.15 M phosphate buffer, while minor amounts of preformed rennin were eluted with 0.2 and 0.25 M phosphate buffer. In all these experiments, the ratio of mono-

TABLE I
PARTIAL PURIFICATION OF PRORENNIN[a]

Procedure	Volume (ml)	RU/ml Preformed activity	RU/ml After activation	mg N/ml	PRU/mg N	Recovery (%)
Crude extract, pH 8.4	2000	0	220	2.75	80	100
Addition of 100 ml 0.33 M aluminum sulfate; precipitate discarded after centrifugation	1600	0	208	1.85	112	75
Addition of 80 ml 0.33 M aluminum sulfate and 80 ml 1 M disodium phosphate; precipitate discarded after centrifugation	1500	0	184	1.62	114	63
1350 ml clarified extract saturated with sodium chloride; precipitate redissolved in 0.05 M phosphate buffer, pH 6.3	220	6.6	690	1.35	510	34
215 ml saturated with sodium chloride; precipitate redissolved in 100 ml 0.05 M phosphate buffer, pH 6.3	110	12	1350			
110 ml dialyzed during 7 hr against 2 × 2 liter distilled water	145	9	1000	1.77	560	33

[a] Foltmann (1958).

sodium to disodium phosphate was 9:1; due to the increase of ionic strength, the pH dropped from 5.8 to 5.6. The prorennin prepared in this way had a potential rennin activity of 780 PRU/mg N. When corrected for the peptide that is split off during the transformation of prorennin into rennin, this figure becomes identical with the milk-clotting activity of crystalline rennin, which on the average is about 900 RU/mg N (Foltmann, 1959a, 1962).

Rand and Ernstrom (1964) report that they have prepared prorennin by a method similar to that just described. Their chromatographically purified prorennin has a potential milk-clotting activity per milligrams N which corresponds to 91% of the activity of their crystalline rennin.

The basic material used by Bundy et al. (1964) was an acetone powder prepared from frozen calf stomachs. The powder was extracted with 0.1 M tris buffer, pH 7.2. The preparation was centrifuged, and the supernatant was precipitated with ammonium sulfate (0.2 g/ml). The precipitate was spun down, redissolved, dialyzed, and freeze-dried. Further purification took place by chromatography on DEAE-cellulose columns. Elution was performed with tris buffer 0.02 M, pH 7.5 and sodium chloride. Prorennin was eluted when the sodium chloride concentration was increased to 0.28 M. The fractions containing prorennin were pooled, freeze-dried, and gel filtration was carried out on a column of Sephadex G-200. The prorennin fractions were again pooled and freeze-dried before a second ion-exchange chromatography on DEAE-Sephadex A-50. In this operation, prorennin was eluted with 0.02 M tris buffer, pH 7.5, containing 0.3 M sodium chloride. Finally, the preparation was dialyzed and freeze-dried. The authors report that the procedure represents a 15-fold purification of the extract of the acetone powder. The potential activity of the preparation has not been compared to the activity of crystalline rennin, but prorennin prepared in this way exhibits a single schlieren peak in the ultracentrifuge, and the sedimentation coefficient is the same as that reported by Foltmann (1960b).

B. RENNIN

The first crystallizations of rennin were described almost simultaneously by Berridge (1943) and Hankinson (1943). According to the reports, the first crystals were not very well shaped, but two years later, Berridge (1945) published pictures of rennin crystallized as rectangular plates.

Hankinson (1943) used commercial rennet as starting material for his preparation of crystalline rennin. The enzyme was purified by four pre-

TABLE II

CRYSTALLIZATION OF RENNIN[a]

Medium	ml	RU/ml	mg N/ml	RU/mg N	Recovery (%)
Clarified extract	5400	164	2.40	68.4	100
Dissolved					
1st precipitation	500	1240	2.88	431	70
2nd precipitation	140	3830	7.06	542	60
1st crystallization	90	3880	4.30	902	39
2nd crystallization	75	3640	3.84	948	31
Suspension					
3rd crystallization	50	4300	4.59	937	24

[a] Foltmann (1959a).

cipitations with sodium chloride, and after dialysis, crystallization took place by slow addition of acid to a salt-free solution of rennin. The crystals were described as needle-shaped. Crystallization of rennin according to Hankinson's procedure has been reported by de Baun et al. (1953) and by Hostettler and Stein (1954, 1955). Ernstrom (1958) prepared crystalline rennin by a method which includes features of both Hankinson's and Berridge's procedures. The crystals obtained were almost block-shaped. Oeda (1956) has reported the crystallization of needle-shaped crystals of rennin: In this case a salt-free solution of rennet was precipitated at pH 4.4, the precipitate was dissolved in hydrochloric acid and crystallization was effected by the addition of calcium chloride dissolved in hydrochloric acid.

The method published by Berridge in 1945 was rather tedious and has been used by Cherbuliez and Baudet (1950). However, the simplified method described by Berridge and Woodward (1953) and by Berridge (1955) is to be recommended. With slight modifications this method has been used by Alais (1956) and by Foltmann (1959a) for preparation of crystalline rennin directly from calf stomachs.

The results of a preparation of crystalline rennin by Foltmann (1959a) are summarized in Table II. The starting material for this preparation was 28 dried calf stomachs (450 g). The stomachs were cut finely and extracted with 7 liter of 1 M sodium chloride. After 4 hr of stirring at room temperature the tissue was separated by straining. The raw extract was clarified, as suggested by van der Burg and van der Scheer (1937); i.e.

150 ml 0.33 M aluminum sulfate were added to the extract, which caused the pH to fall to 3.65. Immediately after this, the pH was raised to 4.6 by the addition of disodium phosphate. In order to convert prorennin into rennin the mixture was left at pH 4.6 and at room temperature for 24 hr. The pH was then adjusted to 5.6 by the addition of more disodium phosphate. The precipitate was discarded after centrifugation, and the last traces of turbidity were removed by filtration. The clear yellow extract was saturated with sodium chloride, and after 24 hr the precipitate which formed was collected by centrifugation. The precipitate was dissolved in 0.05 M phosphate buffer, pH 6.3, the pH of the resulting solution being 5.5. A second precipitation was carried out in a refrigerator, a slow salting-out through a dialyzing bag, according to the method of McMeekin (1939); 130 g sodium chloride were dissolved in the course of 5 days. The heavy globular precipitate was then collected by centrifugation and dissolved in the minimum amount of 0.05 M phosphate buffer, pH 6.3 (115 ml). A slight insoluble residue was removed by centrifugation. The clear solution (140 ml, with final pH 5.4) was left in a refrigerator to crystallize.

After 18 days the crystals were collected by centrifugation, rapidly washed with ice-cold distilled water and redissolved in 100 ml of distilled water. Recrystallization took place after the addition of 5.4 g sodium chloride. After standing for 3 weeks in the refrigerator, the crystals were harvested and a second recrystallization took place as described above.

As can be seen from Table II, the purity of the enzyme is materially increased after the first salting-out and after the first crystallization. Recrystallization has little or no effect on the activity per milligram N.

It is usually easiest to start the preparation of crystalline rennin from commerical rennet extract or rennet powder. Berridge (1955) has noted that some brands of rennet are more difficult to handle than others. It is this author's experience, however, that these difficulties may be overcome if a clarification according to van der Burg and van der Scheer is carried out at the beginning of the procedure.

Since there is no absolute unit for rennin activity, it is difficult to compare the activity of crystalline rennin prepared in different laboratories. Several of the preparations have, however, been compared with Chr. Hansen's commercial rennet; this has made an indirect comparison possible (Foltmann, 1959a). Activities vary from 660 to about 1070 RU/mg N, but most of the preparations show activities around 900 RU/mg N. By making such a comparison it is assumed that the clotting activity of the commerical rennet is constant, but even if we allow for a variation of ± 5% it is still a reasonable basis for comparing different preparations of rennin.

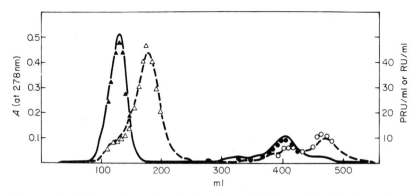

FIGURE 2. Rechromatography of chromatographically purified fractions of prorennin (Foltmann, 1962). Aliquots of the first and second half of the prorennin peak were tested by rechromatography on a column of DEAE-cellulose (0.9 × 25 cm). Elution with linear phosphate gradients: 300 ml, 0.1 M, pH 5.8 plus 300 ml, 0.3 M, pH 5.7. Absorbancy (A) of prorennin A: ----; absorbancy of prorennin B: ——; potential rennin activity (PRU/ml) of prorennin A and B: △, ▲; and rennin activity (RU/ml) of rennin from prorennin A and B: ○, ●.

C. Chromatographic Heterogeneity of Prorennin and Rennin

By chromatography on columns of DEAE-cellulose with stepwise elution, prorennin appears as a single peak (Foltmann, 1960b; Bundy et al., 1964), but if elution is carried out with a shallow gradient of phosphate buffers at pH about 5.75, the prorennin will be more or less resolved into two peaks (Foltmann, 1962). The results of two experiments involving rechromatography of two fractions of prorennin are illustrated in Fig. 2. Two distinct chromatographic peaks are indicated, but at least one of these is contaminated due to tailing in the first fractionation. During the procedure, minor amounts of prorennin are converted into active enzyme, and the elution profile shows that the different fractions of prorennin give rise to chromatographically different fractions of rennin. This fact also appears from experiments in which aliquots of the prorennin preparations have been analyzed by chromatography after complete conversion into rennin (Foltmann, 1962). As described below, similar chromatographic fractions have been observed in crystalline rennin. Bundy et al. (1964, 1967) did not find any chromatographic heterogeneity in their prorennin. The reason might be that the heterogeneity was caused by genetic variants which were not present in the calves used by Bundy et al.

In attempts at purifying rennin for investigations of its optical properties, Jirgensons et al. (1958) observed that crystalline rennin could be separated chromatographically into two or more different components. The chromatographic heterogeneity of rennin has subsequently been studied in detail by Foltmann (1960a, 1962, 1964b). The chromatographic fractionation was carried out on columns of DEAE-cellulose; among the elution systems investigated, the best resolution was obtained with phosphate buffers containing mono- and disodium phosphate in the ratio 9:1. Rennin was eluted at phosphate concentrations from 0.2 to 0.25 M. This system has the advantage that the separation takes place at the pH of optimum stability for rennin and at a pH where prorennin is activated only at a very slow rate. This means that fractionation of prorennin and rennin may be performed in one operation.

Figure 3 illustrates a typical elution profile of crystalline rennin. At the beginning of the chromatogram a negligible amount of nonenzymic material is eluted and all the milk-clotting activity is associated with three peaks, each with a characteristic ratio between milk-clotting activity and absorbancy. The three chromatographic components are designated A-, B-, and C-rennin in order of decreasing specific milk-clotting activity ($RU/ml/A_{278}$). By rechromatography, each of the fractions retains both its position in the chromatogram and its characteristic specific milk-clotting activity. The elution pattern for B-rennin previously purified by chromatography (Fig. 3) is shown in Fig. 4.

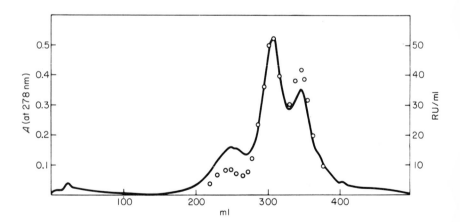

FIGURE 3. Chromatography of crystalline rennin (Foltmann, 1962). Crystalline rennin was dissolved by dialysis. A sample corresponding to about 31 mg was applied to a column of DEAE-cellulose (0.9 × 25 cm). Elution was performed with a linear phosphate gradient: 300 ml, 0.1 M, pH 5.8 plus 300 ml, 0.4 M, pH 5.5. Absorbancy (A) at 278 nm: ———; RU/ml: ○.

13. THE BIOCHEMISTRY OF PRORENNIN AND RENNIN   229

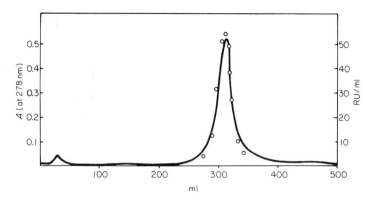

FIGURE 4. Rechromatography of B-rennin (Foltmann, 1965). B-rennin was prepared by a method analogous to that described in Fig. 3. A sample was tested by rechromatography using the same conditions as in Fig. 3. Absorbancy (A) at 278 nm: ——; RU/ml: ○.

It has already been mentioned that two proenzymes corresponding to A- and B-rennin have been isolated. So far there is no indication that A- or B-rennin consist of more than one component; on the other hand, the C-fraction is probably a mixture consisting of a degradation product formed by a limited proteolysis of A-rennin and one or more rennins of unknown origin (Foltmann, 1964b).

Chromatography of rennin has also been described by Schober et al. (1960), who used carboxymethyl cellulose. Rennin was applied to the column in 0.005 M phosphate buffer, pH 5.4, and elution took place by a stepwise increase of pH and buffer concentration. It seems strange that rennin, which has a negative charge at pH above 4.6, could be chromatographed on a cation exchanger when buffers of pH values from 5.4 to 7 are used for the elution; the adsorption probably depends on special properties of the carboxymethyl cellulose used.

It should be emphasized that the heterogeneity of prorennin and rennin described in this section is not disclosed by gel filtration, by ultracentrifugation or by ordinary moving-boundary electrophoresis (Andrews, 1964; Djurtoft et al., 1964; Alais, 1956; Ernstrom, 1958; Foltmann, 1959a). However, by moving-boundary electrophoresis in buffers of low ionic strength, Ernstrom (1958) observed a heterogeneity in the ascending boundary. Payens (1962) has reported the fractionation of rennin by paper electrophoresis, and Schober and Heimburger (1960) found heterogeneity by immunoelectrophoresis.

So far, the individual components of rennin are mainly characterized by their positions in the chromatograms and their specific milk-clotting activities. The amino acid compositions of the individual components are very similar, and we probably will have to await determinations of the primary structures in order to ascertain the chemical differences between the individual fractions.

IV. Formation of Rennin from Prorennin

Prorennin is converted into rennin by a limited proteolysis during which the molecular weight is reduced from about 36,000 (prorennin) to about

TABLE III

AMINO ACID COMPOSITION OF PRORENNIN B, B-RENNIN, AND PEPTIDES SPLIT OFF DURING THE ACTIVATION OF PRORENNIN[a]

Component	Residues per molecule		Molar ratio of amino acid in activation peptides[b]	Difference between proposed formulas of prorennin and rennin[c]
	Prorennin B (residues per 36,200 g)	B-Rennin (residues per 30,700 g)		
Lysine	12.97	8.14	4.91	5
Histidine	5.01	4.10	1.11	1
Arginine	7.01	4.85	2.00	2
Aspartic acid	33.37	29.99	3.40	3
Threonine	20.42	18.45	2.48	2
Serine	30.56	26.68	4.44	4
Glutamic acid	36.31	28.93	7.20	7
Proline	13.98	12.02	1.96	2
Glycine	28.49	24.50	4.89	3
Alanine	15.01	12.82	2.87	2
Half-cystine	(5.08)[d]	(4.88)[d]	[e]	0
Valine	22.80	20.90	1.99	2
Methionine	6.76	6.66	0.14	0
Isoleucine	19.04	15.16	3.42	4
Leucine	25.56	18.54	6.50	7
Tyrosine	18.34	15.38	2.79	3
Phenylalanine	16.45	13.79	2.43	2
Tryptophan	4.05	4.19	[e]	0
Amide N	34.03	30.70	[e]	3

[a] Foltmann (1964a).
[b] Calculated relative to arginine.
[c] Figures derived from the nearest integers of columns 2 and 3.
[d] Sequence studies indicate three SS bridges per molecule of rennin.
[e] Not determined.

31,000 daltons (rennin). Table III shows a balance sheet of the amino acid compositions of prorennin, rennin and peptides liberated during the activation of prorennin (Foltmann, 1964a). Examination of N-terminal residues indicates that the liberation of peptides occurs from the N-terminus of the zymogen. Prorennin has N-terminal alanine (Foltmann, 1960b; Bundy et al., 1964), while rennin has N-terminal glycine (Jirgensons et al., 1958). Bundy et al. (1967) have confirmed that rennin has N-terminal glycine, and from the activation mixture they have isolated a peptide with N-terminal alanine, presumably the N-terminal peptide of prorennin. As the activation peptides have no inhibitory effect on the milk-clotting activity of rennin (Foltmann, 1966), the activation of prorennin may be observed directly through the increase of milk-clotting activity of the reaction mixture.

A. Experimental Observation of the Activation of Prorennin

The activation of prorennin is dependent on pH and salt concentration. From pH 5.3 to 9 solutions of prorennin are relatively stable. At pH 5 the conversion into rennin takes place at a very slow rate, but the rate of the activation process is considerably increased by decreasing pH. This feature has been observed in earlier investigations (Holwerda, 1923; Kleiner and Tauber, 1932, 1934; Ege and Lundsteen, 1934), and although these experiments were performed with rather impure preparations of prorennin, the results are qualitatively consistent with the results obtained using purified preparations of prorennin.

Both the yield of rennin and the course of the activation process change with pH. Rand and Ernstrom (1964) have reported that they obtained maximum activity by an activation over a period of three days at 25°C and pH 5. At pH between 3 and 4 the yield of rennin is relatively low, and this is probably due to an autolytic degradation of the rennin formed (see also Section V.D). In the experiments reported by Foltmann (1966), the maximum yield was obtained by a rapid activation at pH 2.

If the increase of milk-clotting activity is plotted against time, the course of the activation at pH 4.5–5 appears as more or less S-shaped curves. If the activation is carried out in the presence of preformed rennin, the S shape disappears and the initial rate of the activation process increases with increasing concentration of the preformed rennin added (Foltmann, 1966), suggesting that the activation is partly autocatalytic. Rand and Ernstrom (1964) have reported that at pH 5 and in 1.7 M sodium chloride, the course of the activation conforms with that of a purely autocatalytic reaction. At lower values of pH the course of the activation process de-

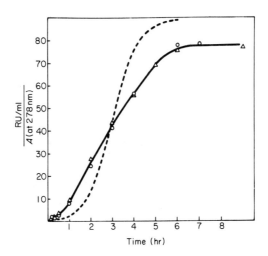

FIGURE 5. Activation of prorennin at pH 4.7 (Foltmann, 1966). The experiments were performed in 0.1 M acetate buffer, pH 4.7 at 25°C. Concentration of prorennin 0.95 mg/ml. The results show the ratio of the milk-clotting activity to the absorbancy of the prorennin solution at 278 nm ($RU/ml/A_{278}$); ------ indicates the course of an autocatalytic reaction (see text).

viates from that of an autocatalytic reaction—the lower the pH value the greater the deviation. The results of two series of experiments performed at pH 4.7 and 25°C are illustrated in Fig. 5.

If the reaction were purely autocatalytic, it would be described by the equation

$$\frac{dx}{dt} = k(PR_0 - x)(R_0 + x) \qquad (2)$$

where x is the amount of rennin formed, and PR_0 and R_0 are the initial concentration of prorennin and rennin, respectively. Assuming that the concentration of prorennin (PR_t) and rennin (R_t) at time t equal ($PR_0 - x$) and ($R_0 + x$) respectively, integration and substitution of the constant will give the equation

$$kt = \frac{2.3}{PR_0 + R_0} \cdot \log\frac{R_t \cdot PR_0}{PR_t \cdot R_0} \qquad (3)$$

The dashed line in Fig. 5 represents the course of an autocatalytic reaction calculated according to (3) on the following assumptions: (a) The initial concentration of prorennin equals the yield of rennin after activation at pH 2; (b) the preformed activity is 0.7% of the potential rennin activity*; and (c) 50% of the prorennin is transformed into rennin after 3 hr.

A comparison of the experimental and calculated curves in Fig. 5 reveals (a) that the initial rate of the activation process is greater than that of a purely autocatalytic reaction, and (b) that the yield at pH 4.7 is significantly lower than that obtained at pH 2. The former observation will be discussed in Section IV.B; the latter may be explained by autocatalytic degradation, which also occurs to a certain extent at pH 4.7.

The course of the activation at pH 2 at both 0° and 25°C is shown in Fig. 6. It was suggested previously that the activation at pH 2 resembles a first-order reaction (Foltmann, 1962; Rand, 1964). The dashed line in Fig. 6 represents the theoretical course of a first-order reaction. As can be seen from the curves in Fig. 6, the rate of the activation process at pH 2 at the beginning of the reaction is obviously considerably higher than that of a first-order reaction. In fact, the course of the activation of the process at pH 2 is reminiscent of a second-order reaction (see Section IV.B).

The experiments reported in Fig. 6 were performed in the absence of

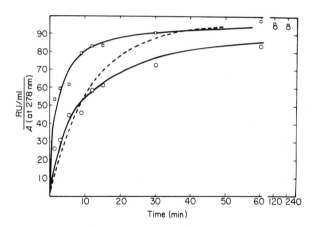

FIGURE 6. Activation of prorennin at pH 2 (Foltmann, 1966). Concentration of prorennin about 1 mg/ml. The pH was adjusted to 2 by addition of hydrochloric acid. The results show the ratio of the milk-clotting activity to the absorbancy of the prorennin solution at 278 nm ($RU/ml/A_{278}$). Points marked ○ (lower solid line) are derived from experiments performed at 0°C; points marked □ (upper solid line) are derived from experiments performed at 25°C; ------ indicates the course of a first-order reaction.

* It is difficult to perform accurate determinations of preformed rennin in prorennin, since a large surplus of potential rennin activity is added to the milk, and a slight transformation of prorennin into rennin may take place during the clotting test.

any salts. At pH 2, small concentrations of salts accelerate the activation. This appears to be a nonspecific effect of the ionic strength since similar results are obtained with solutions of sodium chloride and potassium nitrate. In both cases, the maximum rate of activation is obtained in 0.1 M solutions. At higher salt concentrations the rate of activation is again depressed (Foltmann, 1962). Rand and Ernstrom (1964) have reported that at pH values of 4.7 and 5, sodium chloride at concentrations of up to 2 M will accelerate the activation of prorennin. In a preliminary communication Shukri and Ernstrom (1966) have reported that glycerol reduces the rate of activation of prorennin.

It may be added here that prorennin may also be converted into rennin by the action of other proteolytic enzymes, but in these cases the activation is often accompanied by a considerable general proteolysis of the reaction mixture. Ege and Lundsteen (1934) thus activated a crude extract of calf stomachs (pH 5.4) by adding "pancreatin." By this activation procedure the maximum activity obtained was only half of the yield of acid activation. Foltmann (1966) obtained similar results by the addition of subtilopeptidase to a solution of prorennin at pH 6.

On the basis of experiments on the activation of crude extracts of calf stomachs in 15% sodium chloride, Linklater (1961) arrived at the conclusion that the activation of prorennin under these conditions was primarily due to the presence of pepsin. Rand and Ernstrom (1968) have stated that it is possible to convert prorennin into rennin by means of pepsin at pH as high as between 5 and 6. By adding pepsin to solutions of prorennin, Foltmann (1966) has confirmed that pepsin has a considerable power of activating prorennin, especially at pH 2. It should be pointed out, however, that it has not been possible to demonstrate the presence of pepsin in the preparations of prorennin used for the experiments described in Figs. 5 and 6, using the degradation of ribonuclease as a sensitive test for pepsin (see Section VI.B).

B. A Possible Mechanism for the Activation of Prorennin

In a discussion of the reaction mechanism of the conversion of prorennin into rennin the following features must be borne in mind: At pH 4.5–5 the activation is partly autocatalytic but the initial rate of the reaction is, in most cases, higher than that of a pure autocatalytic reaction. By decreasing the pH, the rate of the initial stage becomes so high relative to the final stage that the course of the reaction resembles that of a second-order reaction.

The autocatalytic part of the reaction may easily be explained as a manifestation of the proteolytic properties of rennin, since a similar autocatalytic activation of zymogens is known from other proteolytic enzymes, e.g., pepsin and trypsin. An increased rate of activation with decreasing pH might possibly be explained by spontaneous hydrolysis of a specially labile peptide bond or by the action of contaminating pepsin. However, both of these possibilities would result in first-order kinetics, perhaps superimposed on an autocatalytic reaction, and would thus not explain the very high rate of the initial stage relative to the final stage of the activation process.

In order to account for the experimental observations in one reaction scheme, a novel hypothesis for the activation of prorennin has been made (Foltmann, 1966). It is assumed that the inactive prorennin molecule is stabilized by the peptide moiety which is split off during the activation and that this peptide branch is at first kept in place by electrostatic interaction between positive and negative groups. If the peptide is removed from its original position, the rest of the molecule may rearrange in such a way that the active center is formed. This rearrangement takes place after the peptide has been removed by the limited proteolysis which causes irreversible activation of prorennin.

By decreasing pH, the carboxyl groups are titrated and become neutral, thereby weakening the electrostatic forces which stabilize the prorennin molecule in its inactive form. It is now postulated that under such conditions, some of the prorennin molecules may rearrange into an active conformation. These molecules may in fact exert activation power without themselves having undergone a limited proteolysis. It must further be assumed that the equilibrium between prorennin molecules in active and in inactive conformations is established very rapidly, so that in a milk-clotting test carried out at pH 6.3 only the molecules which have undergone irreversible activation by the limited proteolysis will exert milk-clotting activity. It should be pointed out however that when nothing else is stated, the term "activation of prorennin" covers the irreversible conversion into rennin.

At pH 4.7 only a small part of the molecules will be found in the active conformation, but a sufficient number will be available to cause the initial phase of the irreversible activation to be higher than that of a purely autocatalytic activation. At pH 2, on the other hand, all the carboxyl groups are almost neutral, and a large part of the prorennin molecules may rearrange into active conformation. A reaction between two molecules of prorennin may explain the very high rate in the beginning of irreversible activation at pH 2.

On the supposition that the autolytic degradation of rennin is negligible, the following equation may be set up for the formation of rennin from prorennin:

$$\frac{dx}{dt} = k_1(PR_0 - x)^2 + k_2(PR_0 - x)x \qquad (4)$$

where x denotes the amount of rennin formed, PR_0 is the initial concentration of prorennin, k_1 is the rate constant for the reaction between two molecules of prorennin, and k_2 is the rate constant for the autocatalytic formation of rennin. Equation (4) may be rearranged to (5):

$$k_1 dt = \frac{dx}{[PR_0 - x][PR_0 + (k_2 - k_1)(x/k_1)]} \qquad (5)$$

From (5), (6) is obtained by integration:

$$k_1 t = \frac{2.3}{PR_0[1 + (k_2 - k_1)(1/k_1)]} \cdot \log \frac{PR_0 + (k_2 - k_1)(x/k_1)}{PR_0 - x} \qquad (6)$$

The course of the solid lines in Fig. 6 is calculated from (6) on the assumption that the ratio between k_1 and k_2 equals 20:1 and that half of the prorennin is converted into rennin after 2 and 8 min at 25° and 0°C, respectively.

On the supposition that at pH 4.7 the ratio between k_1 and k_2 equals 1:10 and that 50% of the prorennin is transformed into rennin during 3 hr, the first half of the solid line in Fig. 5 corresponds to the course of a reaction calculated according to (6). In the second half of the reaction the observed milk-clotting activity deviates from the theoretical activity calculated according to (6). This deviation may be explained by an autolytic degradation of rennin which is not allowed for in the reaction kinetics outlined above.

It should be emphasized that thus far an active conformation of prorennin is only a postulate, but the theory is interesting because it makes it possible to explain in one reaction scheme the course of the activation of prorennin at different pH values.

V. Physical and Chemical Properties

A. MOLECULAR WEIGHTS

The chromatographic heterogeneity which is described in Section III.C does not appear when prorennin and rennin are examined by gel filtration or ultracentrifugation. By gel filtration, Djurtoft et al. (1964) found the molecular weight of prorennin to be about 36,000 daltons. Determinations of the sedimentation coefficient, performed at pH 5.8, showed it to increase with increasing concentrations of prorennin (at least up to 12 mg/ml). The single peak observed in the schlieren pattern over a wide range of concentration and the shape of the s vs. C curve may be taken as indicative of a rapid polymerization reaction. (These are discussed in Chapter 7, Volume I.) On extrapolation to zero concentration, a value of 3.5 S was found (Foltmann, 1960b; Bundy et al., 1964; Djurtoft et al., 1964). From the amino acid composition, the partial specific volume of prorennin may be calculated to be 0.73 ml/g. The diffusion coefficient of prorennin has not been determined in separate experiments, but by comparison of optical patterns from ultracentrifuge studies of prorennin and rennin, it appears that the two proteins have very similar diffusion coefficients ($D = 8.5 \times 10^{-7}$ cm^2/sec). On the basis of the above data, a molecular weight of about 37,000 daltons is obtained for prorennin. This is quite consistent with the value obtained from the amino acid composition; the minimum molecular weights of prorennin were calculated from its contents of lysine, histidine, arginine, proline, and alanine. As mentioned above, the

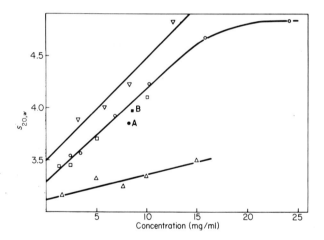

FIGURE 7. Sedimentation coefficients (s_{20}) of prorennin and rennin plotted against concentration (Djurtoft et al., 1964). All experiments, at pH 5.8, were performed in 0.1 M sodium phosphate buffer. The points correspond to the following experimental conditions: ▽: prorennin, pH 5.8 in the Spinco centrifuge; □: rennin, pH 5.8 in the Svedberg centrifuge; ○: rennin, pH 5.8 in the Spinco centrifuge; ●, ■: chromatographically purified preparations of A-rennin and B-rennin, pH 5.8 in the Spinco centrifuge; △: rennin, pH 2, 0.01 M hydrochloric acid plus 0.2 M sodium chloride in the Spinco centrifuge.

molecular weight of prorennin is probably 36,000–37,000 daltons; consequently multiples of the minimum molecular weights closest to 36,500 daltons were calculated, and they varied between only 36,125 and 36,282 daltons (Foltmann, 1964a).

As occurs with prorennin, rennin shows an increasing sedimentation coefficient with increasing protein concentration when studied in the ultracentrifuge at pH 5.8 (Baldwin and Wake, 1959; Djurtoft et al., 1964), but if the sedimentation experiments are carried out at pH 2 the variation of the sedimentation coefficient with concentration is much less pronounced. Figure 7 illustrates the results obtained by Djurtoft et al. (1964). From these experiments a value of $s^{\circ}_{20,w} = 3.2$ S was considered as the best. Djurtoft et al. determined the diffusion coefficient to be 8.5×10^{-7} cm²/sec, and from the amino acid composition and by pycnometer measurement, the partial specific volume ($\bar{v}$) was estimated to be 0.73 ml/g. From these data a molecular weight of 34,000 daltons was calculated. Rennin has also been studied by ultracentrifugation and diffusion by Schwander et al. (1952) and Friedman (1960). An increase of the sedimentation coefficient with increasing rennin concentration was not reported by these authors.

TABLE IV

MOLECULAR WEIGHT OF PRORENNIN AND RENNIN

Method	Mol wt (daltons)	$s_{20,w}$ (S)	$D \times 10^{-7}$ (cm²/sec)	$\bar{v}$ (ml/g)	Reference
Prorennin					
Sedimentation/diffusion	36,500	3.5	~8.5	0.73	Djurtoft et al. (1964)
Gel filtration	36,000				Djurtoft et al. (1964)
Amino acid composition	36,200				Foltmann (1964a)
Rennin					
Sedimentation/diffusion	40,000	4	9.5	0.75	Schwander et al. (1952)
	34,400	2.6	6.8		Friedman (1960)
	34,000	3.2	8.5	0.73	Djurtoft et al. (1964)
Sedimentation/equilibrium	30,000				Cheeseman (1969)
Gel filtration	33,000				Djurtoft et al. (1964)
	31,000				Andrews (1964)
	34,000				de Koning (1967)
Amino acid composition	30,700				Foltmann (1964a)
	30,400				de Koning (1967)

De Koning (1967) has tried to determine the molecular weight of rennin by approach to equilibrium (the Archibald method); the results varied from 46,000 to 49,000 daltons, probably due to associating molecules. By gel filtration, de Koning arrived at a molecular weight of ~34,000 daltons, which is reasonably consistent with the value of 31,000 daltons reported by Andrews (1964) and 33,000 daltons by Djurtoft et al. (1964).

Cheeseman (1969) has reported the determination of the molecular weight of rennin by the sedimentation equilibrium method. At pH 3.1 the observed molecular weight was about 30,000 daltons, while at pH 5.7–6.5 the apparent molecular weight varied from 60,160 to 67,345 daltons, indicating that a dimer is mainly present at pH ~ 6.

As in the case of prorennin, the molecular weight of rennin has been calculated on the basis of the determination of lysine, histidine, arginine, proline, and alanine. These results suggest a molecular weight of 30,700 daltons (Foltmann, 1964a). By similar calculations de Koning (1967) arrived at a molecular weight of 30,400 daltons.

Evaluation of the possible errors in the different methods suggests that the determination which is based on the amino acid composition is probably the most accurate. Rounded off to the nearest thousand, a molecular weight of about 31,000 daltons may therefore be considered to be the best estimate. The results are summarized in Table IV.

B. Isoelectric Points and Solubility

Paper electrophoresis (Foltmann, 1958) and moving-boundary electrophoresis (Foltmann, 1966) both indicate that the isoelectric point of prorennin is somewhat higher than that of rennin and probably about pH 5. By paper electrophoresis, Hankinson (1943) and Schwander et al. (1952) estimated the isoelectric point of rennin to be about 4.6. This value has been substantiated by moving-boundary electrophoresis (Foltmann, 1966).

Rennin crystals are very insoluble in dilute salt solutions. Berridge (1945) has reported that his crystals are only slightly soluble in distilled water, while crystals prepared by Foltmann (1959a) dissolved easily by dialysis against distilled water. The differences might be explained by small differences in the crystallization procedure.

Hankinson and Palmer (1942) observed that amorphous precipitates of rennin have a solubility like a globulin, being soluble in dilute salt solution at pH 5.5 and precipitating in saturated solutions of sodium chloride. Foltmann (1959b) also found that rennin was precipitated in saturated solutions of sodium chloride, but at pH 5.5 a peculiar solubility minimum

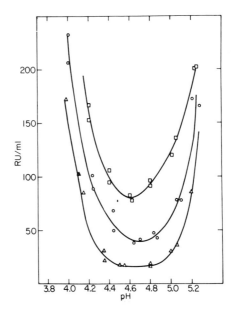

FIGURE 8. Solubility of rennin at pH near the isoelectric point (Foltmann, 1959b). Liquid phase: sodium acetate buffer; △: ionic strength (I) = 0.005; ○: I = 0.025; □: I = 0.05. Solid phase: amorphous precipitates of rennin, previously purified by crystallization. The concentration of rennin in the liquid phase is shown by the milk-clotting activity. The rennin used had a milk-clotting activity of 140 RU/mg.

was observed in buffers of ionic strength about 0.1. Investigations of the solubility relative to pH show that rennin is rather insoluble at pH values near its isoelectric point, the solubility increasing with increasing ionic strength of the solutions. Some solubility curves reported by Foltmann (1959b) are illustrated in Fig. 8.

C. CHEMICAL COMPOSITION

1. *Amino Acid Analyses*

The results currently available are compiled in Table V. On evaluating these analyses the following points should be borne in mind. The results marked (a) represent the first determination of the amino acid composition of rennin; these experiments were performed by means of partition chromatography on starch columns and are probably less reliable than the following determinations which have been obtained by ion-exchange chromatography. The results marked (a) and (c) have been corrected to 100%

13. THE BIOCHEMISTRY OF PRORENNIN AND RENNIN

TABLE V

AMINO ACID COMPOSITION OF RENNIN (moles per 10^5 g)[a]

Amino acid	Total crystalline rennin			B-Rennin
	a	b	c	d
Lysine	23	24.0	24.0	26.5
Histidine	21.5	14.1	13.2	13.4
Arginine	8.5	16.7	17.1	15.8
Aspartic acid	133	103.3	99.7	97.7
Threonine	55.5	61.8	64.2	60.1
Serine	86	80.1	80.0	86.9
Glutamic acid	91	93.3	93.4	94.3
Proline	46.5	43.7	41.8	39.2
Glycine	69.5	75.4	76.0	79.8
Alanine	35	41.1	41.5	41.8
Half-cystine	7.5	19.5	15.8	19.5
Valine	98	66.3	70.7	68.1
Methionine	7	21.4	19.7	21.7
Isoleucine	119[b]	50.6	52.0	49.4
Leucine		64.0	63.2	60.4
Tyrosine	42.5	51.8	52.6	50.1
Phenylalanine	48	45.0	46.4	44.9
Tryptophan	8	12.4	15.8	13.6

[a] Methods used for the determination of amino acids in the individual analyses: column a, partition chromatography on starch columns; columns b–d, ion-exchange chromatography on columns of sulfonated polystyrene. Legend: column a, average of 3 analyses after 24 hr of hydrolysis (Schwander et al., 1952); column b, average of 3 analyses after 24 hr of hydrolysis (Foltmann, 1964a); column c, average or extrapolated values of 3 analyses performed after 24, 48, and 72 hr of hydrolysis (de Koning, 1967); column d, average or extrapolated values of 6 analyses of samples hydrolyzed pairwise for 24, 54, and 72 hr (Foltmann, 1964a).
[b] Leucine plus isoleucine.

recovery; the actual recovery in these experiments was 85–88 g amino acid residue per 100 g of protein. With the exception of the half-cystine content, the results marked (b) and (d) are reproduced directly from experiments which gave 97–98% recovery. In the latter experiments, chromatographic analyses of the acid hydrolyzates indicated a half-cystine content of about 5 mole/30,700 g rennin. Since investigations of the primary structure have shown the existence of three cystine bridges per molecule of rennin (Table VI), the half-cystine content has been corrected according to this. Six half-cystine groups per molecule of rennin are also in agreement with the results published by Cheeseman (1965). Table V shows that the amino acid composition of B-rennin is almost equal to that of crystalline rennin; current investigations in this author's laboratory indicate that this is also the case for A-rennin.

2. Primary Structure

Having purified rennin by chromatography, Jirgensons et al. (1958) determined the N-terminal amino acid by the fluorodinitrobenzene method and found it to be glycine. By hydrazinolysis they found the C-terminal group to be leucine or isoleucine. From this we may assume that the rennin molecule consists of a single peptide chain containing internal SS bridges. After reaction with carboxypeptidase, Bundy et al. (1967) found that asparagine was liberated first followed by the release of glycine or alanine. The results of Jirgensons et al. are consistent with the amino acid sequences reported in Fig. 9, while the C-terminus suggested by Bundy et al. is not. Foltmann and Hartley (1967) have now launched detailed investigations of the primary structure. The first attempts were concentrated on the determination of the sequences around the SS bridges in rennin. As it was difficult to isolate all cystine peptides from one digest, the following degradations were carried out: (a) autolysis at pH 3.5, (b) autolysis at pH 3.5 followed by peptic digestion at pH 2, and (c) tryptic and chymotryptic digestion at pH 9. The digests were analyzed by the so-called diagonal electrophoretic technique according to Brown and Hartley (1966). The amino acid sequence was determined from the N-terminus using the dansyl-Edman method according to Gray and Hartley (1963). The C-terminal amino acids were analyzed after degradation with carboxypeptidase. In this manner, three separate cystine bridges were identified. The results are shown in Fig. 9. Although the starting material in these investigations was total crystalline rennin, the great difference between the three cystine peptides and the similarities in amino acid composition of the individual chromatographic fractions of rennin indicate that all three SS bridges are present in each of the major components of crystalline rennin.

Further analyses of the primary structure were hampered by the fact that after tryptic digestion, most of the molecule was present as insoluble "core peptides." However, in the primary structure there is a cluster of basic amino acids near the C-terminus, and thus a C-terminal amino acid sequence of 28 residues has been compiled from tryptic peptides containing 2–9 residues. The overlapping sequences have been provided by peptides obtained after autolysis and chymotryptic digestion. These peptides were fractionated and purified by two-dimensional paper electrophoresis/paper chromatography as described by Ambler (1963). Recently the N-terminal

N-Terminal sequence

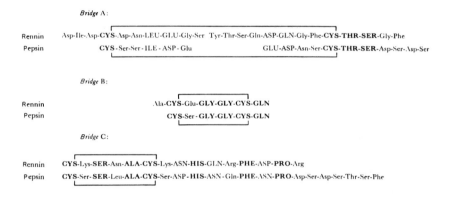

FIGURE 9. Comparison of amino acid sequences of peptides from bovine rennin and porcine pepsin. Data are those of Foltmann and Hartley (1967), Foltmann (1970a), Tang and Hartley (1967), Dopheide et al. (1967), and Stepanov (1968). Chemically similar residues are given in capitals and identical residues are in bold capitals. (See also Chapter 18 for a note on N-terminal sequence.)

amino acid sequence has been established (Foltmann, 1970a). The results are illustrated in Fig. 9. (The homologies with pepsin indicated in Fig. 9 are discussed in Section VII.) Attempts were made to purify the "core peptides" by chromatography on columns of DEAE-cellulose with tris buffer, pH 8, containing 8 M urea. The results have so far not been successful since the peptides have been more or less cross-contaminated, but a few peptides obtained after chymotryptic hydrolysis of partly purified tryptic fragments have been sequenced (for details see Foltmann and Hartley, 1967).

Still very little is known about the tertiary structure of the rennin mole-

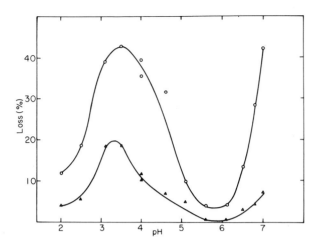

FIGURE 10. Stability of rennin (Foltmann, 1959a). The influence of pH on the stability of rennin expressed as percentage of loss of initial activity after 48 hr at 25°C (○) and 48 hr at 2°C (△). Ionic strength of all solutions 0.05; pH 2–4 citrate buffers; pH 4–5.1 acetate buffers; pH 5.1–7 phosphate buffers. The upper points at pH 4 represent loss in citrate buffers, and the lower points represent loss in acetate buffers.

cule. Bunn et al. (1970) have initiated X-ray diffraction studies on rennin crystals. Great difficulties have been encountered in the preparations of heavy-atom derivatives, and so far the interpretation of the diffraction patterns is ambiguous.

D. STABILITY

As mentioned in Sections III.A and IV.A, the potential milk-clotting activity of prorennin is relatively stable at room temperature from about pH 5.3 to about pH 9. Solutions of rennin have a moderate stability at pH 2. The pH of optimum stability lies between pH 5.3 and 6.3. At neutral or alkaline pH, solutions of rennin rapidly loose their milk-clotting activity. Figure 10 illustrates the observations of Foltmann (1959a), which have been confirmed by Micklesen and Ernstrom (1963, 1967). The latter authors point out that the losses of rennin activity at pH 3.8 are increased considerably when the sodium chloride concentration is raised from 0.03 to 1 M. As rennin exerts maximum proteolytic activity at about pH 3.5 (Section VI.A), the loss of activity in this region of pH is most probably a result of autolysis. This is sustained by Mickelsen and Ernstrom (1967), who find that the loss of rennin activity at pH 3.8 almost parallels the increase of color yield after reaction with ninhydrin.

If solutions of rennin partially inactivated at pH 3.5 are analyzed by chromatography on columns of DEAE-cellulose, it is observed that the A-rennin disappears more rapidly than the other components and that it is transformed into a component which in the chromatograms is eluted at the same position as the C-rennin (Foltmann, 1962). This degradation product has been denoted C_2-rennin (Foltmann, 1964b). The conversion of A-rennin into C_2-rennin is assumed to be a case of limited proteolysis, and by this reaction the ratio of the milk-clotting activity to the absorbancy at 278 nm is reduced from about 120 for A-rennin to about 30 RU/ml/$A_{278\ nm}$ for C_2-rennin. It is remarkable that the corresponding loss of proteolytic activity against hemoglobin amounts to only about 50%.

Schober et al. (1960) have observed that the loss of milk-clotting activity at pH 7 is accompanied by an increase in the color reaction with ninhydrin. These authors also suggest that the inactivation at pH 7 is due to autolysis. Although this pH is far from that of the optimum proteolytic activity of rennin, it is possible that the inactivation at pH 7 is a combination of alkali denaturation and autolysis such that a part of the rennin molecules undergoes an unfolding which makes them easily accessible to the small amount of proteolytic activity still present at pH 7. A similar mechanism has been suggested for the inactivation of pepsin at neutral pH (Bovey and Yanari, 1960).

Cheeseman (1965) studied the denaturation of rennin in urea-containing solutions and found that rennin is more susceptible to urea denaturation than pepsin. He subsequently examined denaturation with urea by comparing the loss of milk-clotting activity with the change in ultraviolet absorbtion of the protein (Cheeseman, 1969); however, the two sets of observations did not correlate. Further experiments with other reagents indicated that tryptophan groups are not specifically associated with enzyme action but are important in maintaining the tertiary structure. One or more tyrosine groups seem to be important for the enzymic activity.

Hill and Laing (1965) have investigated inactivation of rennin by photooxidation, and according to these authors, this inactivation depends on destruction of the histidine groups. It is, however, difficult to evaluate these results quantitatively since the rennin used in these experiments is said to contain 7 mole histidine/32,000 g, which is significantly more than the value reported by other authors (see Table V). Hill and Laing (1966) have also reported that rennin is inactivated by dansyl chloride. Their results suggest that rennin is inactivated by coupling one mole of dansyl chloride to one mole of rennin. In a later paper Hill and Laing (1967) report that the group substituted is lysine.

In a discussion of the stability of rennin, it should be noted that Green and Crutchfield (1969) have attempted the preparation of stable water-insoluble derivatives of rennin by coupling rennin to aminoethyl cellulose according to the method of Habeeb (1967). The aim was not achieved, since the milk-clotting activity of the preparation was due to rennin which dissociated from the insoluble carrier.

VI. Proteolytic Activity

A. GENERAL PROTEOLYTIC ACTIVITY

Having crystallized rennin, Berridge (1945) studied its proteolytic activity with hemoglobin as substrate; a proteolytic optimum of pH 3.7 was observed. Later de Baun et al. (1953) and Fish (1957) confirmed this. Foltmann (1964b) investigated chromatographically purified fractions of rennin and found that each of the individual components has the same pH optimum as assayed by degradation of acid-denatured hemoglobin. Concerning other substrates, it should be mentioned that bovine serum albumin is digested with a rather narrow pH optimum of about 3.4 (Foltmann, 1959a), and that Fish (1957) has reported that the B chain of oxidized insulin is digested with an optimum pH 3.5.

The degradation of casein by rennin is of particular interest from a physiological point of view, because casein is the natural substrate in the stomach of the calf, and from a technological point of view, because this reaction is important for the ripening of cheese. Due to its insolubility at pH around its isoelectric point (pH 4.6), casein is not a suitable substrate for determining the variations of proteolytic activity with pH. However, by performing the digestion in rotating tubes for one hour, the results indicate a proteolytic optimum at about pH 4 (Foltmann, 1959a). Lindqvist and Storgårds (1959, 1960) have studied the degradation of α- and β-caseins during periods of 15–25 days. Under such conditions a maximum liberation of NPN was found to occur at pH 4.5. The electrophoretic patterns of the enzymically transformed casein were also studied and appeared to be rather complex.

Cerbulis et al. (1959, 1960) have investigated the degradation products from α- and β-caseins after reaction at pH 6.4 with rennin and pepsin. The main components were apparently the same, but after the degradation with pepsin a larger amount of low molecular weight peptides was formed. Nitschmann and Varin (1951) have reported that at pH 6.8 rennin is able to split, on the average, 1 peptide bond out of 33 in casein, while at pH 2, 1 bond out of 8 can be hydrolyzed. Fox (1969) has investigated the influence of temperature and pH on the general proteolysis of casein by commercial rennet extract and crystalline rennin. Evaluation of the

TABLE VI
Peptide Bonds of the B Chain of Oxidized Insulin Hydrolyzed by Rennin, Pepsin, Parapepsin I, and Parapepsin II[a]

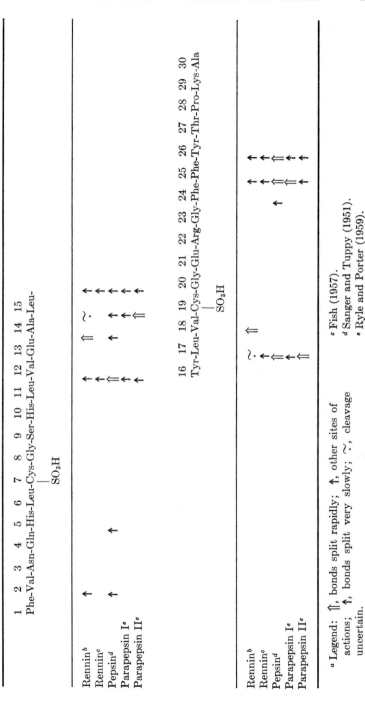

[a] Legend: ⇑, bonds split rapidly; ↑, other sites of actions; ⇈, bonds split very slowly; ∼·, cleavage uncertain.
[b] Bang-Jensen et al. (1964).
[c] Fish (1957).
[d] Sanger and Tuppy (1951).
[e] Ryle and Porter (1959).

experiments by gel electrophoresis of the degraded casein showed a pH optimum of rennin proteolysis at pH ~5.8, while changes in NPN indicated a lower pH optimum.

B. PROTEOLYTIC SPECIFICITY

The specific attack of rennin on κ-casein is described in Chapter 12. In this section the splitting of other peptide bonds are considered.

The proteolytic specificity of rennin has been studied by Fish (1957) and Bang-Jensen et al. (1964) through the degradation of the B chain of oxidized insulin, but the results of the two reports are not quite consistent. The discrepancies stem mainly from degradation products, which were purified only by paper electrophoresis followed by chromatography. Fish purified his fractions of the digests by one-dimensional electrophoresis only, so that he did not observe all of the peptides later found by Bang-Jensen et al. The results are shown in Table VI.

Proteolytic enzymes with an activity similar to that of pepsin have been observed in hog and human gastric mucosa (Ryle and Porter 1959; Tang et al., 1959; Taylor, 1962; Tang and Tang, 1963). For comparison, the specificities of pepsin and the so-called parapepsins are also included in Table VII. It should be pointed out that so far it has not been possible to demonstrate any differences between the proteolytic specificities of the individual components of rennin.

Although there are some similarities between the proteolytic specificities of pepsin and rennin, clear differences are also observed. These differences in proteolytic specificities are seen in the degradation of ribonuclease. Ribonuclease is inactivated by pepsin at pH 2 through liberation of a tetrapeptide from the C-terminus. On the basis of this reaction, Berger et al. (1959) have developed a sensitive method for determination of pepsin: Ribonuclease is first used as a substrate for pepsin, and in a subsequent step the pepsin is assayed indirectly through the remaining ribonuclease activity. Rennin does not inactivate ribonuclease in this way, neither at pH 2, nor at pH 3.5 (Bang-Jensen et al., 1964).

Until now no differences in the specificities of rennin and pepsin have been found by hydrolysis of synthetic peptides. Both Fish (1957) and Friedman (1960) have reported that benzyloxycarbonyl glutamyltyrosine is hydrolyzed by rennin, and Friedman also found hydrolysis of benzyloxycarbonyl glutamylphenylalanine. The pH optimum for these hydrolyses is said to be between pH 5 and 6. This is somewhat above the pH optimum of general proteolysis, but it should be borne in mind that a similar relation between the proteolytic optimum and that of hydrolysis of synthetic peptides is also found for pepsin (Bergmann and Fruton, 1941).

Among the synthetic substrates, Levin (1962) has observed that poly-L-glutamic acid is hydrolyzed by rennin.

Attempting to find suitable synthetic substrates for rennin, Hill (1969) has investigated the hydrolysis of sulfite esters and N-substituted imidazole compounds. Phenyl sulfite esters are good substrates for pepsin (Reid and Fahrney, 1967), and experiments show that they are hydrolyzed by rennin as well. During the milk-clotting a Phe-Met bond is hydrolyzed in κ-casein; this bond seems to be particulary exposed in the casein, since the same bond is not hydrolyzed in a tryptic digest of κ-casein (Jollès et al., 1968). Hill (1969) has also investigated rennin hydrolysis of model peptides similar to the amino acid sequence around the Phe-Met bond in κ-casein, and he arrives at the conclusion that a nearby serine side chain is important for the attack by rennin at the Phe-Met bond.

VII. A Comparison of Rennin and Pepsin

The question of the possible identity of pepsin and rennin, which Dumas (1843) raised more than 100 years ago, is no longer relevant, but a comparison of the two enzymes reveals interesting features. Investigations of the amino acid sequences around the SS bridges of pepsin have been carried out by Tang and Hartley (1970). As indicated in Fig. 9 there is a high degree of homology between the primary structures of rennin and pepsin around the SS bridges marked B and C. Only little resemblance is found in bridge A. The sequences of bridges B and C in pepsin are partially confirmed by the results of Dopheide et al. (1967). These authors have also analyzed the C-terminal sequence of pepsin, which shows a pronounced similarity to that of rennin. In recent years the investigations of primary structures have indicated that many proteins belong to families, each of which has a common ancestor (for a review see Dixon, 1966). The observed homology in the primary structures of calf rennin and porcine pepsin suggests that the gastric proteolytic enzymes also are derived from one common ancestor.

The relationship between rennin and pepsin is likewise reflected in the properties of the zymogens and the way these are transformed into active enzymes. Both of the enzymes are secreted as precursors which are converted into active enzymes in acid medium, and about equal quantities of peptide material are liberated during the activation of the two proenzymes. In both cases the isoelectric point of the proenzyme is higher than that of the active enzyme, the course of the activation being more or less autocatalytic at pH about 4.5 and changing to a much more rapid reaction at lower pH. It is possible that the mechanism of activation suggested for prorennin in Section IV.B also applies to the activation of pepsinogen. The pepsin part of the pepsinogen molecule is more acidic and the peptides liberated are relatively more basic than the corresponding parts of the prorennin molecule. From this one could expect a stronger electrostatic interaction between the pepsin part and the activation peptides in the pepsinogen molecule than is true for the corresponding interaction in the prorennin molecule. This conclusion is consistent with the finding that at pH 4.5 the activation of pepsinogen appears to be more purely autocatalytic than the activation of prorennin (Herriott, 1938, 1939). It is only at lower pH values that pepsinogen would be expected to occur in an active conformation. In this connection it is noteworthy that Perlmann (1967) from physicochemical observations has arrived at the conclusion that the configuration of pepsinogen (at pH above 6) is stabilized by electrostatic bonding between the basic amino acid residues of the activation peptides and some of the dicarboxylic acids of the pepsin moiety. In contrast to this the active pepsin molecule is stabilized by hydrophobic interactions.

It must be stressed that all comparisons between rennin and pepsin are based on experiments performed with bovine rennin and porcine pepsin. Further studies on the properties of bovine pepsin might provide new and interesting aspects of the problem. As mentioned in Section VI.B, several proteolytic enzymes have been isolated from hog gastric mucosa. The differences between these, however, appear to be greater than the differences between the individual components of crystalline rennin. It is perhaps reasonable to assume that the relation between porcine pepsin and the parapepsins corresponds to that of bovine pepsin and rennin. The question remains as to how many species have specific milk-clotting enzymes. Holter and Andersen (1934) analyzed the ratio between the milk-clotting and the proteolytic activities of gastric juice from young and adult mammals. Their investigations comprised the pig, dog, cow, and man. Only in calves was the milk-clotting activity higher in the young than in the adult animals. Malpress (1967) has investigated the gastric secretion of infants during the first six weeks after birth. No rennin was detected in any sample. Henschel et al. (1961) have investigated gastric juice from calves, aged 5 days to 8 weeks, during which period the calves were fed only on milk. A continuous displacement was observed from predominantly rennin-containing to predominantly pepsin-containing gastric juice. There were, however, great variations between the rennin–pepsin ratio for the individual calves. Yamamoto and Takahashi (1952) have prepared rennet from kid stomachs, while Alais et al. (1962) have analyzed rennet from lamb stomachs.

From this scattered information it appears that specific milk-clotting enzymes are characteristic of young ruminants. In this connection attention should be drawn to the distinction, made by Porcher (1930), between different types of milk. Porcher divided the milk into two groups. One, which was called the "casein-like" milk, clotted easily; the milk of ruminants like cow, reindeer, goat, and sheep belonged to this group. The other group was called the "albumin-like" milk. To this group belonged milk from the human, horse, donkey, and dog. It seems very likely that a relation exists between the caseins of the milks and the gastric proteolytic enzymes of the young animals. An investigation along these lines would probably reveal interesting features on the physiological and evolutionary significance of rennin.

REFERENCES

Alais, C. (1956). *Lait* **36**, 26.
Alais, C., Dutheil, H., and Bosc, J. (1962). *Proc. Int. Dairy Congr., 16th, 1962* **B**, 643.
Ambler, R. P. (1963). *Biochem. J.* **89**, 349.
Andrews, P. (1964). *Biochem. J.* **91**, 222.
Baldwin, R. L., and Wake, R. G. (1959). *Abstr. Amer. Chem. Soc. Meeting, 136th*, 35C, No. 74.
Bang-Jensen, V., Foltmann, B., and Rombauts, W. (1964). *C. R. Trav. Lab. Carlsberg* **34**, 326.
Berger, A., Neumann, H., and Sela, M. (1959). *Biochim. Biophys. Acta* **33**, 249.
Bergmann, M., and Fruton, J. S. (1941). *Advan. Enzymol.* **1**, 63.
Berridge, N. J. (1943). *Nature* **151**, 473.
Berridge, N. J. (1945). *Biochem. J.* **39**, 179.
Berridge, N. J. (1951). *In* "The Enzymes" (J. B. Sumner and K. Myrbäck, eds.), Vol. 1, Part 2, pp. 1079-1105. Academic Press, New York.
Berridge, N. J. (1952a). *J. Dairy Res.* **19**, 328.
Berridge, N. J. (1952b). *Analyst (London)* **77**, 57.
Berridge, N. J. (1954). *Advan. Enzymol.* **15**, 423.
Berridge, N. J. (1955). *In* "Methods in Enzymology" (S. P. Colowick and N. O. Kaplan, eds.), Vol. 2, pp. 69-77. Academic Press, New York.
Berridge, N. J., and Woodward, C. (1953). *J. Dairy Res.* **20**, 255.
Bovey, F. A., and Yanari, S. S. (1960). *In* "The Enzymes," 2nd ed. (P. D. Boyer, H. Lardy, and K. Myrbäck, eds.), Vol. 4, pp. 63-92. Academic Press, New York.
British Standards Institution (1963). British Standard 3624 : 1963, London.
Brown, J. R., and Hartley, B. S. (1966). *Biochem. J.* **101**, 214.
Bundy, H. F., Westberg, N. J., Dummel, B. M., and Becker, C. A. (1964). *Biochemistry* **3**, 923.
Bundy, H. F., Albizati, L. D., and Hogancamp, D. M. (1967). *Arch. Biochem. Biophys.* **118**, 536.
Bunn, C. W., Camerman, N , Liang Tung T'sai, Moews, P. C., and Baumber, M. E. (1970). *Phil. Trans. Roy. Soc. London, Ser. B* **257**, 153.
Cerbulis, J., Custer, J. H., and Zittle, C. A. (1959). *Arch. Biochem. Biophys.* **84**, 417.
Cerbulis, J., Custer, J. H., and Zittle, C. A. (1960). *J. Dairy Sci.* **43**, 1725.
Cheeseman, G. C. (1963). *J. Dairy Res.* **30**, 17.
Cheeseman, G. C. (1965). *Nature* **205**, 1011.
Cheeseman, G. C. (1969). *J. Dairy Res.* **36**, 299.

Cherbuliez, E., and Baudet, P. (1950). *Helv. Chim. Acta* **33**, 1673.
de Baun, R. M., Connors, W. M., and Sullivan, R. A. (1953). *Arch. Biochem. Biophys.* **43**, 324.
de Koning, P. J. (1967). "Studies on Rennin and the Genetic Variants of Casein." Thesis, University of Amsterdam, The Netherlands, H. Veenman, Wageningen, The Netherlands.
Deschamps. (1840). *J. Pharm.* **26**, 412.
Dixon, G. H. (1966). *In* "Essays in Biochemistry" (P. N. Campbell and G. D. Greville, eds.), Vol. 2, pp. 147–204. Academic Press, New York.
Djurtoft, R., Foltmann, B., and Johansen, A. (1964). *C. R. Trav. Lab. Carlsberg* **34**, 287.
Dopheide, T. A. A., Mooré, S., and Stein, W. H. (1967). *J. Biol. Chem.* **242**, 1833.
Dumas, J. B. (1843). *In* "Traité de Chimie Appliquée aux Arts," Vol. 6, pp. 379. Bechet Jeune, Paris.
Ege, R., and Lundsteen, E. (1934). *Biochem. Z.* **268**, 164.
Enzyme Commission. (1961). "Report of the Commission on Enzymes of the International Union of Biochemistry," Pergamon Press, London.
Ernstrom, C. A. (1958). *J. Dairy Sci.* **41**, 1663.
Ernstrom, C. A. (1965). *In* "Fundamentals of Dairy Chemistry" (B. H. Webb and A. H. Johnson, eds.), pp. 590–623. Avi, Westport, Connecticut.
Fish, J. C. (1957). *Nature* **180**, 345.
Foltmann, B. (1958). *Acta Chem. Scand.* **12**, 343.
Foltmann, B. (1959a). *Acta Chem. Scand.* **13**, 1927.
Foltmann, B. (1959b). *Acta Chem. Scand.* **13**, 1936.
Foltmann, B. (1959c). *Proc. Int. Dairy Congr., 15th, 1959* **2**, 655.
Foltmann, B. (1960a). *Acta Chem. Scand.* **14**, 2059.
Foltmann, B. (1960b). *Acta Chem. Scand.* **14**, 2247.
Foltmann, B. (1962). *C. R. Trav. Lab. Carlsberg* **32**, 425.
Foltmann, B. (1964a). *C. R. Trav. Lab. Carlsberg* **34**, 275.
Foltmann, B. (1964b). *C. R. Trav. Lab. Carlsberg* **34**, 319.
Foltmann, B. (1965). Unpublished data.
Foltmann, B. (1966). *C. R. Trav. Lab. Carlsberg* **35**, 143.
Foltmann, B. (1970a). *Phil. Trans. Roy. Soc. London, Ser. B* **257**, 147.
Foltmann, B. (1970b). *In* "Methods in Enzymology" (S. Colowick and N. Kaplan, eds.), Vol. 19, pp. 421–436. Academic Press, New York.
Foltmann, B., and Hartley, B. S. (1967). *Biochem. J.* **104**, 1064.
Fox, P. F. (1969). *J. Dairy Sci.* **52**, 1214.
Friedman, L. (1960). *Diss. Abstr.* **20**, 4510.
Fuld, E. (1902). *Ergeb. Physiol.* **1**, 468.
Garnier, J., Mocquot, G., Ribadeau-Dumas, B., and Maubois, J.-L. (1968). *Ann. Nutr. Aliment.* **22B**, 495.
Gray, W. R., and Hartley, B. S. (1963). *Biochem. J.* **89**, 379.
Green, M.-L., and Crutchfield, G. (1969). *Biochem. J.* **115**, 183.
Habeeb, A. F. S. A. (1967). *Arch. Biochem. Biophys.* **119**, 264.
Hammarsten, O. (1872). *Upsala Laekarefoeren. Foerh.* **8**, 63.
Hankinson, C. L. (1943). *J. Dairy Sci.* **26**, 53.
Hankinson, C. L., and Palmer, L. S. (1942). *J. Dairy Sci.* **25**, 277.
Henschel, M. J., Hill, W. B., and Porter, J. W. G. (1961). *Proc. Nutr. Soc. Engl. Scot.* **20**, XL.
Herriott, R. M. (1938). *J. Gen. Physiol.* **21**, 501.
Herriott, R. M. (1939). *J. Gen. Physiol.* **22**, 65.

Hill, R. D. (1969). *J. Dairy Res.* **36**, 409.
Hill, R. D., and Laing, R. R. (1965). *Biochim. Biophys. Acta* **99**, 352.
Hill, R. D., and Laing, R. R. (1966). *Nature* **210**, 1160.
Hill, R. D., and Laing, R. R. (1967). *Biochim. Biophys. Acta* **132**, 188.
Holter, H. (1932). *Biochem. Z.* **255**, 160.
Holter, H. (1941). *In* "Die Methoden der Fermentforschung" (E. Bamann and K. Myrbäck, eds.), Vol. 2, pp. 2081–2090. G. Thieme Verlag, Leipzig.
Holter, H , and Andersen, B. (1934). *Biochem. Z.* **269**, 285.
Holwerda, B. J. (1923). *Biochem. Z.* **134**, 381.
Hostettler, H., and Rüegger, H. R. (1950). *Landwirt. Jahrb. Schweiz* **64**, 669.
Hostettler, H., and Stein, J. (1954). *Landwirt. Jahrb. Schweiz* **68**, 291.
Hostettler, H., and Stein, J. (1955). *Milchwissenschaft* **10**, 40.
Jirgensons, B., Ikenaka, T., and Gorguraki, V. (1958). *Makromol. Chem.* **28**, 96.
Jollès, J., Alais, C., and Jollès, P. (1968). *Biochim. Biophys. Acta* **168**, 591.
King, C. W., and Melville, E. M. (1939). *J. Dairy Res.* **10**, 340.
Kleiner, I. S., and Tauber, H. (1932). *J. Biol. Chem.* **96**, 755.
Kleiner, I. S., and Tauber, H. (1934). *J. Biol. Chem.* **106**, 501.
Lawrence, R. C., and Sanderson, W. B. (1969). *J. Dairy Res.* **36**, 21.
Lea, A. S., and Dickinson, W. L. (1890). *J. Physiol. (London)* **11**, 307.
Levin, Y. (1962). *Bull. Res. Counc. Isr. Sect. A* **11**, 48.
Lindqvist, B. (1963). *Dairy Sci. Abstr.* **25**, 257, 299.
Lindqvist, B., and Storgårds, T. (1959). *Acta Chem. Scand.* **13**, 1839.
Lindqvist, B., and Storgårds, T. (1960). *Acta Chem. Scand.* **14**, 757.
Linklater, P. M. (1961). *Diss. Abstr.* **22**, 533.
McMeekin, T. L. (1939). *J. Amer. Chem.* **61**, 2884.
Malpress, F. H. (1967). *Nature* **215**, 855.
Mickelsen, R., and Ernstrom, C. A. (1963). *J. Dairy Sci.* **46**, 613.
Mickelsen, R. and Ernstrom, C. A. (1967). *J. Dairy Sci.* **50**, 645.
Mulder, H., and Radema, L. (1947). *Neth. Milk Dairy J.* **1**, 128.
Nitschmann, Hs., and Varin, R. (1951). *Helv. Chim. Acta* **34**, 1421.
Oeda, M. (1956). *Proc. Int. Dairy Congr., 14th, 1956,* **2**, Part 2, 367.
Oppenheimer, C. (1926). *In* "Die Fermente und ihre Wirkungen," 5th ed., Vol. 2, pp. 977–1024. G. Thieme Verlag, Leipzig.
Oppenheimer, C. (1939). *In* "Die Fermente und ihre Wirkungen," Suppl. Vol. 2, pp. 863–881. W. Junk, The Hague.
Payens, T. A. J. (1962). *Proc. Int. Dairy Congr., 16th, 1962* **B**, 410.
Perlmann, G. E. (1967). *Advan. Chem.* **63**, 268.
Peters, R. (1894). "Untersuchungen über das Lab und die Labähnlichen Fermente," Dissertation. University of Rostock, Rostock, Germany.
Porcher, C. (1930). *Lait* **10**, 47.
Rand, A. G. (1964). *Diss. Abstr.* **24**, 3479.
Rand, A. G., and Ernstrom, C. A. (1964). *J. Dairy Sci.* **47**, 1181.
Rand, A. G., and Ernstrom, C. A. (1968). *J. Dairy Sci.* **51**, 1756.
Reid, T. W., and Fahrney, D. (1967). *J. Amer. Chem. Soc.* **89**, 3941.
Ryle, A. P., and Porter, R. R. (1959). *Biochem. J.* **73**, 75.
Sanger, F., and Tuppy, H. (1951). *Biochem. J.* **49**, 481.
Schober, R., and Heimburger, N. (1960). *Milchwissenschaft* **15**, 561.
Schober, R., Heimburger, N., and Printz, I. (1960). *Milchwissenschaft* **15**, 506.
Schwander, H., Zahler, P., and Nitschmann, Hs. (1952). *Helv. Chim. Acta* **35**, 553.
Segelcke, T., and Storch, V. (1870). *Ugeskrift Landmaend, Ser. 3* **9**, 341.

Segelcke, T., and Storch, V. (1874). *Milchzeitung.* **3,** 997.
Shukri, N. A., and Ernstrom, C. A. (1966). *J. Dairy Sci.* **49,** 696.
Sommer, H. H., and Matsen, H. (1935). *J. Dairy Sci.* **18,** 741.
Soxhlet, F. (1877). *Milchzeitung.* **6,** 495, 513.
Stepanov, V. M. (1968). *Abstr. 5th FEBS Meeting, Prague* No. 1089.
Storrs, F. C. (1956). *J. Dairy Res.* **23,** 269.
Tang, J , and Hartley B. S. (1970). *Biochem. J.* **118,** 611.
Tang, J., and Tang, K. I. (1963). *J. Biol. Chem.* **238,** 606.
Tang, J., Wolf, S., Caputto, R., and Trucco, R E. (1959). *J. Biol. Chem.* **234,** 1174.
Taylor, W. H. (1962). *Physiol. Rev.* **42,** 519.
van Dam, W. (1912). *Versl. Landbouwk. Onderzoek* **20,** 5.
van der Burg, B., and van der Scheer, A. E. (1937). *Int. Tech. Chim. Ind. Agr. Congr., 5th., 1937* **2,** 321.
Yamamoto, T., and Takahashi, K. (1952). *Bull. Nat. Inst. Agr. Sci. (Japan), Ser. G 4* **27,** No. 7, 1. Cited in (1953) *Dairy Sci. Abstr.* **15,** 411.
Zunz, E. (1911). *In* 'Biochemisches Handlexikon" (E. Abderhalden, ed.), Vol. 5, pp. 618–625. J. Springer Verlag, Berlin.

Part E
Whey Proteins and Minor Proteins

General Introduction

The term *whey protein* is used here in a loose sense to signify a *noncasein* protein occurring in milk in an appreciable amount. *Minor protein* is used to refer to a protein or enzyme occurring in milk in a small amount. In general it will be a "whey" protein, but in some cases it may be intimately connected with a fat globule rather than being in simple aqueous solution.

For many years it was generally believed that β-lactoglobulin is the predominant "whey" protein in the milk of all mammals. While this is true of the milk of ruminants in general, and of cows in particular, it does not appear to be true for other types of mammals. This conclusion may, however, need some qualification, as we shall see in Part G. It is not surprising that β-lactoglobulin has been the most intensively studied of the whey proteins, and our present knowledge of it is discussed by Hugh A. McKenzie in Chapter 14.

α-Lactalbumin is the next most prolific of the ruminant "whey" proteins; it is the dominant one in human milk and occurs in the milk of other mammals. It has long been known as an interesting protein, but it is only since 1966 that an important biological function has been recognized for it. Thus the discussion of this protein by William Cordon, in Chapter 15, has special significance.

One of the most rapidly expanding areas of milk protein research has been the study of the "minor" proteins, especially the enzymes of milk. Their occurrence, isolation and properties, and their relationships to enzymes and proteins with similar functions in other secretions and in blood, are discussed in Chapter 16 by Merton L. Groves.

H. A. McKenzie

14 □ β-Lactoglobulins

H. A. McKENZIE

I. Introduction 258
II. Isolation of β-Lactoglobulins 259
   A. Early Methods 259
   B. Methods of Aschaffenburg and Others 259
   C. Methods of McKenzie and Co-Workers 262
   D. Chromatography of β-Lactoglobulins 266
III. Methods of Zone Electrophoresis of Whey Proteins 271
IV. Species Differences and Genetic Variants 274
   A. Bovine β-Lactoglobulins 274
   B. Ovine β-Lactoglobulins 276
   C. Caprid β-Lactoglobulin 276
   D. β-Lactoglobulins of Other Species 276
V. Amino Acid Composition 277
   A. End-Group Analysis 277
   B. Amino Acid Analysis 279
   C. Peptide Studies and Amino Acid Sequence 280
   D. Location and Reactivity of SS and SH Groups 285
   E. Location of Tyrosine and Tryptophan Groups 293
VI. Electrochemical Properties 294
   A. pH Titration Curves 294
   B. Isoionic Points and Ion Binding 302
   C. Electrophoresis 303
VII. Molecular Size and Conformation 304
   A. Introduction 304
   B. pH Range 1.8–3.5 305
   C. pH Range 3.5–5.4 307
   D. pH Range 5.4–9.2 313
VIII. Denaturation of β-Lactoglobulins 316
   A. Effect of Heat 316
   B. Effect of Detergents 318
   C. Effects of Urea and Guanidine Hydrochloride 319
IX. Interaction of β-Lactoglobulins and κ-Casein 321
X. X-Ray Crystallographic Studies 323
XI. Summary and Conclusions 324
   References 325

I. Introduction

In 1934, during the salt fractionation of skim milk, Palmer (1934) isolated a protein that, although it was salted out with the albumin group, had the characteristic low solubility of a globulin at its isoelectric point in salt solution of very low ionic strength. We have seen in Chapter 1, Volume I, that Pedersen (1936), in the course of an ultracentrifugal examination of milk and whey, observed a number of peaks (designated α, β, γ, etc.) in the sedimentation pattern. He identified the "β-peak" as arising from Palmer's protein. Subsequently Svedberg (1938) called it "β-lactalbumin." However in 1942 Cannan et al. (1942) proposed the name β-lactoglobulin, by which this protein is known today. (The use of the generic term "lactoglobulin" to refer to the immunoglobulin proteins of milk should be avoided. The generic term "lactalbumin" should not be used in referring to whey proteins. The term "α-lactalbumin" refers to a specific protein, originally associated with the "α-peak" in the sedimentation pattern.)

Because of the beautiful crystals of β-lactoglobulin that could be isolated from cow milk, chemists considered that it was a convenient "pure" protein to use in physico-chemical studies. It is not surprising that it was attractive to X-ray crystallographers in their early studies. Some of this early history of the X-ray crystallography of β-lactoglobulin has been discussed by Hodgkin and Riley (1968). Many electrophoretic and ultracentrifugal studies were carried out with the protein. It is ironic that it was in these very studies that several workers began to suspect heterogeneity in their preparations (for example, Ogston and Tilley, 1955; for reviews of this early work see Tilley, 1960; McKenzie, 1967). Aschaffenburg and Drewry (1955, 1957a) made an important discovery that was to have a profound bearing not only on our understanding of the "heterogeneity" and properties of β-lactoglobulin (especially in transport experiments), but on future research in milk proteins generally. They found that cow milk could contain either a mixture of two β-lactoglobulins or only one or other of the two types. It was possible to isolate each type in a state of high purity. They showed that the occurrence of these variants, designated A and B, was genetically determined, the genes controlling the synthesis of the protein being autosomal alleles without dominance. This was the first demonstration of genetic variants in milk proteins. We know now that there are further variants in milk proteins.

Bell and McKenzie (1964) concluded tentatively that the β-lactoglobulins occur only in the milk of ruminants. However there is now evidence that they may occur in the milk of some monogastrics (see Section IV.D). The reasons for the evolution of β-lactoglobulin and its biological function are not yet known. This is one of the fascinating aspects of this group of proteins whose properties we shall be concerned with in the ensuing pages.

II. Isolation of β-Lactoglobulins

A. EARLY METHODS

Soon after Palmer's original isolation of bovine β-lactoglobulin by sodium sulfate fractionation of acid whey, Sørensen and Sørensen (1939) isolated it from an ammonium sulfate whey protein fraction. Jacobsen (1949) modified the Sørensen method, making it more convenient. He obtained a fraction consisting primarily of β-lactoglobulin, α-lactalbumin, serum albumin, and the "red" and "green" proteins. The latter were removed by ammonium sulfate precipitation, and the β-lactoglobulin was crystallized by dialysis of the supernatant protein solution against water. This method has the disadvantage that the protein is crystallized from a solution in which other whey proteins, such as α-lactalbumin and serum albumin, are present. Appreciable contamination of the preparation may result. These and other early procedures have been reviewed by Tilley (1960) and McKenzie (1967).

B. METHODS OF ASCHAFFENBURG AND OTHERS

Aschaffenburg and Drewry (1957b) found that if the pH of a sodium sulfate whey solution from bovine milk is adjusted to 2, α-lactalbumin and serum albumin are precipitated, much of the β-lactoglobulin remaining in solution. They exploited this observation to develop procedures for the isolation of bovine β-lactoglobulins A and B from the milk of dairy breeds of cattle. At the same time they were able to obtain fractions rich in α-lactalbumin. Although our main concern in this chapter is with β-lactoglobulin it is instructive to note the fractions that are enriched in other whey proteins and how these proteins may be isolated. The general procedure of Aschaffenburg and Drewry is outlined in Fig. 1. They followed the course of their fractionation with filter-paper electrophoresis.

Although their procedure was a considerable advance it does have certain disadvantages. The use of anhydrous sodium sulfate is somewhat inconvenient. It involves undesirable heating of the protein solution to 40°C, and the salt does not dissolve readily. Careful control of temperature is required in the subsequent cooling to avoid crystallization of sodium sulfate in the whey protein solution. Furthermore, the final separation of the

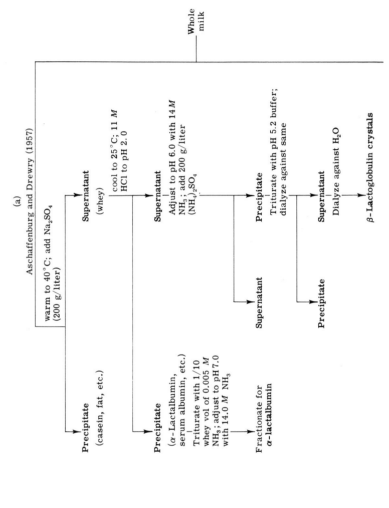

FIGURE 1. (a) Method of Aschaffenburg and Drewry (1957b) for the preparation of β-lactoglobulin, also showing the precipitate suitable for further fractionation to obtain α-lactalbumin.

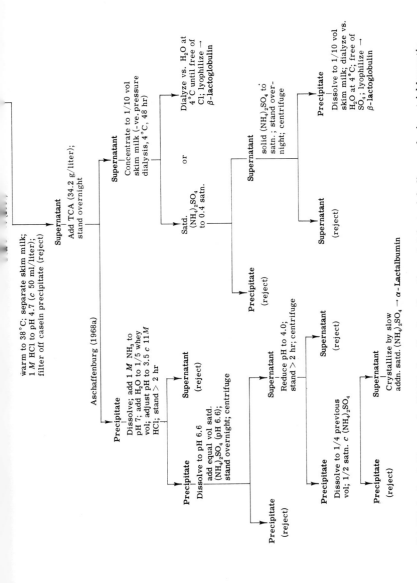

FIGURE 1. (b) Method of Fox et al. (1967) for the fractionation of β-lactoglobulin employing trichloroacetic acid (TCA) precipitation. The modification of Aschaffenburg (1968a) to obtain α-lactalbumin is also shown.

β-lactoglobulin precipitate cannot readily be made by centrifugation owing to its tendency to float on the top of the concentrated salt solution. The use of strong acid and alkali in the pH adjustment is also undesirable due to the chance of local excesses.

Some workers are apprehensive about the use of pH values as low as 2 during the fractionation. There is at present no concrete evidence that irreversible dissociation or other changes occur in β-lactoglobulin under the conditions of pH, protein concentration and ionic strength employed by Aschaffenburg and Drewry. However, it is possible that an irreversible change may occur in α-lactalbumin under these conditions (see Chapter 7, Volume I, and Chapter 15). Fox et al. (1967) have used even stronger acid conditions for the fractionation of β-lactoglobulin. They separated acid whey from skim milk and made it 3% (w/v) with respect to trichloracetic acid. Under these conditions they found a supernatant free of proteins other than β-lactoglobulin. The latter solution was concentrated by negative-pressure dialysis, and the protein solution was dialyzed against water and lyophilized. We would counsel against the use of lyophilization when the protein is to be subsequently used for conformational and other physical studies.

Aschaffenburg (1968a) made use of the observations of Fox et al. to develop a procedure for the isolation of α-lactalbumin. Both of these procedures are shown in Fig. 1.

C. METHODS OF MCKENZIE AND CO-WORKERS

McKenzie and his co-workers (Bell and McKenzie, 1963, 1964; McKenzie, 1967; Armstrong et al., 1967) considered that it is more satisfactory to prepare ammonium sulfate whey rather than sodium sulfate whey. Apart from having the general advantages outlined by Dixon and Webb (1961), ammonium sulfate fractionation has the advantage that, in the case of whey, it is possible to use a temperature of 20°C, and there is no danger of subsequent crystallization of salt in the whey solution or difficulty in the separation of protein by centrifugation. McKenzie's group preferred to use 1 M hydrochloric acid and 1 M ammonia for the pH adjustments to lessen the chance of local excesses of acid or alkali. They avoided extremes of pH adjustment, lowering the pH value to only 3.5, hoping in this way to avoid any irreversible changes in the protein. Their procedures are outlined in Figs. 2 and 3.

14. β-LACTOGLOBULINS

They showed that the solubility of β-lactoglobulin is lower at pH 3.5 than at pH 2 in the whey protein—1.8 M ammonium sulfate solution of method I. Some of the β-lactoglobulin is precipitated from the whey mix-

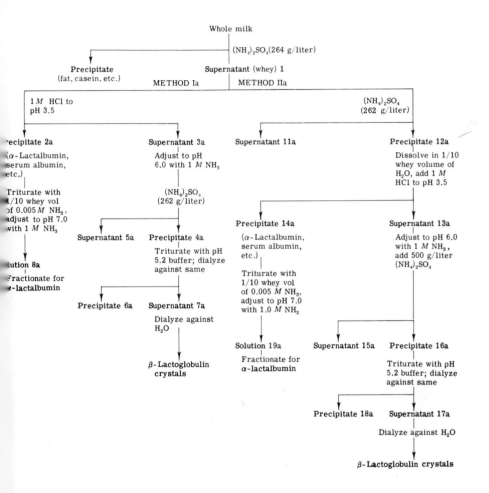

FIGURE 2. Schematic representation of methods Ia and IIa of Armstrong, McKenzie and Sawyer for the isolation of bovine β-lactoglobulins. (From McKenzie, 1967.) The fractions rich in α-lactalbumin are also shown. An error occurs in Fig. VI.2 of McKenzie (1967): In the treatment of precipitate 12a, 0.1 M HCl should read 1 M HCl.

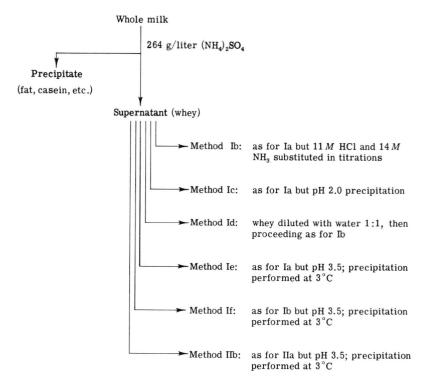

FIGURE 3. Schematic diagram representing variations of Armstrong-McKenzie-Sawyer methods Ia and IIa for the fractionation of bovine whey proteins (McKenzie, 1967).

ture, along with the α-lactalbumin and serum albumin, at both pH values. The amount of this protein in the pH 3.5 supernatant is less than that in the pH 2 supernatant. However, both supernatants contain virtually only β-lactoglobulin. Since solutions of β-lactoglobulin alone show no precipitation at pH 3.5 in 1.8 M $(NH_4)_2SO_4$, the precipitation from the whey protein solution must be due to interaction with one or more of the whey proteins or to some influence of the whey medium. The interaction of β-lactoglobulin and α-lactalbumin in heated solutions has been reported, but no study appears to have been made of interactions under the conditions of the fractionation.

The solubility of β-lactoglobulin A is considerably less than that of β-lactoglobulin B at pH 3.5 under the conditions of method I. Thus the yields of β-lactoglobulin A crystals obtained from method Ia are low (∼0.8 g/liter). However, there is no significant difference in the β-lactoglobulin B yield between method Ia and the Aschaffenburg-Drewry method

(~1.3 g/liter). Bell and McKenzie (1963, 1964, 1967a) found that the solubility of β-lactoglobulin C is greater than that of β-lactoglobulin B, and they obtained somewhat higher yields by method Ia than by the Aschaffenburg-Drewry method.

The question arises as to whether method Ic [$(NH_4)_2SO_4$, pH 2] could be used for the preparation of β-lactoglobulin A to obtain a higher yield without significant dissociation during the period of pH 2, and whether the return to pH 6 is completely reversible. The high ionic strength during the pH 2 period would largely prevent the dissociation. Using present methods for studying conformation and size as criteria of reversibility, it appears that the final crystallized material is not affected by the low pH period. Some workers prefer to avoid this uncertainty.

In method IIa very little β-lactoglobulin is precipitated at pH 3.5 with the α-lactalbumin and serum albumin. However, the β-lactoglobulin supernatant contains an appreciable amount of serum albumin and a small amount of α-lactalbumin. While good yields of β-lactoglobulin of high zone electrophoretic "purity" may be obtained, crystallization is more difficult owing to the presence of other proteins. Method IIa can thus be a valuable method for the preparation of β-lactoglobulin (especially the A variant), particularly if the remaining impurities are removed chromatographically before crystallization is attempted (see Section II.D).

The pH 3.5 precipitate from method IIa is largely free of β-lactoglobulin but contains some serum albumin, so that a solution of this precipitate (19a) provides a good starting point for the isolation of α-lactalbumin. However method IIb, the low-temperature version of the pH 3.5 precipitation (Fig. 3), has the added advantage that the precipitate is largely free of both β-lactoglobulin and serum albumin. The solution of this precipitate (19b) is therefore the best starting point for the further purification of α-lactalbumin by salt fractionation or by chromatography (compare Robbins and Kronman, 1964).

If method Ia is being used for the preparation of β-lactoglobulin and it is also desired to prepare α-lactalbumin, then solution 8a may be used for the latter purpose. Adjustment of the pH of this solution to 3.5 gives a precipitate which may be redissolved at pH 6, forming a solution which has virtually the same composition by starch-gel electrophoresis as solution 19a in method IIa and may be fractionated for α-lactalbumin. The precipitation may also be carried out at low temperature as in method IIb, giving a solution on redissolving which is equivalent to 19b.

Armstrong *et al.* (1967) consider that adjustment of the pH of ammonium sulfate whey to pH 4 at room temperature does not give a satisfactory separation of the β-lactoglobulin from the other whey protein. They examined the Robbins-Kronman (1964) fractionation and found that the precipitation at pH 4 and room temperature, followed by standing at 2°C, did not result in good fractionation.

Method Ia has been applied by Bell and McKenzie (1964, 1967b) to the fractionation of ovine whey proteins. It was possible to isolate two genetic variants of ovine β-lactoglobulin in this way. Some difficulty was found in crystallizing the A variant, on occasion, and attempts to crystallize the B variant were unsuccessful. Subsequently Treacy (1967) made the interesting observation that crystallization of the A variant was facilitated by mixing the final protein solution with Hyflo Super-Cel (Johns-Manville) and filtering before dialyzing against water. This treatment was not successful for the B variant.

Askonas (1954) crystallized caprid β-lactoglobulin from ammonium sulfate solutions. This method of crystallization was also used later by Bell et al. (1968). However Ghose et al. (1968) had no difficulty in crystallizing caprid β-lactoglobulin by dialysis of solutions of pH 5.2 against water. Townend and Basch (1968) found that crystals obtained from final ammonium sulfate solutions in the preparation contained impurities. They fractionated these crystals by precipitation of their solutions with varying amounts of ammonium sulfate. Some fractions contained only caprid β-lactoglobulin by the criterion of acrylamide-gel electrophoresis.

D. CHROMATOGRAPHY OF β-LACTOGLOBULINS

It is possible using the above procedures for fractionation of ammonium sulfate whey to prepare β-lactoglobulin of high purity. In the case of some methods, for example the preparation of the bovine A variant of method Ia of McKenzie et al. (1967), the yields are low. If the milk used is heterozygous in β-lactoglobulin it is not possible by these procedures to isolate the pure single variants composing it. It should be stressed that the procedures do not result in final solutions of α-lactalbumin of high purity. Further fractionation, usually by chromatography, is necessary to achieve an α-lactalbumin preparation of high purity. Also, in the course of the preparations, adjustment of pH is necessary to effect the necessary fractionations; this could result in irreversible changes. Every effort is made to minimize the chances of such changes and it appears to be successful for β-lactoglobulin. The greatest possibility of change in the latter case is in the pH adjustment with alkali. On the other hand, irreversible changes are more likely to occur for α-lactalbumin in the titration with acid. Such possibilities must be considered in preparations intended for conformational studies.

Armstrong et al. (1970) have described studies, which they carried out over the past five years, on the chromatography of whey proteins. When

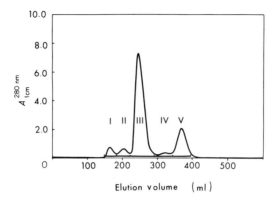

FIGURE 4a. Profile of total ammonium sulfate bovine whey proteins after elution from a Sephadex G-75 column (3.2 cm × 50 cm) at pH 6.3. Sample: homozygous in β-lactoglobulin A and in α-lactalbumin B; dialyzed vs. imidazole–HCl buffer (pH 6.3, $I = 0.043$); 10 ml applied; $A_{1\,cm}^{280\,nm}$ (absorbance), 35.4 (∼0.3 g protein). Elution rate: 0.5 ml/min. Protein recovery: 99 ± 1% (Armstrong et al., 1970).

they began their investigations the main previous attempt at the chromatographic fractionation of whey proteins was that of Préaux and Lontie (1962). The latter group obtained four fractions by chromatography of solutions of freeze-dried, dialyzed whey proteins on Sephadex G-75 columns in pH 4.67 buffer (acetate, ionic strength (I) of 0.05). (For the properties of Sephadex see Chapter 4, Volume I.) Only one of these fractions gave a single band (β-lactoglobulin) on zone electrophoresis in agar. In their initial experiments, Armstrong et al. attempted to fractionate ammonium sulfate whey by chromatography on Sephadex G-75 with pH 7.0 buffer (0.01 M phosphate), followed by further chromatography of individual fractions on DEAE-Sephadex in pH 5.4 buffer (acetate) with a sodium chloride gradient. Although promising fractionations of α-lactalbumin and β-lactoglobulin were obtained it was decided to change the buffer systems. The choice of buffer is a difficult one: It is conditioned by the pH range of the solubility of α-lactalbumin and of its stability, and by the range in which conformational changes occur in β-lactoglobulin. Because of conflict in the pH range required, a compromise choice must be made. It was decided to use a pH 6.3 buffer (0.05 M imidazole—0.043 M hydrochloric acid, $I = 0.043$).

Providing adequate care was taken in the choice of sample size it was possible to fractionate ammonium sulfate bovine whey into five fractions, as shown in Fig. 4a. Zone electrophoretic patterns of the fractions are shown in Fig. 4b. These fractions were studied by zone electrophoresis,

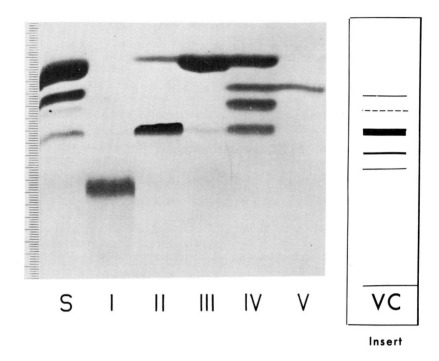

FIGURE 4b. Starch-gel electrophoresis pattern at pH 7.5 of fractions shown in Fig. 4a. The protein concentration in each sample (except fraction V) was adjusted to ∼2 g/dl prior to application to detect more readily minor components of a fraction. Thus the band intensities in a given fraction give some indication of the proportion of the components in that fraction, but they do not reflect the total amount of a given fraction in the sample chromatographed. Buffer system: Ferguson-Wallace semidiscontinuous; 7 V/cm for 4 hr in 20°C room. The pattern marked S is the total whey protein. The fraction V sample is dilute, and the minor components can barely be detected on the gel. The value of concentrating the solution can be seen from the insert, where a typical pattern of concentrated fraction V (V.C) is shown (Armstrong et al., 1970).

chromatography, etc. and their composition was found to be the following:

Fraction I. This peak is eluted in the void volume of Sephadex G-75 (also G-100) and consists of high molecular weight material giving bands that move slower than α-lactalbumin on starch-gel electrophoresis at pH 8.6. This fraction gives positive tests for sialic acid and shows iron (III)-binding capacity; the iron complex formed has the characteristic spectrum of lactoferrins and transferrins, and its behaviour on sedimentation-velocity is consistent with its containing iron-binding protein. This fraction has also been shown to contain the lactose synthetase complex and the "A protein" of this complex (see Armstrong et al., 1970).

Fraction II. From a comparison of the starch-gel electrophoretic patterns at pH 8.6 (or 7.5) with those of pure bovine serum albumin, it was concluded that fraction II is primarily serum albumin. However it is sometimes overwhelmed by the adjacent β-lactoglobulin. When a sharp peak is obtained it may be used as a criterion of good resolution for this chromatographic system.

Fraction III. Irrespective of whether the sample of whey protein is homozygous or heterozygous with respect to β-lactoglobulin, a single peak showing little or no asymmetry is obtained for this fraction. It consists of β-lactoglobulin with some contaminating protein.

Fraction IV. It was concluded in the initial studies that this fraction consists, in part, of α-lactalbumin and an aggregate of it. Recently, more extensive studies of the nature of the protein have been carried out by K. E. Hopper and H. A. McKenzie (see Chapter 18). They now conclude that two of the components of this fraction are α-lactalbumins containing varying amounts of a carbohydrate moiety. It also contains some β-lactoglobulin and serum albumin.

Fraction V. This fraction consists primarily of α-lactalbumin. However it includes appreciable amounts of the α-lactalbumin components of fraction IV and β-lactoglobulin. The trailing edge contains varying amounts of several contaminants, as judged by starch-gel electrophoresis at pH 8.6. These may include, surprisingly, bovine serum albumin. Sacrifice of 30–50% of the yield is necessary if the amount of these contaminants is to be kept minimal. A comprehensive study of this α-lactalbumin fraction is being reported elsewhere by K. E. Hopper and H. A. McKenzie.

If the pH 3.5 precipitate from method Ia of Armstrong et al. (1967) is redissolved by the addition of ammonia to pH 7 and if the protein is reprecipitated with ammonium sulfate and redissolved as a concentrated solution, chromatography of this fraction may be carried out on Sephadex G-75 under similar conditions to the above fractionation of the total whey protein. Resolution into five qualitatively similar fractions is obtained also, but the quantitative proportions of the fractions are naturally different, and because there is less β-lactoglobulin present, resolution is superior. Fraction V may be used for further purification of α-lactalbumin. However, a solution of the pH 3.5 precipitate obtained by method IIb of Armstrong et al. (1967), containing primarily α-lactalbumin with only small amounts of other whey proteins, is a useful starting material for the chromatographic purification of α-lactalbumin on Sephadex G-75 followed by DEAE-Sephadex A-50.

Fraction III from the G-75 chromatography of the total whey protein may be used as starting material for further purification of β-lactoglobulin (for example on DEAE-Sephadex). However this starting material is more prone to contamination by other whey proteins than material obtained from G-75 chromatography of protein that has been fractionated by salt solubility and pH adjustment. The possible disadvantages of the latter material have already been discussed. The supernatant from the pH 3.5 precipitate of method IIa is β-lactoglobulin containing some other whey proteins and may be used for chromatography on Sephadex G-75 followed by DEAE-Sephadex. On the other hand, material from the supernatant of the pH 3.5 precipitate of method Ia is virtually pure β-lactoglobulin and may be chromatographed directly on DEAE-Sephadex to obtain β-lactoglobulin. Use of this method results, of course, in lower overall yield from the original whey protein due to loss of some β-lactoglobulin in the pH 3.5 pre-

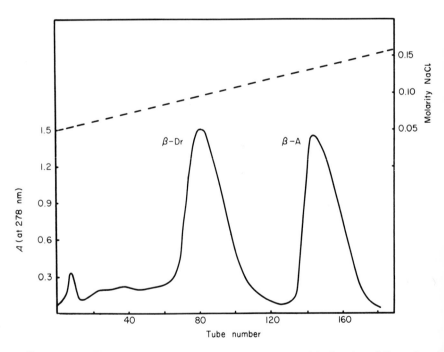

FIGURE 4c. Elution profile for the separation of β-lactoglobulins A and Droughtmaster in a DEAE-Sephadex A-50 column (2.0 cm × 20 cm) at pH 6.3 using a linear gradient of sodium chloride. Load, ~1 g of β-lactoglobulin; yield, ~0.4 g $_{\text{Droughtmaster}}$ and 0.36 g A showing single bands on electrophoresis.

cipitate. It is preferable in general to use starting material homozygous in the particular β-lactoglobulin desired. However in some cases, for example β-lactoglobulin$_{Droughtmaster}$, the homozygous material is not available. An elution pattern for the separation of this protein from the A-Droughtmaster mixture is shown in Fig. 4c (Bell et al., 1966a, 1970).

Basch et al. (1965) separated β-lactoglobulins A and C by chromatography of supernatant fractions from the pH 2 precipitate of Aschaffenburg and Drewry, on DEAE-cellulose at pH 5.8 (0.05 M phosphate buffer) using sodium chloride gradients. Other chromatographic fractionations of whey proteins include those of Schober et al. (1959), Morr et al. (1964), Yaguchi et al. (1961), Tarassuk and Yaguchi (1962), Koike et al. (1964), Szuchet-Derechin and Johnston (1965), Groves (1965), and Mawal (1967). These are mainly concerned with whey proteins other than β-lactoglobulin (see also Chapters 15 and 16).

III. Methods of Zone Electrophoresis of Whey Proteins

The original discovery of genetic variants of bovine β-lactoglobulins was made by Aschaffenburg and Drewry (1955, 1957a) through paper electrophoresis of sodium sulfate whey protein fractions. Their procedure was used widely for some years. However it is not surprising that with the discovery of the method of zone electrophoresis on starch gel by Smithies (1955), various workers applied the method to the study of whey proteins. The first significant contribution was that of Bell (1962), who reported the discovery of a new bovine β-lactoglobulin variant, C, using this method. Bell (1967) described in detail the development of his procedure for the detection of whey protein types. His method offers a number of advantages over earlier methods of typing. It is simple, can be applied directly to skim milk, requires only a small sample (1–2 ml) and gives good resolution. Bell made a careful study of the kind of hydrolyzed starch to be used in preparing the gel. He hydrolyzed a number of commercial samples of potato starch and compared the products with the hydrolyzed starch prepared for electrophoresis by the Connaught Laboratories. Bell found that gels of hydrolyzed starch prepared from potato starch (Drug Houses of Australia, DHA) gave consistently superior resolution for bovine skim milk proteins to that obtained with Connaught starch or starch from other commercial sources. He used a sodium hydroxide–boric acid buffer system of pH 8.5 (see Appendix) and found that careful control of buffer composition was necessary for optimum resolution of β-lactoglobulins in skim milk samples. Subsequently Bell et al. (1966a) resolved bovine β-lactoglobulins A, B, C and $_{Droughtmaster}$ with this buffer system (see Fig. 5).

β-Lg: A-B   A-C   A-Dr   B-C   B-Dr   C

FIGURE 5. Starch-gel electrophoretic patterns illustrating resolution of bovine β-lactoglobulins A, B, C, and Droughtmaster by Bell et al. (1966a, 1970). The samples (reading from left to right) contain the following variants: A–B, A–C, A– Droughtmaster B–C, B– Droughtmaster, C. Gel buffer: 0.028 M H_3BO_3 and 0.0112 M NaOH electrode. Buffer: 0.3 M H_3BO_3 plus 0.06 M NaOH. Electrophoresis: 7 V/cm for 4.5 hr in 20°C room.

McKenzie and Sawyer (1966) have pointed out that care must be exercised in the use of pH 8.5 buffer systems for electrophoresis of whey protein fractions. If solutions containing β-lactoglobulin are allowed to stand at pH 8.5 for any length of time prior to electrophoresis, considerable irreversible changes may occur (see Section VIII). Thus attention has been given by McKenzie and Treacy (1966) to the use of other buffer systems. Bailey and Lemon (1966) have successfully applied a semidiscontinuous buffer system, developed by Ferguson and Wallace (1963),* to the analysis of kangaroo milk. McKenzie and Treacy attempted to resolve bovine β-lactoglobulins A, B, C and $_\text{Droughtmaster}$ with this buffer system. They found that the bands obtained for the isolated proteins were sharper at pH 7.0 than at pH 8.5 (borate system), but the resolution was somewhat lower at pH 7.0, as would be expected. This resolution was achieved, however, with the more readily available Connaught starch (Fig. 5). Larsen and Thymann (1966) resolved β-lactoglobulins A, B, C and D using a similar buffer system and hydrolyzed Danish potato starch. Buffer systems for gel electrophoresis of bovine whey proteins have also been presented by others, for example Arave (1967).

In their original work with ovine β-lactoglobulins Bell and McKenzie (1964, 1967b) achieved resolution of the A and B variants with a phosphate buffer of pH 7.8, but they could not resolve them with the borate system of pH 8.5. Subsequently it was found by H. A. McKenzie and G. B. Treacy that it is desirable when using Connaught starch to double the concentration of phosphate to obtain adequate buffering capacity. If the Ferguson-Wallace system of pH 8.0 is used the A and B variants may also be resolved. Bell (1965) has developed a starch-gel procedure for resolving the ovine A and B variants using the pH 7.5 tris-citrate gel buffer system of Kristjannson (1963). While its resolution is superior to the phosphate system for ovine whey proteins it is inferior to the pH 8.5 borate system for bovine whey proteins. Bell (1968) has found the phosphate buffer system of pH 7.8 very satisfactory for resolving porcine whey proteins.

Peterson (1963) applied the method of acrylamide-gel electrophoresis to the analysis of β-lactoglobulins and α-lactalbumins, using a pH 8.6 borate buffer similar to that employed by Bell (1962) in his starch-gel method. Applying samples of bóvine β-lactoglobulins A and B (supplied by Townend) and C (supplied by Bell), Peterson was able to obtain satisfactory resolution of the variants.

In the Appendix a number of procedures for analysis of β-lactoglobulin and α-lactalbumin variants, in skim milk and in whey protein fractions, are outlined. In some of these methods the authors employ acrylamide gels, others use starch gels. Good results have been obtained with both types. Nevertheless little attention has been given to the systematic design of buffer systems. The need for this and the principles involved are outlined in Chapter 7, Volume I.

Aschaffenburg (1968b) has objected to the use of starch, other than Connaught Laboratories' starch, on the grounds that Connaught starch is a more reproducible product than that prepared by workers in their own

*For a discussion of the principles of this system see Chapter 7, Volume I.

laboratories. While we agree that readily available commercial products should be used in general, nevertheless it is our experience that not all batches of commercial hydrolyzed starch have reproducible properties. Thus careful checking of the properties of each batch is essential to the worker wishing to use commercial hydrolyzed starch. Useful changes in resolution can be obtained by using gels of mixed starches in variable proportions, for example, equal amounts of Connaught and hydrolyzed DHA starch.

Attention must also be given to the properties of chemicals used in the preparation of acrylamide gels. Also, complications can arise from the properties of chemicals remaining after polymerization.

IV. Species Differences and Genetic Variants

A. BOVINE β-LACTOGLOBULINS

Aschaffenburg and Drewry (1957a) designated the first two variants of bovine β-lactoglobulin as A and B. This designation was made on the basis of their mobility on filter-paper electrophoresis at pH 8.6, A having a higher mobility than B. This is also the order of mobility on starch-gel electrophoresis at the same pH value. Aschaffenburg and Drewry examined whey proteins from the milk of ~278 cows of Shorthorn, Friesian, Guernsey and Ayrshire breeds. It was found that the genes controlling the β-lactoglobulin synthesis were autosomal codominant alleles. Cows were described as of phenotype β-lactoglobulin A/A, B/B or A/B and of genotype Lg^A/Lg^B, etc., using Lg as the locus symbol for the single gene involved. This notation is in line with normal genetic usage and has been universally adopted in studies of β-lactoglobulins.

In 1962 Bell (1962, 1967) showed that certain Jersey cows in Australian herds produced milk that had a third variant with a mobility less than that of the A or B variants at pH 8.5 (see also Bell and McKenzie, 1964). He showed that this third type is genetically determined, and he designated it as the C variant, assigning the symbol Lg^C to the gene responsible for its production. The occurrence of this variant has since been confirmed in Jersey cattle in North American, British and European herds. It has been claimed by Osterhoff and Pretorius (1966) that it occurs in South African Nguni cattle. However there has been no confirmation of this claim by other workers.

Bell (1962, 1967) found that the frequency of Lg^C varied markedly with the Jersey herds he studied and ranged from 0.01 to 0.31. Aschaffenburg

(1968b) has criticized Bell's conclusions and considers that the C variant is rare. This may be true of British herds but it does not appear to apply to Australian Jersey herds as confirmed in the figures of Table I for six Jersey herds studied in South Australia by Bailey (1968b).

The occurrence of the C allele in the Jersey and its absence from the Guernsey is a distinguishing feature of the two Channel Island breeds. Aschaffenburg (1961, 1963) has drawn attention to another distinction, β-casein C is found in the Guernsey, but not in the Jersey. Aschaffenburg (1968b) has also stressed the differences in the frequency distribution of κ-casein variants in the two breeds. Thus it seems likely that the two breeds differ in origin and arrived in the Channel Islands by different routes.

In 1966, two other variants were announced. Grosclaude et al. (1966) found a variant, designated β-lactoglobulin D, in Montbéliarde cattle in France. Its occurrence has been confirmed in Danish Jersey herds (Larsen and Thymann, 1966), in Polish Simmental cows (Michalak, 1967) and in two German mountain breeds (Meyer, 1966). The frequency of occurrence of this variant is low in Europe. It has not been found in Western breeds in Australia.

The second new variant was isolated from Australian Droughtmaster beef cattle by Bell et al. (1966a, 1970). This breed is a new one and consists of 3/8–5/8 Brahman and 5/8–3/8 Western breeds of cattle. Most samples of milk studied contained the A or B or AB variants, but 5 out of 200 contained approximately equal amounts of A or B with another variant. The C variant was not observed, this is to be expected since this breed does not contain any Jersey strain. The new variant moves more slowly than the C

TABLE I

OCCURRENCE OF β-LACTOGLOBULIN C IN SIX JERSEY HERDS IN SOUTH AUSTRALIA[a]

Herd	Variant				
	B	AB	A	BC	AC
1	8	13	6	9	2
2	24	33	10	1	0
3	11	7	1	4	3
4	31	20	1	3	0
5	18	3	0	0	1
6	12	4	0	4	0
Total	104	80	18	21	6

[a] Data of Bailey (1968b).

variant on starch-gel electrophoresis in the Bell borate system at pH 8.5 and has been given the name β-lactoglobulin$_{\text{Droughtmaster}}$. This nomenclature was chosen since the variant has the same amino acid composition as the bovine A variant but has a carbohydrate moiety attached to it.

B. OVINE β-LACTOGLOBULINS

Bell and McKenzie (1964, 1967b) found β-lactoglobulin variants in a study of milk from 62 Merino ewes. The variants were designated A and B, the former having greater mobility in starch-gel electrophoresis at pH 7.6 (phosphate buffer). Subsequently Bell and Stormont (1965) confirmed the presence of these variants in sheep in North America. The genes controlling the synthesis of the variants are autosomal codominant alleles. The distribution of ovine β-lactoglobulins has been confirmed in recent studies by King (1969).

C. CAPRID β-LACTOGLOBULIN

Only one caprid β-lactoglobulin has been revealed in studies by Bell and McKenzie (1964; see also Bell et al., 1968). Sen and Chaudhuri (1962) crystallized caprid β-lactoglobulin from an isoelectric salt-free solution. However Phillips and Jenness (1965) were not successful in attempts to crystallize the protein in this way. They did, however, crystallize it from ammonium sulfate solution using a method similar to that described by Askonas (1954). Ghose et al. (1968) and Townend and Basch (1968) also isolated only a single variant (see also Kalan and Basch, 1966).

D. β-LACTOGLOBULINS OF OTHER SPECIES

Bell and McKenzie (1964) concluded from electrophoretic, immunological and physico-chemical studies of whey proteins from a variety of species that β-lactoglobulin is present only in the milk of ruminants. This was contrary to the generally held belief at that time. It seems certain from these and subsequent studies that β-lactoglobulin is not present in the milk of humans (see also Got, 1965; Got et al., 1965). Brew and Campbell (1967) confirmed this conclusion in the case of the guinea pig, isolating the major whey protein, α-lactalbumin. Bell, McKenzie and co-workers have extended their studies to include the horse, kangaroo and echidna. The position with these animals is complex and is discussed further in Chapter 18. They have also made a more extensive study of porcine proteins. The porcine protein that was not originally considered to be a β-lactoglobulin has now been found to occur as genetic variants. Each variant has a molecular weight of 18,000 daltons, has two disulfide groups but no sulfhydryl group present, and from its chemical behavior may be considered as a monogastric analog of the β-lactoglobulin of ruminants. It is not immunologically similar to the ruminant β-lactoglobulins. Further detailed investigation of the milk of a range of species is needed if we are to trace satisfactorily the evolution of β-lactoglobulin.

V. Amino Acid Composition

A. END-GROUP ANALYSIS

Fraenkel-Conrat (1954, 1956) and Niu and Fraenkel-Conrat (1955) found two N-terminal leucine residues and two C-terminal isoleucine residues in pooled bovine β-lactoglobulin per 36,000 daltons. This was indicative of the unit being a dimer. The monomer molecular weight of 18,000 daltons has since been confirmed by a variety of physical and chemical investigations.

Neurath et al. (1954) confirmed these findings by studying the release of amino acids at the C-terminal end using carboxypeptidase A. They also found that histidine is the penultimate residue. Similar studies were made by Davie et al. (1959). These were extended to the A and B variants by Kalan and Greenberg (1961) and to the A, B, and C variants by Kalan et al. (1965). The latter found that all three variants have the same N-terminal residue (leucine) and the same C-terminal residue (isoleucine). In each variant the penultimate residue at the N-terminal end is threonine and at the C-terminal end, it is histidine. The rate of release of amino acids from the carboxyl end of the chain by the action of carboxypeptidase A is the same for the A and B variants, but the rate for C is much less than for A or B. Upon denaturation or rupture of SS bonds and formation of the S-sulfo derivatives, the overall rates of hydrolysis for all three variants becomes the same. In a subsequent publication Greenberg and Kalan (1965) prepared modified crystalline derivatives from the A, B and C variants after reacting the latter with carboxypeptidase A. The derivatives of the A and B variants have one less isoleucine and histidine residue (per monomer) compared with their parent molecules. However the penultimate histidine is incompletely removed in the case of the C variant. Furthermore these derivatives cannot be simply derived from the parent protein (see Section V.D).

TABLE II

AMINO ACID ANALYSES OF β-LACTOGLOBULINS: COMPARISON OF RESULTS[a]

Amino acid	Bovine A PDFG[b]	Bovine A TBK[c]	Bovine A KGW[d]	Bovine A BMS[e]	Bovine B PDFG[b]	Bovine B TBK[c]	Bovine B KGW[d]	Bovine B BMS[e]	Bovine C KGW[d]	Bovine C BMS[e]	Bovine D BRGP[f]	Bovine Dr BMMS[g]	Ovine A BMS[e]	Ovine B BMS[e]	Caprid BMS[e]	Caprid PJ[h]	Caprid TB[i]
Gly	3	3	3	3	4	4	4	4	4	4	4	3	5	5	5	5	5
Ala	13.9	14.2	13.4	14.2	14.9	14.5	14.4	15.2	14.4	15.4	15.1	14.0	15.4	15.7	14.6	15	16.1
Ser	6.9	6.3	6.6	6.1	6.9	6.5	6.6	6.3	6.6	6.4	7.0	6.0	5.5	5.2	5.5	7	6.6
Thr	8.1	7.4	7.8	7.5	8.1	7.4	7.8	7.6	7.6	7.7	8.0	7.4	7.5	7.5	7.2	8	8.1
Pro	8.2	7.9	8.5	8.3	8.2	8.8	8.4	8.3	8.5	8.5	8.0	8.2	8.0	7.9	8.1	8	7.9
Val	9.9	9.6	9.7	9.7	8.9	8.8	8.9	8.9	8.9	8.9	9.1	9.7	9.5	9.7	9.4	10	10.2
Ile	9.7	9.5	9.5	9.0	9.7	9.5	9.6	9.1	9.6	9.1	10.0	9.1	8.1	8.3	8.5	9	9.5
Leu	21.9	20.3	20.5	22.5	21.9	20.3	21.0	22.6	21.0	23.1	21.0	21.5	19.9	20.4	20.0	20	20.7
				(21.2)[j]				(21.4)[j]		(21.9)[j]							
Phe	3.9	3.6	3.9	4.0	3.8	3.6	3.9	3.9	3.9	4.0	4.1	4.0	3.8	3.8	4.1	4	3.9
Typ	3.6	3.5	3.9	4.0	3.6	3.5	3.9	3.8	3.9	4.0	4.1	3.7	3.7	2.9	3.9	4	3.8
												(3.9)[k]					
Try	2.5		2.0	1.9	2.5		2.1	1.8	2.1	2.0	2.1		2.0	2.0		2	2.0
Cys/2	3.9	3.5	3.3	3.9	4.1	3.6	3.3	3.4	3.3	3.5	4.0	3.9	2.9	2.9	3.0	5	3.8
Met	3.7	3.7	3.9	3.8	3.7	3.7	3.9	3.9	3.9	4.0	4.1	3.6	3.5	3.8	3.4	4	3.7
Asp	16.0	15.4	15.2	16.0	15.0	14.6	14.9	15.2	14.8	15.4	15.2	15.9	14.9	14.6	14.5	15	14.7
Glu	25.0	23.7	23.9	26.2	25.0	24.9	24.7	26.4	23.6	25.5	24.9	25.3	24.0	23.6	23.8	25	24.5
				(25.1)[j]				(25.2)[j]		(24.4)[j]							
NH₃				15.6				15.4		14.7	12.4	17.5	17.0	17.2	16.0	14	
Arg	2.8		3.0	2.9	2.9		3.0	2.9	3.0	3.0	3.0	2.9	2.9	2.9	2.8	3	2.9
His	1.9		1.9	1.9	1.8		1.8	1.9	2.9	2.9	2.1	1.9	2.1	2.9	2.0	2	2.0
Lys	14.9		14.4	14.6	14.8		14.4	14.8	14.4	14.9	14.6	14.9	13.6	14.0	14.8	15	15.7

[a] Residues per single-chain monomer.
[b] Piez et al. (1961).
[c] Townend et al. (1965).
[d] Kalan et al. (1965).
[e] Bell and McKenzie (1964), Bell et al. (1966b).
[f] Brignon et al. (1969).
[g] Bell et al. (1970).
[h] Phillips and Jenness (1965).
[i] Townend and Basch (1968).
[j] Corrected (see Bell et al., 1970).
[k] From ultraviolet spectra.

14. β-LACTOGLOBULINS 279

TABLE III
PROBABLE NUMBER OF AMINO ACID RESIDUES PER MONOMER OF β-LACTOGLOBULIN[a]

Residue	Bovine					Ovine		Caprid
	A	B	C	D	Dr[b]	A	B	
Gly	3	4	4	4	3	5	5	5
Ala	14	15	15	15	14	15	15	15
Ser	7	7	7	7	7	6	6	6
Thr	8	8	8	8	8	8	8	8
Pro	8	8	8	8	8	8	8	8
Val	10	9	9	9	10	10	10	10
Ile	10	10	10	10	10	9	9	9
Leu	22	22	22	22	22	20	20	20
Phe	4	4	4	4	4	4	4	4
Tyr	4	4	4	4	4	4	3	4
Trp	2	2	2	2	2	2	2	2
Cystine	2	2	2	2	2	2	2	2
Cysteine	1	1	1	1	1	1	1	1
Met	4	4	4	4	4	4	4	4
Asp	16	15	15	15	16	15	15	15
Glu	25	25	24	25[c]	25	24	24	25
NH₃	15	15	14	15	15	16	16	15
Arg	3	3	3	3	3	3	3	3
His	2	2	3	2	2	2	3	2
Lys	15	15	15	15	15	14	14	15

[a] Modified from McKenzie (1967).
[b] Also contains carbohydrate moiety.
[c] D has 1 Gln substituted for 1 Glu compared with B.

B. AMINO ACID ANALYSIS

The amino acid composition of the bovine A and B variants was first determined by Gordon et al. (1960, 1961) and Piez et al. (1961). Subsequently Bell et al. (1968) determined the composition of the bovine A, B and C variants and the ovine A and B variants (see also Bell and McKenzie, 1964). Kalan et al. (1965) determined the composition of the bovine A, B and C variants. Townend et al. (1965) compared the analyses of samples of the A variant prepared from milk of 24 individual cows (homozygous in A) and of the B variant from 9 individual cows (homozygous in B). There were no differences in analyses for the individual samples of a given variant. Bell et al. (1966b, 1970) have determined the composition of the Droughtmaster variant and Brignon et al. (1969) have analyzed the D variant. Phillips and Jenness (1965), Bell et al. (1968) and Townend and Basch (1968) analyzed caprid β-lactoglobulin.

TABLE IV

Differences in Number of Amino Acid Residues per Monomer Compared with Bovine B

Residue	Bovine				Ovine		Caprid
	A	C	D	Dr	A	B	
Gly	−1		−1		+1	+1	+1
Ala	−1		−1				
Ser					−1	−1	−1
Val	+1		+1		+1	+1	+1
Ile					−1	−1	−1
Leu					−2	−2	−2
Tyr						−1	
Asp	+1		+1				
Gln		−1	*a*		−1	−1	
His		+1				+1	
Lys					−1	−1	
Carbohydrate				*b*			

a D has 1 Gln substituted for 1 Glu.
b Droughtmaster has 1.0 N-acetylneuraminic acid, 4.3 hexosamine, and 2.7 hexose residues.

A comparison is made in Table II of the amino acid analyses of the various workers for the bovine, ovine and caprid variants. All the analyses have been converted to a common basis, namely, an integral number of glycine residues per monomer unit (mol wt 18,000 ± 500 daltons for all variants except Droughtmaster, mol wt 20,000 daltons). The cystine analyses have not been corrected for destruction during hydrolysis. The serine and threonine analyses have been corrected by most workers.

On the basis of these results, McKenzie (1967) has drawn up the most likely composition of the variants. The table has been modified to include the more recent results on the bovine D variant and the data are presented in Table III. A summary of the differences in composition of the variants is given in Table IV.

C. Peptide Studies and Amino Acid Sequence

An important landmark in our understanding of the structure of β-lactoglobulin was achieved at the end of 1967 when Frank and Braunitzer (1967) published their partial amino acid sequence for bovine β-lactoglobulins A and B. Their sequence for β-lactoglobulin A (modified in the light of subsequent investigations, for example the determination of the location of the SH group and SS bridges by McKenzie *et al.*, 1970, and Ralston, 1969) is presented in Fig. 6.

In this section we shall review studies of peptides from variants of β-lactoglobulin in the light of Frank and Braunitzer's sequence. It is important to realize that most of the peptide studies were carried out before Frank and Braunitzer's work was published.

Townend (1965) isolated a pair of tryptic peptides of differing mobility from tryptic digests of S-sulfonated bovine β-lactoglobulins A and B. The peptide from the A variant differed from that of B in the substitution of an aspartic acid residue for a glycine residue. There was a large concentration of carboxylic acid side-chain residues in the neighborhood of the difference residue. This appears to be the area of the peptide chain probably involved in the octamerization reaction of the A variant (see also McKenzie, 1967).

Bell *et al.* (1966b, 1968) examined tryptic digests of the performic acid-oxidized digests of the variants known when they commenced their investigation in 1963. They used a variety of chromatographic and electro-

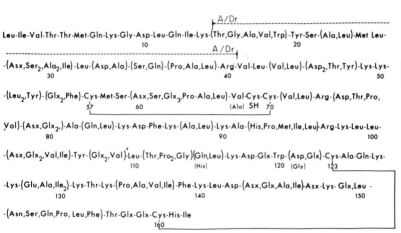

FIGURE 6. Partial amino acid sequence of bovine β-lactoglobulin A. The following substitutions are shown for other bovine variants: B has Ala at residue 69, Gly at 121 or 122; C is similar to B but has His at 115 or 116; the Droughtmaster variant has the same amino acid composition as A but has a carbohydrate moiety, the difference tryptic peptide being indicated by the dashed lines (residues 15–40). The sequence is based on that of Frank and Braunitzer (1967) for the A and B variants and the determination of the location of the SH group and SS bridges by McKenzie *et al.* (1970) and Ralston (1969). The following modifications are made for residues 120–122 (see text): Trp-(Asp, Glx) replaces (Asp, Glx, Trp); for residues 110–117: Leu-(Thr, Pro$_2$, Gly) (Glu, Leu)-Lys replaces Leu-(Thr, Glx, Pro$_2$, Gly, Leu)-Lys.

phoretic methods to separate the peptides. The number of peptides detected is in accord with a monomer molecular weight of 18,000 ± 500 daltons and the postulate that the 36,000 dalton particle (dimer) of a given variant consists of two identical chains.

They found the composition and mobility of the difference peptides, relating to the Asp-Gly difference of bovine A and B, to be similar to those reported by Townend (1965). He concluded that in bovine A there is a sequence Asp-Lys-Lys, in which the acidic residue could hinder hydrolysis by trypsin after the first lysyl residue, and hence cleavage of the second lysyl residue would be favored. On the other hand, hydrolysis by trypsin would be free of interference by an acidic residue in bovine B with the sequence Gly-Lys-Lys. The existence of the Lys-Lys sequence would lead to a low yield of the tryptic peptide, because if the first hydrolysis occurred at the second lysyl residue, subsequent removal of this would be slow since trypsin is most efficient as an endopeptidase. This difference in tryptic digestion was also observed by Bell et al. (1966b, 1968) and by Frank and Braunitzer (1968; see also Préaux et al., 1962).

The composition of the chymotryptic and tryptic difference peptides given by Townend is the following:

Bovine β-A peptides	
Tryptic	↓ (Leu, Trp)-(Val, Ala, Glu$_3$, Asp)-*Asp*-Lys-Lys
Chymotryptic	↓ (Val, Ala, Glu$_3$, Asp)-*Asp*-Lys-Lys-(Glu, Ala$_2$, Lys$_2$, Thr, Pro, Cys, Ile$_2$, Phe) ↓

Bovine β-B peptides	
Tryptic	↓ (Leu, Trp)-(Val, Ala, Glu$_3$, Asp)-*Gly*-Lys ↓ + Lys ↓
Chymotryptic	↓ (Val, Ala, Glu$_3$, Asp)-*Gly*-Lys-Lys-(Glu, Ala$_2$, Lys$_2$, Thr, Pro, Cys, Ile$_2$, Phe) ↓

Bell et al. (1966b), analyzing the tryptic peptides from the material oxidized by performic acid, found cysteic acid present in addition. This is borne out by the sequence published subsequently by Frank and Braunitzer (1967). The latter group found that the sequence near the Lys-Lys residues is different from that given by Townend:

$$\text{Lys-}^{\downarrow\text{T}}\text{Asp-Glx-}\left(\begin{matrix}\text{Asp}\\\text{Gly}\end{matrix}\right)\text{, Glx, Trp)-Cys-Ala-Gln-Lys}^{\downarrow\text{T}}\text{-Lys}^{\downarrow\text{T}}$$
117   120   127

Considering the results of the various groups for both tryptic and chymo-

14. β-LACTOGLOBULINS 283

tryptic peptides it seems to the author that a more likely sequence is

$$\text{Lys}\underset{117}{\downarrow^{\text{T}}}\text{-Asp-Glx-Trp}\underset{120}{\downarrow^{\text{C}}}\text{-}\left(\begin{Bmatrix}\text{Asp}\\\text{Gly}\end{Bmatrix}, \text{Glx}\right)\text{-Cys-Ala-Gln-}\underset{126}{\text{Lys}}$$

Neither Townend nor Bell *et al.* located the Val ↔ Ala difference peptide. However this has been isolated subsequently by Frank and Braunitzer (1968). The valine residue at position 68 in the A variant is substituted by alanine in the B variant.

Bell *et al.* (1966b, 1968) separated and analyzed the difference tryptic peptides for bovine β-lactoglobulin B and C. They found the following partial sequences:

B   (Thr, Glu$_4$, Gly, Val$_2$, Leu$_3$, Lys, Pro$_2$)-(Asp, Ile)-(*Gln*, Leu)-Lys

C   (Thr, Glu$_4$, Gly, Val$_2$, Leu$_3$, Lys, Pro$_2$)-(Asp, Ile)-(*His*, Leu)-Lys

This sequence includes the analyses of some histidine-containing peptides separated from a chymotryptic hydrolyzate of the tryptic core. One peptide containing equal amounts of Asp, Ile, His, Leu and Lys, and another peptide with only the last three residues, were both isolated in low yield.

This sequence has been considered by Ralston (1969) in the light of the sequence work of Frank and Braunitzer (1967). He identified the latter's partial sequence as

-Arg-$\underset{97}{\downarrow}$Lys-Leu-Leu-(Asx, Glx$_2$, Val, Ile)-Tyr-(Glx$_2$, Val)-Leu-(Thr, Glx, Pro$_2$, Gly, Leu)-
       100$$110

$\underset{117}{\text{Lys}}$

Bell *et al.* postulated a Lys-Pro sequence to explain the evidence of two lysine residues in this peptide. Ralston has pointed out that the presence of the Arg-Lys sequence at the N-terminal end would be consistent with cleavage between arginine and lysine, leaving lysine as the N-terminal residue. Comparing the above results of the two groups Ralston concluded that the difference residue for B/C is located as follows:

$$\underset{110}{\text{Leu-(Thr, Pro}_2\text{, Gly)}}\text{-}\left(\begin{Bmatrix}Gln\\His\end{Bmatrix}, \text{Leu}\right)\text{-}\underset{117}{\text{Lys}}$$

Bell *et al.* concluded that the substitution is His/Gln rather than His/Glu on the basis of the difference peptides having the same mobility at pH 8.9. McKenzie (1967) stated incorrectly that the genetic code allowed a His/Glu as well as His/Gln as a single-base exchange. The current code (Crick, 1967) allows only His/Gln (compare Jukes, 1963).

On the basis of an examination of chymotryptic peptides from bovine β-lactoglobulin B and D, Brignon *et al.* (1969) have proposed that the B ↔ D mutation, Glu ↔ Gln, occurs at position 109 in Frank and Braunitzer's (1967) partial sequence for the B variant. It would seem to the author that the substitution could occur at 108 or 109 on the basis of their data.

Bell et al. (1966b, 1970) located the difference tryptic peptides from the bovine A and Droughtmaster variants. These peptides differ only in that there is a carbohydrate moiety attached in the Droughtmaster variant. An examination of their data in the light of the partial sequence of Frank and Braunitzer (1967) leads one to suggest that the partial sequence involved is

(Thr, Gly, Ala, Val, Trp)-Tyr-Ser-(Ala, Leu)-Met-Leu-(Asx, Ser$_2$, Ala$_2$, Ile)-Leu-
15   20   30
   (Asp, Ala)-(Ser, Gln)-(Pro, Ala, Leu)-Arg
                        40

Bell et al. also examined the peptides from ovine β-lactoglobulin A and B. They found that the A variant has two tryptic peptides that do not occur in the same position in the two-dimensional peptide maps (pH 1.9/4.7) from the B variant. The latter has only one peptide occurring in a different position from the A peptides. This B peptide stains strongly for histidine and has the same composition as one of the A peptides, except for the substitution of a histidine residue for a glutamic residue. These two peptides have the same mobility at pH 8.9. Thus the A ↔ B substitution would appear to be Gln ↔ His. Bell et al. concluded that the A ↔ B mutation does not involve a simple substitution, Tyr ↔ His, as would be the simplest interpretation of the amino acid analysis of the proteins. Since the total number of glutamic acid residues is 24 in each variant, it would appear that another one of the glutamic acid residues has been substituted in the B variant for a tyrosine residue of the A variant. The second ovine A difference peptide contains tyrosine but is not a true tryptic peptide (see Bell et al., 1968). It is possible to identify the partial sequence of the Gln ↔ His difference peptides by comparison of the analysis of these peptides with the partial sequence of bovine β-lactoglobulin A determined by Frank and Braunitzer (1967). The region of the bovine chain involved is from positions 15–40. If the results of Bell et al. on ovine A and B and of Frank and Braunitzer on bovine A are correct, there are differences between ovine A and bovine A in this region of the chain, for example there is no tyrosine in position 20 of the bovine chain and this residue must be located in another region of the chain of ovine A. The conclusions of Bell et al. are being reinvestigated in this laboratory.

In conclusion, it is seen that there is good agreement between the peptide studies and the partial sequence of Frank and Braunitzer. There are, however, important differences; these will only be resolved when the complete sequence is elucidated.

D. Location and Reactivity of SS and SH Groups

There have been numerous determinations of the sulfhydryl content of pooled bovine β-lactoglobulin. Many of the workers (Leslie et al., 1962a; Habeeb, 1960; Piez et al., 1961; Fraenkel-Conrat et al., 1952; Stark et al., 1960; Hutton and Patton, 1952; Leach, 1960; Klotz and Carver, 1961) found that there are 2 SH groups/36,000 dalton unit. This figure is in good agreement with the total sulfur content, 1.60 g/100 g (Brand et al., 1945), when the methionine content, 8 residues/36,000 daltons, and cystine content, 4 Cys/2 residues/36,000 daltons, are taken into account. Other workers have obtained higher figures. Christensen (1952), using potassium ferricyanide as oxidant, obtained a value of 4, and Larsen and Jenness (1950), using o-iodosobenzoate, obtained 3.9.

Leslie et al. (1962b) showed that ferricyanide reacts very slowly with bovine β-lactoglobulin AB at pH 7.0. However in the presence of 5 M urea or 3 M guanidine hydrochloride, the reaction proceeds in two stages: a rapid one, followed by a slow one. Sodium dodecylsulfate at high concentrations inhibits the reaction both in the presence and absence of urea (see also Leslie and Varrichio, 1968). There is a simple stoichiometry of 2 mole ferricyanide/protein dimer at low ratios of ferricyanide to protein. The stoichiometry increases to a maximum of 4 at high ratios. Leslie et al. suggested that there is only one SH group on each single-chain monomer, SS groups formed by oxidation being intermolecular, and that at low ferricyanide ratios the stoichiometry is simple. However at high ratios the SH can react with more than one ferricyanide to yield higher oxidation products. This behavior of ferricyanide is quite different from that with ovalbumin where simple stoichiometry is observed at all ratios. In the latter protein, intramolecular disulfide bonds can form readily, as there are two pairs of SH groups present per molecule.

Leslie et al. (1962b) and Habeeb (1960) studied the rate and extent of reaction of the SH groups with N-ethylmaleimide (NEM) and sodium p-hydroxymercuribenzoate (PHMB), incorrectly referred to in early work as p-chloromercuribenzoate or PCMB. The reaction of the native protein was slow with NEM at pH 7.0 and faster with PHMB but very rapid for both reagents with protein that had been denatured in concentrated urea, concentrated guanidine hydrochloride, or alkali at pH 12. Products of denaturation by ethanol or sodium dodecylsulfate appeared to have different reactivity for the SH groups to those formed from urea or alkaline denaturation.

Prolonged exposure of protein to alkali resulted in excessive titers. Habeeb (1960) found that protein kept in urea for 24 hr prior to titration gave low results. It is evident that on prolonged denaturation, cleavage of

SS groups and/or exchange reactions complicate the reaction of the SH groups. In a more recent paper Franklin and Leslie (1968) made a kinetic study of the reaction of NEM with bovine β-lactoglobulin AB in urea and in sodium dodecylsulfate solution. Unfortunately they followed the reaction spectrophotometrically at 300 nm. This led to difficulties under some conditions owing to difference spectral effects arising out of changes in environment of tyrosine and tryptophan groups during the denaturation reaction. It would have been preferable to have followed the reaction at 310 nm. The reaction of the SH group in 8 M urea at pH 4.8 was second order, and the rate would appear to be comparable to that of cysteine at pH 4.8. However the reaction in sodium dodecylsulfate was considerably slower. The latter

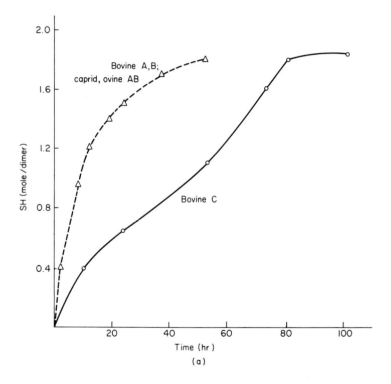

FIGURE 7. (a) Reaction of β-lactoglobulin variants with a twofold excess of DTNB in 0.01 M phosphate buffer—10^{-3} M in EDTA, pH 7.6. The amount of reaction at various times is expressed in terms of moles of SH per protein dimer unit. (b) Reaction of normal and modified β-lactoglobulins with fivefold excess DTNB in the presence and absence of sodium dodecylsulfate (0.5 g/dl). Note time-scale change at 5 hr. (Based on data of Phillips et al., 1967.)

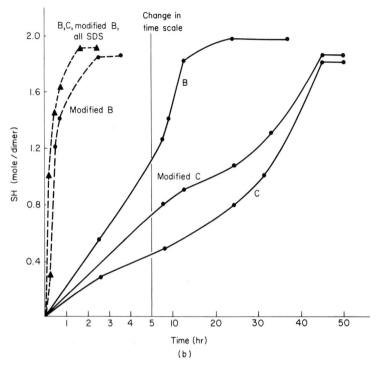

FIGURE 7. (b)

may be contrasted with that of ovalbumin where the reaction in detergent is not simple second order and the rate is comparable to that of cysteine.

Phillips et al. (1967) have made an interesting comparative study of the reactivity of both β-lactoglobulin variants and modified bovine proteins with Ellman's (1959) reagent, 5,5'-dithiobis(2-nitrobenzoic acid) (DTNB). The rates of reaction of bovine A and B, ovine AB(?) and caprid β-lactoglobulins with a twofold excess of DTNB at pH 7.6 (0.05 M phosphate, 0.02 M disodium ethylenediaminetetraacetate) were approximately the same. The rate for the bovine C variant was considerably slower than that for the other variants, as shown in Fig. 7a. When sodium dodecylsulfate was added to the bovine variants in the presence of a fivefold excess, DTNB, the reaction half time increased up to ~0.1 × 10^{-2} g/dl detergent and then decreased. Rates considerably lower than the values in the absence of detergent were attained at 0.33 × 10^{-2} g/dl concentration, and there was only a small further change above this concentration. The reaction in the presence of 0.5 g/dl is shown in Fig. 7b. At the end of the reaction for all variants the consumption of DTNB approached the equivalent of two SH groups per protein dimer of 36,000 daltons.

We have seen in Section V.A that Greenberg and Kalan (1965) prepared crystalline derivatives of β-lactoglobulins after the removal of C-terminal residues by the action of carboxypeptidase A. The rates of reaction of these derivatives with DTNB were compared with those of the native protein by Phillips *et al.* (1967). The rate for the modified bovine β-lactoglobulin B was considerably faster than for the normal B protein and was comparable to the rates attained by both normal and modified proteins when sodium dodecylsulfate (0.5 g/dl) was present (Fig. 7b). The rate for the modified C variant was a little faster than that for the normal C protein, but it was still considerably slower than for the normal B protein. However in the presence of the detergent (0.5 g/dl) the rate for the C variant increased markedly and was similar to that for the B variant with detergent present.

The nature of the modified protein must be regarded with some reservation. While the C-terminal Ile residue (162) and the adjacent His residue (161) were cleaved in the treatment it is uncertain as to what other residues were also cleaved. The derivatives were obtained only in low yield ($\sim 10\%$) and Kalan and Greenberg observed release of considerable Thr (residue 157) and Glx (159), but did not detect Cys (160) and obtained a C-terminal Leu for the derivative (residues 156 in the native protein). It is also uncertain as to what extent the derivatives are denatured since complete optical rotatory dispersion (ORD) curves were not performed on them. Thus detailed interpretation of the data for the modified proteins in structural terms is not justified. However the following general comments may be made.

On the basis of the data for the unmodified proteins, it appears that the mutation His residue (115 or 116) in the C variant sufficiently alters the conformation and/or state of association of the protein to make the SH group less readily available to the DTNB. In the presence of sodium dodecylsulfate greater than $\sim 0.3 \times 10^{-2}$ g/dl, the rate for C increases appreciably, becoming comparable to that for B (or A) in the absence of detergent. The B (or A) variant overcomes the difference in conformation and/or tendency to dissociate. (It will be recalled that at pH 7.5 the tendency for C to dissociate to the monomer is much less than for A or B, McKenzie and Sawyer, 1967.) However the modification of the C variant, in which apparently only approximately half of His (161) is removed, is insufficient (in the absence of detergent) to endow similar reactivity on the SH group to that of the normal B variant.

Lontie, Préaux and their associates have studied extensively the behavior of the cysteine residue in β-lactoglobulin (Binon *et al.*, 1962; Lontie *et al.*, 1964a, b; Préaux *et al.*, 1964a, b, c; Buchet *et al.*, 1965; Roels *et al.*, 1966,

1968a, b; Joniau et al., 1966; Lontie and Préaux, 1966; Bloemmen et al., 1967). They have shown that the titration of the SH group can be followed polarimetrically with the following reagents: mercury(II), monofunctional organic mercurials, NEM, disulfides, and sodium tetrathionate. At pH values of 6.8 and higher, the addition of mercury(II) acetate to bovine β-lactoglobulins A or B results in an increase in levorotation at 436 nm up to one mercury(II) per β-lactoglobulin 36,000 dalton unit. The levorotation remains constant for higher concentrations of mercury(II). The elution volume on Sephadex G-100 at pH 8.2, together with the stoichiometry, are indicative of the linking of two single-chain monomers via one SH group in each to one mercury(II). It was found that in the pH range 6.1–7.3 mercury(II) can also react with a stoichiometry of two mercury(II) per β-lactoglobulin unit of 36,000 daltons. The levorotation for this ratio is less than that for the lower ratio, and its elution volume on Sephadex indicates it can dissociate to a monomer unit.

When PHMB is present in neutral and alkaline solutions of bovine β-lactoglobulin A and B (pH 6.5–8.0) the levorotation increases up to a ratio of 2 moles PCMB/β-lactoglobulin dimer (36,000 daltons). The levorotation of β-lactoglobulin is affected by NEM in alkaline solution only and increases up to a ratio of 2.3 NEM/36,000 dimer unit of β-lactoglobulin. The reactions with organic disulfides and sodium tetrathionate were also studied by Lontie and Préaux (1966) and Préaux and Lontie (1966). They found that cystamine dihydrochloride, like NEM, reacted with the SH group in alkaline solution only. However, the reaction, as measured by the increase in levorotation at 436 nm, was slower than that for NEM. The stoichiometry was 1 mole cystamine/1 mole protein dimer. The reaction of sodium tetrathionate was similar to that of cystamine hydrochloride, but the reaction product differed in that it was present as a dimer only (no monomer).

The marked augmentation of levorotation of β-lactoglobulin on reaction with mercurials was first demonstrated by Pantaloni (1962). It is strongly pH dependent as can be seen strikingly from the data of Lontie and Préaux (1966) in Fig. 8. The reaction with monofunctional mercurials not only affects conformation of β-lactoglobulin but it also may profoundly affect the dissociation of the protein. We shall see in Section VII that the bovine β-lactoglobulins exist predominantly in the dimer (36,000 dalton) form in the pH range 3.6–6.0. The tendency to dissociate to the monomer form is small, but it exists nevertheless. As the pH is lowered below 3.5 the dissociation increases and is nearly complete by pH 2 at *low* protein concentrations. There is little difference between the variants in extent of dissociation at low pH. When the pH is raised above 6, there is an increase in dissociation, but while it is appreciable by pH 7.5, it is not complete except at very low protein concentrations. There is an appreciable difference in the

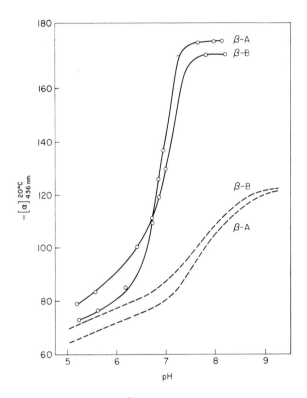

FIGURE 8. Effect of pH on $[\alpha]_{436\,nm}^{20°C}$ for bovine β-lactoglobulins A and B, in the absence (-----) and presence(———) of sodium p-hydroxymercuribenzoate (Lontie and Préaux, 1966).

tendency for dissociation at pH 7.5: A and B ≫ C. Roels et al. (1968b) have attempted to enhance the dissociation by the separate addition of PHMB, 1-[3-(chloromercuri)-2-methoxypropyl]urea (chlormerodrin, ClM) and NEM to the bovine B variant at pH 5.2, 5.6 and 8.1. The elution volumes of the reacted and unreacted protein on Sephadex G-100 at pH 5.6 were the same. At pH 8.2 the elution volume increased in the order PHMB < ClM < NEMI derivative. This was interpreted in terms of increasing dissociation. There was no difference in the $s_{20,w}$ vs. concentration curves for the native protein and its derivatives at pH 5.6 ($s°_{20,w} \approx 3.0$ S). However at pH 8.2 both the native protein and the PCMB or ClM derivatives gave the characteristic plot of $s_{20,w}$ vs. concentration for a rapidly associating–dissociating system of monomer–dimer type (see Volume I, Part C, Chapter 7). The tendency for dissociation of the derivatives was much greater than that of the native protein. The $s_{20,w}$ values for the NEM derivative were lower than those for the other derivatives and extrapolated linearly to zero concentration to give a value of 1.97 S. Roels et al. interpreted this to mean that the NEM induced complete dissociation to the monomer.

In an independent investigation Zimmerman et al. (1969) found that if the sulfhydryl group of bovine β-lactoglobulin A is treated with 4-(p-dimethylaminobenzeneazo)phenylmercuric acetate the protein is dissociated completely.

H. A. McKenzie and G. B. Ralston, in the course of an investigation of the urea denaturation and aggregation reactions of the bovine genetic variants, collaborated with D. C. Shaw to determine the location of the SH group and SS linkages in the peptide chain. They used the diagonal peptide mapping method of Brown and Hartley (1963), making digests of bovine β-lactoglobulin B treated with ^{14}C-iodoacetamide. With the aid of the partial sequence of Frank and Braunitzer (1967), they were able to show that the SH group is located at residue 69, and the two cystine disulfide linkages involve residues 57 and 70, and 123 and 160. These positions can be visualized more clearly by referring to the peptide chain shown in Fig. 6.

If the location of the sulfhydryl group is considered in relation to the effect that its blocking with organic mercurials and NEM has on the behavior of the protein, the following inferences can be made. The conformation of the protein is such that when the sulfhydryl group is blocked at pH 7 and above, it is sterically difficult for the monomer units to form a dimer. This implies that position 69 is reasonably close to the site of contact between the monomer units. However it is not so close that dimerization is prevented at pH values ($\leqslant 6$) where there is no appreciable conformational change in the presence of blocking agent.

This conclusion is at variance with that of Townend et al. (1969) who studied the effect of pH in the region 3.8–5.0 on the molecular size of carboxymethyl, 2-hydroxyethyl and 2-aminoethyl derivatives of bovine β-lactoglobulin A and B (prepared via the sulfenyl iodide derivatives, according to the method of Cunningham and Nuenke, 1959, 1960a, b, 1961). They found that all derivatives were present at pH 5.0 as dimers. At pH 4.6 and low temperature, the derivatives of the A variant showed somewhat less tendency to octamerization than the unsubstituted protein, as found earlier by Cunningham and Nuenke. Below pH 3.5 all derivatives tended to dissociate to the monomer. Townend et al. concluded that the SH group is possibly in the vicinity of the site of interaction of dimers to form the octamer (i.e., residues 117–126) and that it cannot be closer than 6 Å to the site of monomer-monomer contact.

It will be seen in Section X that Green and Aschaffenburg (1959) found that the distance between the SH groups of the dimer is 27 Å. Dunnill and

Green (1965) found that the second-order rate constant k', for the reaction of bovine β-lactoglobulin A ($\sim 3 \times 10^{-5}$ M) with PHMB ($\sim 0.6 \times 10^{-5}$ M), was approximately constant (~ 6 liter/mole/sec) from pH 2.4 to pH 6.8. At pH 7.1 it became 91 liter/mole/sec; at pH 7.8, 430 liter/mole/sec; at pH 7.8, 1030 liter/mole/sec; and at pH 8.5, $> 2.7 \times 10^4$ liter/mole/sec. Their conclusion, that the large increase in k' over the small pH change near 7 involves both dissociation and conformational change, is in agreement with our conclusions concerning the location of the SH group. The above studies of Phillips *et al.* (1967) on comparative reactivity of the SH groups in the A, B and C variants are also in accord with our findings.

The finding that Cys 160 is linked to Cys 123 is of interest in connection with the action of carboxypeptidase on the C-terminal end of the peptide chain. The C-terminal Ile, and penultimate His residues, appear to be readily removed from the bovine A and B variants but the action becomes complex thereafter, no doubt being affected by the fact that residue 160 is cystine and not cysteine. This residue is linked to residue 123, located in the COOH-rich residues of the dimer contact region involved in the octamerization reaction of the A variant. (It is also the region of the A ↔ B, Asp ↔ Gly substitution site.)

The location of the SH group and SS bridges has implications with respect to the exchange reactions and SH oxidation observed in alkaline solutions and in urea solutions of β-lactoglobulin (see Sections VII, VIII). There are also implications concerning the location of other groups in the molecule. On the basis of hybridization experiments Townend *et al.* (1961) concluded that the Val ↔ Ala, bovine A ↔ B substitution is at the area of monomer-monomer contact (Timasheff and Townend, 1962). However doubt was subsequently cast on this on the basis of the interpretation of electrophoresis patterns in which association-dissociation reactions are involved (see Chapter 7, Volume I; Hill, 1964; Townend *et al.*, 1969). However, Val 68 being adjacent to SH residue 69 could well be reasonably near the site of monomer-monomer contact.

We shall see in Section V.E that in each chain of the bovine β-lactoglobulins, one of the tryptophans is "buried" and the other is "exposed" or both are partly "buried." Two of the tyrosines are reasonably exposed, the third partly exposed, and the fourth buried. In the sequence of Fig. 6, residue 120 is a tryptophan residue near Cys 123 and is located in the region rich in carboxyl. Thus it is concluded that Trp 123 is located near the dimer-dimer contact region in the octamerization of A (see Townend *et al.*, 1969). The other Trp (residues 15–19) is considered to be buried or partially buried, and Tyr 20 is also likely to be partly buried or buried.

One wonders whether the SH group is hydrogen bonded $-SH \cdots O$, or $-SH\cdots N$ or $-SH\cdots S$, and when the abnormal COOH group is titrated and the conformational transition occurs near pH 7, whether this bond is broken, both leading to increased reactivity of the SH group (for the possibility of such hydrogen bonds see Mukherjee et al., 1970, and Donohue, 1969; for a contrary view see Edsall, 1965).

E. LOCATION OF TYROSINE AND TRYPTOPHAN GROUPS

Townend et al. (1969) attempted to determine the degree of exposure of the two tryptophan residues in bovine β-lactoglobulin A, B and C, at various states of association, by solvent perturbation spectroscopy (see Chapter 8, Volume I). They found that the degree of exposure to sucrose for the monomer (pH 2.0) and dimer of each of these variants was similar and equal to ~0.5. There was no marked difference in the exposure of the monomer and dimer of the A variant to sucrose, ethylene glycol, methanol, and glycerol, the degree of exposure being in the range 0.4–0.5. The octamer for the A variant had a degree of exposure of 0.5 to sucrose and 0.4 to ethylene glycol. It was concluded that the tryptophan residues are located in regions of the molecule remote from those involved in association-dissociation reactions over the pH range 2–6, or the conformational transitions in the pH range 4–6. It could not be determined by this method, of course, whether one residue is buried from, and one exposed to, solvent, or whether both are partly exposed to solvent.

The availability of the four tyrosine residues in the bovine A and caprid variants was examined by spectrophotometric titration (see Chapter 8, Volume I) and reaction with cyanuric fluoride and N-acetylimidazole (see Chapter 5, Volume I). Townend et al. found from spectrophotometric titration that three tyrosine residues ionize only above pH 9.3 with a pK_{app} of 10.9, the fourth ionizing above pH 11, with a pK_{app} of 12.3. Reaction with cyanuric fluoride results in two tyrosines of bovine A being cyanurated at pH 9.3 and 25°C, and a third at pH 10.8. Two of the groups in the caprid variant reaction are cyanurated at pH 9.3, and a third at pH 10.0. Two of the tyrosines in bovine A protein are acetylated with N-acetylimidazole as the mole ratio of reagent to protein is increased from 20:1 to 300:1. Twice this ratio is needed for the caprid protein to attain the same degree of acetylation. After reaction of the first two groups, the reaction continues very slowly and the degree of acetylation approaches 3 asymptotically. The caprid reaction proceeds at a much slower rate for the acetylation of the second and third residues. Thus two of the tyrosine residues in both variants are more reactive with both reagents than the third, whereas the fourth is totally unreactive.

The conclusions of Townend et al. concerning the tyrosine and tryptophan residues are discussed in Section V.D with relation to the location of the cysteine residue.

VI. Electrochemical Properties

A. pH Titration Curves

1. Introduction

Cannan et al. (1942) were the first to determine the pH titration curve for bovine β-lactoglobulin AB. Nozaki et al. (1959) obtained a similar curve from pH 2–9.7, and Tanford et al. (1959) studied the curve for protein denatured at pH 12.5. By use of a pH-stat Nozaki et al. obtained two curves above pH 9.7: one derived by extrapolation to zero time (regarded as the curve for native protein), the other representing infinite time (regarded as the curve for alkali denatured protein). Subsequently Tanford and Nozaki (1959) studied the titration of the A and B variants. Basch and Timasheff (1967) have compared the titration curves of the A, B and C variants, and Brignon et al. (1969) have studied the D variant. Ghose et al. (1968) made a very careful study of the hydrogen ion equilibria of caprid β-lactoglobulin (see Townend and Basch, 1968). The findings from these studies will now be summarized and discussed (see also the general discussion in Volume I, Chapter 8, Section VII.B).

2. Bovine A and B Variants

The following important conclusions can be reached from the work of Tanford and his colleagues concerning the bovine A and B variants:

The maximum acid-binding capacity (see Fig. 9) is 40 cationic groups per dimer (36,000 daltons). This is equal to the sum of the α-amino, ϵ-amino, imidazole and guanidinium groups interacting with the solvent. Tanford and Nozaki (1959) and Basch and Timasheff (1967) state that this agrees well with the number (40 per dimer unit or 20 per monomer unit) found on amino acid analysis of both variants. However it can be seen on

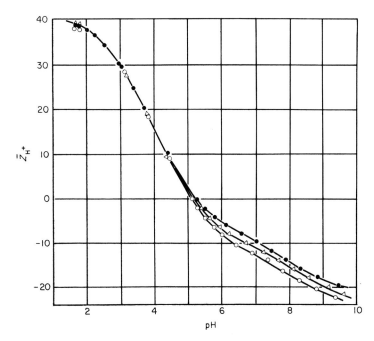

FIGURE 9. Titration curves for bovine β-lactoglobulins A, B and C in 0.15 M potassium chloride at 25°C; ○, A variant; △, B variant; ●, C variant; $\bar{Z}_{H^+}$ is expressed in terms of the number of groups per dimer unit of 36,000 daltons. (Data of Basch and Timasheff, 1967.)

examination of Table III that the total number of these groups is 42 (21 per monomer, there being 15 Lys not 14). The discrepancy between the titration and analytical figures (drawn to my attention by G. B. Ralston) probably arises from an error in the determination of the titration curve. Ghose et al. (1968) in their titration study of the caprid protein (see Section VI.A.5) determined the curve for bovine B and obtained a figure of 42 for the total acid binding in bovine B. While the titration of the B protein needs reinvestigation, the conclusion that all of the cationic groups are "normal" is probably valid (although the position may be more complex).

From pH 6.6 to the acid end point the total number of COOH groups titrated is 52 for the A variant and 50 for the B variant. The number of COOH groups is two less for the dimer of each variant than the number calculated from amino acid analysis (total number of aspartic + glutamic residues − amide groups + α-carboxyl groups).

A characteristic feature of the titration curves for the native protein is the steepening of the curves near pH 7.5. A count of the number of groups titrated in this region is eight for both variants. This is two more than the number of histidine (imidazole) groups plus α-amino groups from amino acid analysis. If denatured protein is titrated this feature disappears, and the total number of carboxyl groups titrated becomes equal to the analytical figure. This is also true for the number of imidazole plus α-amino groups.

A conformational change has been found to accompany the titration of the anomalous groups having a pH of 7.3 in the native protein (see below). Tanford (1961) has developed a theory of conformational transitions related to the ionization of buried groups in protein molecules. A transition involving the ionization of a single protein is described by the equation

$$y = \frac{K'[H^+]}{1 + K'[H^+]} \quad (1)$$

where y is the apparent extent of the transition, $[H^+]$ is the hydrogen ion activity and K' is the equilibrium constant for the reaction; K' is related to pK_{int}, the normal pK of the buried group, and to pK^*_{int}, the anomalous pK of the group by

$$pK^*_{int} = pK_{int} + \log(1 + K') \quad (2)$$

If two ionizing groups of identical pK^*_{int} are involved, then

$$y = \frac{K^1/[H^+]^2}{1 + K^1/[H^+]^2} \quad (3)$$

Theoretical curves for transitions involving one and two protons are shown in Fig. 10a. If either of the forms involved in the transition is a mixture of conformations the curve obtained is intermediate between the curves shown. If the pK^*_{int} values of the buried groups are sufficiently different from one another, each group is titrated independently and then the transition exhibits two steps as shown in Fig. 10b, curve 1. When the pK^*_{int} values are close, but not identical, the two transitions coalesce. (Fig. 10b, curve 2).

Tanford and Taggart (1961) applied this theory to the above transition in bovine β-lactoglobulin AB and found that the group involved is an anomalous carboxyl group of $pK^*_{int} = 7.3$. There is one anomalous group on each polypeptide chain; hence two such groups are involved in the above titration of the dimer. It should be stressed that the COOH group involved is present in both the A and B variants and hence is not the additional aspartic acid residue of the A variant.

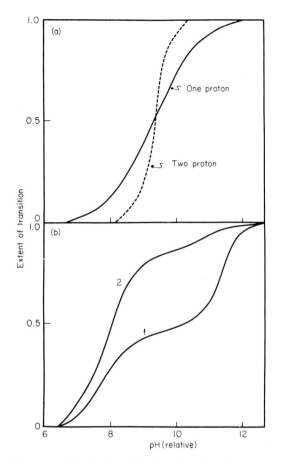

FIGURE 10. The course of ionization-linked conformational changes. (a) Theoretical curves for one- and two-proton transitions according to the theory of Tanford (1961). (b) Curve 1: conformational transition involving two ionizable groups of differing pK value. Curve 2: conformational changes involving two ionizable groups of similar, but not identical, pK value. (After Tanford, 1961.)

3. Bovine C Variant

Basch and Timasheff (1967) have considered the titration curves of the bovine C variant (see Figs. 10 and 11). Their results and conclusions may be summarized as follows:

The total acid-binding capacity from the isoionic point to the acid end point is equivalent to 40 cationic groups, i.e., the same as for A and B. However a total of 44 groups is obtained from amino acid analysis due to

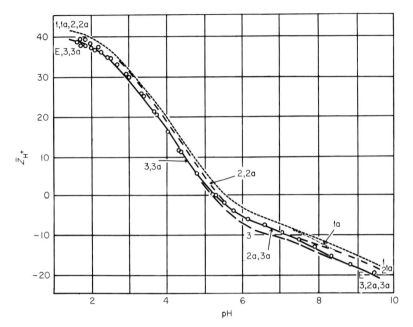

FIGURE 11. Comparison of calculated and experimental titration curves for bovine β-lactoglobulin C, in 0.1 M potassium chloride at 25°C. Curve E: experimental curve; O signifies experimental point. The other curves are calculated assuming, for 1: 6 Im, 2 α-COOH, 46 β,γ-COOH; for 1a: 6 Im, 2 α-COOH, 46 β,γ-COOH, 2 abnormal COOH; for 2: 6 Im, 2 α-COOH, 48 β,γ-COOH; 2a: 6 Im, 2 α-COOH, 48 β,γ-COOH, 2 abnormal COOH; for 3: 4 Im, 2 Im not titrated, 2 α-COOH, 50 β,γ-COOH; and for 3a: 4 Im, 2 abnormal Im, 2 α-COOH, 48 β,γ-COOH, 2 abnormal COOH. (Curves based on data of Basch and Timasheff, 1967.)

the presence of two extra histidine residues in the dimer, compared with the A and B variants. Thus the discrepancy between the titration curve figure and amino acid analysis value for the sum of the imidazole, α-amino, ε-amino, and guanidine groups is four and not two as considered by Basch and Timasheff (see the discussion for A and B in Section VI.A.2).

Titration curves were calculated in the range pH 2–9 making use of the known amino acid composition of C and pK_{int} values obtained by Tanford's group for the A and B variants (α-COOH, 3.75; β,γ-COOH, 4.66; anomalous β,γ-COOH, 7.25; imidazole (Im), 7.25; α-NH₂, 7.80). Several combinations of imidazole and carboxyl groups were used in the calculations, and the resultant curves are compared in Fig. 11 with the experimental curve E.

Curve 1, involving the combination 6 Im, 2 α COOH, 46 β,γ-COOH, runs above the experimental curve throughout the pH range 2–9.5. The latter follows curve 2 (6 Im, 2 α-COOH, 48 β,γ-COOH) in the pH range ~5.5–7.0, then it deviates above pH appreciably from curve 2. However, on comparison of the experimental curve with curve 2a (6 Im, 2 α-COOH, 48 β,γ-COOH, 2 anomalous β,γ-COOH) it is seen that there is close agreement between these curves over the pH range 5.9–9.5. This behavior is in accord with the C variant possessing two anomalous carboxyl groups (per dimer) that are unmasked in an ionization-linked transition near pH 7.0, as is the case for the A and B variants. Thus C has a total of 50 carboxyl groups per dimer. This is the same as in B and is in accord with the conclusion reached in Section V.C from peptide studies by Bell et al. (1968) that the His ↔ Glu substitution involves the substitution of histidine in C for glutamine in B and not for glutamic acid.

On examination of the experimental curve and curves 3a and 2a between pH 5.9 and the acid end point, we can see the following. As the pH falls below 5.9 the experimental curve gradually falls below curve 2a for 6 Im, 2 α-COOH, 48 β,γ-COOH, 2 anomalous β,γ-COOH. However it joins curve 3a at pH 5 and follows it closely to the acid end point. Thus in this region the titration behavior is similar to a curve for Im, 2 Im not titrated, 2 α-COOH, 48 β,γ-COOH, 2 anomalous β,γ-COOH.

When the protein was denatured by exposure to pH 11.1 for 10 min or by solution in 8 M urea, the maximum acid-binding capacity increased. It became consistent with six imidazole groups being titrated normally, whereas denatured A and B showed only four imidazole groups, as expected. Above pH 6 there was a small difference between the B and C variants and a larger difference between A and B variants.

Basch and Timasheff have attempted to explain the anomalous behavior of the histidine residues in the native C variant. They considered the possibility that there is a conformational transition in which one protonated histidine residue (per monomer) is transferred from the surface of the molecule to a hydrophobic interior position. This transfer would have to be accompanied by the forced dissociation of the imidazolium group to its neutral form. Otherwise it would involve a prohibitive amount of stabilization energy, say 70–100 kcal/mole (compare the total free energy of stabilization for the protein of 15 kcal/mole). This dissociation would be accompanied by the protonation of one carboxyl group. In the second mechanism the imidazole would remain protonated and would be transferred to the interior of the protein as a $Im^+ \cdots COO^-$ ion pair. The carboxyl would remain unprotonated even at low pH. These two mechanisms involve only a reasonable amount of energy. Basch and Timasheff were unable on the basis of their analysis to decide between the two mechanisms.

The author would agree with Basch and Timasheff that the substitution histidine of the C variant does behave in an anomalous fashion. However he considers that the pH titration curve for C needs reinvestigation. There are discrepancies in the curves given by Basch and Timasheff for C in the various figures (for example their Figs. 2 and 4) and there is the discrepancy already mentioned for the apparent maximum acid-binding capacity. Furthermore, since the B ↔ C substitution appears to be Gln ↔ His, the titration curves for B and C should agree more closely in the alkaline region than they do in their paper. The mobility of the two variants in electrophoresis should be the same at pH 8.5. However, as first shown by Bell (1962), the variants can be resolved in zone electrophoresis by appropriate choice of hydrolyzed starch. The separation of B and C can also be achieved at pH 7.4 in starch-gel electrophoresis and at low loading in paper electrophoresis at pH 8.5. In all cases the separation of B from C is considerably less than that of A and B. It would appear that the mobility difference in alkaline solution arises because of a shape and/or size factor rather than a simple charge effect. The C variant shows less tendency to dissociation and to undergo conformational change than B at pH 7–8. This would seem to arise out of the difference histidine residue of C. It occurs at position 115 or 116 in the polypeptide chain and in the charged condition would appear to be linked to a sidechain carboxyl group. It is considered that this ion pair bond extends over a wider pH range than the one envisaged by Basch and Timasheff and that it influences considerably the lack of dissociation of the C variant near pH 7.

4. *Bovine D Variant*

Brignon *et al.* (1969) have shown that the bovine D variant has an identical maximum acid-binding capacity to the C variant. The titration curves for both variants are identical up to ∼pH 4 when the D curve moves above the B curve. At pH 6.5 the D variant dissociates less protons than B (per dimer). This is consistent with B ↔ D substituted Glu ↔ Gln. Analysis of the curves also reveals the presence of two abnormal carboxyl groups per dimer as in the other bovine variants.

5. *Caprid Variant*

The titration curve obtained by Ghose *et al.* (1968) for caprid β-lactoglobulin is shown in Fig. 12. It can be seen on comparison of this curve with their curve for the bovine B variant, that the caprid protein binds two more groups than bovine B (44 vs. 42) on titration from the isoionic point to the acid end point. Thus their titration value of the sum of the imidazole, α-amino, ε-amino, and guanidinium groups is in agreement with the analytical data of Table III in the case of the bovine B variant (compare Section VI.A.2) but not in that of the caprid variant. Near pH 4 the curve for caprid becomes closer to the bovine curve but flattens after this region,

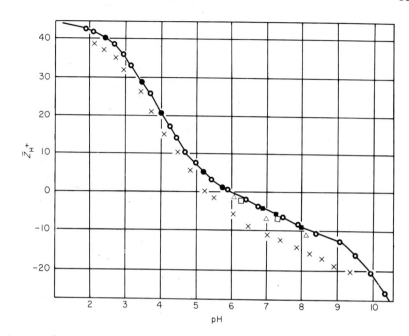

FIGURE 12. Titration curve of caprid β-lactoglobulin in 0.15 M potassium chloride at 25°C. Direct titration signified by ○. Reverse titration: ●, from pH 2; ■, from pH 8.4; □, from pH 8.7; △, from pH 9.7. A direct titration curve of the bovine B variant is shown for comparison, signified by ×. (From Ghose et al., 1968.)

and from ∼pH 6–9 the two curves differ in acid-binding capacity by seven groups. It also shows the steepening near pH 7.5, observed in all other variants, where it involves the titration of two anomalous carboxyl groups.

The number of ε-amino groups revealed by formaldehyde titration is 32, in agreement with the amino acid analysis of Townend and Basch (1968). However this is at variance with the lysine value of 30 (15 per monomer) given in Table III. From the determination of heats of ionization obtained from titration curves at 10.5° and 25°C, it was concluded that there are four imidazole groups and two anomalous carboxyl groups present.

The total number of normal carboxyl groups titrated from the acid end point to pH 6.6 is 46, including two α-carboxyls. However it is known from amino acid analysis that the bovine B and caprid variants both have 50 glutamic and 30 aspartic residues per dimer. There are 54 titratable carboxyl groups (50 + 2 anomalous + 2 α-COOH) in bovine B, and this is six more than for caprid (44 + 2 anomalous + 2 α-COOH). Ghose et al. found that there are 30 amide groups in the caprid protein. Thus there should be 52 (including 2 α-COOH) titratable carboxyl groups in caprid, i.e., four more than have been found in the titration of the native protein. On denaturation at pH 12 and titrating with acid, the number of carboxyl groups titrated with normal pK_{int} becomes equal to 52. Two of these are the groups that have a pK_{int} of ~7.0 in the native protein, the remaining four are apparently completely inaccessible in the native protein.

On thermodynamic analysis of the titration curves, using the Linderstrøm-Lang equation (see Volume I, Part C, Chapter 8, Section VII) it was found that the "normal" side-chain carboxyl groups are not intrinsically identical. Taking a constant value of the electrostatic interaction factor, $w = 0.036$ (see Chapter 8, Volume I), a pK_{int} of 4.4 for 18 β,γ-carboxyls and of 4.8 for the remaining 26, satisfactory agreement is obtained between the computed curve and experimental curve down to pH 4. There is a decrease in w to 0.030 below pH 4, caused presumably by either dissociation into monomers and/or expansion of the protein. The dissociation of caprid protein is negligible in the region pH 2.5–4 and only becomes appreciable below pH 2.5, according to Basch and Timasheff (1967). They have shown from optical rotatory dispersion studies that the caprid variant undergoes a conformational change below pH 4. The lowering of w presumably reflects this change.

The work of Ghose et al. is an excellent example of the value of careful pH titration studies of protein variants. The most significant conclusion from it is that, although the contents of aspartic and glutamic acid residues are similar in the caprid and bovine B variants, there are four buried carboxyl groups in the caprid protein of a type that is not present in the bovine variants. There are probably two of these groups in each monomer chain. They presumably reflect a difference in conformation of the relevant formation of the relevant portion of the caprid polypeptide chain.

B. ISOIONIC POINTS AND ION BINDING

When solutions of β-lactoglobulins are passed through a Dintzis deionization column (see Volume I, Chapter 4, Section I) the bovine A variant crystallizes almost immediately. In the case of other variants, oils tend to separate or the solution tends to become opalescent. This makes determination of the isoionic point of all but dilute solutions impossible. Treece et al. (1964) showed that a rise in pH due to progressive ionization of the bovine protein occurs only at concentrations below 1 g/liter. Isoionic points for several variants found by various workers and the effect of change in ionic strength by addition of potassium chloride are given in Table V.

14. β-LACTOGLOBULINS 303

TABLE V

Isoionic Points of Some β-Lactoglobulin Variants at Several Ionic Strengths[a]

Variant	Ionic strength (KCl)					
	0	0.1	0.2	0.3	0.4	0.5
Bovine A	5.3_5	5.2_6	5.2_2	5.1_9	5.1_8	5.1_7
B	5.4_1	5.3_4	5.3_1	5.2_9	5.2_8	5.2_8
C	5.3_9	5.3_3	5.3_1	5.2_8	5.2_7	5.2_6
Caprid	6.0_8	6.0_3	6.0_2	6.0_0	6.0_0	

[a] Based on data of Nozaki et al. (1959), Basch and Timasheff (1967) and Ghose et al. (1968).

Nozaki et al. (1959) concluded that for the bovine A and B variants, the fall in isoionic point with increasing potassium chloride concentration is due to potassium ion binding. This conclusion is probably valid for the other variants. Baker and Saroff (1965) have carried out direct ion-binding experiments with 0.5 M sodium chloride above pH 5.8 and have questioned the interpretation of Nozaki et al. Basch and Timasheff (1968) have attempted to reconcile the differences in interpretation.

Wishnia (1964) and Wishnia and Pinder (1966) have made a study of the binding of small alkanes to β-lactoglobulin. They found that alkane binding is to a localized, interior, hydrophobic site with high affinity and nontrivial stereospecificity. Wishnia (1969) has attempted to explain this type of behavior. He finds that the heat of dissociation of butane and pentane from β-lactoglobulin is 3–4 kcal higher than from dodecylsulfate micelles and is the source of high alkane binding to the protein; local rearrangements occur in the binding sites ("induced fit").

The binding of organic ions to bovine β-lactoglobulin B has been discussed by Lovrien and Anderson (1969). Ray and Chatterjee (1967) and Seibles (1969) have studied the interactions of dodecyl ions with β-lactoglobulins.

C. Electrophoresis

Early electrophoretic studies of bovine β-lactoglobulin were of great importance from two points of view. First, Ogston and his collaborators observed the behavior in transport experiments of samples of β-lactoglobulin prepared from individual cows. They found nonenantiographic moving-boundary electrophoretic patterns with bimodal descending boundaries (Ogston and Tilley, 1955; Tilley, 1960). These patterns were interpreted in terms of Gilbert's theory for the behavior of associating-dissociating systems in transport experiments (Ogston and Tombs, 1957; Tombs, 1957). This work provided one of the first examples of Gilbert's theory. It has subsequently been shown that at least part of the bimodal descending boundary for the B variant is due to rapid isomerization and that the A variant polymerizes to a unit of 144,000 daltons rather than 72,000 daltons, as originally postulated by Ogston and co-workers (for a review see McKenzie, 1967). The importance of Gilbert's theory in explaining the electrophoretic behavior of β-lactoglobulins and other milk proteins is of such great importance that the theory is discussed at considerable length in Volume I, Chapter 7, Section I.

Second, it was from studies of zone electrophoresis of samples from individual cows that Aschaffenburg and Drewry (1955) first showed genetic polymorphism in β-lactoglobulins. Their work enabled part of the apparent heterogeneity in samples of pooled β-lactoglobulin to be explained, and it provided the first example of genetic variants in a milk protein. Further studies of the zone electrophoresis of β-lactoglobulins are considered in in the next section.

VII. Molecular Size and Conformation

A. INTRODUCTION

Pedersen (1936) found that the sedimentation coefficient of bovine β-lactoglobulin (AB) varied between 2.7 and 3.2 S over the pH range 1–10, decreasing from 3.2 to 2.7 S over the pH range 6–8. He obtained a value of 40,700 daltons for the molecular weight, in this pH range, from sedimentation coefficient (s) and diffusion coefficient (D) measurements. However on the basis of sedimentation equilibrium studies, he obtained a value for the weight-average molecular weight ($\bar{M}_w$) of 37,600 daltons at pH 6.4, but 50,200 daltons at pH 9.8. A much longer time was required for sedimentation-equilibrium at pH 9.8 than at pH 6.4, since a slow aggregation occurred at pH 9.8. This aggregation was not noticed during the short time required for his sedimentation-velocity experiments. Pedersen also noted, from electrophoresis experiments, that a transition occurred near pH 7.

Most of the points raised in these studies have been confirmed or further investigated by other workers.

Positive evidence of the dissociation of β-lactoglobulin into units of molecular weight of 17,000 daltons was obtained by Bull (1946) from surface film studies. He also demonstrated that the presence of copper(II) could prevent the dissociation to monomers. A decade after Bull's work Townend and Timasheff (1957) noted that the s value of bovine β-lactoglobulin (AB) became less as the pH was lowered below 3.5. This was shown to be due to dissociation by molecular weight measurements.

Subsequently other workers have shown that as the pH increases above 6.5, the ruminant β-lactoglobulins undergo conformational transitions and show increasing dissociation. At higher pH they may aggregate and/or be denatured irreversibly. We shall now give a summary of the more important findings arising out of the various workers' investigations.

B. pH RANGE 1.8–3.5

Following their earlier studies on pooled β-lactoglobulin at low pH, Timasheff and Townend (1961b) studied the dissociation of the bovine A and B variants at low pH. Both variants gave similar "spread" patterns in sedimentation-velocity, and similar s vs. concentration curves, to those exhibited by the AB protein. It was concluded that each variant undergoes a rapid monomer-dimer equilibrium, with increasing dissociation as the pH is lowered below 3.5. Timasheff and Townend (1962) confirmed these conclusions by light-scattering measurements and determined the equilibrium constants and thermodynamic parameters for the reaction. Albright and and Williams (1968) made a very careful study of the reaction by the sedimentation-equilibrium method. This work has been discussed in Volume I, Chapter 7, Section III.C. They found that the second virial coefficient was appreciable and that the third final coefficient had to be taken into account at higher protein concentrations (>10 g/liter). Their plots of $\bar{M}_w$ vs. concentration of the B variant at pH 2.58 (I = 0.1, 0.15) and 2.20 (I = 0.15) are shown in Fig. 13a. (These plots should not be confused with the plots involving the *apparent* weight-average molecular weight shown in Fig. 9, Chapter 7, Volume I.) Plots of degree of dissociation (α) of the A and B variants at pH 2.7 (I = 0.1) based on the light-scattering data of Timasheff and Townend (1961b, 1962), are shown in Fig. 13b. It is important to note that even at pH 2.7, the protein does not approach complete dissociation to the monomer until the concentration is quite low.

The conformation of β-lactoglobulin in the pH range 1.8–3.5 is considered in the next part of this section (VIII.C).

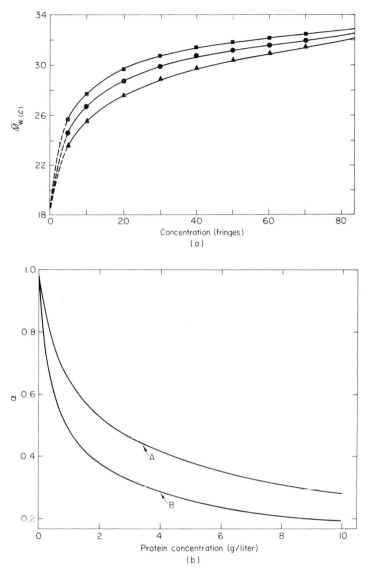

FIGURE 13. The dissociation of bovine β-lactoglobulins at low pH. (a) Idealized curves for the weight-average molecular weight ($\bar{M}_w$) as a function of concentration (c) of bovine β-lactoglobulin B at pH 2.20 and 2.58 (■—pH 2.58, $I = 0.15$; ●—pH 2.20, $I = 0.15$; ▲—pH 2.58, $I = 0.10$). Based on the molecular weight data obtained by sedimentation-equilibrium measurements (Albright and Williams, 1968). The concentration in fringes is related to the concentration in g/liter by 40.2 fringes = 10 g/liter. (b) Curves showing relation between protein concentration and extent of dissociation (α) to monomer of bovine β-lactoglobulins A and B, at pH 2.7 ($I = 0.1$, NaCl–HCl). (Based on light-scattering data of Timasheff and Townend, 1962.)

C. pH Range 3.5–5.4

1. Size

As has been mentioned previously, the ruminant β-lactoglobulins were long considered to exist as a molecular weight unit of 36,000 daltons near pH 5.2, and this was considered to be the "monomer" unit. However, evidence accumulated to indicate that the monomer unit is 18,000 daltons and that the 36,000 dalton unit is a dimer of two identical chains held together by noncovalent forces. It is now apparent that even at pH 5.2, the dimer is very weakly dissociated to the monomer. If the pH is raised or lowered from 5.2, there is an increasing tendency to dissociation, and this becomes appreciable at pH 3.5 and below, and at pH 7.5 and above. This type of behavior has been observed for all variants discovered to date, but the bovine A variant exhibits an additional type of reaction. It behaves anomalously in the region of pH 4.6 and this anomaly becomes greater as the temperature is lowered. At pH 4.65, low temperature and moderate protein concentration, the A variant is largely associated beyond the dimer. In fact, the octamer appears to be preferentially formed under these conditions. The dimer-octamer equilibrium is rapid and its behavior in sedimentation-velocity experiments has been used as a test of Gilbert's theory for the behavior of such systems in transport experiments. This work has been discussed in detail in Volume I, Chapter 7, Section I. In Gilbert's (1963) theoretical treatment of the sedimentation-velocity experiments of Timasheff and Townend (1961a) he assumes that the equilibrium is predominantly one involving the dimer and octamer ("monomer" and "tetramer" in the treatment). This approach was also used by Townend and Timasheff (1960) and Timasheff and Townend (1961a) in their measurement of the equilibrium constant of the reaction by light-scattering. Although some of the arguments used by Townend and Timasheff to justify this are not correct, others, based on the observation that polymers higher than the octamer ("tetramer") cannot be induced to form in appreciable amount on account of the stereochemistry of the polymerization, are strong ones.

Green and Aschaffenburg (1959) proposed a model for the β-lactoglobulin dimer consisting of two identical spheres (of 18,000 daltons) and having an axial ratio of 2:1. On this basis Timasheff and Townend (1964) and Green (1964) proposed the models for the octamer shown in Fig. 14.

It is not implied in this discussion that association proceeds by the simultaneous collision of four dimers but that at equilibrium, the dimer and octamer are the prime species present, other n-mers being present in negligible amount. However, in a recent study of the reaction at 16°C, Adams and Lewis (1968), as discussed in Volume I, Chapter 7, Section III.C, concluded that the association is an indefinite one. This difference has not

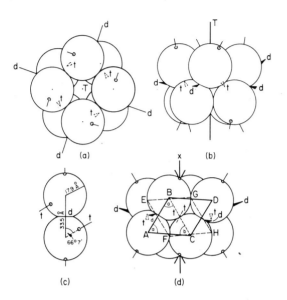

FIGURE 14. Staggered structures for the octamer of bovine β-lactoglobulin A. (a) Top view; 422 symmetry; d, diad axis of symmetry; t, octamer bond. (b) Side view; 422 symmetry; T, tetrad axis of symmetry. (c) Dimer structure. (d) 222 symmetry; X, overall diad axis of symmetry. The preferred structure is 422. (From Green, 1964; Timasheff and Townend, 1964.)

been resolved; nevertheless it would seem that the dimer and octamer are the predominant forms.

The association constants and thermodynamic parameters for the octamerization of the bovine A variant at pH 4.65 in the range 1°–30°C obtained by light-scattering, approach to sedimentation-equilibrium, and optical rotation are compared in Table VI. They are in good agreement. The association is minimal at pH 4.6 and falls to low levels at pH 3.8 and 5.2. From the nature of this pH dependence, it was first suggested by Timasheff and Townend (1961b) that carboxyl groups play an important role in the association. The amino acid composition and peptide studies of the variants discussed in Section V lend strong support to this hypothesis. The COOH blocking studies of Armstrong and McKenzie (1967; see also Chapter 5, Volume I) are in accord with it. Mechanisms involving hydrogen bond formation between the COOH groups have been reviewed elsewhere (McKenzie, 1967).

It is of interest to note here that the Droughtmaster variant does not octamerize although it has the same amino acid composition as the A variant. Apparently the carbohydrate moiety attached to the peptide chain

TABLE VI

ASSOCIATION CONSTANTS AND THERMODYNAMIC PARAMETERS FOR THE OCTAMERIZATION OF BOVINE β-LACTOGLOBULIN A AT pH 4.65[a]

Temp. (°C)	log K_a[b]				Archibald, $\bar{M}_w$ ($\Delta H = -40$ kcal/mole)		α_0 ($\Delta H = -64$ kcal/mole)		Light-scattering ($\Delta H = -53$ kcal/mole)	
	Archibald, $\bar{M}_w$	$[\alpha]_{578}$	α_0	Light-scattering[c]	$-\Delta G°$	$-\Delta S°$	$-\Delta G°$	$-\Delta S°$	$-\Delta G°$	$-\Delta S°$
1.0		13.3	12.6				15.8	178		
2.2				11.7						
3.0		12.3	12.0				15.2	179		
4.0	11.1				14.0	96				
4.5									14.4	138
5.0		11.8	11.6	11.1			14.8	179		
6.1										
7.0		11.4	11.2				14.4	179		
10.0		10.8	10.7				13.9	179		
10.5				10.5						
11.0	9.9				12.9	100				
13.0		10.2	10.1				13.3	179		
15.5				9.8						
16.0		9.8	9.8				12.9	178		
20.0		9.2	9.1				12.2	178		
20.1	9.4			9.2						
25.0				8.5						
30.0				7.9						

[a] From McKenzie (1967).
[b] K_a liter[3] (base moles)$^{-3}$.
[c] Data of Townend and Timasheff (1960). All other data of McKenzie et al. (1967).

sterically prevents the reaction (Bell et al., 1966a, 1970). The B variant can form mixed octamers with the A variant, but B, by itself, only octamerizes weakly (Kumosinski and Timasheff, 1966). No other variant appears to octamerize to any measurable extent.

2. Conformation

The conformation of β-lactoglobulin has long been a matter of controversy. Early work has been reviewed by McKenzie (1967). It is now generally agreed that there are no major changes in conformation over the pH range 2–5.4. There appears to be little change in conformation when the dimer dissociates to the monomer at low pH. However, there are subtle changes in conformation between pH 4 and 6. The optical rotation and a_0 parameter (see Volume I, Chapter 8, Section IX) for the bovine B and C variants change significantly in this pH region, but there is only a small change in the b_0 parameter. There is a change in the titration curve for the C variant during this transition, but none is detectable for the B variant, as has been discussed in Section VI.A. These transitions have been considered in terms of Tanford's theory (Section VI.A) of ionization-linked transitions by Timasheff et al. (1966b), Basch and Timasheff (1967) and McKenzie and Sawyer (1967). There is reasonable agreement between the experimental data of the Canberra and Philadelphia groups for the B and C variants. However there is lack of agreement for the A variant. Curves, obtained by McKenzie and Sawyer (1967), for the pH dependence of $[\alpha]_{578}$ and the a_0 parameter for the bovine A, B and C variants at 20°C are shown in Fig. 15a, b. The extremum in $[\alpha]_{578}$ and a_0 for the A variant near pH 4.6 is apparent. This feature is not apparent in the curves of Timasheff et al. (1966b). However the extremum has also been found by Préaux et al. (1962, 1965) and Lontie and Préaux (1966). It is temperature dependent as can be seen in Fig. 15c. The pH dependence of optical rotatory parameters near pH 4.6 appears to reflect not only the presence of octamer (see Volume I, Chapter 7, Section III.H) but also a small conformational change that is a necessary prelude to the association reaction (McKenzie et al., 1967).

The structure of β-lactoglobulin has been variously considered to contain 0–100% α-helix. This controversy has been reviewed by McKenzie (1967), who considered, on the basis of ORD data, that β-lactoglobulin in the pH region 2–6 consists broadly of 33% α-helix, 33% β-conformation and 33% disordered chain. Analysis of ORD curves for this protein is made difficult, inter alia, by side-chain Cotton effects in the 280–300 nm region. Also there is considerable uncertainty in the parameters relevant to the β-conformation (see Volume I, Chapter 8, Section IX). Analyses of circular dichroism (CD) curves is on a firmer basis but is still difficult. Typical CD

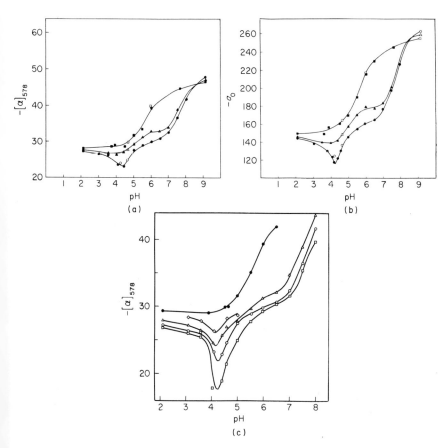

FIGURE 15. Conformational transitions dependent on pH for bovine β-lactoglobulins A, B and C. (a) The specific rotation at 578 nm ($[\alpha]_{578}$) at 20°C. (b) The parameter, a_0, of the Moffitt-Yang equation (see Volume I, Chapter 8, Section IX.H), at 20°C. Symbols for (a) and (b): ●, A (variant); ▲, B; ■, C; solid symbols, direct pH titration; open symbols, buffered solutions (from McKenzie and Sawyer, 1967). (c) The effect of temperature on $[\alpha]_{578}$ for the A variant near pH 4.5. The C variant is shown for comparison. For β-C: ●, 20°C. For β-A: ◇, 45°C; △, 30°C; ○, 20°C; and □, 10°C. (From McKenzie et al., 1967.)

curves obtained by Townend et al. (1967) for bovine β-lactoglobulin B are shown in Fig. 16. Their analyses of the side-chain effects is shown in Fig. 16b. (see also Volume I, Chapter 8, Section XIV). Townend et al. (1967) conclude that in the pH region 2–6 there is probably <10% α-helix, 45% β-conformation, 45% disordered chain in β-lactoglobulin (see also Timasheff et al., 1966a). The presence of β-conformation has been confirmed by infrared spectroscopy (McKenzie, 1967; Timasheff and Susi, 1966).

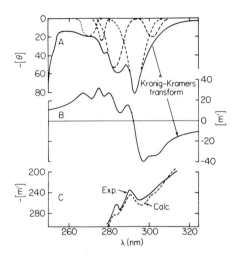

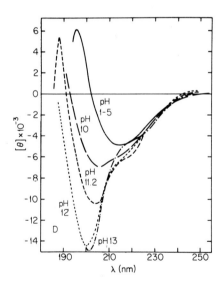

FIGURE 16. A: ———, circular dichroic spectrum of bovine β-lactoglobulin B; -----, empirical decomposition of spectrum into five symmetrical bands. B: optical rotatory dispersion calculated by Kronig-Kramers transforms (see Volume I, Chapter 8, Section IX) of decomposed CD spectrum. C: ———, experimental ORD spectrum; -----, curve of part B added to smooth base line obtained as described by Townend et al. (1967). D: circular dichroic spectra given by β-lactoglobulin B as a function of pH in 0.1 I NaCl–NaOH solutions. (From Townend et al., 1967.)

D. pH RANGE 5.4–9.2

As the pH increases above 5.4, an ionization-linked transition is observed in all variants. This is much greater than the one occurring near pH 4 and discussed above. Analysis of the transition for the A and B variants centered around pH 7.5 is comparatively straightforward (Tanford, 1961; Tanford and Taggart, 1961; Pantaloni, 1961, 1965; Timasheff et al., 1966b; McKenzie and Sawyer, 1967). All workers are now agreed that the titration of one abnormal COOH in each monomer accompanies the transition, but other features of the transition are still somewhat uncertain. The transition would appear to proceed via the monomer. It can be seen in Fig. 15 that the curve for the C variant is displaced somewhat from the A and B curves. It is difficult to analyze the curve for C; this arises partly from the overlap of the transition near pH 4.5. Nevertheless, as shown in Section VI.A, there is an abnormal COOH titrated in the pH 7.5 transition of the C variant.

The conformational changes and increasing dissociation near pH 7.5 are established immediately on mixing. There is very little change with time (during 24 hr) near room temperature. The dissociation has been studied by many workers, for example Georges and Guinand (1960), Georges et al. (1962), Préaux and Lontie (1966), and McKenzie and Sawyer (1967). The latter found that the tendency for dissociation at pH 7.5 and 20°C is in the order A > B > C. The dissociation constants for A and B dimers obtained by sedimentation-equilibrium measurements are $10^{-4.2}$ mole/liter and $10^{-5.1}$ mole/liter, respectively. The latter value is in reasonable agreement with the value of $10^{-4.2}$ mole/liter obtained by Georges et al. (1962) from light-scattering measurements.

During the pH 7.5 transition, an abnormal side-chain carboxyl group is "released," and this is generally considered to be the "ionization linkage" of the transition, as discussed in Section VI.A. Nevertheless, imidazole groups of histidine residues are being titrated in this region. Thus it is possible that the transition could result from the titration of an imidazole group, and the release of the abnormal carboxyl group could be a consequence of the transition. Pantaloni (1963, 1965) considers that a tyrosine residue is "unmasked" during the transition. However it is apparent from the magnitude of the change in the ultraviolet difference spectra (Volume I, Chapter 8, Section X.E) observed by Bell and McKenzie (1967a) that this group cannot be transferred from a hydrophobic interior to the solvent. The situation would seem to be one involving only a change in the environment of the tyrosine residue.

As discussed in Section V.D, the transition is accompanied by an increase in reactivity of the SH group. The increase in reactivity for bovine A, B

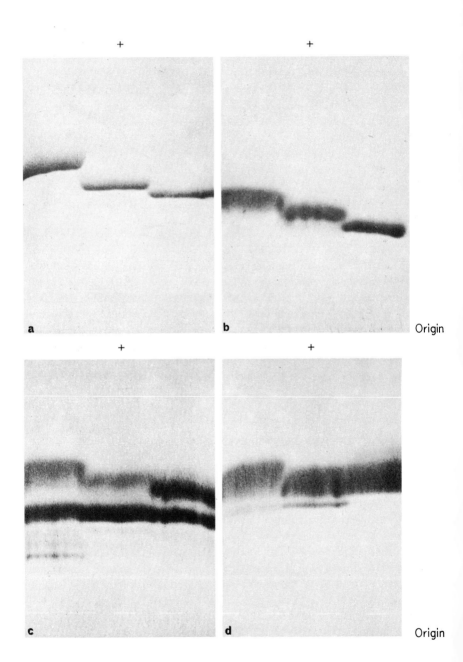

and C variants parallels the degree of dissociation of dimer to monomer (A > B > C). It was also seen in Section V.D that a more profound conformational change and increased dissociation can be effected by reacting the SH group with mercurials and NEM. Such changes can also be effected by increasing the pH. During the additional conformational change Pantaloni (1965) has found evidence for change in the environment of a tryptophan residue.

Above pH 8, time-dependent changes occur (slowly) following the immediate transition. These have been studied by the Danish, American, Belgian, French and Australian groups. Linderstrøm-Lang and Jacobsen (1940) found that bovine β-lactoglobulin denatured rapidly at pH 8.3 and 0°C. Groves et al. (1951) detected the presence of both reversible and irreversible changes above pH 8. Macheboeuf and Robert (1953) and Calvin (1954) concluded that the irreversible changes involved, at least in part, the oxidation of SH groups and the rupture of SS bridges. The formation of the irreversible products of the reactions appears to be inhibited by the presence of SH-specific reagents. This has been demonstrated in the zone electrophoretic studies of McKenzie and Sawyer (1966), typical examples being shown in Fig. 17. They have also demonstrated aggregation of the β-lactoglobulin by ultracentrifugal measurements, and Roels et al. (1966) and Préaux and Lontie (1966) have done so by chromatography on Sephadex G-100. Buchet et al. (1966) have found that the reaction is slowed down by the presence of 10^{-3} M disodium ethylenediaminetetraacetate. This results presumably from the binding of ions, such as copper(II), that accelerate the oxidation of SH groups (see also Buchet et al., 1969b). Thus some of the aggregation reactions involve oxidation of SH groups.

Above pH 8, the optical rotatory dispersion of the A, B and C variants changes slowly with time, at 20°C, following the immediate transition (see, for example, McKenzie and Sawyer, 1967; Lontie and Préaux, 1966; Buchet et al., 1969a). The rate of change is in the order A > B > C, and the rate for each variant is greater at 3°C than at 20°C. The overall reaction (over several days) is essentially first order (although not without some

---

FIGURE 17. Starch-gel electrophoretic patterns of β-lactoglobulin showing effect of pH. For the sodium hydroxide–boric acid gel buffer a voltage gradient of 8 V/cm was applied for 6 hr. For the tris-citrate gel buffer the initial voltage gradient was 5 V/cm; the duration of electrophoresis was 5 hr. (a) β-Lactoglobulins A, B, and C in tris-citrate buffer at pH 7.5; (b) β-lactoglobulins A, B, and C in sodium hydroxide–boric acid buffer at pH 8.5; (c) variants A, B and C in sodium hydroxide–boric acid buffer after 64 hr at pH 8.5; (d) variant A in sodium hydroxide–boric acid buffer after 64 hr at pH 8.5 in the presence of (i) p-chloromercuribenzoate, (ii) phenylmercuric acetate and (iii) n-ethylmaleimide. (From McKenzie and Sawyer, 1966.)

complications). The change in levorotation is primarily caused by the conformational change rather than the aggregation, and the former is the rate-limiting process near pH 9.

Bodanszky et al. (1965; see also Bell and McKenzie, 1967a) have found that the side-chain Cotton effects (involving at least some of the tryptophan residues) persist at pH 9 when the immediate conformational transition has occurred but disappear slowly with time. Other features of the ORD curve (for example, the 230 nm trough) change with time also. Immediate and far more reaching changes in ORD occur with increasing pH (similar changes occur in CD; see, for example, Fig. 16). It has been stressed that β-lactoglobulin is believed to contain α-helical, β- and disordered chain segments near pH 5. During the changes at alkaline pH, the β-segments would appear to unfold more readily than the α-helical portion of the chain (see, for example, the analysis of ORD curves by Timasheff et al., 1966a).

VIII. Denaturation of β-Lactoglobulins

A. Effect of Heat

The essential features of the effect of heat on proteins were realized by Hardy (1899), who considered that there are two main stages. The first stage involves a preliminary change in the structure of the protein—the *denaturation* stage. The second and distinctly different stage involves the process of *aggregation*, and this may be followed by *coagulation*. This distinction is an important one in considering not only the action of heat, but also the action of organic solutes, such as urea, guanidine hydrochloride, detergents, etc. Unfortunately not all investigators have made the distinction, and it is important to keep it in mind in this section. It is the author's experience that the "heat" denaturation step appears to involve generally less unfolding than that wrought by the action of concentrated urea or guanidine hydrochloride. These general features appear to be true for the ruminant β-lactoglobulins.

Among the first to investigate the effect of heat on bovine β-lactoglobulin AB were Briggs and Hull (1945). They examined the effect of heat at pH 7 and $I = 0.1$ and followed the effects by carrying out moving-boundary electrophoresis on the products of reactions occurring at different temperatures. Their procedures were such that they were only able, in effect, to detect irreversible products. At temperatures above 65°C they found that a product was formed with an s value of 5.6 S (compared with 3.1 S for the native dimer). A second product ($s = 15.2$ S) could form readily if the temperature was then lowered. If the temperature was above 75°C the rate of formation of this second product fell off, becoming negligible at 99°C. Also it could not be formed unless the protein was first exposed to a temperature sufficient to enable the first product to form, i.e., the aggregations were considered to be sequential.

Larson and Jenness (1952) found increasing activity of the SH group as they heated bovine β-lactoglobulin AB. Stauff and Ühlein (1955) concluded from light-scattering studies that the dimer dissociates to the monomer on heating and that aggregation occurs with increasing time. Later Stauff and Ühlein (1958) showed that the aggregation involves the SH group reacting to form SS bridges, and Stauff et al. (1961) proposed a theory to explain the aggregation of globular proteins with disulfide bridges. Georges et al. (1962) also demonstrated by light-scattering measurement that bovine β-lactoglobulin B shows increasing dissociation to the monomer at a given pH in the region 6–9 as the temperature is increased from 20° to 45°C.

Dupont (1965a, 1965b) made a study of the effect of heat on bovine β-lactoglobulin A at pH 5–7.5 and also compared the effects of heat on the A and B variants at pH 6.85. On the basis of optical rotation measurements ($[\alpha]_{436}$), Dupont showed that the variants went through immediate transition above 40°C at pH 6.85, the temperature of the midpoint of the transition being 66.5°C for A and 67.5°C for B. The immediate change in rotation was essentially reversible, but at higher temperatures this change was followed by a slow irreversible one. The immediate transition was followed through ultraviolet difference spectral measurements by Pantaloni (1965), with comparable results to the optical rotation measurements.

Sawyer (1968) has examined the irreversible products of the effect of heat on bovine β-lactoglobulins A, B and C, using the technique of starch-gel electrophoresis. He confirmed the earlier observations of Briggs and Hull (1945) that there are essentially two classes of irreversible aggregation reactions involved. Sawyer refers to these as the "primary" and "secondary" denaturation reactions. This author prefers not to use this terminology as it can cause confusion in relation to the overall mechanism of the effects of heat (outlined at the beginning of this section) in which the unfolding (denaturation) and aggregation steps are separated. Since the methods used by Briggs and Hull and Sawyer detect the irreversible products, this author prefers to use the terms "type I aggregation" and "type II aggregation." Sawyer has shown that in the type I aggregation, SH groups are intimately involved. The aggregates are small ($s \approx 3.7$ S after 97°C, 150 min, pH 7.0, $I = 0.1$) and are not formed in the presence of NEM. Type II aggregation results in large aggregates ($s \approx 29$ S) that do not involve SS bridges and are not formed when NEM is present. However he found that there was a third type of aggregation that occurrs even when NEM is present. The type III aggregation is nonspecific and in the case of the C variant is a major part of the overall aggregation. Sawyer concluded that the overall effects of heat for the variants are in the order C > B > A (see Gough and Jenness, 1962; Dupont, 1965b).

Considering the various workers' results it seems to the author that the effects can be summarized in the following way:

$$(R_\beta)_2 \rightleftharpoons 2R_\beta \rightleftharpoons 2R_\gamma + 2H^+ \rightleftharpoons 2R_\sigma \rightleftharpoons (R_\sigma)_2 \xrightarrow{III} (R_\sigma)_z$$

$$-SH, -SS \bigg| I$$

$$(R_\sigma)_x$$

$$\bigg\downarrow II$$

$$(R_\sigma)_y$$

where $(R_\beta)_2$ is the dimer prevalent near pH 5.2 and 20°C and in rapid reversible equilibrium with the monomer (R_β) that undergoes an ionization-linked transition near pH 7 to the form R_γ. This reaction is driven to the right with increasing temperature near pH 7 and is in equilibrium with form R_σ. The latter can undergo type I aggregation involving SH oxidation and/or –SH–SS exchange, giving aggregate $(R_\sigma)_x$, which in turn can undergo aggregation of type II to $(R_\sigma)_y$. Form R_σ can also undergo nonspecific aggregation (III) to form aggregates $(R_\sigma)_z$. The importance of type III should not be underestimated.

The aggregations involving the SH group will be affected by the presence of copper(II) or other heavy metal ions (see the discussion on SH groups) and the presence of calcium(II) (Zittle et al., 1957). Nonspecific ionic-strength effects will play a role in type II and III aggregations (Zittle and DellaMonica, 1957; Georges and Guinand, 1968).

B. EFFECT OF DETERGENTS

The effect of detergents on the reactivity of sulfhydryl groups in ruminant β-lactoglobulins has been discussed in Section V.D. Groves et al. (1951) compared the denaturation of β-lactoglobulin AB and its dodecyl-sulfate derivative in alkaline solution and also compared the optical rotation of the protein in alkaline and 0.17 M detergent solutions. In the course of their study of the effect of sodium dodecylsulfate on the reactivity of the SH group, Leslie et al. (1962b) made parallel measurements of the optical rotation.

C. EFFECTS OF UREA AND GUANIDINE HYDROCHLORIDE (see also p. 285)

Early studies of the effects of urea on bovine β-lactoglobulin were made in the Carlsberg Laboratory by Jacobsen and Christensen (1948), Linderstrøm-Lang (1949), Christensen (1952), Johansen (1953) and Schellman (1958). These workers found that the reaction at pH 5 was complex and that irreversible products, involving the SH group, could be formed. Kauzmann and Simpson (1953) found that levorotation at 589 nm in urea at pH 6.7–7.2 increased immediately on mixing and then exhibited a slow increase. Morr (1967) studied the products of urea denaturation of bovine β-lactoglobulin A and B and concluded that there was no aggregation following unfolding. Pace and Tanford (1968) studied the unfolding of bovine β-lactoglobulin A at pH 2.5–3.5 by optical rotation measurements. They concluded that the unfolding was a two-state process and that the kinetics of the reaction were simple first order.

McKenzie and Ralston (1970) and Ralston (1969) have made a study of the urea denaturation of the bovine A, B and C variants over the pH range 3–9. Their conclusions may be summarized as follows. At pH 5.2 the rate and extent of increase in levorotation at 578 nm is strongly dependent on urea concentration and, for a given urea concentration, is in the order A > B > C. The kinetics are not simple first order, the change may be divided into primary and secondary steps. The primary step can be described by two exponential terms. The half time of the primary process increases slightly with increasing protein concentration. In the early stages of reaction the change in rotation is largely reversible but becomes increasingly irreversible with increasing time of reaction. The rate of change of rotation in the primary stage does not parallel that of molar absorbancy difference at 293 and 286 nm. In the presence of NEM, the rate and extent of optical rotation change are much greater.

At pH 3.5 the rate of change in levorotation is much greater than at pH 5.2. The very rapid initial increase is followed by a small and very slow change. The initial change is so rapid that a kinetic analysis could only be made in a few cases. The C variant is less stable than the A or B variants in urea at this pH. Evidence was obtained from optical rotatory dispersion curves that intermediate states of β-lactoglobulin exist in urea solution at low pH (see Fig. 18). It was found, surprisingly, that with increasing time

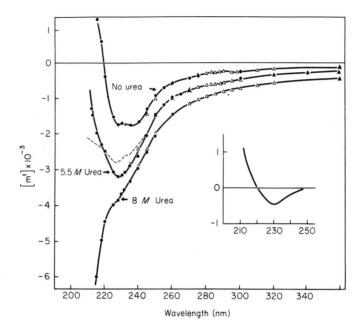

FIGURE 18. Optical rotatory dispersion in the ultraviolet region for bovine β-lactoglobulin B at pH 3.0 and 27°C (a) in the absence of urea, (b) in the presence of 5.5 M urea, and (c) in 8 M urea. The broken line represents the calculated value for 50% denatured protein, based on measurements made at wavelengths greater than 260 nm. The triangles represent data for 1 cm path length and the circles represent data for 0.1 cm. The inset shows the difference between the experimental and calculated curves of the protein in 5.5 M urea. The Cotton effects of 280–300 nm are too small to appear on the scale of the diagram (McKenzie and Ralston, 1970).

of reaction at low pH (3.5), irreversible products involving –SH–SS exchange are formed.

Complex kinetics were found for the reaction near pH 7, and the C variant was more stable than A. Irreversible products formed to a greater extent than at pH 5.2 or 3.5. Isolation of these products indicated that a trimer of molecular weight 54,000 was an important aggregate.

We have seen in Section VII.D that at pH 8.9 β-lactoglobulins A, B and C undergo changes in conformation and state of association in the absence of urea. The changes are more extensive in the presence of urea. There is considerably more aggregation in the presence of urea at pH 8.9 than at the lower pH values. The C variant is the most stable of the three variants at pH 8.9.

In general, the behavior of bovine β-lactoglobulin in urea is complex, irreversible aggregation reactions accompanying the denaturation process at all pH values. However aggregation is not the only cause of complexity; the primary unfolding reaction itself cannot be described adequately in terms of a two-state process. The kinetics of the denaturation and renaturation reactions are in general not simple first order, except for a limited range of denaturation conditions. In most cases two exponential terms are needed to describe the primary stage of denaturation. At the end of this stage, under pH conditions where the onset of aggregation is negligible, the ORD of β-lactoglobulin does not correspond to a simple mixture of native and fully denatured protein. Different parts of the native protein (for example α- helical and β- structure segments) would appear to unfold at different rates. The kinetics of the unfolding reactions are consistent with parallel or sequential unfolding of these segments (although a general mechanism may involve varying contributions from "branching" reactions according to denaturation conditions). The denaturation appears to proceed via the monomer. The unfolded protein from the second step of the unfolding stage undergoes –SH–SS exchange reactions and/or SH oxidation to yield irreversible products. There are other complexities that will be reported in detail elsewhere.

IX. Interaction of β-Lactoglobulins and κ-Casein

The first intimation of the interaction of casein and β-lactoglobulin appears to be in the papers of Tobias et al. (1952) and Slatter and van Winkle (1952), who relied on moving-boundary electrophoresis to detect the interaction. The interpretation of such patterns is fraught with difficulty, as has been seen in Chapter 7, Volume I. Nevertheless other workers have used this method, DellaMonica et al. (1958) and Zittle et al. (1962) for example. The latter workers found that the complex formed involved κ-casein and had an s value of 45 S, compared with 5 S for β-lactoglobulin and 15 S for κ-casein, when heated separately under similar conditions. They also found that β-lactoglobulin, after heat treatment, could react with κ-casein at room temperature (see also Long et al., 1963).

Trautman and Swanson (1958) suggested that SH groups are involved in the reaction of β-lactoglobulin with κ-casein. This was confirmed by Sawyer et al. (1963) who concluded, on the basis of moving-boundary electrophoresis, that complex formation did not take place in the presence of NEM (blocking of SH groups) and that the complex could be disrupted by 2-mercaptoethanol (reduction of SS bridges). Purkayastha et al. (1967) showed by means of polyacrylamide-gel electrophoresis that no complex involving covalent linkages was formed when β-lactoglobulin and S-carboxyamidomethyl κ-casein were heated together. Krescheck et al. (1964) demonstrated interaction between whole, or α-, or κ-casein, and β-lactoglobulin by means of light-scattering and prevented the reaction by addition of PHMB.

Freimuth and Krause (1968) have studied the interaction of κ-casein with β-lactoglobulin after heating and have detected carbohydrate-containing components in the complex formed.

DellaMonica et al. (1958) found that calcium(II) plays a role in the interaction of casein and β-lactoglobulin (see also Morr and Josephson, 1968). It is difficult at present to know what role calcium plays in the reaction in various media, for example, milk. Tessier et al. (1969) isolated heat-induced complexes formed between β-lactoglobulin (β) and κ-casein (κ) by zonal ultracentrifugation. (It will be realized that only irreversible complexes are detected by this method since the ultracentrifugation was carried out after cooling to 5°C.) When the ratio of $\beta:\kappa \leqslant 1.6:1$, all the β reacted with κ after 10 min at 80°C, pH 6.65 (cacodylate buffer). At higher mixing ratios the β reacted with κ in the ratio 3:1, in reasonable agreement with Long et al. (1963), who obtained a maximum of 2.2:1 after 20 min at 85°C. The product of reaction had an s value of 21.4–40.7 S. No complex appeared to form at 110° or 140°C. The reaction in artificial milk serum (90°C, 10 min) was complex, and it was difficult to isolate products other than one with s = 130 S. There was also no evidence of a 25–48 S component from zonal profiles of skim milk and milk free of colloidal calcium phosphate, after heating.

Sawyer (1969) in a comprehensive review of the interaction of κ-casein and β-lactoglobulin points out that there is no unequivocal evidence that the product of interaction involves disulfide bridge formation between the β-lactoglobulin and κ-casein. It may well be that the interaction involves noncovalent interaction between disulfide bridged aggregates of the β-lactoglobulin and κ-casein (see Long et al., 1963, and the beginning of this section.)

Great stress has been laid on the interaction of κ-casein and β-lactoglobulin as playing an important role in the heat stability of milk and in the increased rennin-clotting time for heated milk. This is discussed further in Chapter 17 by Beeby et al., who doubt the fundamental importance of –SH–SS exchange in the interaction. Rose (1963, 1965) has reviewed problems of heat stability of milk proteins, and on the basis of his group's recent zonal ultracentrifugal studies, considered above, has concluded that the "formation in milk of a complex corresponding to that found in heated mixtures of β-lactoglobulin and κ-casein has not been demonstrated."

In considering this problem it is important to realize that other heat-induced interactions can occur in milk, for example between α-lactalbumin and β-lactoglobulin (Hunziker and Tarassuk, 1965). Subclinical mastitis plays an important role as shown by Feagan et al. (1966a, b) and Feagan and Hehir (1967). The author stresses the possible role of immunoglobulins and Bailey (1968a) has considered the role of serum albumin. These matters, including a discussion of heat stability of porcine and human milk, are considered further in Chapter 18.

X. X-Ray Crystallographic Studies

In view of the ease with which bovine β-lactoglobulin AB can be crystallized, it is not surprising that there have been many investigations of the crystals since the protein was first isolated. An example of these studies is the careful work of McMeekin et al. (1954) on the water in the crystals and the partial specific volume of the protein. X-Ray studies date back to the period before World War II, and their history has been well summarized by Hodgkin and Riley (1968), as pointed out in the introduction to this chapter. With this long history it is ironic that the high resolution structure of the protein is still unknown.

Green and Aschaffenburg (1959) prepared crystals of a cadmium derivative of the bovine B variant and of a PHMB derivative of the A variant. They showed, by an analysis of X-ray diffraction patterns that the molecule, of molecular weight 36,000 daltons, consists of two spheres of 36 Å diameter in contact with one another and with a twofold axis of symmetry. The spheres impinge on each other by 2.3 Å at their surface of contact, giving a distance from center to center of 33.5 Å.

Aschaffenburg et al. (1965) have prepared crystals of bovine β-lactoglobulins and some heavy-metal derivatives at low ionic strength and by ammonium sulfate precipitation. A triclinic type (X) was obtained below pH 6.9, whereas two crystal types (Y and Z) formed above 7. Crystal types Y and Z appear to contain identical molecules which are a conformational variant distinct from type X. Both the Y and Z crystals have a single unit of molecular weight 18,000 daltons as the asymmetric unit of the crystal. Aschaffenburg et al. consider whether or not the dimer molecule is dissociated into the monomer under the conditions of crystallization and discuss some possible amino acid residues involved in the area of contact between the units. They point out that the Y and Z crystals are favorable for detailed study because of the 18,000 dalton asymmetric unit. Green has since studied suitable heavy-metal derivatives. These crystals are formed in a pH region where β-lactoglobulin is known to undergo an ionization-linked transition in solution that is driven further by the presence of mercurial, as has been discussed in Sections V.D and VII.D.

Dunnill et al. (1966) have discussed factors affecting isomorphism of native and heavy-atom sulfhydryl derivatives of bovine β-lactoglobulin and hemoglobin. They reached the following conclusion:

> The stability of crystalline β-lactoglobulin in the presence of mercurials, even though mercurials induce a conformational change in solution, suggests that the fixing of the molecules into a crystalline array tends to stabilize the native conformation. This view is supported by the lack of distortion on diffusion of p-iodophenyl mercuric chloride into the crystal. Though none of the mercurials coupled to β-lactoglobulin produced major lattice changes, close examination of photographs indicated changes in dimensions of the order of 1.5% in the most extreme case.

Bishop and Richards (1968), in the course of their studies on the diffusion of small molecules into and out of protein crystals, developed a procedure for estimating the isoelectric point of bovine β-lactoglobulin AB in the crystalline state when cross-linked with glutaraldehyde. They found that in contrast to carboxypeptidase and lysozyme, the cross-linking produced a substantial disordering of the X-ray diffraction pattern of β-lactoglobulin crystals. Thus, this type of cross-linked crystal is not suitable for detailed X-ray studies but probably does not invalidate Bishop and Richards' conclusion that the distribution and nature of the ionizable groups in β-lactoglobulin are similar in dilute solutions and the crystalline state.

XI. Summary and Conclusions

The following variants of β-lactoglobulin have now been isolated from the milk of ruminants: Bovine A, B, C, D and Droughtmaster; Ovine A and B; and Caprid. In solution near pH 5.2 they all exist predominantly as a dimer unit consisting of two single-chain units held together by noncovalent forces. The conformation of the chains is such that there would appear to be a mixture of α-helical, β- and disordered chain structures present. The Droughtmaster variant is unique in that it has a carbohydrate moiety, and in the heterozygote with A or B, it has the same amino acid composition as A. By variation of pH and other factors, all variants can be made to dissociate increasingly to the monomer and under some conditions undergo concomitant ionization-linked transitions. At pH values > 8 these changes may be followed by reactions involving –SH–SS exchange and/or SH oxidation and change in molecular size.

The β-lactoglobulins have proved of great interest in protein chemistry because of their basic structure and the remarkable motility they exhibit in solution. Their interactions with κ-casein in milk are of great technological importance, and these interactions are relevant to allergenicity problems involving β-lactoglobulin. The β-lactoglobulins from a given ruminant appear to be immunologically identical, but there are minor differences between the proteins from different types of ruminant. These properties and their significance in allergenicity reactions have been considered in Chapter 3, Volume I. Now that genetic variants of high purity are available, the stage is set for reinvestigation of some aspects of these reactions. Despite all of the work that has been carried out on this protein we do not yet know what biological function it performs in the milk of ruminants or why an analogous protein of different physical and immunological properties is present in porcine milk (see Chapter 18).

β-Lactoglobulin provides an example of how much information can be obtained about the structure of a protein from its solution chemistry. We await anxiously the determination of the complete amino acid sequence and X-ray crystal structure to see how close the many conclusions, so far drawn, are to the truth.

REFERENCES

Adams, E. T., Jr., and Lewis, M. S. (1968). *Biochemistry* **7**, 1044.
Albright, D. A., and Williams, J. W. (1968). *Biochemistry* **7**, 67.
Arave, C. W. (1967). *J. Dairy Sci.* **50**, 1320.
Armstrong, J. McD., and McKenzie, H. A. (1967). *Biochim. Biophys. Acta* **147**, 93.
Armstrong, J. McD., McKenzie, H. A., and Sawyer, W. H. (1967). *Biochim. Biophys. Acta* **147**, 60.
Armstrong, J. McD., Hopper, K. E., McKenzie, H. A., and Murphy, W. H. (1970). *Biochim. Biophys. Acta* **214**, 419.
Aschaffenburg, R. (1961). *Nature* **192**, 431.
Aschaffenburg, R. (1963). *J. Dairy Res.* **30**, 251.
Aschaffenburg, R. (1968a). *J. Dairy Sci.* **51**, 1295.
Aschaffenburg, R. (1968b). *J. Dairy Res.* **35**, 447.
Aschaffenburg, R., and Drewry, J. (1955). *Nature* **176**, 218.
Aschaffenburg, R., and Drewry, J. (1957a). *Nature* **180**, 376.
Aschaffenburg, R., and Drewry, J. (1957b). *Biochem. J.* **65**, 273.
Aschaffenburg, R., Green, D. W., and Simmons, R. M. (1965). *J. Mol. Biol.* **13**, 194.
Askonas, B. A. (1954). *Biochem. J.* **58**, 332.
Bailey, L. (1968a). "Australian Dairy Board Research Seminar, Brisbane."
Bailey, L. F. (1968b). Private communication.
Bailey, L. F., and Lemon, M. (1966). *J. Reprod. Fert.* **11**, 473.
Baker, H. P., and Saroff, H. A. (1965). *Biochemistry* **4**, 1670.
Basch, J. J., and Timasheff, S. N. (1967). *Arch. Biochem. Biophys.* **118**, 37.
Basch, J. J., Kalan, E. B., and Thompson, M. P. (1965). *J. Dairy Sci.* **48**, 604.
Bell, K. (1900). Private communication.
Bell, K. (1962). *Nature* **195**, 705.
Bell, K. (1965). Private communication.
Bell, K. (1967). *Biochim. Biophys. Acta* **147**, 100.
Bell, K. (1968). Private communication.
Bell, K., and McKenzie, H. A. (1963). Paper read to the Australian Biochemical Society, Melbourne.
Bell, K., and McKenzie, H. A. (1964). *Nature* **204**, 1275.

Bell, K., and McKenzie, H. A. (1967a). *Biochim. Biophys. Acta* **147**, 109.
Bell, K., and McKenzie, H. A. (1967b). *Biochim. Biophys. Acta* **147**, 123.
Bell, K., and Stormont, C. (1965). Private communication.
Bell, K., McKenzie, H. A., and Murphy, W. H. (1966a). *Aust. J. Sci.* **29**, 87.
Bell, K., McKenzie, H. A., and Shaw, D. C. (1966b). *Aust. J. Sci.* **29**, 86.
Bell, K., McKenzie, H. A., and Shaw, D. C. (1968). *Biochim. Biophys. Acta* **154**, 284.
Bell, K., McKenzie, H. A., Murphy, W. H., and Shaw, D. C. (1970). *Biochim. Biophys. Acta* **214**, 427.
Binon, N., Préaux, G., and Lontie, R. (1962). *Arch. Int. Physiol. Biochim.* **70**, 291.
Bishop, W. H., and Richards, F. M. (1968). *J. Mol. Biol.* **33**, 415.
Bloemmen, J., Joniau, M., and Lontie, R. (1967). *Arch. Int. Physiol. Biochim.* **75**, 552.
Bodanszky, A., Kauzmann, W., and McKenzie, H. A. (1965). Unpublished data.
Brand, E., Saidel, L. J., Goldwater, W. H., Kassel, B., and Ryan, F. J. (1945). *J. Amer. Chem. Soc.* **67**, 1524.
Brew, K., and Campbell, P. N. (1967). *Biochem. J.* **102**, 258.
Briggs, D. R., and Hull, R. (1945). *J. Amer. Chem. Soc.* **67**, 2007.
Brignon, G., Ribadeau-Dumas, B., Garnier, J., Pantaloni, D., Guinand, S., Basch, J. J., and Timasheff, S. N. (1969). *Arch. Biochem. Biophys.* **129**, 720.
Brown, J. R., and Harley, B. S. (1963). *Biochem. J.* **89**, 59P.
Buchet, J.-P., Préaux, G., and Lontie, R. (1965). *Arch. Int. Physiol. Biochim.* **73**, 866.
Buchet, J.-P., Préaux, G., and Lontie, R. (1966). *Arch. Int. Physiol. Biochim.* **74**, 721.
Buchet, J.-P., Préaux, G., and Lontie, R. (1969a). *Arch. Int. Physiol. Biochim.* **77**, 143.
Buchet, J.-P., Préaux, G., and Lontie, R. (1969b). *Arch. Int. Physiol. Biochim.* **77**, 145.
Bull, H. B. (1946). *J. Amer. Chem. Soc.* **68**, 746.
Calvin, M. (1954). *U.S. At. Energy Comm. Bull.* U.C.R.L. 2438.
Cannan, R. K., Palmer, A. H., and Kibrick, A. C. (1942). *J. Biol. Chem.* **142**, 803.
Christensen, L. K. (1952). *C. R. Trav. Lab. Carlsberg, Ser. Chim.* **38**, 37.
Crick, F. H. C. (1967). *Proc. Roy. Soc. (London)* **B167**, 331.
Cunningham, L. W., and Nuenke, B. J. (1959). *J. Biol. Chem.* **234**, 1447.
Cunningham, L. W., and Nuenke, B. J. (1960a). *Biochim. Biophys. Acta* **39**, 565.
Cunningham, L. W., and Nuenke, B. J. (1960b). *J. Biol. Chem.* **235**, 1711.
Cunningham, L. W., and Nuenke, B. J. (1961). *J. Biol. Chem.* **236**, 1716.
Davie, E. W., Newman, C. R., and Wilcox, P. (1959). *J. Biol. Chem.* **234**, 2635.
DellaMonica, E. S., Custer, J. H., and Zittle, C. A. (1958). *J. Dairy Sci.* **41**, 465.
Dixon, M., and Webb, E. C. (1961). *Advan. Protein Chem.* **16**, 197.
Donohue, J. (1969). *J. Mol. Biol.* **45**, 231.
Dunnill, P., and Green, D. W. (1965). *J. Mol. Biol.* **15**, 147.
Dunnill, P., Green, D. W., and Simmons, R. M. (1966). *J. Mol. Biol.* **22**, 135.
Dupont, M. (1965a). *Biochim. Biophys. Acta* **102**, 500.
Dupont, M. (1965b). *Biochim. Biophys. Acta* **94**, 573.
Edsall, J. T. (1965). *Biochemistry* **4**, 28.
Ellman, L. (959). *Arch. Biochem. Biophys.* **107**, 449.
Feagan, J. T. (1968). "Australian Dairy Board Research Seminar, Brisbane."
Feagan, J. T., and Hehir, A. (1967). "Australian Dairy Board Research Seminar, Sydney."
Feagan, J. T., Griffin, A. T., and Lloyd, G. T. (1966a). *J. Dairy Sci.* **49**, 933.
Feagan, J. T., Griffin, A. T., and Lloyd, G. T. (1966b). *J. Dairy Sci.* **49**, 1011.
Ferguson, K. A., and Wallace, A. L. C. (1963). *Recent Progr. Horm. Res.* **19**, 1.
Fox, K. K., Holsinger, V. H., Posati, L. P., and Pallansch, M. J. (1967). *J. Dairy Sci.* **50**, 1363.

Fraenkel-Conrat, H. (1954). *J. Amer. Chem. Soc.* **76**, 3606.
Fraenkel-Conrat, H. (1956). *J. Cell. Comp. Physiol.* **47**, 133.
Fraenkel-Conrat, H., Cook, B. B., and Morgan, A. F. (1952). *Arch. Biochem. Biophys.* **35**, 157.
Frank, G., and Braunitzer, G. (1967). *Hoppe-Seyler's Z. Physiol. Chem.* **348**, 1691.
Frank, G., and Braunitzer, G. (1968). *Hoppe-Seyler's Z. Physiol. Chem.* **349**, 1456.
Franklin, J. G., and Leslie, J. (1968). *Biochim. Biophys. Acta* **160**, 333.
Freimuth, V., and Krause, W. (1968). *Nahrung* **12**, 597.
Georges, C., and Guinand, S. (1960). *J. Chim. Phys.* **57**, 606.
Georges, C., and Guinand, S. (1968). *Compt. Rend.* **266**, 1011.
Georges, C., Guinand, S., and Tonnelat, J. (1962). *Biochim. Biophys. Acta* **59**, 739.
Ghose, A. C., Chaudhuri, S., and Sen, A. (1968). *Arch. Biochem. Biophys.* **126**, 232.
Gilbert, G. A. (1963). *Proc. Roy. Soc. (London)* **A276**, 354.
Gordon, W. G., Basch, J. J., and Kalan, E. B. (1960). *Biochem. Biophys. Res. Commun.* **3**, 672.
Gordon, W. G., Basch, J. J., and Kalan, E. B. (1961). *J. Biol. Chem.* **236**, 2908.
Got, R. (1965). *Clin. Chim. Acta* **11**, 432.
Got, R., Goussault, Y., and Marnay, A. (1965). *Clin. Chim. Acta* **11**, 383.
Gough, P., and Jenness, R. (1962). *J. Dairy Sci.* **45**, 1033.
Green, D. W. (1964). Unpublished results, quoted in Timasheff and Townend (1964).
Green, D. W., and Aschaffenburg, R. (1959). *J. Mol. Biol.* **1**, 54.
Greenberg, R., and Kalan, E. B. (1965). *Biochemistry* **4**, 1660.
Grosclaude, F., Pujolle, J., Garnier, J., Ribadeau-Dumas, B. (1966). *Ann. Biol. Animale, Biochim. Biophys.* **6**, 215.
Groves, M. L. (1965). *Biochim. Biophys. Acta* **100**, 154.
Groves, M. L., Hipp, N. J., and McMeekin, T. L. (1951). *J. Amer. Chem. Soc.* **73**, 2790.
Habeeb, A. F. S. A. (1960). *Can. J. Biochem. Physiol.* **38**, 269.
Hardy, W. B. (1899). *J. Physiol.* **24**, 158.
Hill, R. J. (1964). *Brookhaven Symp. Biol.* No. **17**, 210.
Hodgkin, D. C., and Riley, D. P. (1968). In "Structural Chemistry and Molecular Biology" (A. I. Rich and N. Davidson, eds.), p. 16. Freeman, San Francisco.
Hunziker, H. G., and Tarassuk, N. P. (1965). *J. Dairy Sci.* **48**, 733.
Hutton, J. T., and Patton, S. (1952). *J. Dairy Sci.* **35**, 699.
Jacobsen, C. F. (1949). *C. R. Trav. Lab. Carlsberg, Ser. Chim.* **26**, 455.
Jacobsen, C. F., and Christensen, K. L. (1948). *Nature* **161**, 30.
Johansen, G. (1953). *C. R. Trav. Lab. Carlsberg* **28**, 335.
Joniau, M., Bloemmen, J., and Lontie, R. (1966). *Arch. Int. Physiol. Biochim.* **74**, 727.
Jukes, T. B. (1963). *Biochem. Biophys. Res. Commun.* **10**, 155.
Kalan, E. B., and Basch, J. J. (1966). *J. Dairy Sci.* **49**, 406.
Kalan, E. B., and Greenberg, R. (1961). *Arch. Biochem. Biophys.* **95**, 279.
Kalan, E. B., Greenberg, R., and Walter, M. (1965). *Biochemistry* **4**, 991.
Kauzmann, W., and Simpson, R. B. (1953). *J. Amer. Chem. Soc.* **75**, 5154.
King, J. W. B. (1969). *Anim. Prod.* **11**, 53.
Klotz, I. M., and Carver, B. R. (1961). *Arch. Biochem. Biophys.* **95**, 540.
Koike, K., Ariga, H., and Osumi, K. (1964). *J. Biochem. (Tokyo)* **55**, 573.
Krescheck, G. C., van Winkle, O., and Gould, I. A. (1964). *J. Dairy Sci.* **47**, 117.
Kristjannson, F. K. (1963). *Genetics* **48**, 1059.
Kumosinski, T. F., and Timasheff, S. N. (1966). *J. Amer. Chem. Soc.* **88**, 5635.
Larson, B. L., and Jenness, R. (1950). *J. Dairy Sci.* **33**, 890.
Larson, B. L., and Jenness, R. (1952). *J. Amer. Chem. Soc.* **74**, 3090.

Larsen, B., and Thymann, M. (1966). *Acta Vet. Scand.* **7**, 189.
Leach, S. J. (1960). *Aust. J. Chem.* **13**, 520.
Leslie, J., and Varrichio, F. (1968). *Can. J. Biochem.* **46**, 625.
Leslie, J., Williams, D. L., and Gorin, G. (1962a). *Anal. Biochem.* **3**, 257.
Leslie, J., Butler, L. G., and Gorin, G. (1962b). *Arch. Biochem. Biophys.* **99**, 86.
Linderstrøm-Lang, K. U. (1949). *Cold Spring Harbor Symp. Quant. Biol.* **14**, 117.
Linderstrøm-Lang, K. U., and Jacobsen, C. F. (1940). *C. R. Trav. Lab. Carlsberg, Ser. Chim.* **23**, 179.
Long, J. E., van Winkle, Q., and Gould, I. A. (1963). *J. Dairy Sci.* **46**, 1329.
Lontie, R., and Préaux, G. (1966). *In* "Protides of the Biological Fluids: Proceedings of the 14th Colloquium" (H. Peeters, ed.), p. 475. Elsevier, Amsterdam.
Lontie, R., Buchet, J.-P., and Préaux, G. (1964a). *Arch. Int. Physiol. Biochim.* **72**, 524.
Lontie, R., Nagant, D., Ramaekers, C., and Préaux, G. (1964b). *Acta Biol. Med. Ger. Suppl.* **3**, 273.
Lovrien, R., and Anderson, W. (1969). *Arch. Biochem. Biophys.* **131**, 139.
Macheboeuf, M., and Robert, B. (1953). *Bull. Soc. Chim. Biol.* **35**, 399.
McKenzie, H. A. (1967). *Advan. Protein. Chem.* **22**, 55.
McKenzie, H. A., and Ralston, G. B. (1970). *Experientia* (in press).
McKenzie, H. A., and Sawyer, W. H. (1966). *Nature* **212**, 161.
McKenzie, H. A., and Sawyer, W. H. (1967). *Nature* **214**, 1101.
McKenzie, H. A., and Treacy, G. B. (1966). Private communication.
McKenzie, H. A., Sawyer, W. H., and Smith, M. B. (1967). *Biochim. Biophys. Acta* **147**, 73.
McMeekin, T. L., Groves, M. L., and Hipp, N. J. (1954). *J. Polymer Sci.* **12**, 309.
Mawal, R. B. (1967). *Arch. Physiol. Pharmacol.* **14**, 317.
Meyer, H. (1966). *Zuchthygiene* **1**, 49.
Michalak, W. (1967). *J. Dairy Sci.* **50**, 1319.
Morr, C. V. (1967). *J. Dairy Sci.* **50**, 1752.
Morr, C. V., and Josephson, R. V. (1968). *J. Dairy Sci.* **51**, 1349.
Morr, C. V., Kenkare, D. B., and Gould, I. A. (1964). *J. Dairy Sci.* **47**, 621.
Mukherjee, S., Palit, S. R., and De, S. K. (1970). *J. Phys. Chem.* **74**, 1389.
Neurath, H., Gladner, J. A., and Davie, E. W. (1954). *In* "The Mechanism of Enzyme Action" (W. D. McElroy and B. Glass, eds.), p. 50. Johns Hopkins, Baltimore.
Niu, C. L., and Fraenkel-Conrat, H. (1955). *J. Amer. Chem. Soc.* **77**, 5882.
Nozaki, Y., Bunville, L. G., and Tanford, C. (1959). *J. Amer. Chem. Soc.* **81**, 4032.
Ogston, A. G., and Tilley, J. M. A. (1955). *Biochem. J.* **59**, 644.
Ogston, A. G., and Tombs, M. P. (1957). *Biochem. J.* **66**, 399.
Osterhoff, D. R., and Pretorius, A. M. G. (1966). *Proc. S. Afr. Soc. Anim. Prod.* **5**, 166.
Pace, N. C., and Tanford, C. (1968). *Biochemistry* **7**, 198.
Palmer, A. H. (1934). *J. Biol. Chem.* **104**, 359.
Pantaloni, D. (1961). *Compt. Rend.* **252**, 2459.
Pantaloni, D. (1962). *Compt. Rend.* **254**, 1884.
Pantaloni, D. (1963). *Compt. Rend.* **256**, 4994.
Pantaloni, D. (1964). *Compt. Rend.* **259**, 1775.
Pantaloni, D. (1965). Thesis. University of Paris, France.
Pedersen, K. O. (1936). *Biochem. J.* **30**, 948.
Peterson, R. F. (1963). *J. Dairy Sci.* **46**, 1136.
Phillips, N. I., and Jenness, R. (1965). *Biochem. Biophys. Res. Commun.* **21**, 16.
Phillips, N. I., Jenness, R., and Kalan, E. B. (1967). *Arch. Biochem. Biophys.* **120**, 192.

Piez, K. A., Davie, E. W., Folk, J. E., and Gladner, J. A. (1961). *J. Biol. Chem.* **236**, 2912.
Préaux, G., and Lontie, R. (1962). *In* "Protides of the Biological Fluids: Proceedings of the 9th Colloquium" (H. Peeters, ed.), p. 103. Elsevier, Amsterdam.
Préaux, G., and Lontie, R. (1966). *In* "Peptides of the Biological Fluids: Proceedings of the 14th Colloquium" (H. Peeters, ed.), p. 611. Elsevier, Amsterdam.
Préaux, G., Morelle, A., Grosjean, N., and Lontie, R. (1962). *Arch. Int. Physiol. Biochem.* **70**, 754.
Préaux, G., Nagant, D., Ramaekers, C., and Lontie, R. (1964a). *Arch. Int. Physiol. Biochem.* **72**, 526.
Préaux, G., van den Berghe-van Orshoven, M., Leujeune, N., and Lontie, R. (1964b). *Arch. Int. Physiol. Biochim.* **72**, 693.
Préaux, G., Binon, N., and Lontie, R. (1964c). *Acta Biol. Med. Ger. Suppl.* **3**, 277.
Préaux, G., Hasselle, C., Jenard, R., and Lontie, R. (1965). *Arch. Int. Physiol. Biochim.* **73**, 374.
Purkayastha, R., Tessier, H., and Rose, D. (1967). *J. Dairy Sci.* **50**, 764.
Ralston, G. B. (1969). Ph.D. thesis. Australian National University, Canberra.
Ray, A., and Chatterjee, R. (1967). *In* "Conformation of Biopolymers" (G. N. Ramachandran, ed.), Vol. I, p. 235. Academic Press, New York.
Robbins, F. M., and Kronman, M. J. (1964). *Biochim. Biophys. Acta* **82**, 186.
Roels, H., Préaux, G., and Lontie, R. (1966). *Arch. Int. Physiol. Biochim.* **74**, 522.
Roels, H., Préaux, G., and Lontie, R. (1968a). *Arch. Int. Physiol. Biochim.* **76**, 198.
Roels, H., Préaux, G., and Lontie, R. (1968b). *Arch. Int. Physiol. Biochim.* **76**, 200.
Rose, D. (1963). *Dairy Sci. Abstr.* **25**, 45.
Rose, D. (1965). *J. Dairy Sci.* **48**, 139.
Sawyer, W. H. (1968). *J. Dairy Sci.* **51**, 323.
Sawyer, W. H. (1969). *J. Dairy Sci.* **52**, 1347.
Sawyer, W. H., Coulter, S. T., and Jenness, R. (1963). *J. Dairy Sci.* **46**, 564.
Schellman, J. A. (1958). *C. R. Trav. Lab. Carlsberg, Ser. Chim.* **30**, 395.
Schober, R., Heimburger, N., and Enkelmawn, D. (1959). *Milchwissenschaft* **14**, 432.
Seibles, T. S. (1969). *Biochemistry* **8**, 2949.
Sen, A., and Chaudhuri, S. (1962). *Nature* **195**, 286.
Slatter, W. L., and van Winkle, Q. (1952). *J. Dairy Sci.* **35**, 1083.
Smithies, O. (1955). *Biochem. J.* **61**, 629.
Sørensen, M., and Sørensen, S. P. L. (1939). *C. R. Trav. Lab. Carlsberg, Ser. Chim.* **23**, 55.
Stark, G. R., Stein, W. H., and Moore, S. (1960). *J. Biol. Chem.* **235**, 3177.
Stauff, J., and Ühlein, E. (1955). *Kolloid-Z.* **143**, 1.
Stauff, J., and Ühlein, E. (1958). *Biochem. Z.* **329**, 549.
Stauff, J., Barthel, H., Jaenike, R., Krekel, R., and Ühlein, E. (1961). *Kolloid-Z.* **178**, 121.
Svedberg, T. (1938). *Ind. Eng. Chem., Anal. Ed.* **10**, 113.
Szuchet-Derechin, S., and Johnson, P. (1965). *Eur. Polymer J.* **1**, 271.
Tanford, C. (1961). *J. Amer. Chem. Soc.* **83**, 1628.
Tanford, C., and Nozaki, Y. (1959). *J. Biol. Chem.* **234**, 2874.
Tanford, C., and Taggart, V. G. (1961). *J. Amer. Chem. Soc.* **83**, 1634.
Tanford, C., Bunville, L. G., and Nozaki, Y. (1959). *J. Amer. Chem. Soc.* **81**, 4032.
Tarassuk, N. P., and Yaguchi, M. (1962). *J. Dairy Sci.* **45**, 253.

Tessier, H., Yaguchi, M., and Rose, D. (1969). *J. Dairy Sci.* **52**, 139.
Tilley, J. M. A. (1960). *Dairy Sci. Abstr.* **22**, 111.
Timasheff, S. N., and Susi, H. (1966). *J. Biol. Chem.* **241**, 249.
Timasheff, S. N., and Townend, R. (1961a). *J. Amer. Chem. Soc.* **83**, 464.
Timasheff, S. N., and Townend, R. (1961b). *J. Amer. Chem. Soc.* **83**, 470.
Timasheff, S. N., and Townend, R. (1962). *J. Dairy Sci.* **45**, 259.
Timasheff, S. N., and Townend, R. (1964). *Nature* **203**, 517.
Timasheff, S. N., Townend, R., and Mescanti, L. (1966a). *J. Biol. Chem.* **241**, 1863.
Timasheff, S. N., Mescanti, L., Basch, J. J., and Townend, R. (1966b). *J. Biol. Chem.* **241**, 2496.
Tobias, J., Whitney, R. M., and Tracy, P. (1952). *J. Dairy Sci.* **35**, 1036.
Tombs, M. P. (1957). *Biochem. J.* **69**, 491.
Townend, R. (1965). *Arch. Biochem. Biophys.* **109**, 1.
Townend, R., and Basch, J. J. (1968). *Arch. Biochem. Biophys.* **126**, 59.
Townend, R., and Timasheff, S. N. (1957). *J. Amer. Chem. Soc.* **79**, 3613.
Townend, R., and Timasheff, S. N. (1960). *J. Amer. Chem. Soc.* **82**, 3168.
Townend, R., Kiddy, C., and Timasheff, S. N. (1961). *J. Amer. Chem. Soc.* **83**, 1419.
Townend, R., Basch, J. J., and Kiddy, C. A. (1965). *Arch. Biochem. Biophys.* **109**, 325.
Townend, R., Kumosinski, T. F., and Timasheff, S. N. (1967). *J. Biol. Chem.* **242**, 4538.
Townend, R., Herskovits, T. T., Timasheff, S. N., and Gorbunoff, M. J. (1969). *Arch. Biochem. Biophys.* **129**, 567.
Trautman, J. C., and Swanson, A. M. (1959). *J. Dairy Sci.* **41**, 715.
Treacy, G. B. (1967). Private communication.
Treece, J. M., Sheinson, R. S., and McMeekin, T. L. (1964). *Arch. Biochem. Biophys.* **108**, 99.
Wishnia, A. (1964). *Fed. Proc. Fed. Amer. Soc. Exp. Biol.* **23**, 160.
Wishnia, A. (1969). *Biochemistry* **8**, 5070.
Wishnia, A., and Pinder, T. W. (1966). *Biochemistry* **5**, 1534.
Yaguchi, M., Tarassuk, N. P., and Hunziker, H. G. (1961). *J. Dairy Sci.* **44**, 589.
Zimmerman, J. K., Klotz, I. M., and Barlow, G. (1969). *Fed. Proc. Fed. Amer. Soc. Exp. Biol.* **28**, 914.
Zittle, C. A., and DellaMonica, E. S. (1957). *J. Amer. Chem. Soc.* **79**, 126.
Zittle, C. A., DellaMonica, E. S., Rudd, R. K., and Custer, J. H. (1957). *J. Amer. Chem. Soc.* **79**, 4661.
Zittle, C. A., Thompson, M. P., Custer, J. H., and Cerbulis, J. (1962). *J. Dairy Sci.* **45**, 807.

Note added in proof: The partial sequence for bovine β-lactoglobulin A given in Fig. 6 of this chapter has been modified in the light of recent work by Dr. D. C. Shaw in the region of the chain between residues 60 and 72. This revised partial sequence is:

$$(\text{Asx, Ser, Glx}_2, \text{Pro-Ala, Leu})\text{-Val-Cys-Gln-Cys-Leu-Val}$$
$$(60) \qquad\qquad\qquad | \qquad\quad \downarrow \qquad (72)$$
$$\qquad\qquad\qquad\qquad\quad \text{SH} \quad\;\; \text{Cys} \;\; (57)$$

Full details will be given in H. A. McKenzie, G. B. Ralston, and D. C. Shaw (to be published).

15 □ α-Lactalbumin

W. G. GORDON

I. Reports of the Isolation of Crystalline "Albumins" from Cow Milk . . 332
   A. α-Lactalbumin 332
   B. β-Lactoglobulin 332
   C. Crystalline Insoluble Substance 333
   D. Bovine Serum Albumin 333
   E. Equivalence of α-Lactalbumin and Crystalline Insoluble Substance . 333
II. Preparation and Purification of Bovine α-Lactalbumin 334
   A. Isolation and Crystallization 334
   B. Purification by Chromatography and Gel Filtration 338
III. Composition and Structure 339
   A. Amino Acid Composition 339
   B. End-Group Analysis 341
   C. Derivatives and Primary Structure 342
IV. Physico-Chemical Properties 345
   A. Electrophoretic Behavior 345
   B. Solubility and Isoelectric Point 346
   C. Ultracentrifugal Behavior and Molecular Weight 346
   D. Association and Aggregation in Acid and Alkaline Solution . . . 347
   E. Conformational Changes 348
V. Genetic Polymorphism 351
VI. α-Lactalbumins in the Milk of Other Mammals 352
   A. Immunological Relationships 352
   B. Caprid α-Lactalbumin 353
   C. Water Buffalo α-Lactalbumin 353
   D. Human Lactalbumin 354
   E. Guinea Pig Lactalbumin 354
   F. Ovine α-Lactalbumin 354
VII. Crystallography and X-Ray Diffraction 356
VIII. The Biological Function of α-Lactalbumin 356
   References 361

I. Reports of the Isolation of Crystalline "Albumins" from Cow Milk

A. α-LACTALBUMIN

The occurrence in cow milk of a protein, designated "lactalbumin" according to the nomenclature of the time because it was not precipitable from whey by saturation with magnesium sulfate, was first demonstrated by Sebelien (1885). Following the removal of globulin which was so precipitated, Sebelien acidified the filtrate and precipitated lactalbumin. Years later, Wichmann (1899) also prepared lactalbumin by this method. Furthermore, he succeeded in crystallizing the protein by means of ammonium sulfate and acid.

In early studies concerned with the ultracentrifugal behavior of milk proteins, Sjögren and Svedberg (1930) prepared and crystallized lactalbumin by similar methods, but they found it to be inhomogeneous on ultracentrifugation as well as in the newly developed electrophoretic apparatus of Tiselius. Svedberg's investigations with milk proteins continued and, in reviewing their ultracentrifugal experiments with whey proteins, Svedberg and Pedersen attributed the α-peak, one of three major components in the sedimentation diagram of whey, to a lactalbumin isolated by Kekwick (1935). This protein, with $s_{20} = 1.9$ S, $D_{20} = 10.6 \times 10^{-7}$ cm^2/sec and a molecular weight of 17,400 daltons, was referred to as α-lactalbumin (Svedberg, 1937; Svedberg and Pedersen, 1940). Kekwick's method of isolation was not published in subsequent years so that these experiments were not repeated elsewhere.

B. β-LACTOGLOBULIN

A most important discovery in resolving the so-called "lactalbumin" fraction into its constituent parts was made by Palmer in 1934. On long dialysis at pH 5.2 of the salt-precipitated "lactalbumin" fraction, a crystalline protein was obtained. The protein was later named β-lactoglobulin, the Greek prefix again being derived from the sedimentation diagram of skim milk in which the β-peak (see Chapter 1, Volume I) was attributable to Palmer's crystalline protein (Cannan et al., 1942). The crystallization and characterization of this, the most abundant bovine whey protein, greatly facilitated further progress in the isolation and characterization of other protein components in skim milk.

C. CRYSTALLINE INSOLUBLE SUBSTANCE

Following Palmer's crystallization of β-lactoglobulin, Sørensen and Sørensen (1939) began work on the fractionation of lactalbumin. They crystallized β-lactoglobulin in several different ways, prepared red and green protein fractions, and were also able to crystallize a protein which they called "crystalline insoluble substance" on the basis of its insolubility in water and in dilute salt solution at pH 4.6. The protein dissolved easily in dilute ammonia solution at pH 6–7 and could be crystallized from such a solution by the slow addition of saturated ammonium sulfate solution. The microscopic crystals appeared to have the same form as β-lactoglobulin crystals obtained under the same conditions. However, purified crystalline insoluble substance differed from β-lactoglobulin in that it was insoluble in dilute salt solution at pH 4.6 and had a considerably higher content of tryptophan. The new protein was not characterized further.

D. BOVINE SERUM ALBUMIN

The discovery of β-lactoglobulin made possible, some years later, the isolation of still another component of the lactalbumin fraction. By repeated fractionation of the mother liquor remaining after crystallization of β-lactoglobulin from the crude lactalbumin fraction of whey, Polis *et al.* (1950) purified and crystallized a true, water-soluble, milk albumin. This was shown to be identical with the crystalline albumin which had been obtained from bovine blood serum.

E. EQUIVALENCE OF α-LACTALBUMIN AND CRYSTALLINE INSOLUBLE SUBSTANCE

In 1953, Gordon and Semmett fractionated the "lactalbumin" fraction along the lines described by Sørensen and Sørensen and confirmed many of their observations, including the crystallization of "insoluble substance." A method for the routine preparation of the protein was worked out. The purified material was found to exhibit a single peak in moving-boundary electrophoresis under most conditions as well as in the ultracentrifuge. Its sedimentation and diffusion coefficients, $s_{20} = 1.7$ S and $D_{20} = 10.6 \times 10^{-7}$ cm^2/sec, were in such close agreement with those reported by Svedberg and Pedersen for α-lactalbumin that there could be little doubt that the proteins were identical. It was proposed, therefore, that the name "α-lactalbumin," already in the literature, be adopted even though the protein was only slightly soluble in water at pH 4.6. The equivalence of α-lactalbumin prepared by the method of Gordon and Semmett and the "crystalline insoluble substance" actually isolated by the Sørensens was demonstrated by Wetlaufer in 1961 by direct comparisons of crystal form, electrophoretic mobility and ultraviolet spectrum.

II. Preparation and Purification of Bovine α-Lactalbumin

A. ISOLATION AND CRYSTALLIZATION

1. Method of Sørensen and Sørensen

Sørensen and Sørensen (1939) did not present a detailed procedure for the isolation of "crystalline insoluble substance." It was sometimes found as an insoluble precipitate in the course of recrystallization of first- or second-crop β-lactoglobulin and sometimes from another lactalbumin fraction, "gelatinous substance," of indeterminate composition. Nevertheless, however obtained, it could be dissolved in weak ammonia and reprecipitated by the addition of acid to pH 4.6. Also, when saturated ammonium sulfate solution was added slowly to an ammoniacal solution of the protein at pH 6.5–7.0, crystallization of "insoluble substance" ensued. These important observations provided the basic information subsequently developed into the isolation method of Gordon and Semmett.

2. Methods of Gordon, Semmett, and Ziegler

The procedure worked out by Gordon and Semmett (1953) may be outlined as follows. Casein was separated from skim milk by acidification with hydrochloric acid to pH 4.6. The whey proteins were then fractionated at pH 6.0 by salting out with ammonium sulfate. A crude globulin fraction was removed at 2.3 M ammonium sulfate concentration. The remaining proteins, including β-lactoglobulin and α-lactalbumin, were precipitated as the crude albumin fraction at about 3.3 M salt concentration. Salt was removed from the albumin fraction by prolonged dialysis at pH 5.2, whereupon β-lactoglobulin crystallized. The crystals were removed and protein remaining in the mother liquor was again precipitated at 3.3 M ammonium sulfate concentration. This precipitate was dissolved in water and also dialyzed at pH 5.2. A precipitate consisting of crude α-lactalbumin and additional β-lactoglobulin separated slowly and was centrifuged off; the mother liquor could be reworked for larger yields. The centrifuged precipitate was extracted with 0.1 M sodium chloride to remove β-lactoglobulin. The insoluble α-lactalbumin was dissolved in dilute ammonia, reprecipitated by dilute sulfuric acid at pH 4.6, again dissolved at pH 6.6 and crystallized by slowly adding saturated ammonium sulfate solution previously adjusted to the same pH. The amount of salt required for crystallization under these conditions varied from 0.5 to 0.67 saturation, depending on the concentration of protein. The microscopic crystals were purified further by repeating the paired acid-reprecipitation and crystallization steps several times. Four grams of pure α-lactalbumin were obtained from 60 liters of skim milk in a typical preparation.

The preceding method was simplified and improved by Gordon et al. in 1954. Crystallization of β-lactoglobulin by dialysis at pH 5.2 was accomplished as described above. The crystals were centrifuged off but then the mother liquor was adjusted to pH 4.0 and ammonium sulfate was added only to a concentration of 1.3 M. The precipitate in this case was mostly α-lactalbumin, with some residual β-lactoglobulin also present. Reprecipitations of α-lactalbumin to remove β-lactoglobulin were carried out at pH 4.0 instead of 4.6; however, the protein was crystallized and recrystallized as before. These modifications resulted in considerably increased yields (18.5 g of α-lactalbumin from 57 liters of skim milk). The modified method was described in detail by Gordon and Ziegler (1955a).

Two or three paired reprecipitations and recrystallizations usually suffice to yield α-lactalbumin pure enough for most purposes. It has been shown by Larson and Hageman (1963), however, that although the protein prepared in this way is 95% pure by electrophoretic analysis, it is not sufficiently pure to be used as an antigen for eliciting a single immunological response. To obtain such a sample Larson and Hageman recrystallized the α-lactalbumin several times more under modified conditions, accepting rather large losses of protein in order to accomplish their purpose. The protein (crystallized five times) was pure by immunological criteria and was then used to produce antisera. A quantitative immunochemical method for determining α-lactalbumin in complex systems was worked out, by means of which it was estimated that 1 liter of cow milk contains 1.0–1.5 g of this protein; this value is in agreement with previous estimates derived from peak areas in Tiselius electrophoretic patterns of the whey proteins (Larson and Jenness, 1955; Larson and Rolleri, 1955).

3. *Method of Zweig and Block*

A rapid method of concentrating α-lactalbumin and β-lactoglobulin in the form of "ferrilactin," the ferric complex of whey proteins, was developed by Block and his co-workers (Block et al., 1953; Block and Zweig, 1954). Subsequently, Zweig and Block (1954) removed iron from ferrilactin by ion-exchange chromatography of an acid solution of the protein at pH 1.3 and dialyzed the iron-free effluent for two days. When the dialyzed solution was adjusted to pH 5.2, a precipitate formed. The washed precipitate yielded α-lactalbumin, while the mother liquor could be fractionated to give both β-lactoglobulin and more α-lactalbumin. It is noteworthy that both proteins could be crystallized in these experiments, not only because pasteurized skim milk was used as the starting material, but also because the proteins were exposed to pH 1.3.

4. *Method of Aschaffenburg and Drewry*

A simple acidification procedure for separating α-lactalbumin from β-lactoglobulin, which considerably expedited the preparation and crystallization of each protein, was reported by Aschaffenburg and Drewry (1957a). In this method, casein, globulins, other proteins and fat were precipitated together from whole milk by the addition of sodium sulfate. The precipitate was removed and the filtrate, containing the lactalbumin fraction, was acidified at pH 2. Under these conditions α-lactalbumin was precipitated rapidly, whereas β-lactoglobulin remained in solution. The solution, being virtually free of proteins other than β-lactoglobulin, was easily worked up for the crystalline protein. The precipitate consisted largely of α-lactalbumin but also contained blood serum albumin and other whey proteins in small amounts, including residual β-lactoglobulin. The α-lactalbumin was reprecipitated from weakly ammoniacal solutions, first at pH 3.5 to remove β-lactoglobulin and then at pH 4.0 to remove serum albumin; it was then crystallized under the conditions described by Gordon and Semmett. Yields of purified protein entirely comparable to those reported by Gordon and Ziegler were obtained by Aschaffenburg and Drewry. This comparatively simple method for preparing α-lactalbumin and β-lactoglobulin has been used widely since its introduction.

5. *Method of Robbins and Kronman*

Because of the special requirements for preparing a large amount of α-lactalbumin under the mildest of conditions, Robbins and Kronman (1964) wished to avoid the large volumes of solutions and long dialyses characteristic of the method of Gordon and Ziegler, as well as exposure of the protein to pH 2.0 in the method of Aschaffenburg and Drewry. They described a modified Gordon and Ziegler procedure which was applicable to 80–400 liter quantities of milk. The crude albumin fraction obtained at 3.3 M ammonium sulfate was not dialyzed but was dissolved in water. α-Lactalbumin was then precipitated at pH 4.0 by the addition of acid, and the mixture was stored at 1°C overnight. The precipitate was centrifuged off in the cold, redissolved at pH 6.6 and reprecipitated at pH 4.0. This step was repeated once more prior to crystallization of the α-lactalbumin at pH 6.6 by the addition of ammonium sulfate in the usual manner. It was demonstrated by the starch-gel electrophoresis patterns of the crystallized α-lactalbumin that these reprecipitations effectively removed β-lactoglobulin.

6. *Method of Armstrong, McKenzie, and Sawyer*

In reviewing methods for the preparation of α-lactalbumin, McKenzie (1967) mentioned that Armstrong *et al.* (1967) had found in the Robbins-Kronman procedure that the precipitation of α-lactalbumin at pH 4.0 and room temperature, followed by standing at 2°C, did not result in satisfactory fractionation. Armstrong *et al.* prefer to use, as the best starting point for the further purification of α-lactalbumin by salt fractionation or by chromatography, a precipitate (14b) obtained at pH 3.5 and 3°C (see Figs. VI-2 and VI-3 of McKenzie, 1967). This precipitate is prepared (method IIb) by fractionating the milk proteins by means of ammonium sulfate and careful control of pH and temperature (see Armstrong *et al.*, 1970, for details of their chromatographic procedure).

7. *Methods of Dautrevaux and Bleumink*

References to a method reported by Dautrevaux (1963) have appeared in the literature. Details are not available but the method apparently involves fractionation of whey with ammonium sulfate, precipitation of α-lactalbumin at acid pH values and purification by gel filtration.

Biserte *et al.* (1966) preferred to use the method of Robbins and Kronman (1964) to obtain partially purified α-lactalbumin; they then completed the purification by means of preparative paper electrophoresis in tris-citric acid buffer, pH 8.6, at a voltage of 2300 in the Elphor-VaP apparatus (Brinkmann Instruments).

In a method proposed by Bleumink (1966), acetone was used to precipitate all whey proteins except α-lactalbumin (and some β-lactoglobulin) from whey. Thereafter, α-lactalbumin was easily isolated by a modification of the procedure of Aschaffenburg and Drewry. It is pointed out by Bleumink that the acetone precipitation step is potentially deleterious and must be carried out carefully at low temperature.

8. *Method of Aschaffenburg*

A simple method which improves the yield of α-lactalbumin from cow or goat milk has been described by Aschaffenburg (1968). The method is based on the observation by Fox et al. (1967) that β-lactoglobulin is quite soluble in acid whey with 3% trichloracetic acid, while α-lactalbumin is not. The precipitated crude α-lactalbumin is purified by dissolution at pH 7, reprecipitation at pH 3.5, and simple fractionation with ammonium sulfate prior to crystallization in the usual way. By this procedure 0.6–0.7 g of bovine α-lactalbumin per liter of whey can be obtained. Aschaffenburg states that no evidence has ever been presented that exposure of α-lactalbumin to strongly acid conditions is detrimental. In support of this statement it may be pointed out that any effects of acid pH values on the properties of the protein (Kronman et al., Section IV.D, E) appear to be completely reversible.

B. Purification by Chromatography and Gel Filtration

It has been noted previously that α-lactalbumin which has been crystallized several times may not be completely pure. Effective purification, as judged by disc electrophoretic patterns at different pH values, may be accomplished by chromatography on columns of DEAE-cellulose, using phosphate buffers at pH 8.2 (Groves, 1965). Likewise, with vertical acrylamide-gel electrophoretic patterns as criteria of purity, pure α-lactalbumin can be obtained by chromatography on DEAE-cellulose using a lower pH (0.005 M phosphate buffer, pH 7.3) plus an increasing gradient of sodium chloride concentration from 0 to 0.4 M (Thompson, 1965).

Small quantities of quite pure α-lactalbumin can be prepared directly by overall chromatographic fractionation of the "lactalbumin" fraction on DEAE-cellulose, the elution scheme consisting of a shallow gradient of pH decreasing from 7.28 to 5.0, together with increasing ionic strength of phosphate buffers and sodium chloride (Szuchet-Derechin and Johnson, 1965). Starch-gel electrophoresis, paper electrophoresis, moving-boundary electrophoresis and ultracentrifugation were applied in the work of Szuchet-Derechin and Johnson in order to evaluate the purity of the protein and to characterize it further.

Dextran gel filtration is another technique which has been used to remove impurities from crystalline α-lactalbumin as well as to prepare fractions from skim milk that are rich in α-lactalbumin. Thus, Bengtsson et al. (1962) filtered α-lactalbumin (crystallized five times) through Sephadex G-75 in order to remove a trace of serum albumin detectable by immunoelectrophoresis. Separations of whey proteins into fractions by filtration through Sephadex were described by Préaux and Lontie (1962) and Hill and Hansen (1964). A fraction consisting mainly of α-lactalbumin was obtained in each case.

III. Composition and Structure

A. AMINO ACID COMPOSITION

α-Lactalbumin is a simple protein with a molecular weight of about 16,000 daltons. Elementary analysis of the protein showed phosphorus to be absent, nor could carbohydrate be detected by the orcinol method (but see Section IV.A). α-Lactalbumin is unusually rich in tryptophan, approximately 7% being found by the method of Spies and Chambers (1949) applied to the substance in solid form. It may be noted parenthetically that this figure is probably incorrect; by a newly developed method, Spies (1967) found only 5.26% tryptophan in the protein; that is, four residues per molecule. The total sulfur of the protein (1.9%) was accounted for satisfactorily in terms of cystine and methionine. Sulfhydryl groups were not present (Gordon and Semmett, 1953; Gordon et al., 1954).

A complete amino acid analysis of α-lactalbumin by the chromatographic technique of Moore and Stein was made by Gordon and Ziegler (1955b) and the results are shown in Table I. Of the total nitrogen of the protein ($N = 15.86\%$), 97% could be accounted for in terms of amino acids and amide nitrogen, while the summation of amino acid residue weights was about 96%. From the amino acid residue weights a specific volume of 0.729 ml/g was calculated, a value in good agreement with Groves' pycnometric determination of 0.735 ml/g (Gordon and Semmett, 1953). A molecular weight of 15,500 daltons for the protein was also calculated from the data. A molecule of α-lactalbumin comprises about 125 amino acid residues with only single residues of methionine and arginine and two of proline but 22 of aspartic acid, at least four of tryptophan, and four of cystine.

Independent amino acid analyses of α-lactalbumin by Block and Weiss (1955) using quantitative paper chromatography, and by Weil and Seibles (1961b) using the improved chromatographic procedures of Moore, Spackman and Stein, gave results in substantial agreement with those in Table I.

With the renewed interest in α-lactalbumin engendered by the discovery of its biological function (Ebner et al., 1966), other analyses of its composition have been made. Brodbeck et al. (1967) compared α-lactalbumin (crystallized five times and prepared by conventional methods) with the purified "B protein" component of lactose synthetase and obtained essentially the same results which are summarized in Table I. Also shown in Table I are figures derived from the partial amino acid sequence (Structure I) reported by Brew et al. (1967). It will be seen that all of the analytical results are essentially the same. The residue numbers found by Brew

TABLE I
AMINO ACID COMPOSITION OF α-LACTALBUMINS

Amino acid	Bovine α-lactalbumin B			Goat	Guinea pig
	Grams per 100 g protein[b]	Residues[a] per 15,500 daltons[b]	Residues[a] per 15,000 daltons[c]	Residues[a] per 14,176 daltons[d]	Residues[a] per 15,000 daltons[e]	Residues[a] per 15,800 daltons[f]

Amino acid	Grams per 100 g protein[b]	Residues[a] per 15,500 daltons[b]	Residues[a] per 15,000 daltons[c]	Residues[a] per 14,176 daltons[d]	Residues[a] per 15,000 daltons[e]	Residues[a] per 15,800 daltons[f]
Aspartic acid	18.65	22	23	21	25	21
Threonine	5.50	7	7	7	6	6
Serine	4.76	7	7	7	6	8
Glutamic acid	12.85	14	13–14	13	15	13
Proline	1.52	2	3–4	2	2	3
Glycine	3.21	7	6	6	5	4
Alanine	2.14	4	3–4	3	5	6
Cystine	6.4[g]	4		4	4	4
Valine	4.66	6	6	6	6	4
Methionine	0.95[g]	1	1	1	0	1
Isoleucine	6.80	8	8	8	8	13
Leucine	11.52	14	14	13	13	16
Tyrosine	5.37	5	4	4	4	5
Phenylalanine	4.47	4	4	4	4	3
Lysine	11.47	12	13	12	14	11
Histidine	2.85	3	3	3	3	4
Amide N	1.37	15				
Arginine	1.15[g]	1	1	1	1	2
Tryptophan	(7.0)[g], 5.3[h]	(5), 4[h]		4	4	3

[a] Rounded to nearest integer.
[b] Gordon and Ziegler (1955b).
[c] Brodbeck et al. (1967).
[d] Brew et al. (1967); molecular weight recalculated from sequence of Brew et al. (1967).
[e] Sen (1968) and Chaudhuri and Sen (1964).
[f] Brew and Campbell (1967a).
[g] Determined by colorimetric methods.
[h] See text.

et al. may be considered the definitive amino acid analysis of the protein; additional analyses, as yet unpublished in detail (Gordon *et al.*, 1968), confirm them exactly.

Because of the widespread use of absorbancy at 280 nm for the determination of protein concentration, and because ultraviolet absorption of proteins is largely due to their content of tryptophan and tyrosine, it is of interest to consider absorptivity of α-lactalbumin at this point. The absorbancy index, $A_{1\,cm}^{1\%}$, of a solution of the protein in 0.1 M NaCl at pH 6.3 was found to be 21.0 by Gordon and Semmett (1954); values in the literature are 20.9 (Wetlaufer, 1961) and 20.1 in phosphate buffer, pH 6.9 (Kronman and Andreotti, 1964). Wetlaufer (1962) has related the absorptivity of α-lactalbumin to its content of tryptophan, tryosine and cystine as follows: The observed molar absorbancy index of the protein is 32,500; the calculated value, based on five tryptophan, five tyrosine and four cystine residues per molecule, is 35,000; the ratio of observed extinction to calculated extinction is thus 0.93. Since the ratio is less than 1, Wetlaufer was of the opinion that either the observed absorptivity or the compositional analysis was in error. If the revised analytical value for tryptophan of four residues per molecule is used in this calculation, the ratio becomes 1.10. This figure is now in excellent agreement with similar ratios derived for other well-characterized proteins, substantiating Wetlaufer's argument and conclusions.

B. END-GROUP ANALYSIS

By Sanger's dinitrofluorobenzene method, Yasunobu and Wilcox (1958) found 0.91 residue of glutamic acid per molecule of α-lactalbumin. These authors confirmed the finding of a single N-terminal glutamic acid residue in other experiments with phenyl isothiocyanate. The same result was obtained by Weil and Seibles (1961a) with a modification of Sanger's method and also by Wetlaufer (1961).

α-Lactalbumin is rapidly attacked by carboxypeptidase to yield leucine exclusively. This finding by Davie (1954) was checked by Yasunobu and Wilcox (1958), who applied the chemical method of hydrazinolysis and also indentified only leucine as C-terminal in amounts approximating one residue per molecule. Weil and Seibles (1961a) employed carboxypeptidase A in later experiments and confirmed Davie's observations.

Clearly, then, α-lactalbumin consists of a single polypeptide chain with N-terminal glutamic acid and C-terminal leucine.

C. DERIVATIVES AND PRIMARY STRUCTURE

For many years the only information available on the primary structure of α-lactalbumin was that due to the work of the late Dr. Leopold Weil and his collaborators. In an early investigation Weil and Telka (1957) prepared N-acetylated α-lactalbumin, DNP-α-lactalbumin, and guanidinated α-lactalbumin and then compared the unmodified and modified protein with respect to hydrolysis by trypsin. Native α-lactalbumin was hydrolyzed to the extent of about 13 peptide bonds per molecule of protein, in accordance with the total number of arginine and lysine residues (1 + 12). The modified proteins, in which the ε-amino groups of the lysine residues were no longer free, were attacked to the extent of only one peptide bond—presumably that involving the single arginine residue.

As these investigations continued, other derivatives of α-lactalbumin were prepared, including S-sulfo-α-lactalbumin (Weil and Seibles, 1959) and S-cyanoethyl-α-lactalbumin (Weil and Seibles, 1961b). With both native and S-sulfo-α-lactalbumin, Weil and Seibles (1961a) investigated the sequential cleavage of amino acids from the C-terminus resulting from the action of carboxypeptidase A, followed by carboxypeptidase B and carboxypeptidase A. The sequence of six amino acids at this end was determined to be -Ile-Val-Tyr-Thr-Lys-Leu-COOH. A number of other amino acids subsequently released in these experiments were identified, but their positions in the chain were not established. It was also reported in this paper that cleavage of the disulfide bonds in α-lactalbumin, as in the preparation of S-sulfo-α-lactalbumin, did not result in any change in sedimentation characteristics in the ultracentrifuge, thus confirming that α-lactalbumin is a single polypeptide chain with intrachain disulfide bonds.

In their last paper on this subject Weil and Seibles (1964) determined the amino acid sequences of four peptides isolated from a peptic digest of α-lactalbumin. They are as follows:

I. -Tyr-Gly-Ser-Gly-Asp-Thr-Glu-Ala-Ile-Val-
II. -Lys-Val-Gly-Ile-Asn-Tyr-
III. -Leu-Lys-Asp-(Leu,Gly)-Lys-Tyr-
IV. -Leu-Lys-Glu-Arg-Asp-Leu-Lys-Gly-Tyr-

These peptides accounted for about a fourth of the amino acids present in the α-lactalbumin molecule. Unfortunately, further work on the problem was halted by the untimely death of Dr. Weil.

A few years later, however, research on the primary structure of α-lactalbumin was resumed in two laboratories. Dautrevaux et al. (1966a,b) oxidized the protein with performic acid, hydrolyzed it with trypsin, and isolated nine peptides representing 49 amino acids; however, the sequences of the peptides were not worked out. These authors also reported that the protein could be split into two peptides by cyanogen bromide cleavage of the single methionine residue. One of the peptides was isolated. It contained 28 amino acid residues and represented the C-terminal portion of the molecule.

The problem was also taken up by Brew et al. following the investigations of Brew and Campbell (1967a,b) on guinea pig α-lactalbumin and their observation that certain α-lactalbumins and lysozymes from higher animals have similar molecular weights, identical or similar numbers of some amino acids, the same number of disulfide bonds, and similar or identical NH_2- and COOH-terminal residues. These similarities had in fact been pointed out in 1958 by Yasunobu and Wilcox. In a remarkably short time Brew et al. (1967) were able to determine the partial amino acid sequence of bovine α-lactalbumin and to demonstrate the surprising degree of homology between its structure and that of hen egg white lysozyme when the two sequences are aligned. Subsequently Brew et al. (1970) (Brew and Hill, 1970; Vanaman et al., 1970) were able to determine the complete amino acid sequence of α-lactalbumin. Its homology with hen egg white lysozyme and human lysozyme can be seen in the comparison of sequences in Fig. 1. This aspect of the structure of α-lactalbumin will be discussed further in Section VIII.

Elucidation of the sequence in Fig. 1 was made possible by a combination of two procedures, aminoethylation of reduced α-lactalbumin with ethyleneimine and cleavage of the modified peptide chain at Met 90 by cyanogen bromide. The fragments, one of 90 residues and the other of 33, were easily separable by gel filtration. Skillful use, thereafter, of conventional methods led to the sequence illustrated. It is made up of 123 residues. The molecular weight calculated from the sequence is 14,176 daltons (the value 14,437, published by Brew et al., 1967, appears to be incorrect).

Of the peptides isolated by Weil and Seibles, peptide II is identical in sequence with residues 98–103; peptide III seems to involve residues 12–18, although Lys-Gly are in reverse order. Peptide IV appears to come from a similar portion of the sequence, involving residues 10–20, and peptide I may be related to the sequence 33–42, although there are inconsistencies between the sequences of peptides I and IV and the relevant sequences of Brew et al. (1970).

Establishment of the entire sequence of the molecule may well depend on further careful chemical work because the X-ray diffraction approach has been hampered by the lack of suitable crystals. It is likely that the work in progress in several laboratories on goat α-lactalbumin, which is more amenable to crystallization, will result in the rapid determination by

```
α-Lactalbumin  -                        Glu-Gln-Leu-Thr-Lys-CYS-GLU-Val-Phe-ARG-Glu-LEU-LYS              Asp-LEU-Lys-GLY-TYR-Gly-GLY
Lysozyme                                                                     10                                              20
Chicken        -        Lys-Val-Phe-Gly-Arg-CYS-GLU-Leu-Ala-Ala-Met-LYS-Arg-His-Gly-LEU-Asp-Asn-TYR-Arg-GLY
                          1                                 10                                 20
Human          -        Lys-Val-Phe-Gly-Arg-CYS-GLU-Leu-Ala-ARG-Thr-LYS-Arg-Leu-Gly-Met-Asp-GLY-TYR-Arg-GLY

α-Lactalbumin  -        Val-SER-LEU-Pro-Glu-TRP-VAL-CYS-Thr-Thr-                PHE-His-Thr-SER-GLY-TYR-Asp-THR-Glu-ALA-Ile-Val
Lysozyme                              30                                                              40
Chicken        -        Tyr-SER-LEU-Gly-Asn-TRP-VAL-CYS-Ala-Ala-Lys-PHE-Glu-    SER-Asn-Phe-Asn-THR-Gln-ALA-Thr-Asn
                                  30                                                          40
Human          -        Ile-SER-LEU-Ala-Asn-TRP-Met-CYS-Leu-Ala-Lys-Trp-Glu       SER-GLY-TYR-Asn-THR-Arg-ALA-Thr-Asn

α-Lactalbumin  -Glu-ASN-         Asn-Gln-SER-THR-ASP-TYR-GLY-Leu-PHE-GLN-ILE-ASN-Asn-Ile-TRP-CYS-Lys-Asn
Lysozyme                                      50                                 60
Chicken        -Arg-ASN-Thr      -Asp-Gly-SER-THR-ASP-TYR-GLY-Ile-Leu-GLN-ILE-ASN-Ser-Arg-Trp-TRP-CYS-Asn-Asp
                                          50                                 60
Human          -Tyr-ASN-Ala-Gly-Asp-ARG-SER-THR-ASP-TYR-GLY-Ile-PHE-

α-Lactalbumin  -        Asp-Gln-Asp-PRO-His-SER-Ser-ASN-Ile-CYS-ASN-ILE-SER-CYS-Asp-Lys-PHE-LEU-Asn-Asn-ASP-Leu
Lysozyme                                  70                                 80
Chicken        -        Gly-Arg-Thr-PRO-Gly-SER-Arg-ASN-Leu-CYS-ASN-ILE-Pro-CYS-Ser-Ala-Leu-LEU-Ser-Ser-ASP-Ile
                                      70                                 80

α-Lactalbumin  -        THR-Asn-Asn-Ile-Met-CYS-Val-LYS-LYS-ILE-Leu        ASP-Lys-Val-GLY-ILE-ASN-Tyr-TRP-Leu-ALA
Lysozyme                              90                                          100                        110
Chicken        -        THR-Ala-Ser-Val-Asn-CYS-Ala-LYS-LYS-ILE-Val-Ser-ASP-Gly-Asp-GLY-Met-ASN-Ala-TRP-Val-ALA
                                          90                                 100

α-Lactalbumin  -        His-Lys-Ala-Leu-CYS-Ser-Glu-Lys-Leu-Asp-GLN        TRP-Leu            CYS-Glu-Lys-LEU
Lysozyme                          110                          120                                      123
Chicken        -        Trp-Arg-Asn-Arg-CYS-Lys-Gly-Thr-Asp-Val-GLN-Ala-TRP-Ile-Arg-Gly-CYS            Arg-LEU
                                                              120                                      129
```

FIGURE 1. A comparison of the amino acid sequences of bovine α-lactalbumin B (Brew et al., 1970), hen (chicken) egg white lysozyme (Canfield and Liu, 1965; see also Jollès, 1967) and the partial sequence for human lysozyme (R. Canfield, cited in Brew et al., 1970). For the alignment of the sequences shown, identical residues at corresponding positions in the peptide chain are shown in capital letters. The disulfide linkages are shown in Fig. 3. (Figure 1 is taken from Brew et al., 1970.)

15. α-LACTALBUMIN

combined chemical and X-ray diffraction techniques of the complete structure of this homologous α-lactalbumin.

IV. Physico-Chemical Properties

A. ELECTROPHORETIC BEHAVIOR

It was reported by Gordon and Semmett (1953) that α-lactalbumin gave a single peak in moving-boundary electrophoresis under most conditions, in pH 2 and 3 glycine buffer, pH 6.6 and 7.7 phosphate and pH 8.5 veronal; however, in pH 3.3 lactate buffer, two peaks were observed. Klostergaard and Pasternak (1957) found the protein to exhibit more than one peak not only in lactate buffer, but also in buffers above pH 7. Wetlaufer (1961) also demonstrated multiple peaks in alkaline solutions in his electrophoretic experiments. Although these experimental results differed in certain respects, they all provided some evidence for electrophoretic heterogeneity of the protein. Furthermore, such heterogeneity was observed in paper electrophoresis. Yet ultracentrifugal experiments under comparable conditions in these investigations failed to demonstrate inhomogeneity. Temporarily disregarding the possible occurrence of genetically controlled polymorphism, one might postulate that the apparent electrophoretic heterogeneity of α-lactalbumin involves strong reversible interactions with buffer ions (Klostergaard and Pasternak, 1957; Wetlaufer, 1961) of the kind originally suggested by Zittle (1956) in his investigation of the "solubility transformation" of α-lactalbumin.

One aspect of the heterogeneity of α-lactalbumin recrystallized several times, originally seen in paper electrophoresis by the author as well as by Aschaffenburg, can now be explained readily. Aschaffenburg and Drewry (1957a) mentioned a minor protein impurity (noticeable as a faint, faster-moving band in paper electrophoresis at pH 8.6) as being tenaciously held by α-lactalbumin. This material, designated satellite α-lactalbumin, has been isolated by Aschaffenburg (1967) using column chromatography. I have analyzed his preparation and have found its amino acid composition to be the same as that of bovine α-lactalbumin B. However, Brew and Hill (1967) have discovered that it contains one residue of hexosamine per molecule of protein. There appears to be, therefore, a rational basis for the difference in electrophoretic mobility. Whether this discovery of another form of α-lactalbumin may be relevant to its role in the biosynthesis of lactose (see Section VIII) is an interesting question.

Other aspects of the behavior of α-lactalbumin in transport experiments are discussed in Chapter 7, Volume I.

B. SOLUBILITY AND ISOELECTRIC POINT

Zittle and DellaMonica (1955) found that a sample of crystalline α-lactalbumin, thought to be homogeneous on the basis of the electrophoretic and ultracentrifugal evidence then available, could be separated into two components with different solubilities by a solvent-gradient extraction procedure. Subsequently, Zittle (1956) reported that one component could be transformed, reversibly, to the other. He showed that in dialyzed solutions at pH 6.6, α-lactalbumin exists to the extent of 75% in a form insoluble in 2 M ammonium sulfate. In the presence of 0.1 M sodium chloride or a variety of other salts it is transformed largely (89%) to a form soluble in 2 M ammonium sulfate. The two forms did not differ in the ultracentrifuge nor in electrophoresis at 0.1 ionic strength; however, at 0.01 ionic strength electrophoretic heterogeneity was demonstrable. Zittle concluded that the transformation depended on the binding of anions by the protein.

Zittle also investigated the effect of sodium chloride on the acid titration curve and the solubility of α-lactalbumin in the isoelectric region. The isoelectric point in the absence of salt was pH 4.8 as determined by titration-curve inflection and by maximal precipitation. In 0.5 M sodium chloride the pH of minimal solubility was 3.6, indicating a shift of isoelectric point presumably due to binding of chloride ions.

In more extensive investigations of the solubility of α-lactalbumin in the isoelectric region, Kronman and his co-workers referred to the isoelectric point of the protein as pH 4.2–4.5. They pointed out that certain characteristics of α-lactalbumin, such as its limited solubility near its isoelectric point, its tendency to bind anions and to undergo an association reaction, as well as difficulties in approaching equilibrium conditions, make determinations of solubilities in this region empirical quantities without thermodynamic significance (Kronman and Andreotti, 1964; Kronman et al., 1964). Because of such considerations, the isoelectric point of α-lactalbumin has not been determined with great accuracy.

C. ULTRACENTRIFUGAL BEHAVIOR AND MOLECULAR WEIGHT

Early determinations of the sedimentation coefficient of α-lactalbumin have been mentioned previously in this chapter (Section I.A, E). Subsequent and more complete studies related sedimentation coefficients to protein concentration and gave the following values for $s_{20,w}$ extrapolated to zero protein concentration: 1.87 S (Wetlaufer, 1961); 1.92 S (Kronman and Andreotti, 1964); 1.98 S (Szuchet-Derechin and Johnson, 1965). All of these measurements were made on the alkaline side of the isoelectric

TABLE II
MOLECULAR WEIGHT OF α-LACTALBUMIN

Method	Molecular weight (daltons)	Reference
Sedimentation-diffusion	17,400	Svedberg and Pedersen (1940)
Sedimentation-diffusion	15,100	Gordon and Semmett (1953)
Light-scattering	16,500	Gordon and Semmett (1953)
Amino acid analysis	15,500	Gordon and Ziegler (1955)
Osmometry	16,300	Wetlaufer (1961)
Sedimentation-viscosity	14,900	Wetlaufer (1961)
Sedimentation-equilibrium	16,200	Kronman and Andreotti (1964)
Amino acid sequencing	14,176[a]	Brew et al. (1967)

[a] Recalculated from the sequence of Brew et al.; the published value, 14,437, is incorrect.

zone, where only one sedimenting peak is seen. From such determinations, and also by other methods, molecular weights of α-lactalbumin have been calculated. These are summarized in Table II and show that the molecular weight of the protein is about 16,000 daltons.

D. ASSOCIATION AND AGGREGATION IN ACID AND ALKALINE SOLUTION

The heterogeneity of α-lactalbumin which was observed in the electrophoretic and solubility experiments just discussed has also been demonstrated in ultracentrifugal experiments at acid pH by Kronman and his co-workers (Kronman and Andreotti, 1964; Kronman et al., 1964). In the first two papers of a series of investigations on inter- and intramolecular interactions of α-lactalbumin, these authors carried out sedimentation-velocity and equilibrium measurements at pH values (4–2) acid to the isoelectric zone. Under these conditions α-lactalbumin was found to exist largely in associated or aggregated form. In these papers the term "association" refers to the formation of low molecular weight polymers such as dimers, trimers, etc., whereas "aggregation" denotes the formation of a heavy component having a sedimentation constant of 10–14 S. The association process at acid pH was rapid, stronger at 10° than at 25°C, and reversible. Little or no association was observed at pH values alkaline to the isoelectric region. The aggregation process at low pH values was time dependent. The rate of formation of the heavy component decreased with decreasing temperature and with decreasing pH and ionic strength. Aggregation, too, was reversible and did not occur at alkaline pH values. Protein concentration markedly influenced the extent of aggregation, with little aggregation being noticeable below 1%, although association might be appreciable at that concentration. On the basis of these and other experiments involving studies of solubility in ammonium sulfate solutions, Kronman and his associates have advanced the hypothesis that the increased tendency of α-lactalbumin to aggregate below its isoelectric point is the consequence of a "denaturation-like" process whereby certain groups become available for intermolecular interaction. Conformational changes in the protein molecule parallel the aggregation, and experimental evidence of such changes in acid solution have been observed, thus supporting the hypothesis.

The investigations by Kronman and his co-workers have provided reasonable explanations for many of the puzzling aspects of the apparent heterogeneity of α-lactalbumin, especially in acid solutions of the protein. Apparent heterogeneity at alkaline pH may be explained, at least in part, by protein–ion interactions, as previously noted. But there is also an alkaline conformational change which has been demonstrated by the later experiments of Kronman *et al.* (1967). Though little or no association to species of low molecular weight had been noticed previously, as mentioned above, titration curve (Robbins *et al.*, 1967) and optical rotatory dispersion measurements (Kronman *et al.*, 1966) indicated the presence of a structural change above pH 10. The later ultracentrifuge experiments of Kronman *et al.* (1967) showed that α-lactalbumin exists in a somewhat expanded state above pH 9.5, although the degree of expansion is somewhat less than that observed at low pH values. This association to low molecular weight species was demonstrated as occurring above pH 9.5. The reversible time-dependent aggregation seen at low pH values, however, was indeed absent in alkaline solution.

E. CONFORMATIONAL CHANGES

The experimental evidence for the reversible "denaturation-like" process in acid solutions of α-lactalbumin reported by Kronman and his associates has been obtained in various ways. Kronman *et al.* (1965b) showed that the acid denaturation was accompanied by blue shifts of the absorption spectrum in the region of 270–300 nm, characteristic of changes in the environment of tryptophan groups. The difference extinction coefficient, ΔE_{293}, was strongly dependent on pH but insensitive to changes in ionic strength, indicating the probable absence of charge perturbations. In the transition region (pH 3–4), ΔE_{293} was strongly dependent on temperature.

15. α-LACTALBUMIN   349

The conformational change which takes place in alkaline solution is also accompanied, as demonstrated by measurement of ultraviolet difference spectra, by a tryptophan blue shift which underlies the usual tyrosine ionization red shift above pH 10 (Kronman et al., 1967). The spectral changes are comparable to those observed during the acid transition.

Additional evidence for the alkaline conformational change at pH 9.5 is provided by the experiments of Gorbunoff (1967) in which the tyrosine residues of α-lactalbumin were modified by treatment with cyanuric fluoride at pH 10 and 9.3. The expanded α-lactalbumin molecule at pH 10 contains four reactive tyrosine groups, only one of which is ionized; at pH 9.3, the native molecule contains three reactive tyrosines, none of which is ionized. Thus, the change in conformation imparts reactivity to an additional tyrosine residue.

Whether the transition effected by changing the pH from 6 to below 3 at 25°C involved the exposure of tryptophan groups in α-lactalbumin was determined by Kronman and Holmes (1965) under a variety of conditions using the solvent perturbation method of difference spectrophotometry. It was concluded from these experiments that the "denaturation blue shift" could not be accounted for on the basis of increased exposures of tryptophan groups but must have other origins.

In another approach, Robbins et al. (1965) introduced new nonpolar groups into the α-lactalbumin molecule by amidination of the free amino groups and compared the modified and native proteins with respect to such properties as sedimentation-velocity, ultraviolet fluorescence, optical rotatory dispersion and electrophoretic behavior. They found that although no major structural changes resulted from the modification, the modified protein was much more susceptible to association and aggregation than α-lactalbumin. They interpreted their observations as supporting the hypothesis that hydrophobic interactions play a role in the association and aggregation of α-lactalbumin, and that the differences in behavior of the "alkaline" and "acid" forms of the protein are due to a high density of nonpolar groups at the molecular surface of the "acid" form.

Other observations on the conformation of α-lactalbumin as revealed by the optical rotatory dispersion method have been reported by Herskovits and Mescanti (1965) and, more extensively, by Kronman et al. (1965a, 1966). Judging from the b_0 parameter $(-235°)$ calculated from measurements of pH 7 aqueous solutions of the protein, Herskovits and Mescanti estimated that α-lactalbumin contains a fair amount of α-helix. A similar estimate was made by Kronman et al., who found further that acid denaturation of α-lactalbumin below pH 4 was accompanied by a decrease in the value of b_0 of the order of 75°. It was thought at first that the pH 4 conformational change might involve some melting out of helical regions of the molecule. However, other considerations led to the conclusion that the changes observed in the parameter b_0 are probably due to alterations in the freedom of rotation of side chains, such as tryptophans. Kronman (1967) has summarized the results of his investigations on the conformational changes as follows:

> There appear to be two rather distinct molecular changes for this protein, both of which involve alteration of the environment of tryptophan residues but in strikingly different ways: (a) below pH 4 molecular swelling occurs; a tryptophan difference spectrum is generated; changes in optical rotation dispersion properties are observed and the protein becomes increasingly prone to aggregation and dissociation. Although a denaturation blue shift of the absorption spectrum is observed, solvent perturbation measurements indicate that no change in exposure of tryptophan groups occurs during the acid conformational change, *i.e.* three groups are completely buried and two are completely exposed. Comparable changes in physicochemical properties of α-lactalbumin suggest that an identical process occurs above pH 10. (b) A more subtle conformational alteration is observed if the temperature of α-lactalbumin solution is lowered from 25 to 0–2°C at pH 6. Solvent perturbation measurements indicate that at the lower temperature the two exposed groups are no longer accessible to large perturbants such as sucrose molecules but remain fully available for small perturbants such as heavy water. This process, which we have called "crevice contraction" is not accompanied by any change in absorption spectrum.

Kronman (1968) has made a comparison of the conformation of native bovine α-lactalbumin, "acid-denatured" α-lactalbumin, and hen egg white lysozyme by circular dichroism (CD) measurements over the range 185–300 nm. He found that there are only small differences in the CD spectra in the region below 240 nm and concluded that these differences arise from side-chain effects, the backbone conformations being essentially similar. He found differences in the 260–295 nm region arising from side-chain transitions (tryptophanyl, tyrosyl or cystinyl residues; see Chapter 8, Volume I). Kronman's conclusion from the CD spectra that the native and "acid-denatured" forms of α-lactalbumin do not differ greatly in conformation is, of course, in agreement with his earlier findings based on other measurements. Indirect support for these views comes from a possible three-dimensional model, proposed by Browne *et al.* (1969), for bovine α-lactalbumin based on the X-ray structure of lysozyme. They made a careful assessment of the consequences of side-chain replacements in α-lactalbumin. There were, however, some regions in which they were unable to deduce the conformation unequivocally; Glu 35, which acts as a proton donor in lysozyme, is not present in α-lactalbumin where a neighboring His residue may assume such a function. Evidence against this model has been obtained for α-lactalbumin in solution by Maes et al. (1969) and Krigbaum and Kügler (1970). On the other hand, Castellino and Hill (1970) have obtained evidence from the relative rates of carboxymethylation of certain residues in bovine α-lactalbumin that they consider supports the predicted conformation of α-lactalbumin. They found that Met 90 was the most reactive residue; His 68 reacted with iodoacetate at a slower rate but more rapidly than His 32. The carboxymethylation of His 107 was much slower than the other histidyl residues. The structural model is considered further in Section VIII.

V. Genetic Polymorphism

In the 278 milk samples from British breeds of cows typed individually for β-lactoglobulin polymorphism by the paper electrophoresis method of Aschaffenburg and Drewry (1957b), only a single kind of α-lactalbumin was found (Aschaffenburg, 1963). In similar experiments with milks from Icelandic cattle, Blumberg and Tombs (1958) likewise observed only a single α-lactalbumin band. But when milks from Nigerian White Fulani or Lyre-Horned Zebu were examined, Blumberg and Tombs discovered two forms of α-lactalbumin, one corresponding in mobility to the α-lactalbumin previously found in the British and Icelandic cattle, while the other moved more rapidly. One milk sample contained only the fast-moving α-lactalbumin (the A form), 12 milk samples contained both forms, and 33 milk samples contained only the slower B form. Because the distribution of the three types agreed with the predictions of the Hardy-Weinberg law, Blumberg and Tombs suggested that the production of α-lactalbumin is genetically determined.

The observations of Blumberg and Tombs were extended by Aschaffenburg (1963) to milk from animals of another breed of African cattle, Boran or Short-Horned Zebus from Kenya. Aschaffenburg confirmed the existence of genetic polymorphism of α-lactalbumin in African Zebu cattle. Also, in a similar investigation of Indian Zebu cattle, Bhattacharya et al. (1963) found that these animals, too, produced milk in which the forms of α-lactalbumin were under genetic control, two variants (A and B) being distinguished. Bhattacharya et al. reported that the polymorphs have approximately the same electrophoretic mobility (Tiselius), crystalline form, nitrogen content, specific extinction coefficient at 280 nm, and sedimentation coefficient.

Through the cooperation of S.K. Ghosh and Dr. A. Sen, the author has been able to carry out amino acid analyses of a sample of highly purified α-lactalbumin A prepared from the milk of an Indian Zebu cow. No arginine has been found in α-lactalbumin A. Arginine 10 in the B variant is replaced by a glutamic acid residue in A, and this is presumed to be in the form of glutamine, a substitution permitted by the genetic code. Other differences in composition could not be detected (Gordon et al. 1968).

It has been reported previously (McKenzie, 1967) that both α-lactalbumins are present in milks from Droughtmaster beef cattle. Bell et al. (1970) have independently determined the amino acid composition of both variants from milks of this breed. They have found the B from Droughtmaster to be the same as B from Western dairy breeds and also that Droughtmaster A differs from B by the simple substitution of glutamine for Arg 10.

There is, at present, no reason to believe that α-lactalbumin prepared from milks from Western breeds of dairy cattle is other than α-lactalbumin B. The apparent heterogeneity discussed in the preceding sections is, in all probability, not a consequence of genetic polymorphism. Yet the possibility of the existence of such polymorphism in Western cattle, demonstrable perhaps by amino acid analysis, fingerprinting of peptides, analysis for carbohydrate, or other typing procedures not yet applied, cannot be dismissed.

VI. α-Lactalbumins in the Milk of Other Mammals

A. IMMUNOLOGICAL RELATIONSHIPS

The quantitative immunochemical method of Larson and Hageman (1963) for estimating α-lactalbumin, which has been noted previously (Section II.A.2), has been applied further in studying milks of various species (Johke et al., 1964). These investigators found that the antiserum to cow α-lactalbumin reacted equally strongly with cow and sheep milk, about twice as strongly with goat and water buffalo milk, but not at all with the nonruminant milks tested: camel, horse, rat, mouse, pig, dog and rabbit. The *apparent* contents of α-lactalbumin in the ruminant milks estimated from quantitative cross-reactions were cow, 1,2; sheep, 1.5; goat, 2.2; and water buffalo, 2.7 g/liter.

Of other investigations along these lines, only the extensive survey of Lyster et al. (1966) will be cited. Positive cross-reactions with bovine α-lactalbumin antiserum were obtained not only with sheep, goat and water buffalo milks but with milk from other ruminants of the families, *Bovidae, Giraffidae, Cervidae* and *Antilocapridae*. Milk from nonruminants, man and guinea pig among others, did not react.

B. CAPRID α-LACTALBUMIN

A crystalline "lactalbumin" was prepared from goat milk by Sen and Chaudhuri (1962) using a modification of the Aschaffenburg and Drewry method. Preliminary measurements of its properties showed that it was very much like bovine α-lactalbumin. This "lactalbumin" was crystallized in the usual way, by the addition of ammonium sulfate to a solution of the protein at pH 6, and it was also crystallized by storing a salt-free concentrated solution at 2°–4°C. It may be noted that it has not yet been possible to crystallize bovine α-lactalbumin in the absence of salt. Milk from individual goats was used in this work but no evidence for genetic polymorphism was found.

Chaudhuri and Sen (1964) later completed their characterization of this protein and, on the basis of additional evidence for similarity between goat and cow proteins, designated it α-lactalbumin. Measurements of mobility in moving-boundary electrophoresis, sedimentation and diffusion coefficients, partial specific volume, frictional ratio, etc., all substantiated this designation. Furthermore, amino acid analysis revealed a distribution of amino acids strikingly like that of bovine α-lactalbumin, as shown in Table I. The chemical evidence for similarity is corroborated by the immunological cross-reactions already noted.

C. WATER BUFFALO α-LACTALBUMIN

Sen and Sinha (1961) noted that lactalbumin could be crystallized from buffalo milk by a modified Aschaffenburg and Drewry procedure. In 1963 Bhattacharya et al. found that buffalo α-lactalbumin and Zebu cow α-lactalbumin A, purified and crystallized by the usual methods, had virtually the same electrophoretic mobilities at three different pH values, the same crystalline form, similar nitrogen contents, similar absorbancy indexes at 280 nm, and similar tyrosine and tryptophan contents. The sedimentation coefficients of both proteins were within the range 1.87–1.99 S, and the molecular weight of buffalo α-lactalbumin was calculated to be 16,200 daltons. Only a single buffalo α-lactalbumin was described.

Preliminary amino acid analyses by the author reveal that buffalo α-lactalbumin is virtually indistinguishable in composition from bovine α-lactalbumin B, differing only in that it contains five glycine residues per molecule rather than six as in B and possibly 22 rather than 21 aspartic acid residues (Gordon et al., 1968). Again, immunological cross-reactions point up the close relationship between the bovine and buffalo α-lactalbumins.

D. HUMAN LACTALBUMIN

The occurrence in human milk whey of a protein similar in some respects to α-lactalbumin was reported by Johansson in 1958. He separated the proteins in the albumin fraction of the whey by chromatography on calcium phosphate gel columns. The protein in the first of three main fractions was essentially homogeneous in electrophoresis and in the ultracentrifuge. Its sedimention constant was 1.73 S.

Some years later Maeno and Kiyosawa (1962) isolated a human lactalbumin by acidification to pH 4.8 of a solution of the crude lactalbumin fraction. The protein was purified by reprecipitations. It, too, had a sedimentation coefficient of 1.7 S, and its molecular weight was calculated to be 23,000 daltons. Reference has also been made to the presence of a human lactalbumin in the gel electrophoresis experiments of Bell and McKenzie (1964). No evidence of polymorphism of the protein was found.

This lactalbumin appears to be the major protein of human whey. Whether human lactalbumin and bovine α-lactalbumin can be considered to be homologous proteins cannot yet be decided on the basis of the experimental evidence at hand. It has already been pointed out that bovine α-lactalbumin antiserum does not react with human milk.

E. GUINEA PIG LACTALBUMIN

In research concerned with the biosynthesis of milk proteins, Brew and Campbell (1967a,b) found an α-lactalbumin to be the major protein in guinea pig milk whey. They isolated it by chromatography on carboxymethyl cellulose and by gel filtration through Sephadex G-100. Its amino acid composition (shown in Table I), taken in conjunction with its other properties, indicated homology with bovine α-lactalbumin. Its molecular weight was 15,800 daltons. Lysine was found to be N-terminal and glutamine C-terminal.

Immunological cross-reaction of guinea pig milk with bovine α-lactalbumin antiserum was not observed in the experiments of Lyster *et al.* (1966).

F. OVINE α-LACTALBUMIN

Identification of an α-lactalbumin in sheep milk has been reported by Bell and McKenzie (1964).

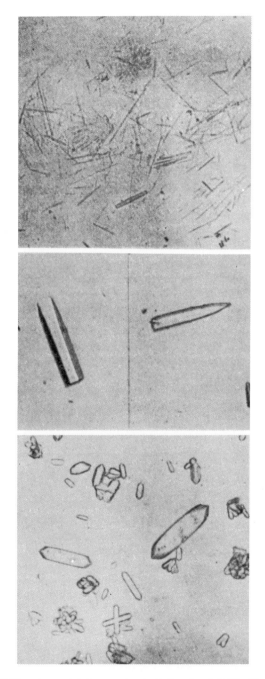

FIGURE 2. Different crystal forms of α-lactalbumin (×160) (Aschaffenburg and Drewry, 1957a).

VII. Crystallography and X-Ray Diffraction

Bovine α-lactalbumin crystals have been obtained only by salting out solutions of the protein at about pH 6.6 with ammonium sulfate. Ordinarily, microscopic crystals are formed under these conditions but their shape may vary. Figure 2, reproduced from the paper of Aschaffenburg and Drewry (1957a), shows three such crystal habits. Crystals of this size are not suitable for X-ray diffraction. Incidentally, it may be noted that Weil and Seibles (1961a) were able to prepare small crystals of α-lactalbumin from which the C-terminal leucine had been removed by treatment with carboxypeptidase.

α-Lactalbumin appears to be more soluble in strong ammonium sulfate solutions at 0°–4°C than at 25°C. On the basis of this observation Gordon (1964) worked out a method for growing large crystals of the protein, some at least 3 mm long. The method was not always successful in producing crystals considered perfect for X-ray diffraction, although some pictures, preliminary in nature, of crystals prepared in this manner were indeed taken by Dr. David W. Green of the Royal Institution, London.

Crystals suitable for X-ray work were also grown by Inman and Bryan (1966). They allowed a solution of highly purified α-lactalbumin, initially adjusted to 51% saturation with ammonium sulfate, with 1.56% protein, at pH 6.64, to stand for many months in a stoppered vial in a refrigerator. The crystals were examined by X-ray diffraction and dimensions of the unit cell were determined. Because there are discrepancies in the dimensions reported by Inman and Bryan and those found by Green (1964), and since the available information is still preliminary in nature, the actual figures will not be recorded here. However, it is apparent that the unit cell has the symmetry A2 and contains four molecules, or some multiple thereof.

VIII. The Biological Function of α-Lactalbumin

In research on pathways for the biosynthesis of lactose, Hassid and his co-workers demonstrated that an enzyme present in particulate preparations from lactating guinea pig or cow mammary glands catalyzed the reaction

$$\text{UDP-D-galactose} + \text{D-glucose} \rightarrow \text{lactose} + \text{UDP}$$

The enzyme, lactose synthetase (E.C. 2.4.1c), was subsequently found in soluble form in bovine milk and was purified 70-fold (Watkins and Hassid, 1962; Babad and Hassid, 1966).

Brodbeck and Ebner (1966) purified the soluble form further and discovered that it could be resolved into two protein components, A and B, which individually did not possess lactose synthetase activity. (To avoid possible confusion with the genetic polymorphs of α-lactalbumin, also designated A and B, the protein components of lactose synthetase will be discussed as "A protein" and "B protein.") Recombination of "A protein" and "B protein" restored activity and the suggestion was made that the two proteins are subunits of the enzyme. Investigation of the properties of purified "B protein" revealed a close similarity to α-lactalbumin (Ebner et al., 1966). Highly purified recrystallized α-lactalbumin served equally well as purified "B protein" in recombining with "A protein" to give active enzyme. Furthermore, immunological assay showed that "B protein" was primarily α-lactalbumin. On the basis of this evidence, Ebner et al. suggested that an enzymatic role may be assigned to α-lactalbumin as one of the subunits of lactose synthetase.

Further convincing evidence supporting the conclusion that the "B protein" of lactose synthetase is α-lactalbumin and, conversely, that α-lactalbumin is one of the subunits of lactose synthetase was reported subsequently by Brodbeck et al. (1967) and summarized by Ebner and Brodbeck (1968) as follows: (a) α-Lactalbumin substitutes for the "B protein" of lactose synthetase at identical protein concentration in the enzymatic assays to give identical rates. (b) The ultraviolet spectra, amino acid composition (see Table I) and molecular weights are identical. (c) The "B protein" and α-lactalbumin are immunologically indistinguishable. (d) The mobilities of the "B protein" and α-lactalbumin in starch-gel electrophoresis are the same and they co-chromatograph on DEAE-cellulose and Sephadex G-100.

The interesting question of whether α-lactalbumins obtained from different species can substitute for each other in the lactose synthetase rate assays when mixed with bovine "A protein" was studied by Tanahashi et al. (1968). The enzymic activities of the following purified α-lactalbumins in the assays were all comparable tô that of bovine α-lactalbumin B: bovine A, water buffalo, sheep, goat, pig, guinea pig and human. However, in agreement with previous findings, only the α-lactalbumins from ruminants gave immunological cross-reactions with antisera to bovine α-lactalbumin. Incidentally, the "satellite α-lactalbumin" isolated by Aschaffenburg is also active in the role of "B protein" (Aschaffenburg and Andrews, 1967).

The discovery of the role of α-lactalbumin as a subunit of lactose synthetase by Ebner and his co-workers gave added importance to the elucidation of its chemical structure. As described earlier in this chapter, Brew *et al.* (1967, 1970) succeeded in working out the sequence shown in Fig. 1 and drew attention to the similarities in the primary structures of α-lactalbumin and hen egg white lysozyme. They suggested that the integral role of α-lactalbumin in lactose synthesis implies a functional as well as a structural similarity in the enzymes; one involves the synthesis and the other the cleavage of a $\beta 1 \to 4$ glucopyranosyl linkage. It is true that lysozyme does not participate in lactose synthesis nor does α-lactalbumin act upon lysozyme substrates. Still, as the sequences for α-lactalbumin and hen egg white lysozyme are aligned in Fig. 1, 49 of the residues of α-lactalbumin are identical with corresponding residues in lysozyme. Many of the residues are structurally similar at corresponding positions. It can also be seen that there are many similarities with human lysozyme. Because of the close homology in primary structure, Brew *et al.* believe that the structural genes for α-lactalbumin and lysozyme have evolved from a relatively recent common ancestor. A useful diagrammatic comparison of their covalent structures has been made by Hill *et al.* (1968) in their review of the structure and function of α-lactalbumin. A revised version is shown in Fig. 3.

As mentioned earlier, this group has concluded that the two proteins are similar in conformation. They found that it is possible to fit the side chains of bovine α-lactalbumin to the established lysozyme polypeptide backbone (Blake *et al.*, 1967) and thereby generate a structure which retains the major structural features of the lysozyme molecule (Brew *et al.*, 1968; Browne *et al.*, 1969). As mentioned in Section IV some aspects of this model have been challenged. However, one of its very interesting aspects is the surface cleft. It is the site for substrate binding in lysozyme, but it is shorter in the α-lactalbumin model. Browne *et al.* (1969) point out that while this is consistent with α-lactalbumin having a mono- or disaccharide as substrate, the biochemical evidence indicates that the role of α-lactalbumin in lactose synthesis is a complex one. Interaction with the "A protein" is necessary in order for α-lactalbumin to function in the lactose synthetase system.

It is interesting to note here that Atassi *et al.* (1970) have found that there is a lack of immunochemical cross-reaction between lysozyme and α-lactalbumin. They have also found differences in their conformation in solution.

Concomitant with the work on α-lactalbumin structure, there have been attempts to characterize the role of the "A protein" in the lactose synthetase system. Progress in this area during 1969–1970 was very rapid; thus attention is drawn to the most pertinent advances that have been made and to the review by Ebner (1970) of the biological role of α-lactalbumin. For

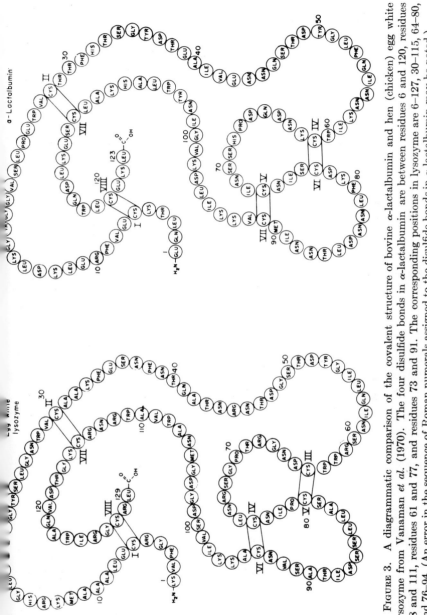

FIGURE 3. A diagrammatic comparison of the covalent structure of bovine α-lactalbumin and hen (chicken) egg white lysozyme from Vanaman et al. (1970). The four disulfide bonds in α-lactalbumin are between residues 6 and 120, residues 28 and 111, residues 61 and 77, and residues 73 and 91. The corresponding positions in lysozyme are 6–127, 30–115, 64–80, and 76–94. (An error in the sequence of Roman numerals assigned to the disulfide bonds in α-lactalbumin may be noted.)

details of the latest developments the reader should refer to the papers cited and to others that will be forthcoming in this rapidly changing area of research.

Much of the more recent work has been directed toward testing various aspects of the hypothesis of the Oklahoma (Ebner *et al.*) and North Carolina (Hill *et al.*) schools and in improving the quality of "A protein" preparations. Brew *et al.* (1968) found that the "A protein" alone behaves as a galactosyl transferase, catalyzing the reaction

UDP-galactose + N-acetyl-D-glucosamine→N-acetyllactosamine + UDP

In other words, in the absence of α-lactalbumin, the "A protein" is primarily a N-acetyllactosamine synthetase. (This activity in partially purified "A protein" is perhaps identical to that found earlier by McGuire *et al.*, 1965, in goat colostrum and in particulate fractions of several rat tissues.)

When α-lactalbumin is present, the "A protein" becomes a lactose synthetase, the N-acetyllactosamine activity apparently becoming inhibited. It should be noted that Turkington *et al.* (1968) have shown that, during pregnancy, the activity of the galactosyl transferase is high, whereas lactose synthetase activity is virtually nonexistent. However at parturition, when α-lactalbumin is formed, galactosyl transferase activity is inhibited and lactose synthesis occurs (see also Turkington and Hill, 1969). Brew *et al.* postulated that the α-lactalbumin modifies the substrate (acceptor) specificity and proposed that α-lactalbumin be termed a specifier protein. They pointed out that "The mechanism by which it effects a change in substrate specificity appears to be complex, but it is possible that it represents a new type of molecular control of a biological reaction." Most of the recent work has generally been in accord with their hypothesis. It is now known that there are several enzymes in which two proteins participate (a list of them is given in Table I of Ebner's 1970 review). However, the unusual point about α-lactalbumin is that it is able to change the acceptor specificity of a galactosyl transferase. (The only other analogous types of behavior occur with sucrose synthetase and the PEP-dependent system forming fructose-1-phosphate; see Ebner, 1970, Table I.) The various reactions involved in lactose biosynthesis are shown schematically in Chapter 18, where some of the more recent methods of assay of the "A protein" and α-lactalbumin are discussed (for example, the methods of McGuire, 1969 and Fitzgerald *et al.*, 1970).

In their original work on the resolution of the lactose synthetase system into two proteins, Brodbeck and Ebner (1966) attempted to purify the enzyme from bovine mammary microsomes. Their efforts were not very successful and they turned to bovine skim milk (see also Babad and Hassid; 1964, 1966). After the removal of casein, an ammonium sulfate fraction was subjected to gel filtration several times on Bio-Gel P-30 (BIO-RAD Laboratories) columns; the result was separation of the fraction into the now well-known "A protein" and "B protein" (α-lactalbumin). It was concluded in 1966 that the "A protein" was mainly associated with the "microsomal fraction," the "B protein" being distributed between microsomal fractions and soluble fractions. However, Coffey and Reithel (1968a,b) have concluded in more recent studies that the "A protein" is associated with the Golgi apparatus and particles comparable in size to lysosomes. Nevertheless conclusions concerning the size distribution of the "A protein" should be regarded with caution as they could be based on observations that are sensitive to different experimental procedures. In their earlier work the Duke University group used essentially the method for isolating "A protein" developed by Brodbeck and Ebner (1966). They have now published an abstract of a new procedure (Trayer et al., 1970). The "A protein" from cow milk is partially purified by chromatography of whey on DEAE-Sephadex followed by chromatography on cellulose phosphate. The active fraction is further purified on columns of Sepharose 4B (Pharmacia) to which α-lactalbumin has been covalently linked. The Oklahoma group has published a very detailed procedure for chromatographic purification of the "A protein" from an ammonium sulfate fraction of skim milk (Fitzgerald et al., 1970). Because the lactose synthetase activity of human milk is about ten times higher than that of bovine milk, Andrews (1969) has preferred to isolate the "A protein" from human milk. He has used ammonium sulfate fractionation, chromatography on DEAE-cellulose and gel filtration on Sephadex G-150.

Brew (1969) has speculated further on the relationship of α-lactalbumin secretion with the organization and control of lactose synthetase. Palmiter (1969c) has discussed factors regulating the lactose content of milk. In two other articles Palmiter (1969a,b) studies the lactose synthetase system of mouse mammary gland. He complicates the lactose synthetase issue by proposing that a third component must be invoked to explain his observations.

Ebner and Schanbacher (1970) have summarized their more recent observations on the galactosyl acceptor specificity of the "A protein." They make the important point that lactose and certain glycoprotein syntheses are compatible and carried out by the same galactosyl transferase.

References

Andrews, P. (1969). *Biochem. J.* **111,** 14 P.
Armstrong, J. McD., McKenzie, H. A., and Sawyer, W. H. (1967). *Biochim. Biophys. Acta* **147,** 60.

Armstrong, J. McD., Hopper, K. E., McKenzie, H. A., and Murphy, W. H. (1970). *Biochim. Biophys. Acta* **214**, 419.
Aschaffenburg, R. (1963). In "Man and Cattle" (A. E. Mourant and F. E. Zeuner, eds.), pp. 50–54. Royal Anthropological Institute Occasional Paper No. 18, London.
Aschaffenburg, R. (1967). Private communication.
Aschaffenburg, R. (1968). *J. Dairy Sci.* **51**, 1295.
Aschaffenburg, R., and Andrews, P. (1967). Private communication.
Aschaffenburg, R., and Drewry, J. (1957a). *Biochem. J.* **65**, 273.
Aschaffenburg, R., and Drewry, J. (1957b). *Nature* **180**, 376.
Atassi, M. Z., Habeeb, A. F. S. A., and Rydstedt L. (1970). *Biochim. Biophys. Acta* **200**, 184.
Babad, H., and Hassid, W. Z. (1964). *J. Biol. Chem.* **239**, PC946.
Babad, H., and Hassid, W. Z. (1966). *J. Biol. Chem.* **241**, 2672.
Bell, K., and McKenzie, H. A. (1964). *Nature* **204**, 1275.
Bell, K., Hopper, K. E., McKenzie, H. A., Murphy, W. H., and Shaw, D. C. (1970). *Biochim. Biophys. Acta.* **214**, 437.
Bengtsson, C., Hanson, L. Å., and Johansson, B. G. (1962). *Acta Chem. Scand.* **16**, 127.
Bhattacharya, S. D., Roychoudhury, A. K., Sinha, N. K., and Sen, A. (1963). *Nature* **197**, 797.
Biserte, G., Dautrevaux, M., Crouwy, F., and Moschetto, Y. (1966). *Bull. Soc. Chim. Biol.* **48**, 1107.
Blake, C. C. F., Mair, G. A., North, A. C. T., Phillips, D. C., and Sarma, V. R. (1967). *Proc. Roy. Soc. (London).* **B**, **167**, 365.
Bleumink, E. (1966). *Neth. Milk Dairy J.* **20**, 13.
Block, R. J., and Weiss, K. W., (1955). *Arch. Biochem. Biophys.* **55**, 315.
Block, R. J., and Zweig, G. (1954). *Arch. Biochem. Biophys.* **48**, 386.
Block, R. J., Bolling, D., Weiss, K. W., and Zweig, G. (1953). *Arch. Biochem. Biophys.* **47**, 88.
Blumberg, B. S., and Tombs, M. P. (1958). *Nature* **181**, 683.
Brew, K. (1969). *Nature* **222**, 671.
Brew, K., and Campbell, P. N. (1967a). *Biochem. J.* **102**, 258.
Brew, K., and Campbell, P. N. (1967b). *Biochem. J.* **102**, 265.
Brew, K., and Hill, R. L. (1967). Private communication.
Brew, K., and Hill, R. L. (1970). *J. Biol. Chem.* **245**, 4559.
Brew, K., Vanaman, T. C., and Hill, R. L. (1967). *J. Biol. Chem.* **242**, 3747.
Brew, K., Vanaman, T. C., and Hill, R. L. (1968). *Proc. Natl. Acad. Sci. U. S.* **59**, 491.
Brew, K., Castellino, F. J., Vanaman, T. C., and Hill, R. L. (1970). *J. Biol. Chem.* **245**, 4570.
Brodbeck, U., and Ebner, K. E. (1966). *J. Biol. Chem.* **241**, 762.
Brodbeck, U., Denton, W. L., Tanahashi, N., and Ebner, K. E. (1967). *J. Biol. Chem.* **242**, 1391.
Browne, W. J., North, A. C. T., Phillips, D. C., Brew, K., Vanaman, T. C., and Hill, R. L. (1969). *J. Mol. Biol.* **42**, 65.
Canfield, R. E., and Liü, A. K. (1965). *J. Biol. Chem.* **240**, 1997.
Cannan, R. K., Palmer, A. H., and Kibrick, A. C. (1942). *J. Biol. Chem.* **142**, 803.
Castellino, F. J., and Hill, R. L. (1970). *J. Biol. Chem.* **245**, 417.
Chaudhuri, S., and Sen. A. (1964). Private communication; "Report of the Bose Institute (1962–1963)," p. 20; "Report of the Bose Institute (1963–1964)," p. 22. Bose Institute, Calcutta, India.

Coffey, R. G., and Reithel, F. J. (1968a). *Biochem. J.* **109,** 169.
Coffey, R. G., and Reithel, F. J. (1968b). *Biochem. J.* **109,** 177.
Dautrevaux, M. (1963). *Ann. Biol. Animale,Biochim. Biophys.* **3,** 125; *Dairy Sci. Abstr.* **27,** 603 (1965).
Dautrevaux, M., Crouwy, F., Moschetto, Y., and Biserte, G. (1966a). *Bull. Soc. Chim. Biol.* **48,** 1111.
Dautrevaux, M., Crouwy, F., Han, K., and Biserte, G. (1966b). *Bull. Soc. Chim. Biol.* **48,** 1119.
Davie, E. W. (1954). Ph.D. thesis. University of Washington, Seattle.
Ebner, K. E. (1970). *Accounts Chem. Res.* **3,** 41.
Ebner, K. E., and Brodbeck, U. (1968). *J. Dairy Sci.* **51,** 317.
Ebner, K. E., and Schanbacher, F. L. (1970). *Fed. Proc., Fed. Amer. Soc. Exp. Biol.* **29,** 873.
Ebner, K. E., Denton, W. L., and Brodbeck, U. (1966). *Biochem. Biophys. Res. Commun.* **24,** 232.
Fitzgerald, D. K., Brodbeck, U., Kiyosawa, I., Mawal, R., Colvin, B., and Ebner, K. E. (1970). *J. Biol. Chem.* **245,** 2103.
Fox, K. K., Holsinger, V. H., Posati, L. P., and Pallansch, M. J. (1967). *J. Dairy Sci.* **50,** 1363.
Gorbunoff, M. J. (1967). *Biochemistry* **6,** 1606.
Gordon, W. G. (1964). *Biochim. Biophys. Acta* **82,** 613.
Gordon, W. G., and Semmett, W. F. (1953). *J. Amer. Chem. Soc.* **75,** 328.
Gordon, W. G., and Semmett, W. F. (1954). Unpublished results.
Gordon, W. G., and Ziegler, J. (1955a). *Biochem. Prep.* **4,** 16.
Gordon, W. G., and Ziegler, J. (1955b). *Arch. Biochem. Biophys.* **57,** 80.
Gordon, W. G., Semmett, W. F., and Ziegler, J. (1954). *J. Amer. Chem. Soc.* **76,** 287.
Gordon, W. G., Aschaffenburg, R., Sen, A., and Ghosh, S. K. (1968). *J. Dairy Sci.* **51,** 947.
Green, D. W. (1964). Private communication.
Groves, M. L. (1965). *Biochim. Biophys. Acta* **100,** 154.
Herskovits, T. T., and Mescanti, L. (1965). *J. Biol. Chem.* **240,** 639.
Hill, R. D., and Hansen, R. R. (1964). *J. Dairy Res.* **31,** 291.
Hill, R. L., Brew, K., Vanaman, T. C., Trayer, I. P., and Mattock, P. (1968). *Brookhaven Symp. Biol.* **21,** 139.
Inman, J. K., and Bryan, R. F. (1966). *J. Mol. Biol.* **15,** 683.
Johansson, B. (1958). *Nature* **181,** 996.
Johke, T., Hageman, E. C., and Larson, B. L. (1964). *J. Dairy Sci.* **47,** 28.
Jollès, P. (1967). *Proc. Roy. Soc. (London)* **B, 167,** 350.
Kekwick, R. A. (1935). Cited by Pedersen, K. O. *Biochem. J.* **30,** 948.
Klostergaard, H., and Pasternak, R. A. (1957). *J. Amer. Chem. Soc.* **79,** 5674.
Krigbaum, W. R., and Kügler, F. R. (1970). *Biochemistry* **9,** 1216.
Kronman, M. J. (1967). *Biochim. Biophys. Acta* **133,** 19.
Kronman, M. J. (1968). *Biochem. Biophys. Res. Commun.* **33,** 535.
Kronman, M. J., and Andreotti, R. (1964). *Biochemistry* **3,** 1145.
Kronman, M. J., and Holmes, L. G. (1965). *Biochemistry* **4,** 526.
Kronman, M. J., Andreotti, R. E., and Vitols, R. (1964). *Biochemistry* **3,** 1152.
Kronman, M. J., Blum, R., and Holmes, L. G. (1965a). *Biochem. Biophys. Res. Commun.* **19,** 227.
Kronman, M. J., Cerankowski, L., and Holmes, L. G. (1965b). *Biochemistry* **4,** 518.

Kronman, M. J., Blum, R., and Holmes, L. G. (1966). *Biochemistry* **5,** 1970.
Kronman, M. J., Holmes, L. G., and Robbins, F. M. (1967). *Biochim. Biophys. Acta* **133,** 46.
Larson, B. L., and Hageman, E. C. (1963). *J. Dairy Sci.* **46,** 14.
Larson, B. L., and Jenness, R. (1955). *J. Dairy Sci.* **38,** 313.
Larson, B. L., and Rolleri, G. D. (1955). *J. Dairy Sci.* **38,** 351.
Lyster, R. L. J., Jenness, R., Phillips, N. I., and Sloan, R. E. (1966). *Comp. Biochem. Physiol.* **17,** 967.
Maeno, M., and Kiyosawa, I. (1962). *Biochem. J.* **83,** 271.
Maes, E. D., Dolmans, M., Vincentelli, J. B., and Léonis, J. (1969). *Arch. Int. Physiol. Biochim.* **77,** 388.
McGuire, E. J., Jourdian, G. W., Carlson, D. M., and Roseman, S. (1965). *J. Biol. Chem.* **240,** PC4112.
McGuire, W. L. (1969). *Anal. Biochem.* **31,** 391.
McKenzie, H. A. (1967). *Advan. Protein Chem.* **22,** 55.
Palmer, A. H. (1934). *J. Biol. Chem.* **104,** 359.
Palmiter, R. D. (1969a). *Biochem. J.* **113,** 409.
Palmiter, R. D. (1969b). *Biochim. Biophys. Acta* **178,** 35.
Palmiter, R. D. (1969c). *Nature* **221,** 912.
Polis, B. D., Shmukler, H. W., and Custer, J. H. (1950). *J. Biol. Chem.* **187,** 349.
Préaux, G., and Lontie, R. (1962). *In* "Protides of Biological Fluids: Proceedings of the 9th Colloquium, 1961" (H. Peeters, ed.), p. 103. Elsevier, Amsterdam.
Robbins, F. M., and Kronman, M. J. (1964). *Biochim. Biophys. Acta* **82,** 186.
Robbins, F. M., Kronman, M. J., and Andreotti, R. E. (1965). *Biochim. Biophys. Acta* **109,** 223.
Robbins, F. M., Andreotti, R. E., Holmes, L. G., and Kronman, M. J. (1967). *Biochim. Biophys. Acta* **133,** 33.
Sebelien, J. (1885). *Z. Physiol. Chem.* **9,** 445.
Sen, A. (1968). Private communication.
Sen, A., and Chaudhuri, S. (1962). *Nature* **195,** 286.
Sen, A., and Sinha, N. K. (1961). *Nature* **190,** 343.
Sjögren, B., and Svedberg, T. (1930). *J. Amer. Chem. Soc.* **52,** 3650.
Sørensen, M., and Sørensen, S. P. L. (1939). *C. R. Trav. Lab. Carlsberg, Ser. Chim.* **23,** 55.
Spies, J. R. (1967). *Anal. Chem.* **39,** 1412.
Spies, J. R., and Chambers, D. C. (1949). *Anal. Chem.* **21,** 1249.
Svedberg, T. (1937). *Nature* **139,** 1051.
Svedberg, T., and Pedersen, K. O. (1940). "The Ultracentrifuge," p. 379. Oxford University Press, London.
Szuchet–Derechin, S., and Johnson, P. (1965). *Eur. Polymer. J.* **1,** 271.
Tanahashi, N., Brodbeck, U., and Ebner, K. E. (1968). *Biochim. Biophys. Acta* **154,** 247.
Thompson, M. P. (1965). Unpublished results.
Trayer, I. P., Mattock, P., and Hill, R. L. (1970). *Fed. Proc., Fed. Amer. Soc. Exp. Biol.* **29,** 597.
Turkington, R. W., and Hill, R. L. (1969). *Science* **163,** 1458.
Turkington, R. W., Brew, K., Vanaman, T. C., and Hill, R. L. (1968). *J. Biol. Chem.* **243,** 3382.
Vanaman, T. C., Brew, K., and Hill, R. L. (1970). *J. Biol. Chem.* **25,** 4583.
Watkins, W. M., and Hassid, W. Z. (1962). *J. Biol. Chem.* **237,** 1432.

Weil, L., and Seibles, T. S. (1959). *Arch. Biochem. Biophys.* **84,** 244.
Weil, L., and Seibles, T. S. (1961a). *Arch. Biochem. Biophys.* **93,** 193.
Weil, L., and Seibles, T. S. (1961b). *Arch. Biochem. Biophys.* **95,** 470.
Weil, L., and Seibles, T. S. (1964). *Arch. Biochem. Biophys.* **105,** 457.
Weil, L., and Telka, M. (1957). *Arch. Biochem. Biophys.* **71,** 473.
Wetlaufer, D. B. (1961). *C. R. Trav. Lab. Carlsberg, Ser. Chim.* **32,** 125.
Wetlaufer, D. B. (1962). *Advan. Protein Chem.* **17,** 378.
Wichmann, A. (1899). *Z. Physiol. Chem.* **27,** 575.
Yasunobu, K. T., and Wilcox, P. E. (1958). *J. Biol. Chem.* **231,** 309.
Zittle, C. A. (1956). *Arch. Biochem. Biophys.* **64,** 144.
Zittle, C. A., and DellaMonica, E. S. (1955). *Arch. Biochem. Biophys.* **58,** 31.
Zweig, G., and Block, R. J. (1954). *Arch. Biochem. Biophys.* **51,** 200.

16 ☐ Minor Milk Proteins and Enzymes

M. L. GROVES

I. Introduction 367
II. Minor Milk Proteins. 368
   A. Lactoferrin (Red Protein) 368
   B. Serum Transferrin (Milk) 375
   C. Lactollin 376
   D. Serum Albumin 378
   E. Fat Globule Membrane Proteins 380
   F. Glycoprotein-a 383
   G. Kininogen 384
   H. M-1 Glycoproteins 384
III. Milk Enzymes 385
   A. Nucleases 385
   B. Lactoperoxidase 387
   C. Xanthine Oxidase 396
   D. Lipases and Esterases 399
   E. Amylases 402
   F. Phosphatases 403
   G. Lysozyme 406
   H. Miscellaneous Enzymes 407
   References 411

I. Introduction

More refined methods of fractionation and analysis of proteins have led to a renewed interest in the minor milk proteins. Although they are of minor importance nutritionally, they can affect the stability and flavor of milk since many of them are enzymes and others form complexes with metal ions. Some of the minor proteins are associated with the casein fraction of milk; some of the enzymes are concentrated on the surface of the fat globule membrane. McMeekin (1954) and Whitney (1958) reviewed the early work in this field. Subsequent reports by Jenness and Patton (1959), Garnier (1964), and Corbin and Whittier (1965) have also dealt with many of these proteins. Shahani (1966) listed 19 enzymes reported to be found in normal milk and summarized information concerning these milk proteins. During 1968–1970 there was a tremendous upsurge of interest in the secretory enzymes and proteins. While an attempt is made in this chapter to include the more important of the most recent publications, they are not dealt with in as much detail as the earlier publications.

II. Minor Milk Proteins

A. LACTOFERRIN (RED PROTEIN)

Certain proteins have the unique ability to bind iron at alkaline pH values, producing a salmon-red complex in solution. Schade and Caroline (1944) first recognized the specific iron-binding capacity of an egg white component which was later identified by Alderton *et al.* (1946) as conalbumin. A similar iron-binding protein was found in human plasma (Schade and Caroline, 1946) and has been called β_1-metal-binding globulin, siderophilin or transferrin, the latter name finding more usage in recent years. In 1960, iron-binding proteins were isolated from human milk and were called red protein (Johansson, 1960) and lactotransferrin (Montreuil *et al.*, 1960). For the purpose of this discussion the transferrins in blood and milk are identical. As has been pointed out in Section III, Chapter 2, Volume I, they are referred to as serum transferrin (blood) or serum transferrin (milk) to designate their source. The other iron-binding protein found in milk but absent in blood is called (in this book) lactoferrin.

Sørensen and Sørensen (1939) noted in cow milk a red protein which they partially purified from the whey fraction. In separating alkaline phosphatase from buttermilk, Morton (1953) obtained a red fraction as a by-product. Also, Polis and Shmukler (1953) and Morrison *et al.* (1957) partially purified a red protein while isolating lactoperoxidase from rennet-treated skim milk.

Several methods are now available for isolating lactoferrin from bovine milk in relatively pure form. It can be separated from the acid-precipitated casein fraction (Groves, 1960) by acid extraction of the casein, followed by ammonium sulfate fractionation and finally by elution from a

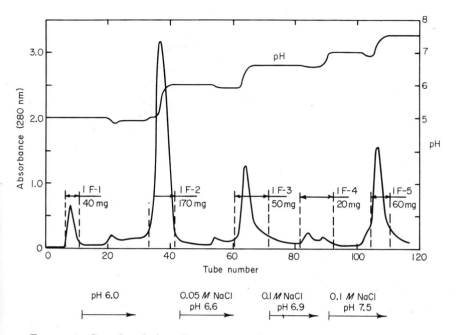

FIGURE 1. Stepwise elution diagram of a whey protein fraction from phosphocellulose: 1F-2—IgG globulin and glycoprotein-a (Section II.F); 1F-3, 1F-4—lactoperoxidase and ribonuclease (Section III.A.1, B); 1F-5—lactoferrin (Groves, 1965). Buffer systems: pH 6.0–6.9, 0.1 M phosphate (Na); pH 7.5, 0.2 M phosphate (K).

DEAE-cellulose column. Lactoferrin is also found in the whey fraction of milk (Groves, 1965); it can be concentrated by chromatography on a DEAE-cellulose column, which retains most of the whey proteins while the basic proteins such as lactoperoxidase and the red protein are eluted near the front. The proteins in this fraction can then be resolved by a phosphocellulose column as shown in Fig. 1. The first major peak (1F-2) contains immunoglobulins and glycoprotein-a (Section II.F). Lactoperoxidase and ribonuclease are found in the combined 1F-3 and 1F-4 fractions while the red protein, lactoferrin (1F-5), is eluted at the higher salt concentration and pH. Yields of about 18 mg of lactoferrin per liter of milk can be expected from the casein fraction. The amount of this protein in the whey fraction has not been estimated.

The original Morrison method (Morrison et al., 1957) for the isolation of lactoperoxidase and a modified procedure (Morrison and Hultquist, 1963), both of which also yield a red protein, involve the direct extraction of these proteins from skim milk with a weak cation-exchange resin. The adsorbed proteins are eluted and are then fractionated by column chromatography on cation-exchange resin. The red protein is more firmly bound and requires ammonium hydroxide for elution.

Szuchet-Derechin and Johnson (1965a,b) describe a method for isolating from the "albumin" fraction of pooled milk two red proteins, F1 and F2, which are very similar to lactoferrin and serum transferrin, respectively. The albumin fraction is the fraction that remains in solution after the cream, casein, immunoglobulins and several other components have been precipitated from milk by the addition of sodium sulfate to the extent of 200 g/liter (Aschaffenburg and Drewry, 1957). The F1 protein (lactoferrin) in 0.005 M phosphate buffer, pH 7–8, is not retained by a DEAE-cellulose column, while the F2 fraction (transferrin) and other proteins are adsorbed. At pH 6.0 and higher salt concentration the F2 fraction is eluted. By controlling the relative amounts of total protein and adsorbent to give the required competition among the different proteins for the adsorbent, the less strongly adsorbed F2 protein can be displaced and remains in solution with the F1 protein. In this way, proteins other than the F1 and F2 proteins remain on the column. Finally, F1 and F2 proteins are resolved by chromatography on a DEAE-cellulose column equilibrated at pH 7.0. Szuchet-Derechin and Johnson (1966a) further resolved the red protein on DEAE-cellulose (initially at pH 9) into several fractions, some of which were artifacts while others gave reproducible elution patterns on rechromatography. The first peak, which was the major fraction, was designated *red protein A*. These fractions showed similar absorption spectra and extinction coefficients but differed in their sedimentation coefficient vs. concentration curves; also, small differences were noted in their starch-gel electrophoretic behavior. A sample of lactoferrin from an individual cow was examined in which fewer fractions were obtained; however, the results were not conclusive since the fractions were not analyzed in detail. Since lactoferrin isolated from the milk of individual cows has been shown to give a number of bands on gel electrophoresis (Groves et al., 1965), some fractionation on DEAE-cellulose might be expected.

It is of interest to note that Gordon and his associates (1962) isolated a colorless iron-binding protein from cow milk using the Morrison procedure (Morrison et al., 1957) followed by chromatography on hydroxyapatite. The colorless protein is free of iron but is otherwise similar to the red protein in physical properties and chemical composition. The iron-binding protein is present in milk in two forms—the iron-complexed red protein and the iron-free protein. The same is true for human blood in which only a third of the transferrin is complexed with iron (Laurell, 1960). Iron-binding proteins in rat milk are also found to be only partially saturated; however, it is possible that they are derived from the blood transferrin (Ezekiel, 1965). Rabbit milk contains a relatively high concentration of unsaturated iron-binding protein which appears to differ from the serum transferrin only in its sialic acid content (Jordan et al., 1967; Baker et al., 1968).

The red protein of cow milk is distributed between the casein, whey, and, apparently, the fat fractions of milk. This distribution is probably a consequence of its ability to form protein complexes which, in some instances, are strong enough to change its chromatographic behavior (Groves, 1965).

Groves (1960) observed a single symmetrical peak for bovine lactoferrin in moving-boundary electrophoretic patterns at 0.1 ionic strength in the pH range of 4.5–10.0. Szuchet-Derechin and Johnson's value (1965b) of pH 8.0 ± 0.2 for the isoelectric point of their preparation is in good agreement with Groves' earlier value of pH 7.8. However, at alkaline pH they found a small amount of faster-moving material. Zone electrophoresis of lactoferrin at an alkaline pH value on polyacrylamide gels results in a number of closely related sharp bands which vary depending on the cow. This led Groves et al. (1965) to suggest that polymorphism in the red protein is genetically controlled. Three closely related bands and a minor slow-moving zone which might be due to aggregation are obtained on disc electrophoresis at pH 4.3 (Groves, 1965). The electrophoretic pattern of protein at this pH does not vary with the cow.

A molecular weight of 86,100 daltons has been calculated for lactoferrin (Groves, 1960) from sedimentation and diffusion coefficients of $s^o_{20,w}$ = 5.55 S and D = 5.75 × 10^{-7} cm²/sec, respectively, and a partial specific volume of 0.725 ml/g. The iron-binding protein studied by Gordon et al. (1962) was found to bind 0.12% iron, a value in agreement with the iron content of lactoferrin (Groves, 1960). This is equivalent to a minimum molecular weight of 46,500 daltons or a molecular weight of 93,000 daltons based on two moles of iron per mole of protein. Lactoferrin, at a concentration of 10 g/liter has a sedimentation coefficient of 3.0 S at pH 2.2 compared to a value of 5.2 S at pH 7.0 (Groves, 1963). This is consistent with the finding of Roberts et al. (1966) that the sedimentation coefficient of human transferrin decreases below pH 5.0 from 4.9 to 2.8 S at pH 2.1, yet no change in molecular weight is observed.

For their red protein A, Szuchet-Derechin and Johnson (1966a,b) obtained the molecular weight of 93,000 ± 3000 daltons ($s^o_{20,w}$ = 5.73 S, D = 5.63 × 10^{-7} cm²/sec, $\bar{v}$ = 0.736 ml/g). The molecule behaves like a compact prolate ellipsoid of axial ratio 6 and is only weakly hydrated. The molecular weight of 93,000 daltons is in good agreement with the molecular weight based on iron analysis.

The composition of the red protein is summarized in Table I. Alanine

TABLE I

AMINO ACID COMPOSITION OF SOME BOVINE MILK PROTEINS

Amino acid	Lactoferrin residues per 86,100 mol wt[a]	Lactollin residues per 43,000 mol wt[b]	Lactoperoxidase residues per 77,500 mol wt[c]	Xanthine oxidase residues per 275,000 mol wt[d]
Aspartic acid	64	40	71	199
Threonine	34	7	28	169
Serine	40	30	30	154
Glutamic acid	66	43	60	243
Proline	32	34	42	130
Glycine	48	11	41	196
Alanine	64	4	40	179
½ Cystine	36	8	16	62
Valine	44	18	29	163
Methionine	5	0	12	47
Isoleucine	16	22	28	120
Leucine	62	29	68	206
Tyrosine	20	22	15	61
Phenylalanine	26	15	31	118
Lysine	49	33	33	162
Histidine	10	14	14	55
Arginine	36	18	39	105
Tryptophan	15	8	16	9
Hexose	22			
Hexosamine	11		26	
N-Acetylneuraminic acid	1		0	

[a] Gordon et al. (1963).
[b] Groves et al. (1963).
[c] Rombauts et al. (1967).
[d] Bray and Malmström (1964).

is the only N-terminal amino acid and amounts to approximately one residue per 86,100 molecular weight (Gordon et al., 1963).

Comparative amino acid analysis between bovine lactoferrin and bovine transferrin (blood) shows significant differences (Gordon et al., 1963). The proteins also differ in electrophoretic mobility and chromatographic and sedimentation behavior. The bovine lactoferrin is not related immunologically to transferrin (blood) or to any other bovine serum component (Szuchet-Derechin and Johnson, 1962). Human lactoferrin and bovine milk red protein also differ immunologically (Blanc and Isliker, 1961b) and in amino acid composition (Blanc et al., 1963).

Human transferrin and human lactoferrin show differences in their contents of carbohydrate and amino acids and in their peptide maps (Montreuil et al., 1965; Spik and Montreuil, 1966). Johansson (1969) crystallized lactoferrin from a fraction of human milk prepared by chromatography on CM-Sephadex C-50. The iron content was 0.137–0.147 g/100 g, indicating a minimum molecular weight of 38,000–41,000 daltons, while permeation studies on Sephadex G-150 were indicative of a molecular weight of 80,000 daltons (i.e. two Fe/molecule). Schade et al. (1969) compared various methods of isolating lactoferrin, as well as its physical properties as reported by different workers. Using a procedure similar to that of Szuchet-Derechin and Johnson (1966a), they found an $s^\circ_{20,w}$ of 8.8 S for lactoferrin before isolation from human milk and a value of 5.3 S after isolation. They concluded that it is bound to another protein in human milk.

It is possible that transferrin and lactoferrin are composed of two similar subunits since they contain two moles of iron per mole of protein; in human transferrin, the two iron-binding sites are apparently separate and independent (Aasa et al., 1963; Aisen et al., 1966). On the basis of physical studies on the reduced and alkylated protein, Jeppsson (1967) suggests that human transferrin consists of two subunits, each with a molecular weight of about 40,000 daltons. Two subunits are also suggested by the finding that peptide maps of reduced and alkylated transferrin contain only half the expected number of tryptic peptides. Furthermore, evidence from C-terminal amino acid analysis indicates two subunits (Spik et al., 1969). Results from peptide maps of oxidized rabbit transferrin are consistent with this finding (Baker et al., 1968). However, Bezkorovainy and Grohlich (1967) and Greene and Feeney (1968) have made physical studies on the reduced and alkylated transferrin from human and rabbit serum and they conclude that the molecule consists of a single polypeptide chain. Mann et al. (1970) also found only one polypeptide chain for human serum transferrin on the basis of sedimentation-equilibrium determinations and gel filtration in guanidine HCl.

It has been suggested that the synthesis of proteins similar to milk iron-binding protein is not limited to the mammary gland. An iron-binding protein resembling human lactoferrin by immunoelectrophoresis has been reported for saliva, semen, and in bronchial and other external secretions (Masson et al., 1965, 1966a). Loisillier et al. (1967) demonstrated the presence of lactoferrin in the stomach and colon tissues and showed that cancerous tumors of these tissues contain more lactoferrin than normal tissues. Masson et al. (1969b) found it in the intestine, genital tract and kidneys and more recently in leucocytes (Masson et al., 1969a). The occurrence of both red protein and lactoperoxidase (cow milk) in bovine salivary and lacrimal glands has been demonstrated by immunodiffusion analysis (Morrison and Allen, 1966). The presence of these proteins in lacrimal glands has been confirmed by their isolation. The wide distribution of lactoferrin in biological fluids is discussed by Masson et al. (1966a,b; Masson and Heremans, 1967), who suggest that the iron-binding property of this protein is of value in defense of the epithelial surface against infection. Also, Reiter and Oram (1967) list lactoferrin among other bacterial inhibitors found in milk and other biological fluids. Oram and Reiter (1968) later studied the bacteriostatic effects of iron-saturated and unsaturated lactoferrin, the saturated form being far less toxic to bacteria. Blanc (1967) also found that lactoferrin acts as a bacteriostatic agent towards certain pathogenic organisms requiring iron for growth and reproduction. He studied the close relationship between the physiological role of lactoferrin and iron metabolism, and concluded that during the formation of milk the lactoferrin being synthesized absorbs its iron directly from the blood transferrins. During the digestion of milk and proteolysis of milk proteins the gastric pH favors dissociation of the iron protein. Nevertheless, lactoferrin remains sufficiently unchanged so that the iron complex can reoccur in the duodenum. Jordan and Morgan (1969) found that the synthesis of lactoferrin (incorporation of ^{14}C-leucine in tissue homogenates) in rat mammary gland decreased from a high value at the beginning of lactation to a minimum on the fifth day and then increased to a maximum on the twentieth day when the rate of synthesis was greater than in liver slices.

The iron-complexed lactoferrin is salmon-red with a broad absorption maximum near 470 nm. Titration of the colorless iron-binding protein (Gordon et al., 1962) with a standard solution of iron at alkaline pH to the point of maximum absorbancy gives a product with a well-defined absorption maximum that corresponds to two atoms of iron per molecule of protein. This is also found for human blood transferrin (Aasa et al., 1963), conalbumin (Warner and Weber, 1953; Fraenkel-Conrat and Feeney, 1950), and human lactoferrin (Johansson, 1960). Montreuil et al. (1960) report an unusually high value of six iron atoms per molecule of human lactoferrin. Wishnia et al. (1961) found that three tyrosines are chelated to one ferric atom in the iron–conalbumin complex. With lactoferrin, the absorption for tyrosine is apparently affected by the iron. Absorptivity at 280 nm for lactoferrin is 15.1, a value considerably higher than the 12.0 value for the iron-free apoprotein (Gordon et al., 1962); this is consistent with the difference in absorptivity for the iron-complexed and iron-free conalbumin (Glazer and McKenzie, 1963; Warner and Weber, 1953). Windle et al. (1963) found the electron paramagnetic resonance spectra for iron–protein complexes of human transferrin (blood), human lactoferrin, and bovine lactoferrin to be essentially the same. Warner and Weber (1953) found that bicarbonate is involved in the formation of the metal complex of conalbumin (for a discussion on this and other aspects of the structure of conalbumin, lactoferrins, etc., see the review of Feeney and Komatusu, 1966). Masson and Heremans (1968) confirmed that bicarbonate is also involved in the formation of the metal complexes of human lactoferrin, one bicarbonate being taken up per iron(III) or copper(II). Other studies on transferrin and conalbumin by electron paramagnetic resonance indicate, however, that specific iron binding can occur in the absence of bicarbonate (Aisen et al., 1967). The iron complex of lactoferrin in both human (Montreuil et al., 1960; Johansson, 1960) and bovine milk is apparently more stable at acid pH values than the corresponding complex of transferrin in blood. The iron bound to human transferrin (blood) begins to dissociate below pH 6 and the protein loses all color at about pH 4 (Schade et al., 1949), while lactoferrin does not completely lose its color until pH 2 is reached. Electrophoretic comparisons by Warner and Weber (1951) on the iron-complexed and iron-free conalbumin show the complex to be more negatively charged than the metal-free protein. Also, a similar relationship is found for human serum transferrin (Inman, 1956; Roop, 1963). However, with lactoferrin the reverse is found. The apoprotein prepared by the dissociation of iron at pH 2 is more negatively charged at pH 10 than the iron-complexed lactoferrin. This is also shown to be true by a comparison of the mobility of the colorless iron-binding protein (Gordon et al., 1962) with lactoferrin at pH 8.4. The explanation of these differences in mobility at alkaline pH and also in stability of the iron complex of transferrin and lactoferrin at acid pH will require further work on the nature of the iron-binding sites.

It has been suggested that since the human milk iron-binding protein has a greater affinity for iron than serum transferrin does, it may be involved in the transfer of iron through the mammary gland (Blanc and Isliker, 1961a). According to Blanc and Isliker (1963), when the exchange of radioactive iron among the proteins was studied by equilibrium dialysis, the affinity of the iron-binding proteins for iron decreased in the following order: human lactoferrin > bovine lactoferrin > human serum transferrin. Studies of iron-binding proteins in the marsupial, quokka, and rat showed no exchange of radioactive iron when samples of plasma and milk were incubated (Ezekiel, 1963).

B. SERUM TRANSFERRIN (MILK)

A small amount of transferrin, which appears to be identical to blood serum transferrin by electrophoresis (Szuchet-Derechin and Johnson, 1962; Groves, 1965) and immunological techniques (Szuchet-Derechin and Johnson, 1962; Gahne, 1961), is found in bovine milk. Apparently it is present in colostrum in somewhat larger amounts (Gahne et al., 1960). Zone electrophoresis of bovine transferrin (blood) shows polymorphism which is genetically controlled (Smithies and Hickman, 1958; Ashton, 1958). Disc electrophoretic patterns of the transferrin in milk also reflect these differences, as illustrated in Fig. 2 which compares the pattern of the transferrin from blood serum and the milk of individual cows. Three transferrin bands can be seen in the patterns from the blood and milk of one cow, and four bands can be seen in the case of the second animal. Milk transferrin has been partly purified from the albumin fraction by chromatography on DEAE-cellulose or displacement chromatography (Szuchet-Derechin and Johnson, 1965a,b). Chromatography of whey proteins on DEAE-cellulose and then on phosphocellulose has also been used to isolate transferrin from bovine milk (Groves, 1965). Very few physical or chemical measurements have been made on transferrin isolated from milk, since blood is a better source of the protein. A crude fraction of milk transferrin gave an absorption spectrum similar to that of lactoferrin; its mean sedimentation coefficient was about 5 S (Szuchet-Derechin and Johnson, 1965b). A preparation of transferrin from bovine blood at a concentration of 10 g/liter showed a sedimentation coefficient of 4.7 S. A mobility of -2.9×10^{-5} cm²/sec/V was obtained on electrophoresis in the Tiselius apparatus at pH 8.5 (Gordon et al., 1963). Comparative amino acid analysis of bovine transferrin and lactoferrin showed significant differences, as mentioned in Section II.A.

Baker et al. (1968) isolated transferrin in crystalline form from the milk and blood of a rabbit. The two proteins appear to be identical by molecular weight, light-absorption spectra, amino acid composition, peptide mapping of tryptic digests and by double diffusion in agar against specific antibodies. They differ in electrophoretic mobility because of differences in the N-acetylneuraminic acid content. The finding that rabbit milk contains a relatively large amount of transferrin appears to be unique since the major iron-binding proteins in bovine and human milks are lactoferrins. Further physiological studies and comparison of the rabbit milk and serum transferrins have been reported by Baker et al. (1969).

C. LACTOLLIN

A crystalline protein, lactollin, is found to be associated with the red protein in very small amounts when the red protein is isolated from bovine milk. It appears to be present in colostrum in a significantly higher amount than in normal milk. Chromatography on DEAE-cellulose of the red fraction, obtained by acid extraction of casein followed by ammonium sulfate

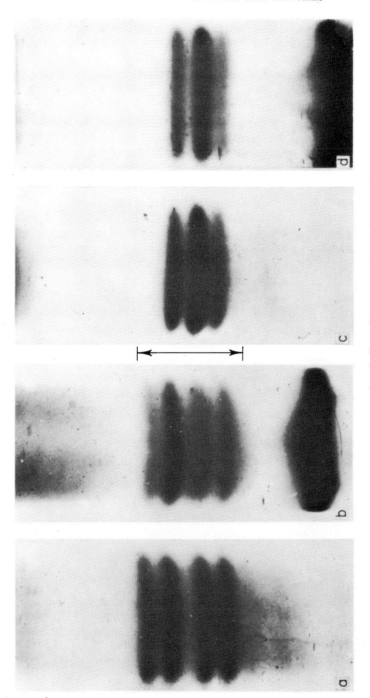

FIGURE 2. Gel electrophoretic comparisons of transferrins (indicated by vertical arrow) found in milk and blood of individual cows: (a) cow A, blood serum; (b) cow A, milk whey fraction; (c) cow B, blood serum; (d) cow B, milk whey fraction (Groves, 1965).

fractionation, gives a small fraction immediately following the red peak. This fraction crystallizes on standing overnight at 3°C (Groves, 1960). The crystalline protein has minimum solubility at pH 8 and is soluble on both the acidic and alkaline side of this pH at low salt concentrations. Typical crystals of lactollin are shown in Fig. 3. The yield of this protein amounts to about 2 mg/liter of milk.

The amino acid composition of lactollin is shown in Table I. Lactollin has no methionine and only small amounts of alanine and cystine. The high proportion of aromatic amino acids is reflected in the absorptivity of 16.5 at 280 nm. A minimum molecular weight based on alanine and cystine contents and then multiplied by four gives a molecular weight of 43,000 daltons, which is consistent with the value 43,000 ± 5000 calculated from the sedimentation pattern.

Lactollin has an apparent isoelectric point of pH 7.1 in 0.1 ionic strength buffer. It shows a single peak by moving-boundary electrophoresis at pH 9.5 but complicated patterns at acid pH values; starch-gel electrophoresis in 5 M urea at pH 3.7 shows only one band.

Ultracentrifuge measurements on lactollin at pH 5.0 and 25°C give a sedimentation coefficient of 3.21 S, which is unchanged on dilution. At pH 10.1, 6°C and a concentration of 10 g/liter, a $s^\circ_{20,w}$ value of 3.51 S is obtained. Lactollin undergoes slow irreversible polymerization at pH 5.0 in 0.1 ionic strength acetate. After storage at 3°C for several days, ultracentrifugation shows rapidly sedimenting boundaries together with slow-moving, fast-spreading boundaries. This suggests the simultaneous occurrence of aggregating and disaggregating phenomena (Timasheff, 1964).

D. SERUM ALBUMIN

Cow milk serum albumin amounts to about 1% of the total proteins and 6% of the whey proteins as determined by electrophoretic area analysis (Rolleri et al., 1956; Larson and Kendall, 1957). The amount of serum albumin in cow milk is highest on the day of parturition.

On salt fractionation of the milk whey proteins, serum albumin is found in a fraction rich in α-lactalbumin and β-lactoglobulin. It has been isolated from this fraction in crystalline form by ammonium sulfate and alcohol fractionation and is identical to blood serum albumin by physical, chemical (Polis et al., 1950), and immunological measurements (Coulson and Stevens, 1950).

In studying the levels of incorporation of radioactive carbon in milk proteins, Larson and Gillespie (1957) found evidence that milk serum al-

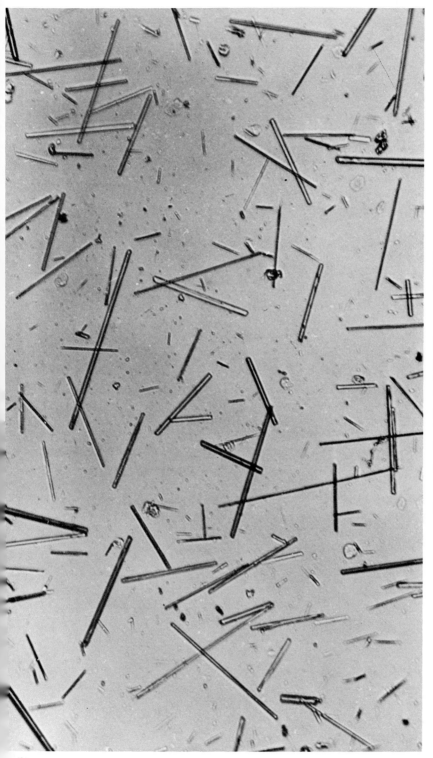

FIGURE 3. Crystalline lactollin (Groves et al., 1963).

bumin, together with the immunoglobulins and γ-casein, enters the milk preformed from the blood serum.

E. Fat Globule Membrane Proteins

The nature of the interphase between fat globules and milk plasma is of major importance in milk chemistry and has been the subject of several review articles (King, 1955; Jenness and Patton, 1959; Brunner, 1965; Prentice, 1969). Studies by Palmer (1944) and his colleagues established that the stability of the dispersed fat globules is controlled by aggregates of colloidal particles oriented at the surface of milk fat globules. These particles, often called "membrane" material, are made up of protein–phospholipid complexes, together with a neutral high-melting glyceride. The protein part of this material consists of a number of individual proteins, some of which are present in small quantities, such as the enzymes, xanthine oxidase, phosphatase and aldolase. Isolation of the membrane material is usually accomplished by churning cream which has previously been washed several times with water to remove the milk plasma proteins, and then separating the membrane proteins from the resulting buttermilk and butter serum. Some of the enzymes loosely associated with the membrane are also removed from the membrane material on washing the cream. Washed cream contains about 0.5–0.9 g membrane protein and 0.2–0.4 g phospholipids per 100 g fat (Jenness and Palmer, 1945). The membrane proteins are sensitive to heat and liberate sulfhydryl groups when cream or buttermilk are heated momentarily to 90°C (Townley and Gould, 1943).

Morton (1954) found that the lipoprotein complex of the globule membrane is similar in many respects to microsomes from the lactating mammary gland and other tissues; they suggested that the microsomes in milk are derived from the mammary gland during the normal secretory process. The microsomes are brown lipoprotein particles containing 22% lipids, largely phospholipids, nucleic acid and a number of enzymes (alkaline phosphatase, xanthine oxidase, diaphorase, and DPN-cytochrome c reductase). Swope and Brunner (1965) have confirmed the presence of ribonucleic acid in the fat globule membrane and support the conclusion that microsomes are located in the fat globule material. Evidently the ribonucleic acid is protected by other membrane components from the action of the ribonuclease in milk (Bingham and Zittle, 1964). Ribonuclease and phosphodiesterase activities have also been detected in the membrane microsome fraction (Matsushita et al., 1963); the phosphodiesterase has been solubilized from the microsomes and partly purified (Matsushita et al., 1965).

Bargmann and Knoop (1959) examined the lactating cell with the electron microscope. They suggested that the fat droplet is encased with plasma membrane when it reaches a critical size at the apex of the cell. Finally, it is completely enclosed by the membrane as it is pinched off from the cell. Dowben *et al.* (1967) and Patton and Fowkes (1967) have presented more evidence in support of this theory. Dowben *et al.* (1967) found that most of the enzyme activities associated with fractions of other tissues which contain membranous material are present in the fat globule membrane. These include alkaline phosphomonoesterase, acid glucose-6-phosphatase, ATPase activated by Mg^{2+} and Na^+–K^+–Mg^{2+}, true cholinesterase, xanthine oxidase and aldolase. They also found that antisera prepared by immunization of rabbits with fat globule membranes produce agglutination and hemolysis of bovine erythrocytes. They concluded that this result supports the view that the globule membrane is a cell-wall derivative. Patton and Fowkes (1967) found that the fat droplets within the cell are relatively devoid of certain membrane constituents such as phosphatidylethanolamine and carotenoids characteristically present in the surface coat of secreted milk fat globules. The origin of these components is thought to be the plasma membrane of the cell, as estimated by a combination of compositional analysis and turnover studies with ^{14}C-fatty acid. Milk proteins appear to be secreted from the vacuoles through the plasma membranes, making the membranes around the vacuole available to coat more fat particles. Stewart and Irvine (1969) have concluded from electron microscopic studies that the fat globule is enveloped by a loose interfacial membrane-like zone and, on secretion, this is enclosed by the cell plasma membrane which then disintegrates, leaving the original intracellular membrane in a more condensed state. Further electron microscopic observations have been made by Hood and Patton (1968) and Keenan *et al.* (1969). The latter have found distinct morphological differences between the plasma membrane of the lactating mammary gland cell and milk fat globule membrane. The plasma membrane was vesicular, while the globule membrane had a plate-like structure. They also concluded that the globule membrane is derived from the plasma membrane with some structural rearrangement. Brunner *et al.* (1969) have proposed a model for fat globule membrane consisting of a protein matrix with adsorbed micelle-like lipoproteins.

Hayashi and Smith (1965) found that water-soluble lipoproteins were released from the membranous fractions of the intact fat globules by sodium deoxycholate. On centrifugation, 45% of the total protein nitrogen of the membrane was solubilized. In a control experiment without deoxycholate, the cream was churned, and the buttermilk, after centrifugation, yielded soluble lipoproteins in smaller amounts. When deoxycholate was added to the buttermilk, soluble lipoproteins were obtained in amounts comparable to those released from intact fat globules. The lipoproteins obtained by either method were similar in gross chemical composition and sedimentation behavior. Hayashi and Smith (1965) proposed a fat globule membrane model that contains two types of lipoprotein complexes, approximately equal in amount and classified by their solubility. The water-soluble lipoproteins are thought to be adsorbed on the water-insoluble lipoprotein complex which borders the triglyceride core. Xanthine oxidase and alkaline phosphatase are found principally with the soluble lipoproteins (Hayashi et al., 1965).

Chien and Richardson (1967a) fractionated the fat globule membrane into two parts of about equal amount. An outer lipoprotein fraction was obtained from washings of the cream and an inner lipoprotein fraction was obtained by melting and washing the butter after the cream phase inverted. These results are compatible with the idea of an outer and inner lipoprotein layer for the fat globule membrane, as proposed by Hayashi and Smith (1965). Both fractions were further subdivided by centrifugation into pellet and supernatant fractions. The pellet fraction was high in protein and low in lipid, while the supernatant was low in protein and high in lipid. The amino acid composition of the various fractions after lipid extraction was uniform and was in general agreement with the analyses of Herald and Brunner (1957). Hexosamine and ribonucleic acid were distributed throughout the membrane fractions, but the iron distribution was highest in the outer membrane layer (Chien and Richardson, 1967b).

The membrane protein freed from lipids consists of approximately equal portions of water-soluble and water-insoluble fractions (Brunner, 1962; Herald and Brunner, 1957). The soluble fraction before lipid extraction contains about 6.5% carbohydrate and has a sedimentation coefficient of from 8 to 17 S. With the removal of lipids there is a corresponding reduction in the sedimentation coefficient, which reflects the role played by the lipids in the state of aggregation of the lipoproteins. Amino acid values for the two fractions are similar, with the exception of larger amounts of arginine, aspartic acid, glutamic acid, methionine, and valine in the insoluble fraction. The soluble fraction shows a single peak by electrophoresis (free-boundary) at alkaline pH but several components at acid pH. It can be dissociated by sodium dodecylsulfate into smaller units. Guanidine hydrochloride with 2-mercaptoethanol is effective in dissociating both soluble and insoluble fractions, yielding molecular weights for the smallest components of 20,000 and 50,000 daltons, respectively (Harwalkar and Brunner, 1965).

Jackson et al. (1962) isolated a soluble membrane mucoprotein by a

16. MINOR MILK PROTEINS AND ENZYMES   383

large-scale preparation following the method of Herald and Brunner (1957). The protein shows a single symmetrical peak by free-boundary electrophoresis at all pH values except 2.0 and an isoelectric point of pH 4.2 in 0.1 ionic strength buffer. It shows a strong tendency to aggregate in high gravitational fields but has a sedimentation coefficient of 4.8 S when determined at reduced centrifugal speeds. The membrane protein has a molecular weight of 123,000 daltons and contains 2% lipid, 5.5% hexose, 3.9% hexosamine, and 4.5% sialic acid. It is highly antigenic and differs immunologically from all other recognized milk proteins (Coulson and Jackson, 1962).

Swope and Brunner (1968) reassessed the isolation procedures and mineral composition of the fat globule membrane of bovine milk. They found that molybdenum, iron and copper are the principal minerals present in the globule membrane and suggest that the presence of molybdenum is due to xanthine oxidase.

Swope et al. (1968) isolated and partially characterized a milk fat globule membrane glycoprotein. It has 7.8% hexose and a high concentration of serine and threonine.

The relationship of various aspects of the fat globule membrane to some milk-processing problems have been considered by Cheeseman and Mabbitt (1968), Chien (1968), and Copius Peereboom (1969b).

F. GLYCOPROTEIN-a

A new glycoprotein-a has been isolated from milk (Groves and Gordon, 1967). It is prepared from the whey proteins by chromatography on DEAE-cellulose and phosphocellulose. The glycoprotein-a and IgG globulins are eluted from phosphocellulose in one peak (Fig. 1, 1F-2; Section II.A). The glycoprotein and IgG globulins are then separated on a Sephadex G-200 column. The glycoprotein-a shows a single band by gel electrophoresis at pH 4.3 and a number of closely spaced bands at alkaline pH values. It has a minimum molecular weight of about 48,000 daltons based on the presence of a single methionine residue, and it contains 3.12% hexose. Smith (1946) found that the immunoglobulins amount to 10% of the total proteins in normal milk whey as determined by free-flowing electrophoresis at pH 8.5. Electrophoretic determinations for the glycoprotein-a and IgG globulin fraction eluted from phosphocellulose indicate that a significant amount of the protein corresponding in mobility to the slower immunoglobulin is contributed by glycoprotein-a. Yields of glycoprotein-a and IgG globulin are approximately 50 and 30 mg, respectively, per liter of milk.

Although glycoprotein-a is present in milk in the free form, Butler *et al.* (1968) reported that some glycoprotein-a is bound to an immunoglobulin of bovine milk and colostrum; they proposed that the immunoglobulin is secretory IgA and that glycoprotein-a is its "secretory piece" (Butler, (1969). Mach *et al.* (1969) have also identified a secretory piece in bovine colostrum IgA and found that the free form of the secretory piece is present in both colostrum and mature milk.

G. KININOGEN.

Cow milk contains a protein, kininogen, which when incubated with trypsin or snake venom, releases a material with the kinin-like ability to contract smooth muscle (Leach *et al.*, 1967). It is found in the whey fraction of milk and has been concentrated more than 40-fold by chromatography on DEAE-cellulose.

H. M-1 GLYCOPROTEINS

In human serum, acid glycoproteins that are negatively charged at pH 4.5 have been designated M-1 and M-2 fractions in order of their decreasing acidity. Two acid glycoprotein fractions of bovine serum have been isolated and they too show electrophoretic mobilities corresponding to the M-1 and M-2 human fractions. Bovine milk and colostrum also contain M-1 acid glycoproteins, which Bezkorovainy (1965, 1967) has isolated and characterized (see also Chapter 10). These glycoproteins contain several N-terminal amino acids and in some instances, show several bands on gel electrophoresis. He concludes that they represent a family of closely related molecular species with an average molecular weight of 10,000 daltons. The M-1 glycoproteins obtained from colostrum contain galactose, glucosamine, galactosamine and sialic acid. Bezkorovainy and Grohlich (1969) further fractionated the colostrum M-1 glycoproteins and isolated one fraction (containing 28.4% carbohydrate) with a molecular weight of 7200 daltons and a heavier glycoprotein of 12,000 daltons. The heavier glycoprotein contains 39.0% carbohydrate and shows no absorption maximum between 240–300 nm. Both proteins have relatively large amounts of glutamic acid, threonine and proline, but no cystine or tryptophan. Tyrosine, arginine and histidine are absent in the larger glycoprotein.

III. Milk Enzymes

A. NUCLEASES

1. Ribonuclease (E.C. 2.7.7.16)

Ribonucleases (RNases) are enzymes that catalyze the hydrolysis of phosphodiester linkages in ribonucleic acid. Although they are widely distributed in nature, only the RNase isolated from bovine pancreas has been thoroughly studied (for a review of this subject see Scheraga and Rupley, 1962, and Shahani, 1966).

Since RNase activity in cow milk is high compared with that of other body fluids (Bingham and Zittle, 1962), milk appears to be a good source of the enzyme. RNase remains in the whey fraction of milk after the casein is precipitated. Isolation of ribonuclease from the whey is accomplished by ammonium sulfate precipitation, acid precipitation and carboxylic acid resin fractionation (Bingham and Zittle, 1964). Two fractions with enzymatic activity are eluted from the resin—a minor peak, ribonuclease B, followed by a major peak, ribonuclease A (Fig. 4). Yields of

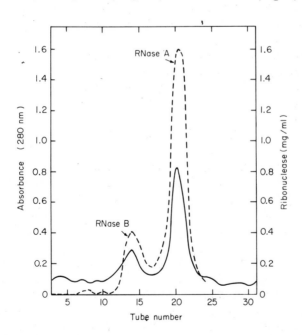

FIGURE 4. Chromatography of milk ribonuclease on IRC-50. Solid line, absorbance at 280 nm; dashed line, ribonuclease activity (Bingham and Zittle, 1964).

the A and B fractions are 148 and 38 mg, respectively, from 95 liters of milk. Milk RNase A is identical or very similar to pancreatic RNase A in amino acid composition, chromatographic behavior, specific activity and electrophoretic mobility. Milk RNase A is serologically identical to pancreatic RNase A (Coulson and Stevens, 1964), and the hydrolyzing specificity of a partially purified milk RNase is also quite similar to that of pancreatic RNase (Ibuki et al., 1965).

The final purification of RNase B has been accomplished by gel filtration on Sephadex G-75 and by chromatography on a carboxylic acid resin (Bingham and Kalan, 1967). It has an amino acid composition and specific activity identical to milk RNase A and pancreatic RNase A and B. In contrast to RNase A, it contains carbohydrates. Milk RNase B contains 4.2% hexosamine (glucosamine and galactosamine) and 5.2% mannose. It differs from pancreatic RNase B, which has 2.2% glucosamine (no galactosamine) and 5.7% mannose (Plummer and Hirs, 1963).

Ribonuclease has also been isolated from cow milk by chromatography of the whey proteins on DEAE-cellulose and phosphocellulose columns. Two enzymes, RNase and lactoperoxidase, are eluted from the phosphocellulose (Fig. 1, 1F-3, 1F-4; Section II.A) and can be resolved on a Sephadex G-200 column (Groves, 1966). The purified milk RNase obtained from Sephadex fractionation has been crystallized. It shows a slightly lower activity than crystalline pancreatic ribonuclease. The amino acid composition is in agreement with values obtained for milk RNase A (Bingham and Zittle, 1964) and pancreatic RNase A (Plummer and Hirs, 1963). Dalaly et al. (1968) have isolated human milk ribonuclease.

2. *Phosphodiesterase (E.C. 3.1.4.1)*

Although phosphodiesterase activity is found in skim milk (Bingham and Zittle, 1962), a major portion of the enzyme is concentrated in the microsome fraction of the fat globule membrane (Matsushita et al., 1963, 1965). When milk is fractionated according to the method of Bailie and Morton (1958), the microsomes are sedimented from the buttermilk fraction and then washed by resuspension and centrifugation. The phosphodiesterase is separated from the microsomes by extraction with a mixture of water and *tert*-amyl alcohol followed by centrifugation, ammonium sulfate fractionation, and finally DEAE-cellulose chromatography.

The enzyme has a specific activity 1500 times that of the original milk. It is inhibited by ethylenediaminetetraacetate but reactivated by the addition of magnesium and calcium ions; reducing agents also inhibit the activity.

B. Lactoperoxidase

Lactoperoxidase (E.C. 1.11.1.7) was first isolated from milk in crystalline form by Theorell and his co-workers (Theorell and Åkeson, 1943; Theorell and Paul, 1944). Polis and Shmukler (1953) isolated and crystallized the enzyme after salt fractionation and displacement chromatography of the whey proteins, which had been freed of casein by rennet treatment. By electrophoresis and spectrophotometry they found two lactoperoxidases distinguished by different mobilities at pH 5.0 and by different absorbancy ratios (at 412 nm/280 nm) of 0.90 and 0.77. Later a method was developed (Morrison et al., 1957; Morrison and Hultquist, 1963) employing a carboxylic acid resin in the ammonium or sodium form to adsorb preferentially lactoperoxidase from rennet whey or whole milk. The crude protein fraction obtained from the resin was then chromatographed on the cation-exchange resin and finally passed through a Sephadex G-100 column. Absorbancy ratios, A_{412}/A_{280}, of 0.91–0.95 were obtained for these preparations. By this method two lactoperoxidases were generally found that differed in elution sequence from the resin columns and in mobility on paper electrophoresis at pH 8.6; however, several preparations, using the modified procedure, gave only a single form of lactoperoxidase. When antiserum to crude lactoperoxidase was employed in the agar diffusion analysis of a preparation in which the two forms of enzyme were present, both forms were immunologically identical (Allen and Morrison, 1963). It is of interest that some fractions from the resin column with no significant absorption at 412 nm also showed immunological identity with the active enzyme. This suggests that large hemin-free polypeptides derived from lactoperoxidase might still react with antibody although they lack enzyme activity. In all the fractionation procedures described the red protein is found to be closely associated with lactoperoxidase. The two proteins are not immunologically identical nor is the enzyme related to bovine transferrin or lactollin (Allen and Morrison, 1966).

Rombauts et al. (1967) further modified the method of isolating lactoperoxidase by substituting for the Sephadex fractionation a chromatographic procedure using an intermediate base anion-exchange resin. By this method, only one lactoperoxidase was found and they suggested that the second lactoperoxidase observed in earlier experiments was an artifact of extraneous proteolytic action during the preparative procedure. Attempts at further purification of the enzyme using reverse salting-out chromatography with ammonium sulfate gave only a slight increase in purity, as determined by an increase in the absorbancy ratio.

In another study of the heterogeneity of lactoperoxidase Carlström

(1965) used a cation-exchange resin, carboxymethyl cellulose, and Sephadex G-200 to isolate milk peroxidase from rennet whey. Chromatography of the purified enzyme on DEAE-Sephadex G-50 resulted in five active fractions. The major portion of the enzyme was found in fractions 1 and 2, while relatively little was found in fractions 3, 4, and 5. Rechromatography of fractions 1 and 2 showed that fraction 1 was homogeneous but 2 was slightly contaminated with fraction 1. The A_{412}/A_{280} ratio for fractions 1 and 2 was 0.96 and 0.85, respectively. The iron content and specific activity of these two major fractions were the same. Lactoperoxidase prepared from a single cow was also heterogeneous when chromatographed on DEAE-Sephadex. Carlström and Vesterberg (1967) subdivided lactoperoxidase into six fractions by the isoelectric focusing method and found that the isoelectric points of the six fractions varied between pH 9.2 and 9.9. These values are consistent with the isoelectric point of pH 9.6 reported by Polis and Shmukler (1953) as determined by moving-boundary electrophoresis, in phosphate buffer, when extrapolated to low ionic strength.

Using a different isolation procedure, Groves (1965, 1966) chromatographed milk whey from single cows on DEAE-cellulose and phosphocellulose and found that lactoperoxidase and ribonuclease were eluted in fractions 1F-3 and 1F-4 of Fig. 1. The two enzymes were then resolved by Sephadex G-200 into fractions A and B, which contain lactoperoxidase, and fraction C, which contains ribonuclease (Fig. 5). The 412 nm ab-

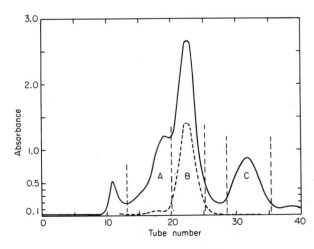

FIGURE 5. Gel filtration of lactoperoxidase fractions 1F-3 and 1F-4 (Fig. 1) on Sephadex G-200. Solid line, absorbance at 280 nm; dashed line, absorbance at 412 nm (Groves, and Kiddy, 1964).

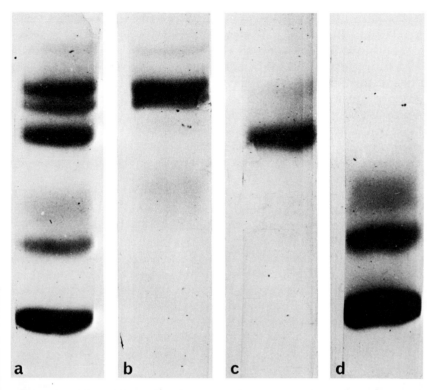

FIGURE 6. Disc-gel electrophoresis, pH 4.3, of fractions eluted from Sephadex G-200 (Fig. 5). (a) Material before Sephadex filtration; (b) fraction A; (c) fraction B, lactoperoxidase; (d) fraction C, ribonuclease (Groves and Kiddy, 1964).

sorbancy is indicated by the dashed line, which shows that the major portion of lactoperoxidase is in fraction B with a ratio A_{412}/A_{280} of 0.5. Disc-gel electrophoresis at pH 4.3 of the proteins in the Sephadex fractions A, B, and C are shown in Fig. 6. Electrophoresis of the proteins before fractionation on Sephadex is also shown (Fig. 6a). Other lactoperoxidase preparations from single cows give similar patterns, although slower-moving bands appear when some fractions are reworked or when the method of isolation is varied. This indicates that aggregation of the enzyme has occurred. Rombauts et al. (1967) observed five bands with peroxidase activity when lactoperoxidase was subjected to disc electrophoresis at an acid pH value. The major fraction corresponded to the band with the greatest mobility while the other four bands varied in relative concentration. It was thought that the slower-moving bands were not aggregates of increasing degrees of polymerization but rather that the multiple components were artifacts of the electrophoretic procedure, since it was found that when the major band was cut out and subjected to electrophoresis again, all five zones were obtained. The multiple bands did not appear to result from differential oxidation states or altered ligands of the hemoprotein during polymerization of the sample in the gel, since samples layered on the gel in a high-density solution also gave five bands. A single band was obtained under conditions where the spacer and sample gels were omitted and the enzyme was layered directly onto a 15% gel. (In this author's opinion, results of zone electrophoresis under the latter system might be misleading if the enzyme contains aggregates too large to move into the gel.)

Gel electrophoretic patterns of lactoperoxidase fractions obtained at pH 9.1 on vertical gels are shown in Fig. 7 (Groves and Kiddy, 1964). All samples are from individual cows except one (Fig. 7e), which is a purified sample from pooled milk. (This lactoperoxidase sample was kindly furnished by Dr. M. Morrison.) The ratio A_{412}/A_{280} of about 0.5 for lactoperoxidase from individual cows indicates a lack of purity, based on the ratio for purified preparations; however, the same gel electrophoretic bands are found on staining for peroxidase activity using benzidene reagent as on staining for protein.

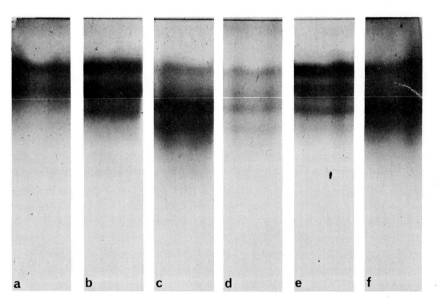

FIGURE 7. Vertical gel electrophoresis of lactoperoxidase fractions at pH 9.1 and 5% gel: (a), (b), (c), (d), (f) from milk of individual cows; (e) from pooled milk (Groves and Kiddy, 1964).

16. MINOR MILK PROTEINS AND ENZYMES 391

Morrison and Hultquist (1963) suggested that the heterogeneity of lactoperoxidase might result from proteolytic activity, since casein is precipitated by rennet, or that it represents chemically different proteins produced by genetically different animals. Carlström and Vesterberg (1967) have separated lactoperoxidase into subcomponents by the isoelectric focusing method and have found no difference in patterns for preparations of lactoperoxidase made with and without rennet in the isolation procedure. Since milk from both pooled and single cows gives lactoperoxidase that can be fractionated into several fractions, Carlström (1965) suggested that either the cow produces several peroxidases or there is one native peroxidase that is converted into several others during isolation. The gel electrophoretic patterns at alkaline pH of lactoperoxidase from individual cows (Fig. 7) do show differences. Whether they result from changes during the fractionation procedure or from reaction with reagents in the gel is not known. If changes do take place they appear to be consistent. Further work on lactoperoxidase from a number of related cows will be required to establish whether genetic polymorphism does exist.

Swope et al. (1966) found that starch-gel electrophoresis of a whey fraction enriched in lactoperoxidase showed only one band with peroxidase activity at both acid and alkaline pH values. It would be of interest to know if a relatively pure sample of lactoperoxidase would give a single band under these conditions. It is possible that polymorphism in lactoperoxidase, when polyacrylamide gels are used, might result from reactions of the protein with residual catalyst or by-products formed during polymerization of the acrylamide. This has been suggested to explain artifacts observed with some enzymes (Brewer, 1967; Fantes and Furminger, 1967; Mitchell, 1967), although Rombauts et al. (1967) observed no difference in the electrophoretic pattern of lactoperoxidase on polyacrylamide at an acid pH value when the sample was applied either before or after gel photopolymerization. Carlström and Vesterberg (1967) found that the polyacrylamide bands obtained at an alkaline pH value correspond to the zones isolated by the method of isoelectric focusing of ampholytes. This also appears to discount the implication of the acrylamide system in the formation of artifacts.

Carlström (1969a,b,c), using ion-exchange chromatography, moving-boundary electrophoresis, disc electrophoresis and isoelectric focusing, made further studies on the heterogeneity of lactoperoxidase prepared from pooled milk and the milk of individual cows. He found that the proportion of certain fractions of the enzyme varied among individual cows. He was unable to explain the heterogeneity on the basis of differences between the A and B fractions containing four and six subfractions, respectively, in amino acid analysis and molecular weight studies (see

TABLE II

PROPERTIES OF DIFFERENT LACTOPEROXIDASE PREPARATIONS

Property	Preparation				
	Theorell[a,e]	Polis[b]	Morrison[c,d]	Carlström[f]	
				A Fractions	B Fractions
Fe (%)	0.070	0.069	0.0729	0.0747	0.0680–0.0709
N (%)		15.56		15.9	15.4–15.9
Absorbancy ratio (A_{412}/A_{280})	0.77	0.90	0.95	0.92	0.92–0.98
Molecular weight (daltons)	92,700	82,000	77,500	76,500	76,400

($s^0_{20,w}$ = 5.37 S; 1% protein; pH 4–11; D_{20} = 5.95 × 10⁻⁷; sp vol = 0.764)	(Light-scattering)	(Sedimentation-equilibrium; 0.8% protein)	(From s, D and sedimentation-equilibrium) 74,800 (From Fe content)[g]	(From s, D; $s^0_{20,w}$ = 5.19 S) 78,000 (From sedimentation-equilibrium) 78,800–82,100 (From Fe content)
Isoelectric point				
7.7 (Phosphate buffer, 0.1 ionic strength)		8.05 (Lactoperoxidase A; Veronal buffer) 9.6 (Lactoperoxidase interacts with phosphate buffer; extrapolated value)		
			[e] Theorell and Paul (1944). [f] Carlström (1969b). [g] In earlier preparations (Carlström, 1965) Fe, 0.062; absorbancy ratio, 0.99.	

[a] Theorell and Pederson (1944).
[b] Polis and Shmukler (1953).
[c] Morrison and Hultquist (1963).
[d] Rombauts et al. (1967).

below). He concluded that the heterogeneity arises from catabolic processes *in vivo*.

Variation of lactoperoxidase activity in milk is influenced by factors such as breed, season, feed, and especially stage of lactation. Peroxidase activity is high in colostrum, reaching a maximum about four days after the calf is born; then it gradually decreases to the level of normal milk (Kiermeier and Kayser, 1960a). Skim milk contains about 30 mg of the enzyme per liter (Polis and Shmukler, 1953).

A number of workers (Paul, 1963; Polis and Shmukler, 1953; Morrison *et al.*, 1957) find that lactoperoxidase tends to become less stable with increasing purity. Reduction in activity can occur on dilution or dialysis of enzyme solutions. Also, denaturation can occur in salt precipitation of the protein if the salt addition is not carefully controlled.

Heat inactivation of peroxidase in milk is markedly sensitive to temperature changes around 80°C but much less affected by variation in heating time at a particular temperature. Depending on the temperature of inactivation, some lactoperoxidase activity is restored after storage (Kiermeier and Kayser, 1960b, c; Woerner, 1961).

Several properties of lactoperoxidase are shown in Table II. Molecular weight calculations, based on the iron content and one iron atom per mole, are in good agreement with the physical measurements shown by Polis and Shmukler (1953) and Rombauts *et al.* (1967). The low absorbancy ratio of 0.77 (but normal iron content) shown by Theorell and Pedersen (1944) could result from contamination of their preparation with lactoferrin. Carlström (1969b) obtained a molecular weight of about 77,000 daltons and an iron content higher than he had previously reported (see Table II). The amino acid composition of lactoperoxidase is shown in Table I. Rombauts *et al.* (1967) found that lactoperoxidase is a glycoprotein in which 16 of the carbohydrate residues tabulated are glucosamine and 10 are galactosamine. They also found neutral carbohydrate in the amount of 1.5% but no acetylneuraminic acid. The enzyme contains 10 acetyl residues per mole which could be present as acetylated hexosamines and possibly acetylated N-terminal groups. Leucine, in the amount of 0.5 residue per mole, is the only significant N-terminal amino acid found. They concluded that lactoperoxidase is composed of two very similar subunits, since the total number of spots found on peptide maps of tryptic peptides of the enzyme is about equal to the number that would be expected if lactoperoxidase had two identical subunits. This conclusion is supported by preliminary sedimentation-equilibrium studies in 8 M urea and 6 M guanidine hydrochloride in which a molecular weight of about 40,000 daltons was obtained. The low value obtained for N-terminal leucine would argue against two subunits, although it is possible that the N-terminal residue of one subunit is blocked (Rombauts et al., 1967).

Carlström (1969b) obtained amino acid analyses similar to those of Rombauts et al. (1967), but he found considerable differences in carbohydrate content which he attributed to differences in analytical procedures between the two groups. He found 14 of the carbohydrate residues to be glucosamine and 4 to be galactosamine; neutral sugars were 5.37%, and there was no sialic acid. Carlström found that his B and A fractions had similar amino acid compositions but A was lower than B in carbohydrate content. He suggested that deamination of asparagine or glutamine may be responsible for interconversion of A and B.

The prosthetic group of lactoperoxidase has been studied by Hultquist and Morrison (1963) and Morell and Clezy (1963). They agree that an ester, or possibly an amide bond (Morell and Clezy, 1963), links the hemin and protein together. Carlström (1969c) has suggested that the prosthetic group is protoheme. There is a lack of agreement on the other types of substituent groups conjugated to the porphyrin ring.

The possibility that lactoperoxidase might be found in tissues other than the mammary gland has been investigated. Morrison and Allen (1963) and Morrison et al. (1965), using immunodiffusion analysis for detecting lactoperoxidase in bovine tissues, found a protein antigenically identical to lactoperoxidase in sublingual, submaxillary, and parotid glands. The enzyme purified from the submaxillary gland was shown to be identical to lactoperoxidase by a number of physical-chemical measurements. Lactoperoxidase has since been identified and isolated from the harderian and lacrimal glands of both cow and steer (Morrison and Allen, 1966). The nonheme iron-containing red protein found in milk has also been identified in these ectodermal glands. The function of lactoperoxidase in these glands is not known. It may be protective, since studies (Jago and Morrison, 1962; Klebanoff and Luebke, 1965; Steele and Morrison, 1969) have shown that it can inhibit certain bacteria under aerobic conditions. The presence of a natural antimicrobial factor in cow milk exhibiting the properties of a protein has been studied by other workers (Jones and Simms, 1930; Wright and Tramer, 1958) and has been reviewed by Reiter and Oram (1967). During the purification of lactoperoxidase, Portmann and Auclair (1959) found a direct relationship between the lactoperoxidase activity of the fractions and their antibacterial activity. They concluded that lactoperoxidase is the inhibitor. The inhibiting properties of lactoperoxidase appear to be specific since horseradish peroxidase does not inhibit bacterial growth (Stadhouders and Veringa, 1962; Jago and Morrison, 1962).

Lactoperoxidases isolated from the milk of goats and sheep are immunochemically indistinguishable from cow milk lactoperoxidase (Allen and Morrison, 1966).

C. XANTHINE OXIDASE

Schardinger (1902) discovered an enzyme in milk which catalyzed the decolorization of methylene blue by formaldehyde. Subsequently, this enzyme has been identified as xanthine oxidase (E.C. 1.2.3.2.), which also catalyzes the oxidation of the purines, hypoxanthine and xanthine to uric acid. Xanthine oxidases are found in animal tissues, particularly the liver, and also in bovine milk, which is a convenient source for isolating the enzyme. It is associated with the fat globule membrane and gradually increases in amount during lactation (Jenness and Patton, 1959). Whole milk is estimated to contain about 120 mg xanthine oxidase per liter (Corran et al., 1939). Bray (1963) has reviewed work on xanthine oxidase, and Zittle (1964) has summarized data on the heat stability of the enzyme. Zikakis and Treece (1969) have concluded that there is codominant polymorphism of the enzyme resulting in two variants of different activity. However, Hart et al. (1970) suggest that nutritional (availability of molybdenum) rather than genetic factors determine the relative amounts of active and inactive xanthine oxidase.

Xanthine oxidase is usually isolated from milk by a modification of the method of Ball (1939). The enzyme is concentrated in the buttermilk fraction on churning cream. Casein and turbid material are removed from the buttermilk after the action of crude trypsin. Subsequently, the enzyme is further purified by salt fractionation and by chromatography on a calcium phosphate column. Avis et al. (1955) crystallized xanthine oxidase from phosphate solutions containing ethanol, and Uozumi et al. (1967) crystallized an iron-free xanthine oxidase from phosphate buffer.

A significant amount of the xanthine oxidase associated with particulate matter is lost during fractionation when the buttermilk is clarified after enzyme digestion. A modified procedure that gives 70–90% yields involves the use of cysteine to release the xanthine oxidase associated with the particulate complex. By this procedure, yields of 400 mg/liter of buttermilk are obtained (Gilbert and Bergel, 1964).

Xanthine oxidase in low yields has also been isolated from buttermilk without the addition of proteolytic enzymes (Mackler et al., 1954). The characteristics of xanthine oxidase isolated from milk may depend on whether the step involving proteolysis is used, since Carey and his associates (1961) observed differences in mobilities and certain enzymatic activities in xanthine oxidases prepared by the two methods.

The crystalline protein has a $s^{\circ}_{20,w}$ of 11.4 S (Avis et al., 1956) and a molecular weight of 275,000 daltons, the average of three values determined from sedimentation-diffusion, approach to sedimentation-equilibrium and gel filtration (Andrews et al., 1964). Between pH 3.5 and 9.8 there is no indication of dissociation of the protein, nor is it dissociated by urea or urea and thioglycolate. The isoelectric point of xanthine oxidase is about pH 5.4 in 0.2 ionic strength buffer (Avis et al., 1956). The active enzyme consists of a high molecular weight protein which firmly binds flavine adenine dinucleotide (FAD) and also the metals molybdenum and iron. One mole of the enzyme contains 8 moles iron, 2 moles FAD, and probably 2 moles molybdenum (Bray, 1963). The amino acid composition of milk xanthine oxidase is shown in Table I. Tryptophan is significantly low compared to the other amino acids.

Under proper storage conditions, xanthine oxidase solutions are reasonably stable. The enzyme is inactivated by heavy metals, photooxidation, and thermal denaturation (Bergel and Bray, 1959).

The reddish appearance of milk xanthine oxidase, which shows a visible absorption maximum around 450 nm, is due in part to FAD. The balance of the absorption in this region appears to be due to the iron–protein complex (Bray, 1963; Rajagopalan and Handler, 1964). Bayer and Voelter (1966) selectively removed the iron from xanthine oxidase without denaturing the protein and found that the iron–protein complex does not contribute to the visible spectrum but that a molybdenum complex and perhaps an unknown bond to the FAD are responsible. The iron-free protein still retains catalytic activity but it denatures more easily than the iron-containing enzyme. Uozumi et al. (1967) also found that the absorption spectra and enzymatic activity of xanthine oxidase remain essentially the same when the iron content is reduced to 0.3 mole per mole of protein. Although the binding site of iron to xanthine oxidase is not known, the iron-containing flavoproteins are reported to have acid-labile sulfide groups in amounts stoichiometric with the iron (Rajagopalan and Handler, 1964). Labile sulfide groups equivalent to the iron (4 moles per mole FAD) have been found for xanthine oxidase isolated from both pig liver and bovine milk (Brumby et al., 1965).

Roussos and Morrow (1967) fractionated a commercial preparation of xanthine oxidase using care in controlling contamination by extraneous metals. After chromatography on DEAE-Sephadex, the eluates with the highest specific activities were pooled and applied to a weakly acidic chelating resin. Several fractions isolated by this method showed specific activities which are significantly higher than those reported previously. Bovine intestinal xanthine oxidase is found to be devoid of molybdenum but to contain iron, copper, and FAD (Roussos and Morrow, 1966). Results of studies on milk xanthine oxidase indicate that molybdenum and possibly a major portion of the iron is not necessary for enzymatic activity. Furthermore, milk xanthine oxidase contains a significant amount of copper (Roussos and Morrow, 1967). Avis et al. (1956) determined that two moles of FAD are associated with one mole of enzyme and that molecular weight calculations based on the FAD content are consistent with the value, 275,000 daltons, obtained by physical measurements. Roussos and Morrow (1967) found considerably less FAD in their preparations and calculated a minimum molecular weight, based on FAD content, of about 354,000 daltons. They also suggested that previous investigations on this enzyme have been carried out on samples which had lost over half of their activity during preparation, storage or other treatment. This finding is consistent with studies by Hart and Bray (1967) which indicate that earlier preparations of xanthine oxidase contained no more than half the active enzyme. They found that milk xanthine oxidase is normally contaminated by two inactive forms of the enzyme, one of which is devoid of molybdenum and can be removed by selective denaturation with high concentrations of sodium salicylate. The other appears to be a preparation artifact that can be minimized by the use of chelating agents and rapid fractionation procedures. In contrast to the finding of Roussos and Morrow (1967), this inactive fraction of xanthine oxidase contains no molybdenum, while the active enzyme, prepared by the salicylate method, contains the molar ratios of approximately 2:2:8 for molybdenum, FAD and iron, respectively, as reported earlier by Bray (1963). Further work by Hart et al. (1970) has confirmed that the amount of active and inactive enzyme present (hence, the FAD/Mo ratio) is influenced greatly by the method of isolation and purification. Their preparation is claimed to have a higher specific activity than previous preparations and has 2 moles FAD, 2 g-atoms of Mo and 8 g-atoms of Fe/mole protein ($\sim$275,000 daltons).

Coughlan et al. (1969) have investigated the role of molybdenum in the enzyme and found that certain inhibitors, such as cyanide, arsenite and methanol, bind to the molybdenum which is located at the active center of the enzyme.

Nelson and Handler (1968) prepared xanthine oxidase without exposure of the enzyme to proteases and obtained a particle weight of 300,000 daltons. This was reduced to 150,000 in guanidine and acid but could not be resolved into smaller units.

A number of thiol groups in milk xanthine oxidase react relatively fast with sodium p-chloromercuribenzoate without affecting the enzyme activity (Bergel and Bray, 1959), while in the presence of substrate (hypoxanthine) the enzyme is rapidly inactivated by the mercurial (Fridovich and Handler, 1958). Bray and Watts (1966), using iodoacetamide,

16. MINOR MILK PROTEINS AND ENZYMES 399

found that inactivation occurs when approximately one thiol group per mole of xanthine-reduced enzyme reacts. Green and O'Brien (1967) found that hydrogen peroxide and o-iodosobenzoate are similar to iodoacetamide in that inactivation of the enzyme occurs only in the presence of substrate. Inactivation by p-chloromercuribenzoate differs since it is more rapid, is reversed by cysteine, is less in the presence of FAD and is formally competitive with the substrate.

A summary of catalytic reactions and the mechanism of the action of xanthine oxidase has been made by Bray (1963). Electron paramagnetic resonance spectroscopic studies on the enzyme have been reported by Beinert and Palmer (1965), Palmer and Massey (1969), Orme-Johnson and Beinert (1969), Gibson and Bray (1968), Bray and Vänngård (1969), Pick and Bray (1969), Nakamura and Yamazaki 1969). Spectral and rapid-reaction studies have been reported by Massey et al. (1969). Arneson (1970) has studied chemiluminescence produced when milk xanthine oxidase operates on its substrate, and Komai et al. (1969) present evidence that flavine is the prosthetic group of the enzyme responsible for reaction with oxygen.

D. LIPASES AND ESTERASES

Milk contains esterase and lipase activities. Lipases (E.C. 3.1.1.3) are important to the milk industry since they act on milk fats and release volatile short-chain fatty acids which can produce undesirable milk flavors; yet they produce desirable flavors in certain cheeses. Reviews of the work on milk lipases have been made by Jensen (1964), Chandan and Shahani (1964) and Shahani (1966).

Two lipases have been identified in bovine milk by Tarassuk and his co-workers (Tarassuk and Frankel, 1957; Tarassuk et al., 1964). One is associated with the casein fraction of milk and requires for activation certain treatments such as homogenization, agitation or foaming, conditions which disturb the fat globule membrane and expose the enzyme to substrate. All normal cow milk contains this type of enzyme. The other lipase becomes irreversibly adsorbed on the fat globule membrane material by cooling fresh milk. It is found in the milk of a relatively small number of cows.

Lipase in raw milk is also activated by variation in temperatures such as cooling, then warming and recooling the milk. If raw milk is rapidly cooled to below 10°C and held there until pasteurization, lipolysis can be minimized (Krukovsky, 1961).

When casein is isolated from skim milk by acid precipitation, the lipase activity is significantly lower than that of casein obtained by centrifugation. The casein, on electrophoretic fractionation, is resolved into three fractions. Only one, the "α-casein" fraction, contains the lipase (Skean and Overcast, 1961). Evidently this "α-casein" fraction consists of both $α_s$- and κ-casein since the enzyme is associated with the κ-casein fraction when skim milk is chromatographed on DEAE-cellulose (Yaguchi et al., 1964). Lipase can be dissociated from κ-casein by the use of dimethylformamide (Fox et al., 1967).

Water extracts of rennet casein were examined for lipase fractionation with DEAE-cellulose by Gaffney et al. (1966). Stepwise elution of the extracted protein on DEAE-cellulose gave eight fractions, all of which contained lipolytic activity, suggesting that lipase is associated with proteins other than κ-casein. The total lipase activity of the various fractions represented about twice that in the original milk and indicates either the liberation of additional enzyme or the unmasking of active sites. The extract of rennet casein gave a 10-fold enrichment of lipase, while one fraction from the DEAE-cellulose produced another 100-fold purification. All the activity was lost, however, when the protein in these fractions was recovered by lyophilization.

Fox and Tarassuk (1968) have coagulated fresh skim milk with rennin, separated the curd and solubilized the lipase from the curd by maceration with 1 M NaCl. After centrifugation the supernatant was half saturated with ammonium sulfate, the precipitate collected, dissolved, dialyzed and fractionated on DEAE-cellulose. The lipase-rich fraction was made 30% (v/v) in dimethylformamide and half saturated with ammonium sulfate. The lipase remains soluble under these conditions and may be further purified by Sephadex gel filtration. Such a preparation was subsequently characterized by Patel et al. (1968). It has 0.6% sialic acid, 14.8% nitrogen, 0.16% phosphorus, but its amino acid composition is unlike any casein (or other milk proteins).

The properties of this lipase may be contrasted with those of a lipase prepared from milk clarifier sediment by Chandan and Shahani (1963a,b). Their enzyme was extracted from the sediment followed by fractionation with salt and acetone and finally by filtration on a Sephadex G-50 column. The material exhibited a single peak in sedimentation-velocity patterns with $s°_{20,w}$ = 1.14 S. A molecular weight of about 7000 daltons was determined by both sedimentation-diffusion and by osmotic pressure measurements. This lipase also showed a single band by starch-gel electrophoresis at alkaline pH and contained 14.33% nitrogen, 1.04% sulfur and 0.26% phosphorus. This was the first milk lipase to be purified sufficiently for extensive physical-chemical measurements. Chandan and Shahani (1963a) suggested that the material isolated from clarifier slime is the same as the lipase associated with the casein fraction. This does not now seem likely in view of the properties of the above preparation of Tarassuk's group. Based on earlier work, the preparation from separator slime was assumed to originate from milk. There is evidence (Gaffney and Harper, 1965) that the somatic cells also present in separator slime contain an endolipase which appears to differ from the lipase in skim milk.

Downey and Andrews (1965a,b) found that lipase is associated with the casein micelles in milk and that most of it can be dissociated from the micelles by adding sodium chloride to skim milk to a concentration of 0.75 M and then removing most of the undissolved casein by centrifugation at 80,000 g for 1 hr. Partial separation of lipase is accomplished by gel filtration of the supernatant on Sephadex G-200. Five active peaks are obtained and much of the activity is in three fractions corresponding to molecular weights, based on gel filtration, of between 62,000 and 112,000 daltons. One active peak eluted in the vicinity of low molecular weight protein might represent the lipase isolated from clarifier slime (Chandan and Shahani, 1963a). It is possible that one enzyme of low molecular weight exists both in the free form and in association with casein or other proteins. All the fractions have low substrate specificities, but differences in their relative activity toward various substrates between higher and lower molecular weight fractions are pronounced. This suggests that there are several enzymes. Further evidence that this is so has been obtained subsequently by Downey and Andrews (1969) (see also Gaffney et al., 1968). Downey and Murphy (1970) have concluded that micelle structure is not necessary for the binding of lipase to casein.

Solutions of purified lipase (Chandan and Shahani, 1963b) are quite stable in the frozen state and exhibit a pH optimum of about 9.1 and a temperature optimum of 37°C. A distinction has been made between lipases and esterases based on the solubility of the substrate (Desnuelle, 1961); the lipase splits ester bonds of emulsion substrates while esterase acts on esters in true solution. The purified milk lipase is found to be a true lipase by this definition. Milk lipase is inactivated by such factors as aging, heat and light. It is also sensitive to heavy metals, which act as inhibitors. Although iron and copper are inhibitors, they also increase the heat stability of the enzyme (Chandan and Shahani, 1964; Frankel and Tarassuk, 1959). Patel et al. (1968) found that their purified preparation was inactivated on storage at 37°C for 24 hr or by freeze-drying.

Studies by Robertson et al. (1966) on the influence of selected inhibitors on milk lipase suggest that the hydroxyl group of serine and the imidazole group of histidine are primarily involved in the action of milk lipase and that sulfhydryl groups are located near the active site but are not necessarily associated with the active center.

Essentially similar conclusions have been reached by Patel et al. (1968) (see also the earlier work from this group, Tarassuk and Yaguchi, 1959). Based on sulfur analysis, the purified lipase preparation has two moles of sulfur per mole of enzyme (Chandan and Shahani, 1965). Studies with sulfhydryl reagents indicate that the enzyme contains a free and a masked sulfhydryl group, both of which are necessary for full activity of the enzyme. The sulfhydryl groups are implicated in the inactivation of the enzyme by oxidation or by interaction with certain heavy metals.

Milk also contains a lipoprotein lipase similar in properties to those of postheparin plasma, adipose tissue and heart (Korn, 1962). Most of its activity is found in skim milk, although some of the enzyme with relatively high specific activity has been isolated from cream. Its function in milk is unknown.

On the basis of the specificity towards different esters and the effect of inhibitors, three classes of esterases have been identified in cow milk (Forster et al., 1959, 1961): A-esterase (aryl esterase), which hydrolyzes phenyl acetate, B-esterase (aliphatic esterase or lipase), which hydrolyzes tributyrin, and C-esterase (cholinesterase) which hydrolyzes phenylpropionate. Although the source of milk A-esterase is not known, it is probably derived from the blood (Marquardt and Forster, 1966). The amount of A-esterase in milk, as determined with phenyl acetate, was corrected for nonenzymic hydrolysis of the substrate by a control containing heated milk (Forster et al., 1961). Downey and Andrews (1965c) found that a number of nonenzymic proteins of milk and serum will hydrolyze p-nitrophenylacetate and that the true esterase amounts to only about 20% of the activity in cow milk. The B-esterase has been concentrated by adsorption of the enzyme on a magnesium hydroxide suspension (Montgomery and Forster, 1961). It possesses properties similar to milk lipase in hydrolyzing triglycerides (Jensen et al., 1961). Studies by Korn (1962) with different substrates and inhibitors on lipoprotein lipase indicate that it is not an A-, B-, or C-esterase.

E. AMYLASES

The α-1,4-D-glucosidic linkages of starch and glycogen are hydrolyzed by amylase. α-Amylase (E.C. 3.2.1.1) splits central glucosidic bonds in the polysaccharide chain while β-amylase (E.C. 3.2.1.2) acts on the nonreducing end of the polysaccharide. Although several workers have found amylase activity in milk, Richardson and Hankinson (1936) demonstrated that milk contains α-amylase, the liquifying and dextrinizing enzyme, and some saccharifying activity, which would indicate the presence of β-amylase.

Guy and Jenness (1958) partially purified α-amylase of cow milk. A major portion of the enzyme remains with the soluble whey proteins after acid precipitation of the casein from skim milk and is subsequently precipitated with the lactoglobulin fraction at 43% saturation in ammonium sulfate. Further purification is attained by dissolving the globulin fraction in a solution containing ethylene glycol, ethanol and water, adsorbing the proteins on rice starch and eluting with a saturated calcium sulfate solution. Milk α-amylase requires both calcium(II) and chloride ions for activity and is inhibited by iodine. The enzyme has optimum activity at pH 7.4 and is heat labile, being inactivated after 30 min at 45°–52°C. Although there is evidence that milk contains some β-amylase, the results are not conclusive.

Goussault et al. (1967) have isolated α-amylase from human colostrum by Sephadex gel filtration. Got et al. (1968) have concluded that there are six isoamylases in these fractions.

F. PHOSPHATASES

1. *Alkaline Phosphatase (E.C. 3.1.3.1)*

Stadtman (1961) has reviewed alkaline phosphatases in general, while Whitney (1958), Jenness and Patton (1959), Zittle (1964) and Shahani (1966) have summarized work on the milk enzyme. Alkaline phosphatase splits phosphomonoesters, with a maximum activity around pH 9–10. About 30–40% of the enzyme is found with the fat globule membrane and can be concentrated in the buttermilk fraction by churning cream. The phosphatase is associated with lipids and can be released from the complex by butanol. By this method, together with chemical fractionation, Morton (1953) prepared a purified alkaline phosphatase free of any diesterase or pyrophosphatase activity. Fractions with specific activities several times greater than Morton's most active preparation have been obtained using repeated acetone fractionation and then charcoal and alumina-gel or ion-exchange chromatography in the final steps; however, details of the methods are not given (Kresheck, 1963; Lyster and Aschaffenburg, 1962). Alkaline phosphatase has also been purified from skim milk by dissociation of the lipid–enzyme complex with butanol and then fractionation with acetone; this preparation contains some phosphodiesterase activity (Zittle and DellaMonica, 1952).

There is not a great deal of physical or chemical information on the enzyme. Earlier preparations were only partly characterized by electrophoresis (Morton, 1953; Zittle and DellaMonica, 1952). The molecular weight has been estimated to be 180,000 (Andrews, 1965) and 190,000 daltons (Barman and Gutfreund, 1966) using Sephadex G-200. Alkaline phosphatase contains 16.2% nitrogen but no phosphorus, carbohydrate or nucleotide (Morton, 1955).

The specific activity of the purified milk enzyme, using β-glycerophosphate as a substrate, is one-fifth that of bovine intestinal phosphatase purified by a similar procedure (Morton, 1955). Intestinal phosphatase reacts equally with the substrates phenyl phosphate and o-carboxyphenylphosphate. The milk phosphatase is less reactive with the o-carboxyphenylphosphate compared to phenyl phosphate (Zittle and Bingham, 1960). The difference in reactivity is attributed to differences in enzyme-substrate dissociation constants.

The inactivation of alkaline phosphatase is used as a measure of the efficiency of pasteurization (Kay et al., 1949) (For a recent study of a new alkaline phosphatase assay system for milk see Kleyn and Lin, 1968; Copius Peereboom and Beekes, 1969.) Nevertheless, with high-temperature, shorttime processing methods, the milk phosphatase under certain conditions may recover some of its activity. In studying this reactivation process Lyster and Aschaffenburg (1962) found that milk contains a dialyzable heat-labile inhibitor and a nondialyzable heat-stable activator for alkaline phosphatase. Reactivation of the enzyme in a simple system containing magnesium ions, β-glycerophosphate and β-lactoglobulin was considerably greater than in milk. An activator in the aqueous phase of milk was found to be capable of replacing β-lactoglobulin in the simple system. Magnesium(II) was the best divalent cation for enzyme reactivation. It was also concluded that thiol groups were not essential for reactivation. Using another model system, Kresheck (1963) showed that sulfhydryl groups were effective but not necessary components of the system. Native and reactivated forms of the enzyme were not susceptible to N-ethylmaleimide and p-chloromercuriphenylsulfonate, while the inactivated enzyme was susceptible, so that the reactivation process was blocked. The chelating action of ethylenediaminetetraacetate, depending on whether it was added before or after the incubation period, was effective in stopping or inhibiting the reactivation of the enzyme; magnesium had its maximum effect on reactivation when added prior to incubation (O'Sullivan and Shipe, 1963). Richardson and his associates (1964) have reviewed work in this field and suggest that since magnesium and calcium are shifted to the colloidal suspension during the brief period of heat processing, the enzyme, free of magnesium ions, may have an increased heat stability. Subsequent storage allows sufficient calcium and magnesium to ionize and recombine with the enzyme, resulting in enzyme reactivation. Kresheck and Harper (1967), in a review of the reactivation process of alkaline phosphatase, suggested that the sulfhydryl groups liberated from the enzyme by heat treatment are affected by the natural inhibitors in milk, such as heavy metal, zinc, copper, and calcium. Under proper conditions, other sulfhydryl groups can compete for the inhibitors and enhance reactivation of the enzyme by allowing a reversible change in the tertiary or quaternary structure of the molecule.

Copius Peereboom (1968, 1969a) has shown that there are three isozymes of alkaline phosphatase in milk: the α form is present in skim milk, β is on the fat globule membrane, and γ is present in cream. He found that only the β form is reactivated under practical storage conditions. Using Sephadex gel thin-layer chromatrography Copius Peereboom and Beekes (1969) have been able to distinguish between native and reactivated alkaline phosphatase. They exploit this observation as the basis of a test for effective pasteurization.

Purified milk alkaline phosphatase splits inorganic phosphate from both casein and phosphoserine (Zittle and Bingham, 1959). The enzyme shows maximum activity at about pH 9.5 for phosphoserine. It is much less reactive toward casein, with greatest activity near neutrality. When hydrolysis of phosphoserine and α-casein is carried out near pH 9.5, 25% of the phosphoserine phosphorus is hydrolyzed in 2 min, while 24 hr are required for equivalent hydrolysis of the phosphorus of α-casein (Bellinzona and Lanzani, 1960).

Alkaline phosphatases from sources other than milk are found to have one reactive serine hydroxyl group per molecule (Schwartz, 1963) which is readily phosphorylated by incubation with orthophosphate in slightly acid solutions (Engström and Ågren, 1958; Schwartz and Lipmann, 1961). Studies on the phosphorylation of bovine milk alkaline phosphatase (Barman and Gutfreund, 1966) indicate that the active center of the enzyme is involved and that maximum phosphorylation of the enzyme occurs at pH 5.0. The phosphorylation is found to be specific and the phosphoryl enzyme is readily hydrolyzed at alkaline pH values.

Buruiana and Marin (1969) fractionated milk samples from four main breeds of cow in Roumania by gel filtration on Sephadex G-200 at pH 8.2 (tris-HCl-NaCl). They found the alkaline phosphatase activity patterns of the eluants to be characteristic of each breed. Friesian (Holstein) cows had four isozymes; others had less.

2. *Acid Phosphatase (E.C. 3.1.3.2.)*

Mullen (1950) investigated acid phosphatase in milk and found it to be relatively heat stable compared to alkaline phosphatase. It is distributed in both cream and skim milk. A partially purified preparation from either fraction showed no significant difference in enzymatic properties (Bingham *et al.*, 1961). Using a resin that preferentially adsorbs the enzyme from skim milk, Bingham and Zittle (1963) purified acid phosphatase 40,000-fold. It is a basic protein with an optimum pH of 4.7, and it acts on aromatic phosphates, casein, and pyrophosphates but is inactive toward serine phosphate and glycerol phosphate. Acid phosphatase concentrated about 70-fold by acetone fractionation is stable between pH 3 and 9, while the enzyme in milk is rapidly inactivated at pH 4 (Hansson and Rasmusson, 1962).

The activities of both alkaline and acid phosphatases are influenced greatly, but in opposite directions, by the stage of lactation. The total alkaline phosphatase activity increases and acid phosphatase activity declines as lactation proceeds (Kiermeier and Meinl, 1961).

G. LYSOZYME

Shahani (1966) has reviewed the work on milk lysozyme (E.C. 3.2.1.17). The presence or absence of lysozyme in bovine milk was debated by several early workers. With the successful purification of lysozyme from human milk (Jollès and Jollès, 1961) the question of the presence of the enzyme in cow milk was reexamined. Analysis of a large number of milk samples from different breeds shows the considerable variation in lysozyme content of from 0 to 2600 μg/liter. The average value of 130 μg/liter is about 3000 times lower than that found in human milk (Chandan *et al.*, 1964; Shahani *et al.*, 1962). Chandan *et al.* (1968) have compared the lysozyme contents of human milk (400,000 μg/liter), goat milk (250 μg/liter), cow milk (100 μg/liter) and sow milk (nil). Jollès *et al.* (1968) have compared the activities of human milk lysozyme and egg white lysozyme in the presence of N-acetylglucosamine as inhibitor. Human milk lysozyme was more strongly inhibited than hen egg white lysozyme (see also Charlemagne and Jollès, 1967). Saint-Blancard *et al.* (1969) reported that lysozymes with different chemical composition and different reaction velocities have different affinities for *Micrococcus lysodeikticus* cells. All the human secretory lysozymes have similar affinity but the affinities of the bird egg white lysozymes differ from one another. Several microorganisms have been found to be susceptible to bovine and human milk lysozymes by Vakil *et al.* (1969), who suggested that the lysozymes may play a significant role in the inherent antibacterial activity of milk.

Chandan *et al.* (1965a) purified lysozyme from the whey fraction of bovine milk using a modification of the Jollès fractionation procedure. Further purification was obtained with ammonium sulfate fractionation and gel filtration on Sephadex. Approximately 40 mg of lysozyme was obtained from 230 liters of milk. The isolated enzyme shows a 16,300-fold purification and a specific activity of 0.35 compared with 3.5 for human milk lysozyme and 1.0 for egg white lysozyme. It is homogeneous by zone electrophoresis but differs from egg white lysozyme in mobility and apparently has a lower isoelectric point. The isoelectric points for bovine and human lysozyme are near pH 9.5 and 11.0, respectively. The ultraviolet absorption of the milk lysozyme is significantly less than egg white lysozyme and differences are found in the pH optimum of the two enzymes. Bovine milk lysozyme has a pH optimum of 7.90 compared to 6.35 for human milk. The sedimentation behavior of the bovine milk enzyme indicates that it is homogeneous, with a molecular weight similar to that of egg white lysozyme. The sedimentation coefficient of cow milk lysozyme is 2.00 S compared to 2.19 S for the purified human milk lysozyme (Chandan et al., 1965a,b; Parry et al., 1967; Shahani, 1966).

A comparison of the primary structure of bovine α-lactalbumin with that of hen egg white lysozyme indicates that the structure of the two is quite similar (Brew et al., 1967). A comparison with bovine milk lysozyme would be desirable. Unfortunately, cow milk contains very little lysozyme.

However, Jollès and Jollès (1967) and Parry et al. (1969) have determined the amino acid composition of human milk lysozyme and found it to be similar to that of hen egg white lysozyme. Jollès and Jollès (1968) have determined the amino acid sequence in 13 of the 18 tryptic peptides they obtained from human milk lysozyme. There are 124 ± 3 amino acid residues, 11–12 being arginine; the C-terminal residue is valine. Mouton and Jollès (1969) have found human milk lysozyme and the lysozymes from normal and abnormal tissues and secretions to be identical in amino acid composition of tryptic peptides. Jollès et al. (1969) have considered the relationships between chemical structure and biological activity of the lysozymes.

A preliminary report of the X-ray structure of human lysozyme from the urine of leukemia patients has been published by Osserman et al. (1969). The human lysozyme crystals, so far prepared, are not related to the many forms of hen egg white lysozyme crystals. The implications of the relationship between lysozyme and α-lactalbumin are discussed in Chapters 15 and 18.

H. MISCELLANEOUS ENZYMES

1. *Aldolase*

Aldolase is concentrated on the fat globules in milk, although some remains in the skim milk. Polis and Shmukler (1950) obtained a 50-fold enrichment of the enzyme from the latter. The whey proteins, after 1.5 M salt precipitation of the casein, were fractionated at 3°C, and the aldolase was precipitated between 2.4 and 2.8 M ammonium sulfate. In milk, the enzyme was inactivated at 37°C, while the partially purified sample was stable at this temperature. Aldolase is found in both the milk and blood serum of cows in about the same concentration.

2. Catalase (E.C. 1.11.1.6)

Catalase has not been purified from milk, although normal milk does contain catalase activity. The catalase content of milk varies among cows and is quite high in colostrum and mastitic milks. There are conflicting data on the distribution of catalase in milk (Jenness and Patton, 1959). Van Maele and Vercauteren (1962) found that catalase and also xanthine oxidase are closely associated with the milk microsomes.

3. *Protease and Trypsin Inhibitor*

A proteolytic enzyme in milk was found to be associated with the casein fraction by Warner and Polis (1945), who studied the slow hydrolysis of casein solutions by changes in viscosity and soluble nitrogen. They concentrated the enzyme about 150 times more than that in the original casein and showed that it had a pH optimum of 8.5. Enzymatic activity was demonstrated in a sterile system suggesting that the enzyme was a milk component and not the result of bacterial contamination. Harper *et al.* (1960) added further evidence that the enzyme was not bacterial in origin and showed that raw milk usually contained a small and variable amount of the protease. McMeekin and his associates (1959) found that acid extraction of casein at pH 4.0 solubilized the proteolytic enzyme, although no quantitative measurements were made. Livrea *et al.* (1964) have since measured the activity of milk fractions and, as might be expected, found that the casein fraction was higher in specific activity than the skim milk or whey fraction. The acid-extracted casein was, however, slightly higher in activity than the casein before extraction. Fractions of the acid extract precipitated at pH 4.7 and 6.0 gave the highest specific activity. The greater activity found in the extracted casein compared to the original casein is puzzling but might be explained in part by the finding (Kiermeier and Semper, 1960a) that milk contains a trypsin inhibitor which could complicate analysis of this type. Zittle (1965) and Zittle and Custer (1963) found that the protease associated with acid-precipitated casein was extracted quantitatively with the κ-casein when casein was extracted with sulfuric acid; a 20-fold concentration of the protease activity was obtained when the extraction was made at pH 3.5.

Kiermeier and Semper (1960b,c) found the proteolytic activity in milk to be optimum at pH 6.8 and 37°C. This pH optimum is in contrast to the higher value obtained by earlier workers who made their measurements on the casein fraction. The enzyme was concentrated 10-fold from milk in precipitates from ammonium sulfate fractionation, while the trypsin inhibitor was found in the filtrates. It was suggested that the measurable enzymatic activity in commerical milk would be much greater if there were no trypsin inhibitor in milk.

The proteolytic activity in milk was highest immediately after the calf was born (Kiermeier and Semper, 1960b). Also, the amount of trypsin inhibitor was found by some workers to be high in the colostrum (Laskowski and Laskowski, 1951; Laskowski et al., 1952). A crystalline trypsin inhibitor isolated from colostrum had an isoelectric point of pH 4.2, while that of the trypsin–trypsin inhibitor complex was pH 7.2, values considerably lower than those for the pancreatic inhibitor and its trypsin complex. The molecular weight of the pancreatic complex by light-scattering was approximately 36,000 daltons compared to 89,000 for the colostrum complex. Cechova et al. (1969) have found that bovine colostrum trypsin inhibitor contains 67 amino acid residues vs. 58 for the pancreatic inhibitor, 21 of the residues being in the same positions in each.

The relative amounts of both the proteolytic enzyme and trypsin inhibitor are of importance to the stability of milk (see also Yamauchi et al., 1969). Also, the stability of the purified caseins may be affected by trace impurities of the enzyme, since storage of purified lyophilized samples produces minor changes in gel electrophoretic patterns.

4. *Phosphohexose Isomerase (E.C. 5.3.1.9)*

It has been reported that phosphohexose isomerase occurs in milk (Heyndrickx, 1964).

5. *Lactose Synthetase*

Lactose synthetase is a microsomal enzyme system in the mammary glands of lactating animals. A soluble form of the enzyme was first demonstrated in milk by Babad and Hassid (1964) and has since been purified 70-fold (Babad and Hassid, 1966) by a combination of centrifugation and ammonium sulfate and hydroxyapatite fractionation. The biosynthesis of lactose involves the catalytic action of galactosyl transferase on uracil nucleoside diphosphate D-galactose, and D-glucose to give lactose and uracil nucleoside diphosphate. Milk lactose synthetase shows maximum activation with the divalent manganese ion and is inhibited by ethylenediaminetetraacetate or the divalent mercury ion. It has an optimum pH of 7.5 and optimum temperature of 42°C.

Brodbeck and Ebner (1966) have concentrated lactose synthetase 30-fold by ammonium sulfate fractionation of milk whey. Two peaks (A and B) were obtained, following gel filtration, corresponding to proteins of higher and lower molecular weights. Assays of the two peaks separately gave no activity, but on recombination of fractions A and B the catalytic activity was restored, indicating that the "A" and "B" proteins represent protein subunits of lactose synthetase. Ebner and associates (1966) found that α-lactalbumin is the "B protein" in lactose synthetase. The more recent developments on this enzymatic system are discussed in detail in Chapters 15 and 18.

6. *Ceruloplasmin*

The copper-binding protein, ceruloplasmin, is found in colostrum, normal milk and blood serum as determined by immunodiffusion techniques (Hanson *et al.*, 1967).

7. *β-N-Acetylglucosaminase (E.C. 3.2.1.30)*

Mellors (1968) has found β-N-acetylglucosaminase in bovine milk. It resembles catalase in its distribution within the subfractions of milk and may be derived from milk leucocytes. The cellular material that can be readily sedimented from milk contains the highest specific activity.

8. *β-Glucuronidase*

The occurrence of β-glucuronidase in cow milk has been reported by Kiermeier and Güll (1966) and in human milk by Kiermeier and Güll (1968a,b). The activity depends on the stage of lactation and is particularly high in colostrum, and in milks of animals with mastitis and carcinomas. A relationship was found between the leucocyte count and the enzyme activity in inflammations. Human milk appears to have 20–30 times the activity of cow milk.

9. *α-Mannosidase*

Mellors and Harwalker (1968) found α-mannosidase activity in milk to be associated with the β-casein fraction.

10. *Lysyl Arylamidase*

Mellors (1969) has isolated a lysyl arylamidase from cow milk.

11. *Sulfhydryl Oxidase*

An enzyme that catalyzes the oxidation of sulfhydryl groups has been concentrated 20 times more than that of the milk whey fraction by ammonium sulfate fractionation. The enzyme oxidizes glutathione to the disulfide in a pH range of 4.5–9.0, with an optimum at pH 7.0 (Keirmeier and Petz, 1967a). It is inactivated by heating 7 min at 75°C, by potassium cyanide, ascorbic acid, guanine and partly by urea. The addition of the enzyme preparation to heated milk causes oxidation of the protein-bound sulfhydryls formed (Kiermeier and Petz, 1967b,c).

12. *Lactic and Malic Dehydrogenases*

Lactic and malic dehydrogenases in their multiple molecular forms have been found in bovine milk and it is concluded that these enzymes are synthesized in the mammary gland (Kjellberg and Karlsson, 1966, 1967). A discussion of these enzymes can be found in Chapter 2, Volume I.

REFERENCES

Aasa, R., Malmström, B. G., Saltman, P., and Vänngård, T. (1963). *Biochim. Biophys. Acta* **75**, 203.
Aisen, P., Leibman, A., and Reich, H. A. (1966). *J. Biol. Chem.* **241**, 1666.
Aisen, P., Aasa, R., Malmström, B. G., and Vänngård, T. (1967). *J. Biol. Chem.* **242**, 2484.
Alderton, G., Ward, W. H., and Fevold, H. L. (1946). *Arch. Biochem.* **11**, 9.
Allen, P. Z., and Morrison, M. (1963). *Arch. Biochem. Biophys.* **102**, 106.
Allen, P. Z., and Morrison, M. (1966). *Arch. Biochem. Biophys.* **113**, 540.
Andrews, P. (1965). *Biochem. J.* **96**, 595.
Andrews, P., Bray, R. C., Edwards, P., and Shooter, K. V. (1964). *Biochem. J.* **93**, 627.
Arneson, R. M. (1970), *Arch. Biochem. Biophys.* **136**, 352.
Aschaffenburg, R., and Drewry, J. (1957). *Biochem. J.* **65**, 273.
Ashton, G. C. (1958). *Nature* **182**, 370.
Avis, P. G., Bergel, F., and Bray, R. C. (1955). *J. Chem. Soc.* p. 1100.
Avis, P. G., Bergel, F., Bray, R. C., James, D. W. F., and Shooter, K. V. (1956). *J. Chem. Soc.* p. 1212.
Babad, H., and Hassid, W. Z. (1964). *J. Biol. Chem.* **239**, PC946.
Babad, H., and Hassid, W. Z. (1966). *J. Biol. Chem.* **241**, 2672.
Bailie, M. J., and Morton, R. K. (1958). *Biochem. J.* **69**, 35.
Baker, E., Shaw, D. C., and Morgan, E. H. (1968). *Biochemistry* **7**, 1371.
Baker, E., Jordan, S. M., Tuffery, A. A., and Morgan, E. H. (1969). *Life Sci. N.Y.* **8**, 89.
Ball, E. G. (1939). *J. Biol. Chem.* **128**, 51.
Bargmann, W., and Knoop, A. (1959). *Z. Zellforsch. Mikrosk. Anat.* **49**, 344.
Barman, T. E., and Gutfreund, H. (1966). *Biochem. J.* **101**, 460.
Bayer, E., and Voelter, W. (1966). *Biochim. Biophys. Acta* **113**, 632.
Beinert, H., and Palmer, G. (1965). *Advan. Enzymol.* **27**, 160.
Bellinzona, G., and Lanzani, G. A. (1960). *Biochim. Appl.* **7**, 131.
Bergel, F., and Bray, R. C. (1959). *Biochem. J.* **73**, 182.
Bezkorovainy, A. (1965). *Arch. Biochem. Biophys.* **110**, 558.
Bezkorovainy, A. (1967). *J. Dairy Sci.* **50**, 1368.
Bezkorovainy, A., and Grohlich, D. (1967). *Biochim. Biophys. Acta* **147**, 497.
Bezkorovainy, A., and Grohlich, D. (1969). *Biochem. J.* **115**, 817.

Bingham, E. W., and Kalan, E. B. (1967). *Arch. Biochem. Biophys* **121**, 317.
Bingham, E. W., and Zittle, C. A. (1962). *Biochem. Biophys. Res. Commun.* **7**, 408.
Bingham, E. W., and Zittle, C. A. (1963). *Arch. Biochem. Biophys.* **101**, 471.
Bingham, E. W., and Zittle, C. A. (1964). *Arch. Biochem. Biophys.* **106**, 235.
Bingham, E. W., Jasewicz, L., and Zittle, C. A. (1961). *J. Dairy Sci.* **44**, 1247.
Blanc, B. (1967). *In* "Protides of the Biological Fluids: Proceedings of the 14th Colloquium, 1966" (H. Peeters, ed.), p. 125. Elsevier, Amsterdam.
Blanc, B., and Isliker, H. (1961a) *Helv. Physiol. Pharmacol. Acta* **19**, C13.
Blanc, B., and Isliker, H. (1961b). *Bull. Soc. Chim. Biol.* **43**, 929.
Blanc, B., and Isliker, H. (1963). *Helv. Chim. Acta* **46**, 2905.
Blanc, B., Bujard, E., and Mauron, J. (1963). *Experientia* **19**, 299.
Bray, R. C. (1963). *In* "The Enzymes" (P. D. Boyer, H. Lardy, and K. Myrbäck, eds.), Vol. 7, Chap. 22. Academic Press, New York.
Bray, R. C., and Malmström, B. G. (1964). *Biochem. J.* **93**, 633.
Bray, R. C., and Watts, D. C. (1966). *Biochem. J.* **98**, 142.
Bray, R. C., and Vänngård, T. (1969). *Biochem. J.* **114**, 725.
Brew, K., Vanaman, T. C., and Hill, R. L. (1967). *J. Biol. Chem.* **242**, 3747.
Brewer, J. M. (1967). *Science* **156**, 256.
Brodbeck, U., and Ebner, K. E. (1966). *J. Biol. Chem.* **241**, 762.
Brumby, P. E., Miller, R. W., and Massey, V. (1965). *J. Biol. Chem.* **240**, 2222.
Brunner, J. R. (1962). *J. Dairy Sci.* **45**, 943.
Brunner, J. R. (1965). *In* "Fundamentals of Dairy Chemistry" (B. H. Webb and A. H. Johnson, eds.), Chap. 10. Avi, Westport, Connecticut.
Brunner, J. R., Swope, F. C., and Carroll, R. J. (1969). *J. Dairy Sci.* **52**, 1092.
Buruiana, L. M., and Marin, M. (1969). *Lait* **49**, 1.
Butler, J. E. (1969). *J. Dairy Sci.* **52**, 1895.
Butler, J. E., Coulson, E. J., and Groves, M. L. (1968). *Fed. Proc., Fed. Amer. Soc. Exp. Biol.* **27**, 617.
Carey, F. G., Fridovich, I., and Handler, P. (1961). *Biochim. Biophys. Acta* **53**, 440.
Carlström, A. (1965). *Acta Chem. Scand.* **19**, 2387.
Carlström, A. (1969a). *Acta Chem. Scand.* **23**, 171.
Carlström, A. (1969b). *Acta Chem. Scand.* **23**, 185.
Carlström, A. (1969c). *Acta Chem. Scand.* **23**, 203.'
Carlström, A., and Vesterberg, O. (1967). *Acta Chem. Scand.* **21**, 271.
Cechova, D., Svestkova, V., Keil, B., and Sorm, F. (1969). *F.E.B.S. Letters* **4**, 155.
Chandan, R. C., and Shahani, K. M. (1963a). *J. Dairy Sci.* **46**, 275.
Chandan, R. C., and Shahani, K. M. (1963b). *J. Dairy Sci.* **46**, 503.
Chandan, R. C., and Shahani, K. M. (1964). *J. Dairy Sci.* **47**, 471.
Chandan, R. C., and Shahani, K. M. (1965). *J. Dairy Sci.* **48**, 1413.
Chandan, R. C., Shahani, K. M., and Holly, R. G. (1964). *Nature* **204**, 76.
Chandan, R. C., Parry, R. M., Jr., and Shahani, K. M. (1965a). *Biochim. Biophys. Acta* **110**, 389.
Chandan, R. C., Parry, R. M., Jr., and Shahani, K. M. (1965b). *J. Dairy Sci.* **48**, 768.
Chandan, R. C., Parry, R. M., Jr., and Shahani, K. M. (1968). *J. Dairy Sci.* **51**, 606.
Charlemagne, D., and Jollès, P. (1967). *Bull. Soc. Chim. Biol.* **49**, 1103.
Cheeseman, G. C., and Mabbitt, L. A. (1968). *J. Dairy Res.* **35**, 135.
Chien, H. C. (1968). *Diss. Abstr. B* **28**, 4611.
Chien, H. C., and Richardson, T. (1967a). *J. Dairy Sci.* **50**, 451.
Chien, H. C., and Richardson, T. (1967b). *J. Dairy Sci.* **50**, 1868.

Copius Peereboom, J. W. (1968), *Neth. Milk Dairy J.* **22**, 137.
Copius Peereboom, J. W. (1969a). *Milchwissenschaft* **24**, 266.
Copius Peereboom, J. W. (1969b). *Fette, Seifen, Anstrichm.* **71**, 314.
Copius Peereboom, J. W., and Beekes, H. W. (1969). *J. Chromatog.* **39**, 339.
Corbin, E. A., and Whittier, E. O. (1965). *In* "Fundamentals of Dairy Chemistry" (B. H. Webb and A. H. Johnson, eds.), p. 1. Avi, Westport, Connecticut.
Corran, H. S., Dewan, J. G., Gordon, A. H., and Green, D. E. (1939). *Biochem. J.* **33**, 1694.
Coughlan, M. P., Rajagopalan, K. V., and Handler, P. (1969). *J. Biol. Chem.* **244**, 2658.
Coulson, E. J., and Jackson, R. H. (1962). *Arch. Biochem. Biophys.* **97**, 378.
Coulson, E. J., and Stevens, H. (1950). *J. Biol. Chem.* **187**, 355.
Coulson, E. J., and Stevens, H. (1964). *Arch. Biochem. Biophys.* **107**, 336.
Dalaly, B. K., Vakil, J. R., and Shahani, K. M. (1968). *J. Dairy Sci.* **51**, 940.
Desnuelle, P. (1961). *Advan. Enzymol.* **23**, 129.
Dowben, R. M., Brunner, J. R., and Philpott, D. E. (1967). *Biochim. Biophys. Acta* **135**, 1.
Downey, W. K., and Andrews, P. (1965a). *Biochem. J.* **94**, 642.
Downey, W. K., and Andrews, P. (1965b). *Biochem. J.* **94**, 33P.
Downey, W. K., and Andrews, P. (1965c). *Biochem. J.* **96**, 21c.
Downey, W. K., and Andrews, P. (1969). *Biochem. J.* **112**, 559.
Downey, W. K., and Murphy, R. F. (1970). *J. Dairy Res.* **37**, 47.
Ebner, K. E., Denton, W. L., and Brodbeck, U. (1966). *Biochem. Biophys. Res. Commun.* **24**, 232.
Engström, L., and Ågren, G. (1958). *Acta Chem. Scand.* **12**, 357.
Ezekiel, E. (1963). *Biochim. Biophys. Acta* **78**, 223.
Ezekiel, E. (1965). *Biochim. Biophys. Acta* **107**, 511.
Fantes, K. H., and Furminger, I. G. S. (1967). *Nature* **215**, 750.
Feeney, R. E., and Komatusu, S. K. (1966). *Struct. Bonding (Berlin)* **1**, 149.
Forster, T. L., Bendixen, H. A., and Montgomery, M. W. (1959). *J. Dairy Sci.* **42**, 1903.
Forster, T. L., Montgomery, M. W., and Montoure, J. E. (1961). *J. Dairy Sci.* **44**, 1420.
Fox, P. F., and Tarassuk, N. P. (1968). *J. Dairy Sci.* **51**, 826.
Fox, P. F., Yaguchi, M., and Tarassuk, N. P. (1967). *J. Dairy Sci.* **50**, 307.
Fraenkel-Conrat, H., and Feeney, R. E. (1950). *Arch. Biochem.* **29**, 101.
Frankel, E. N., and Tarassuk, N. P. (1959). *J. Dairy Sci.* **42**, 409.
Fridovich, I., and Handler, P. (1958). *J. Biol. Chem.* **231**, 899.
Gaffney, P. J., Jr., and Harper, W. J. (1965). *J. Dairy Sci.* **48**, 613.
Gaffney, P. J., Jr., Harper, W. J., and Gould, I. A. (1966). *J. Dairy Sci.* **49**, 921.
Gaffney, P. J., Jr., Harper, W. J., and Gould, I. A. (1968). *J. Dairy Sci.* **51**, 1161.
Gahne, B. (1961). *Anim. Prod.* **3**, 135.
Gahne, B., Rendel, J., and Venge, O. (1960). *Nature* **186**, 907.
Garnier, J. (1964). *Ann. Biol. Animale., Biochim. Biophys* **4**, 163.
Gibson, J. F., and Bray, R. C. (1968). *Biochim. Biophys. Acta* **153**, 721.
Gilbert, D. A., and Bergel, F. (1964). *Biochem. J.* **90**, 350.
Glazer, A.N., and McKenzie, H. A. (1963). *Biochim. Biophys. Acta* **71**, 109.
Gordon, W. G., Ziegler, J., and Basch, J. J. (1962). *Biochim. Biophys. Acta* **60**, 410.
Gordon, W. G., Groves, M. L., and Basch, J. J. (1963). *Biochemistry* **2**, 817.
Got, R., Bertagnolio, G., Pradal, M. B., and Frot-Coutaz, J. (1968). *Clin. Chim. Acta* **22**, 545.

Goussault, Y., Got, R., and Marnay, A. (1967). *In* "Protides of the Biological Fluids: Proceedings of the 14th Colloquium, 1966" (H. Peeters, ed.), p. 621. Elsevier, Amsterdam.
Green, R. C., and O'Brien, P. J. (1967). *Biochem. J.* **105**, 585.
Greene, F. C., and Feeney, R. E. (1968). *Biochemistry* **7**, 1366.
Groves, M. L. (1960). *J. Amer. Chem. Soc.* **82**, 3345.
Groves, M. L. (1963). Unpublished data.
Groves, M. L. (1965). *Biochim. Biophys. Acta* **100**, 154.
Groves, M. L. (1966). *J. Dairy Sci.* **49**, 204.
Groves, M. L., and Gordon, W. G. (1967). *Biochemistry* **6**, 2388.
Groves, M. L., and Kiddy, C. A. (1964). Unpublished data.
Groves, M. L., Basch, J. J., and Gordon, W. G. (1963). *Biochemistry* **2**, 814.
Groves, M. L., Peterson, R. F., and Kiddy, C. A. (1965). *Nature* **207**, 1007.
Guy, E. J., and Jenness, R. (1958). *J. Dairy Sci.* **41**, 13.
Hanson, L. Å., Samuelsson, E. G., and Holmgren, J. (1967). *J. Dairy Res.* **34**, 103.
Hansson, E., and Rasmusson, Y. (1962). *Z. Lebensm.-Unters.-Forsch.* **118**, 141.
Harper, W. J., Robertson, J. A., Jr., and Gould, I. A. (1960). *J. Dairy Sci.* **43**, 1850.
Hart, L. I., and Bray, R. C. (1967). *Biochim. Biophys. Acta* **146**, 611.
Hart, L. I., McGartoll, M. A., Chapman, H. R., and Bray, R. C. (1970). *Biochem. J.* **116**, 851.
Harwalkar, V. R., and Brunner, J. R. (1965). *J. Dairy Sci.* **48**, 1139.
Hayashi, S., and Smith, L. M. (1965). *Biochemistry* **4**, 2550.
Hayashi, S., Erickson, D. R., and Smith, L. M. (1965). *Biochemistry* **4**, 2557.
Hellung-Larsen, P. (1968). *Comp. Biochem. Physiol.* **27**, 703.
Herald, C. T., and Brunner, J. R. (1957). *J. Dairy Sci.* **40**, 948.
Heyndrickx, G. V. (1964). *Enzymologia* **27**, 209.
Hood, L. F., and Patton, S. (1968). *J. Dairy Sci.* **51**, 928.
Hultquist, D. E., and Morrison, M. (1963). *J. Biol. Chem.* **238**, 2843.
Ibuki, F., Mori, T., Matsushita, S., and Hata, T. (1965). *J. Agr. Biol. Chem. (Tokyo)* **29**, 635.
Inman, J. K. (1956). Ph.D. thesis, Harvard University, Cambridge, Massachusetts.
Jackson, R. H., Coulson, E. J., and Clark, W. R. (1962). *Arch. Biochem. Biophys.* **97**, 373.
Jago, G. R., and Morrison, M. (1962). *Proc. Soc. Exp. Biol. Med.* **111**, 585.
Jenness, R. (1959). *J. Dairy Sci.* **42**, 895.
Jenness, R., and Palmer, L. S. (1945). *J. Dairy Sci.* **28**, 611.
Jenness, R., and Patton, S. (1959). "Principles of Dairy Chemistry." Wiley, New York.
Jensen, R. G. (1964). *J. Dairy Sci.* **47**, 210.
Jensen, R. G., Gander, G. W., Sampugna, J., and Forster, T. L. (1961). *J. Dairy Sci.* **44**, 943.
Jeppsson, J.-O. (1967). *Acta Chem. Scand.* **21**, 1686.
Johansson, B. G. (1960). *Acta Chem. Scand.* **14**, 510.
Johansson, B. G. (1969). *Acta Chem. Scand.* **23**, 683.
Jollès, J., and Jollès, P. (1967). *Biochemistry* **6**, 411.
Jollès, J., and Jollès, P. (1968). *Bull. Soc. Chim. Biol.* **50**, 2543.
Jollès, P., and Jollès, J. (1961). *Nature* **192**, 1187.
Jollès, P., Saint-Blancard, J., Charlemagne, D., Dianoux, A.-C., Jollès, J., and Le Baron, J. L. (1968). *Biochim. Biophys. Acta* **151**, 532.
Jollès, P., Jollès, J., Dianoux, A.-C., Hermann, J., and Charlemagne, D. (1969). *In* "Protides of the Biological Fluids. Proceedings of the 16th Colloquium, 1968" (H. Peeters, ed.), p. 181. Pergamon, Oxford.

Jones, F. S., and Simms, H. S. (1930). *J. Exp. Med.* **51,** 327.
Jordan, S. M., and Morgan, E. H. (1969). *Biochim. Biophys. Acta* **174,** 373.
Jordan, S. M., Kaldor, I., and Morgan, E. H. (1967). *Nature* **215,** 76.
Kay, H. D., Aschaffenburg, R., and Mullen, J. E. C. (1949). *Proc. Int. Dairy Congr., 12th, 1949* **2,** 743.
Keenan, T. W., Morre, D. J., Olson, D. E., and Patton, S. (1969). *J. Dairy Sci.* **52,** 918.
Kiermeier, F., and Güll, J. (1966). *Naturwissenschaften* **53,** 613.
Kiermeier, F., and Güll, J. (1968a). *Z. Lebensm.-Unters. -Forsch.* **138,** 205.
Kiermeier, F., and Güll, J. (1968b). *München Med. Wochenschr.* **110,** 1813.
Kiermeier, F., and Kayser, C. (1960a). *Z. Lebensm.-Unters. -Forsch.* **112,** 481.
Kiermeier, F., and Kayser, C. (1960b). *Z. Lebensm.-Unters. -Forsch.* **113,** 22.
Kiermeier, F., and Kayser, C. (1960c). *Z. Lebensm.-Unters. -Forsch.* **113,** 97.
Kiermeier, F., and Meinl, E. (1961). *Z. Lebensm.-Unters. -Forsch.* **114,** 189.
Kiermeier, F., and Petz, E. (1967a). *Z. Lebensm.-Unters. -Forsch.* **132,** 342.
Kiermeier, F., and Petz, E. (1967b). *Z. Lebensm.-Unters. -Forsch.* **134,** 97.
Kiermeier, F., and Petz., E. (1967c). *Z. Lebensm.-Unters. -Forsch.* **134,** 149.
Kiermeier, F., and Semper, G. (1960a). *Z. Lebensm.-Unters. -Forsch.* **111,** 373.
Kiermeier, F., and Semper, G. (1960b). *Z. Lebensm.-Uters. -Forsch.* **111,** 282.
Kiermeier, F., and Semper, G. (1960c). *Z. Lebensm.-Unters. -Forsch.* **111,** 483.
King, N. (1955). "The Milk Fat Globule Membrane." Commonwealth Agriculture Bureau, Farnham Royal, Bucks, England.
Kjellberg, B., and Karlsson, B. W. (1966). *Biochim. Biophys. Acta* **128,** 589.
Kjellberg, B., and Karlsson, B. W. (1967). *Comp. Biochem. Physiol.* **22,** 397.
Klebanoff, S. J., and Luebke, R. G. (1965). *Proc. Soc. Exp. Biol. Med.* **118,** 483.
Kleyn, D. H., and Lin, S. H. C. (1968). *J. Ass. Offic. Anal. Chem.* **51,** 802.
Komai, H., Massey, V., and Palmer, G. (1969). *J. Biol. Chem.* **244,** 1692.
Korn, E. D. (1962). *J. Lipid Res.* **3,** 246.
Kresheck, G. C. (1963). *Acta Chem. Scand.* **17,** Suppl. 1, S295.
Kresheck, G. C., and Harper, W. J. (1967). *Milchwissenschaft* **22,** 72.
Krukovsky, V. N. (1961). *J. Agr. Food Chem.* **9,** 439.
Larson, B. L., and Gillespie, D. C. (1957). *J. Biol. Chem.* **227,** 565.
Larson, B. L., and Kendall, R. A. (1957). *J. Dairy Sci.* **40,** 377.
Laskowski, M., Jr., and Laskowski, M. (1951). *J. Biol. Chem.* **190,** 563.
Laskowski, M., Jr., Mars, P. H., and Laskowski, M. (1952). *J. Biol. Chem.* **198,** 745.
Laurell, C. B. (1960). In "The Plasma Proteins" (F. W. Putnam, ed.), Vol. I. Academic Press, New York.
Leach, B. E., Blalock, C. R., and Pallansch, M. J. (1967). *J. Dairy Sci.* **50,** 763.
Livrea, G., Campanella, S., and Fama Cambria, M. (1964). *Quad. Nutr.* **24,** 1.
Loisillier, F., Got, R., Burtin, P., and Grabar, P. (1967). In "Protides of the Biological Fluids: Proceedings of the 14th Colloquium, 1966" (H. Peeters, ed.), p. 133. Elsevier, Amsterdam.
Lyster, R. L. J., and Aschaffenburg, R. (1962). *J. Dairy Res.* **29,** 21.
Mach, J. P., Pahud, J. J., and Isliker, H. (1969). *Nature* **223,** 952.
Mackler, B., Mahler, H. R., and Green, D. E. (1954). *J. Biol. Chem.* **210,** 149.
McMeekin, T. L. (1954). In "Proteins" (H. Neurath and K. Bailey, eds.), Vol. II, Part A, p. 389. Academic Press, New York.
McMeekin, T. L., Hipp, N. J., and Groves, M. L. (1959). *Arch. Biochem. Biophys.* **83,** 35.
Mann, K. G., Fish, W. W., Cox, A. C., and Tanford, C. (1970). *Biochemistry* **9,** 1348.
Marquardt, R. R., and Forster, T. L. (1966). *J. Dairy Sci.* **49,** 19.

Massey, V., Brumby, P. E., Komai, H., and Palmer, G. (1969). *J. Biol. Chem.* **244**, 1682.
Masson, P. L., and Heremans, J. F. (1967). *In* "Protides of the Biological Fluids: Proceedings of the 14th Colloquium, 1966" (H. Peeters, ed.), p. 115. Elsevier, Amsterdam.
Masson, P. L., and Heremans, J. F. (1968). *Eur. J. Biochem.* **6**, 579.
Masson, P. L., Heremans, J. F., and Brignot, J. (1965). *Experientia* **21**, 604.
Masson, P. L., Heremans, J. F., and Dive, C. (1966a). *Clin. Chim. Acta* **14**, 735.
Masson, P. L., Heremans, J. F., Prignot, J. J., and Wauters, G. (1966b). *Thorax* **21**, 538.
Masson, P. L., Heremans, J. F., and Schonne, E. (1969a). *J. Exp. Med.* **130**, 643.
Masson, P. L., Heremans, J. F., Schonne, E., and Crabbe, P. A. (1969b). *In* "Protides of the Biological Fluids: Proceedings of the 16th Colloquium, 1968" (H. Peeters, ed.), p. 633. Pergamon, Oxford.
Matsushita, S., Ibuki, F., Mori, T., and Hata, T. (1963). *Agr. Biol. Chem. (Tokyo)* **27**, 736.
Matsushita, S., Ibuki, F., Mori, T., and Hata, T. (1965). *Agr. Biol. Chem. (Tokyo)* **29**, 436.
Mellors, A. (1968). *Can. J. Biochem.* **46**, 451.
Mellors, A. (1969). *Can J. Biochem.* **47**, 173.
Mellors, A., and Harwalkar, V. R. (1968). *Can. J. Biochem.* **46**, 1351.
Mitchell, W. M. (1967). *Biochim. Biophys. Acta* **147**, 171.
Montgomery, M. W., and Forster, T. L. (1961). *J. Dairy Sci.* **44**, 721.
Montreuil, J., Tonnelat, J., and Mullet, S. (1960). *Biochim. Biophys. Acta* **45**, 413.
Montreuil, J., Spik, G., Monsigny, M., Descamps, J., Biserte, G., and Dautrevaux, M. (1965). *Experimentia* **21**, 254.
Morell, D. B., and Clezy, P. S. (1963). *Biochim. Biophys. Acta* **71**, 157.
Morrison, M., and Allen, P. Z. (1963). *Biochem. Biophys. Res. Commun.* **13**, 490.
Morrison, M., and Allen, P. Z. (1966). *Science* **152**, 1626.
Morrison, M., and Hultquist, D. E. (1963). *J. Biol. Chem.* **238**, 2847.
Morrison, M., Hamilton, H. B., and Stotz, E. (1957). *J. Biol. Chem.* **228**, 767.
Morrison, M., Allen, P. Z., Bright, J., and Jayasinghe, W. (1965). *Arch. Biochem. Biophys.* **111**, 126.
Morton, R. K. (1953). *Biochem. J.* **55**, 795.
Morton, R. K. (1954). *Biochem. J.* **57**, 231.
Morton, R. K. (1955). *In* "Methods in Enzymology" (S. P. Colowick and N. O. Kaplan, eds.), Vol. II, p. 533. Academic Press, New York.
Mouton, A., and Jollès, J. (1969). *F.E.B.S. Letters* **4**, 337.
Mullen, J. E. C. (1950). *J. Dairy Res.* **17**, 288.
Nakamura, S., and Yamazaki, I. (1969). *Biochim. Biophys. Acta* **189**, 29.
Nelson, C. A., and Handler, P. (1968). *J. Biol. Chem.* **243**, 5368.
Oram, J. D., and Reiter, B. (1968). *Biochim. Biophys. Acta* **170**, 351.
Orme-Johnson, W. H., and Beinert, H. (1969). *Biochem. Biophys. Res. Commun.* **36**, 337.
Osserman, E. F., Cole, S. J., Swan, I.D.A., and Blake, C.C.F. (1969). *J. Mol. Biol.* **46**, 211.
O'Sullivan, A. C., and Shipe, W. F. (1963). *J. Diary Sci.* **46**, 596.
Palmer, L. S. (1944). *J. Dairy Sci.* **27**, 471.
Palmer, G., and Massey, V. (1969). *J. Biol. Chem.* **244**, 2614.
Parry, R. M., Jr., Chandan, R. C., and Shahani, K. M. (1967). *J. Dairy Sci.* **50**, 943.
Parry, R. M., Jr., Chandan, R. C., and Shahani, K. M. (1969). *Arch. Biochem. Biophys.* **130**, 59.
Patel, C. V., Fox, P. F., and Tarassuk, N. P. (1968). *J. Dairy Sci.* **51**, 1879.

Patton, S., and Fowkes, F. M. (1967). *J. Theor. Biol.* **15**, 274.
Paul, K. G. (1963). *In* "The Enzymes" (P. D. Boyer, H. Lardy, and K. Myrbäck, eds.), 2nd ed., Vol. 8, Chap. 7. Academic Press, New York.
Pick, F. M., and Bray, R. C. (1969). *Biochem. J.* **114**, 735.
Plummer, T. H., Jr., and Hirs, C. H. W. (1963). *J. Biol. Chem.* **238**, 1396.
Polis, B. D., and Shmukler, H. W. (1950). *J. Dairy Sci.* **33**, 619.
Polis, B. D., and Shmukler, H. W. (1953). *J. Biol. Chem.* **201**, 475.
Polis, B. D., Shmukler, H. W., and Custer, J. H. (1950). *J. Biol. Chem.* **187**, 349.
Portmann, A., and Auclair, J. E. (1959). *Lait* **39**, 147.
Prentice, J. H. (1969). *Dairy Sci. Abstr.* **31**, 353.
Rajagopalan, K. V., and Handler, P. (1964). *J. Biol. Chem.* **239**, 1509.
Reiter, B., and Oram, J. D. (1967). *Nature* **216**, 328.
Richardson, G. A., and Hankinson, C. L. (1936). *J. Dairy Sci.* **19**, 761.
Richardson, L. A., McFarren, E. F., and Campbell, J. E. (1964). *J. Dairy Sci.* **47**, 205.
Roberts, R. C., Makey, D. G., and Seal, U. S. (1966). *J. Biol. Chem.* **241**, 4907.
Robertson, J. A., Harper, W. J., and Gould, I. A. (1966). *J. Dairy Sci.* **49**, 1386.
Rolleri, G. D., Larson, B. L., and Touchberry, R. W. (1956). *J. Dairy Sci.* **39**, 1683.
Rombauts, W. A., Schroeder, W. A., and Morrison, M. (1967). *Biochemistry* **6**, 2965.
Roop, W. E. (1963). Ph.D. thesis. University of Florida, Gainesville, Florida.
Roussos, G. G., and Morrow, B. H. (1966). *Arch. Biochem. Biophys.* **114**, 599.
Roussos, G. G., and Morrow, B. H. (1967). *Biochem. Biophys. Res. Comm.* **29**, 388.
Saint-Blancard, J., Locquet, J.-P., and Jollès, P. (1969). *In* "Protides of the Biological Fluids: Proceedings of the 16th Colloquium, 1968" (H. Peeters, ed.), p. 191. Pergamon, Oxford.
Schade, A. L., and Caroline, L. (1944). *Science* **100**, 14.
Schade, A. L., and Caroline, L. (1946). *Science* **104**, 340.
Schade, A. L., Reinhart, R. W., and Levy, H. (1949). *Arch. Biochem.* **20**, 170.
Schade, A. L., Pallavicini, C., and Wiesmann, U. (1969). *In* "Protides of the Biological Fluids: Proceedings of the 16th Colloquium., 1968" (H. Peeters, ed.), p. 619. Pergamon, Oxford.
Schardinger, F. (1902). *Z. Unters. Nahr. Genussm. Gebrauchsgegenstaende* **5**, 1113.
Scheraga, H. A., and Rupley, J. A. (1962). *Advan. Enzymol.* **24**, 161.
Schwartz, J. H. (1963). *Proc. Natl. Acad. Sci. U.S.* **49**, 871.
Schwartz, J. H., and Lipmann, F. (1961). *Proc. Natl. Acad. Sci. U.S.* **47**, 1996.
Shahani, K. M. (1966). *J. Dairy Sci.* **49**, 907.
Shahani, K. M., Chandan, R. C., Kelly, P. L., and Macquiddy, E. L., Sr. (1962). *Proc. Int. Dairy Congr., 16th, 1962* **8**, 285.
Skean, J. D., and Overcast, W. W. (1961). *J. Dairy Sci.* **44**, 823.
Smith, E. L. (1946). *J. Biol. Chem.* **165**, 665.
Smithies, O., and Hickman, C. G. (1958). *Genetics* **43**, 374.
Sørensen, M., and Sørensen, S.P.L. (1939). *C. R. Trav. Lab. Carlsberg., Ser. Chim.* **23**, 55.
Spik, G., and Montreuil, J. (1966). *C. R. Soc. Biol.* **160**, 94.
Spik, G., Monsigny, M., and Montreuil, J. (1969). *Bull. Soc. Chim. Biol.* **50**, 2186.
Stadhouders, J., and Veringa, H. A. (1962). *Neth. Milk Dairy J.* **16**, 96.
Stadtman, T. C. (1961). *In* "The Enzymes" (P. D. Boyer, H. Lardy and K. Myrbäck, eds.), 2nd ed., Vol. 5, Chap. 4. Academic Press, New York.
Steele, W., and Morrison, M. (1969). *J. Bacteriol.* **97**, 635.
Stewart, P. S., and Irvine, D. M. (1969). *J. Dairy Sci.* **52**, 917.
Swope, F. C., and Brunner, J. R. (1965). *J. Dairy Sci.* **48**, 1705.
Swope, F. C., and Brunner, J. R. (1968). *Milchwissenschaft* **23**, 470.

Swope, F. C., Kolar, C. W., Jr., and Brunner, J. R. (1966). *J. Dairy Sci.* **49**, 1279.
Swope, F. C., Rhee, K. C., and Brunner, J. R. (1968). *Milchwissenschaft* **23**, 744.
Szuchet-Derechin, S., and Johnson, P. (1962). *Nature* **194**, 473.
Szuchet-Derechin, S., and Johnson, P. (1965a). *Eur. Polym. J.* **1**, 271.
Szuchet-Derechin, S., and Johnson, P. (1965b). *Eur. Polym. J.* **1**, 283.
Szuchet-Derechin, S., and Johnson, P. (1966a). *Eur. Polym. J.* **2**, 29.
Szuchet-Derechin, S., and Johnson, P. (1966b). *Eur. Polym J.* **2**, 115.
Tarassuk, N. P., and Frankel, E. N. (1957). *J. Dairy Sci.* **40**, 418.
Tarassuk, N. P., and Yaguchi, M. (1959). *J. Dairy Sci.* **42**, 864.
Tarassuk, N. P., Nickerson, T. A., and Yaguchi, M. (1964). *Nature* **201**, 298.
Theorell, H., and Åkeson, Å. (1943). *Arkiv Kemi, Mineral. Geol.* **17B**, No. 7, 1.
Theorell, H., and Paul, K. G. (1944). *Arkiv Kemi, Mineral. Geol.* **18A**, No. 12, 10.
Theorell, H., and Pedersen, K. O. (1944). *In* "The Svedberg" (A. Tiselius and K. O. Pedersen, eds.), p. 523. Almqvist and Wiksell, Uppsala.
Timasheff, S. N. (1964). *In* "Symposium on Foods—Proteins and Their Reactions" (H. W. Schultz and A. F. Anglemier, eds.), p. 179. Avi, Westport, Connecticut.
Townley, R. C., and Gould, I. A. (1943). *J. Dairy Sci.* **26**, 843.
Uozumi, M., Hayashikawa, R., and Piette, L. H. (1967). *Arch. Biochem. Biophys.* **119**, 288.
Vakil, J. R., Chandan, R. C., Parry, R. M., and Shahani, K. M. (1969). *J. Dairy Sci.* **52**, 1192.
van Maele, A., and Vercauteren, R. (1962). *Naturwissenschaften* **49**, 14.
Warner, R. C., and Polis, E. (1945). *J. Amer. Chem. Soc.* **67**, 529.
Warner, R. C., and Weber, I. (1951). *J. Biol. Chem.* **191**, 173.
Warner, R. C., and Weber, I. (1953). *J. Amer. Chem. Soc.* **75**, 5094.
Whitney, R. M. (1958). *J. Dairy Sci.* **41**, 1303.
Windle, J. J., Wiersema, A. K., Clark, J. R., and Feeney, R. E. (1963). *Biochemistry* **2**, 1341.
Wishnia, A., Weber, I., and Warner, R. C. (1961). *J. Amer. Chem. Soc.* **83**, 2071.
Wright, R. C., and Tramer, J. (1958). *J. Dairy Res.* **25**, 104.
Woerner, F. (1961). *Kiel. Milchwirt. Forschungsber.* **13**, 361.
Yaguchi, M., Tarassuk, N. P., and Abe, N. (1964). *J. Dairy Sci.* **47**, 1167.
Yamauchi, K., Kaminogawa, S., and Tsugo, T. (1969). *Jap. J. Zootech. Sci.* **40**, 67.
Zikakis, J. P., and Treece, J. M. (1969). *J. Dairy Sci.* **52**, 916.
Zittle, C. A. (1964). *J. Dairy Sci.* **47**, 202.
Zittle, C. A. (1965). *J. Dairy Sci.* **48**, 771.
Zittle, C. A., and Bingham, E. W. (1959). *J. Dairy Sci.* **42**, 1772.
Zittle, C. A., and Bingham, E. W. (1960). *Arch. Biochem. Biophys.* **86**, 25.
Zittle, C. A., and Custer, J. H. (1963). *J. Dairy Sci.* **46**, 1183.
Zittle, C. A., and DellaMonica, E. S. (1952). *Arch. Biochem. Biophys.* **35**, 321.

Part F
Milk Proteins and Technology

General Introduction

It is important that those concerned with the technological problems of milk are aware of the basic information that has now become available on milk proteins. At the same time, those who are actually working on the structures and properties of the proteins should be cognizant of some of the applied problems. The interrelationships of milk protein research and milk technology are discussed by Beeby, Hill, and Snow in Chapter 17.

H. A. McKenzie

17 □ Milk Protein Research and Milk Technology

R. BEEBY, R. D. HILL, AND N. S. SNOW

I. Introduction 422
II. Cheese Manufacture 422
    A. Manufacturing Processes and Associated Problems 422
    B. Action of Rennin on Casein 424
    C. Coagulation and Syneresis of Casein 426
    D. Natural Variations in Milk Composition and Their Effect on Cheese Manufacture 427
    E. Effects of Milk Processing on the Cheesemaking Properties of Milk . . 428
    F. New Cheesemaking Processes 430
    G. Research and Technology 432
III. Concentrated Milk and Milk Powder 434
    A. Processes and Problems 434
    B. Effect of Heat on Milk. 435
    C. Evaporated Milk 438
    D. Sweetened Condensed Milk 442
    E. Frozen Concentrated Milk 443
    F. Milk Powder 444
IV. The Manufacture of Casein, Coprecipitate, and Whey Proteins . . . 446
    A. Casein Manufacture 446
    B. Coprecipitate Manufacture 447
    C. Research Aspects of Casein and Coprecipitate Manufacture 448
    D. Whey Protein Manufacture 449
    E. Research Aspects of Whey Protein Manufacture 451
V. Specialized Products 452
    A. Use of Milk Products in Other Foods and as Special-Purpose Foods . . 452
    B. Use of Milk Products to Replace Other Foods: Modification of Proteins . 453
    C. Investigations Related to Fat–Protein Interactions 455
VI. Conclusion 457
    References 459

I. Introduction

Although liquid milk is the main source of milk proteins in the diet, a substantial proportion is provided in developed countries by products manufactured from milk, such as cheese, powdered whole and skim milks, sweetened and unsweetened concentrated milks, and casein. An impression of the importance of these commodities is given by the statistics for output in the main producing countries of the world during 1968. In millions of tons, these were cheese, 4.2; concentrated milks, 3.1; milk powders, 3.2; and casein, 0.15 ("Dairy Produce," 1969). In 13 countries of Western Europe, North America and Australasia, about 13% of the milk supply was used in cheese manufacture and about 10% for the manufacture of milk concentrates and ice cream. In addition to these well-established products, new products such as butter powder and coprecipitated milk protein are being developed for special purposes; though not important at present, they may become so in time.

The manufacture of all these products results in, or may depend on, some change in the state of the milk protein. In some cases, the need to control these changes determines manufacturing techniques; in other cases, alterations of the milk protein are related to difficulties which occur in manufacture. A full understanding of the nature of milk proteins and their interactions in various conditions is therefore clearly desirable. At the same time, the complexity of the milk protein system—which has only recently begun to be appreciated—makes the gaining of this understanding a difficult task. Much research has necessarily been concerned with the isolation and characterization of the individual milk proteins and their many variants. While this work is essential to ultimate understanding, its results may not have an immediate application to the problems of milk technology. Because of the size of the field to be covered, this chapter is restricted to short descriptions of the methods of production and attendant problems and to summaries of recent research that seems to bear on these problems. We have also attempted to indicate some lines of research that may fruitfully be followed in the future.

II. Cheese Manufacture

A. Manufacturing Processes and Associated Problems

Except for some types of cottage cheese or quarg, in which the milk is clotted mainly by acid, and except for some whey cheese, all cheese manufacture depends upon the formation of a curd by the action of rennin or similar enzymes. Typical stages in the manufacture are, briefly, as follows:

(1) *Pasteurization.* The milk is pasteurized under conditions that do not alter the casein sufficiently to affect significantly its later reaction with rennet.

(2) *Adjustment of composition.* The desired fat–protein ratio is obtained by adding cream or skim milk.

(3) *Coagulation.* Bacterial starter culture and rennet are added to the milk in the vat, usually at a temperature of about 30°C. The conversion of lactose to lactic acid by the bacteria and the alteration of the casein by the rennet proceed simultaneously. Usually, sufficient rennet is added to cause coagulation of the milk in about half an hour to one hour.

(4) *Cutting the curd and expulsion of whey.* When the coagulum is sufficiently firm the curd is cut, usually into cubes of about 12 mm, and the contents of the vat are stirred to accelerate the expulsion of the whey from the curd particles (syneresis). This process may be expedited by raising the temperature and by dry-stirring the curd in the vat after the bulk of the whey has been drained off. The rate of syneresis is also increased as the whey becomes more acid.

(5) *Addition of salt and fusion of the curd particles.* Salt is added either to the curd particles in solid form or, later, to the cheese by immersion in brine. It acts as a preservative, inhibiting the growth of undesirable organisms. The curd particles are formed into blocks of a size suited to the requirements of the maturing stage.

(6) *Maturing.* The flavor and texture of the cheese are developed by the action of the starter bacteria, residual rennet, incidental flora and externally added ripening agents such as molds. These act on the fat, protein and residual lactose in the cheese.

The great variety of cheese types arises through differences in such things as the composition of the curd (the proportions of fat, protein and water), the extent of acid production, the structure of the curd, the salt content, and the degree and mode of fat and protein breakdown during ripening. For the manufacture of a hard, pressed cheese such as cheddar the fat content is about 50% of the dry weight, and a final water content of about 36% is sought. The maturing agents act in an acid environment, at approximately pH 5. In producing a softer cheese, several factors may be altered. The fat content may be increased to 60–70% of the dry matter as in cream cheeses, or the water content may be greater as in Livarot and Camembert cheeses, in which it is 46–50%. Further differences between varieties of cheese occur as a result of differences in conditions of maturing. In cheeses ripened with the aid of molds (Camembert) or bacterial smears (Limburger), there is a characteristic mode of fat and protein breakdown, which affects both the flavor and texture of the cheese. The temperature and humidity conditions of storage, the salt content, composition and the acidity of the cheese are adjusted to favor the desired course of ripening.

This brief account indicates that the manufacture of cheese, from the addition of rennet to the final maturing, is closely concerned with alterations in the state of the protein. Research into the nature and behavior of milk proteins should therefore be helpful in understanding and overcoming some of the problems of the cheese industry. Some examples of problems associated with the behavior of protein in cheese manufacture are (1) slow coagulation with rennet, (2) failure to form a coagulum (soft curd), (3) slow, or insufficient, syneresis of the curd, (4) poor fusion of the curd particles, and (5) undesirable changes in flavor and texture of the cheese during maturing. Information relevant to problems 1–4 may be obtained by conducting research on (a) the nature of the substrate of rennin, (b) the reaction between rennin and that substrate, both in isolated systems and in milk, (c) the nature of the coagulation and syneresis of the casein in milk following rennin action, (d) the influence of milk composition on b and c, and (e) the influence of processing treatments on b and c. Although problems connected with maturing of the cheese are important in cheese manufacture, they are less directly related to milk protein research than the others mentioned and will therefore not be considered in this chapter.

B. ACTION OF RENNIN ON CASEIN

1. *Course of the Reaction*

Although it had been realized earlier (Berridge, 1942) that the alteration of casein by rennet and the subsequent coagulation are two separate reactions, Alais *et al.* (1953) first made the important observation that the action of rennin causes the release from casein of a glycopeptide soluble in 12% trichloracetic acid. In the presence of calcium, the casein then polymerizes to form a gel, presumably because the release of the peptide exposes the functional groups required for the polymerization. The presence of calcium is not needed for the enzymic reaction, the rate of which increases with increasing temperature up to 42.5°C (Foltmann, 1966) and decreases as the pH value is raised from 5.5 to 6.7 (Nitschmann and Bohren, 1955). Berridge (1942) pointed out that the two reactions (enzymic alteration and coagulation) can be separated because of the relatively high temperature coefficient of the latter. At temperatures of 30°C and higher, coagulation follows almost immediately after the enzymic alteration. At low temperatures, the enzymic reaction still proceeds at a useful rate, whereas the coagulation may be delayed indefinitely. This fact has been made the basis of several cheese manufacturing processes.

2. Nature of the Substrate of Rennin

The isolation of the κ-casein fraction and its identification as the substrate of rennin are dealt with in Chapter 12, and it is sufficient to give here only the broad results of this work. These are the separation of the κ-casein fraction and the demonstration of its ability to protect α_s-casein from precipitation by calcium (Waugh and von Hippel, 1956), and the action of rennin on this fraction to release peptide material (Wake, 1959) which has a composition similar to that released by rennin from whole casein (Nitschmann and Beeby, 1960). As a result of the release of this material the κ-casein loses its power to protect the α_s-casein, and in the presence of calcium(II), coagulation occurs. κ-Casein is subject to genetic variation and its sialic acid and carbohydrate contents may also vary, although variants without sialic acid still possess the ability to protect α_s-casein and are acted on by rennin (Mackinlay and Wake, 1965). There is however still some uncertainty as to the nature of κ-casein (vs. Beeby, 1965), and it appears likely that the κ-caseins isolated by Mackinlay and Wake (1965) may not be the only rennin-sensitive material in whole casein, but this does not affect the general concept outlined above. Schmidt and Koops (1965) have reported that differences in the type of κ-casein in milk are associated with differences in the heat stability of the concentrated product, but it is not known if there is any relation between the properties of milk for cheesemaking and the type of κ-casein it contains.

3. Functional Groups Concerned in the Reaction between Rennin and Casein

Detailed information on the bond in casein that is split by rennin and the nature of the active center of rennin is not yet available. The results of Delfour et al. (1965) suggest that this bond should be phenylalanyl-methionine, and this is in reasonable accord with the specificity of the enzyme (Fish, 1957; Foltmann, 1966). However, a number of aspects of the reaction suggest that the bond may possess unusual features. The specific action of rennin on casein is much faster than its normal proteolytic action, and it can proceed quite rapidly at a pH of 6.5 which is rather higher than the optimum for proteolytic action, pH 4 (Fish, 1957); the same bond is also split relatively rapidly by other proteolytic enzymes (Dennis and Wake, 1965). From studies of the action of rennin on modified casein and on model peptides, Hill (1968) has suggested that the attack on the rennin-sensitive bond is catalyzed by nearby serine and histidine side chains in the κ-casein.

Concerning the functional groups in the active center of rennin, Cheeseman (1963) and Mocquot and Garnier (1965) reported that rennin is not inactivated by diisopropyl fluorophosphate, and from this it would appear that the active center does not contain a functional serine residue. As in many enzymes, histidine side chains appear to be essential for activity (Hill and Laing, 1965b). Rennin may also contain essential lysine (Hill and Laing, 1966, 1967) and carboxyl functions (Stepanov et al., 1968).

The preparation of rennin, its relation to prorennin, its amino acid composition, molecular weight and variability have been the subjects of a thorough study by Foltmann, who has also commenced the determination of its amino acid sequence (Foltmann, 1966 and Chapter 13). It is to be hoped that all these lines of study may eventually be integrated to explain the mechanism of the reaction between rennin and its substrate.

C. COAGULATION AND SYNERESIS OF CASEIN

1. *Factors Affecting the Rate and Extent of Coagulation and Syneresis*

This stage is most important in cheese manufacture, as the properties of the curd have an important influence on the mature cheese. For example, the amount of whey retained in the curd has an effect on the texture and acidity of the cheese and may influence flavor by affecting the growth of bacteria or other ripening agents. In most of the studies of factors that affect the rate of coagulation, no distinction has been made between the primary enzymic reaction and the coagulation proper. As previously stated, calcium(II) is necessary for the coagulation, and the rate of coagulation increases rapidly with temperature (Berridge, 1962). In milk, the rate is reduced as the content of colloidal calcium phosphate is lessened (Pyne and McGann, 1962), in spite of the fact that the stability of the casein toward calcium ion is also reduced (McGann and Pyne, 1960). The firmness of the coagulum (curd tension) is reduced by decreasing casein, total calcium and phosphate contents; it is little affected by serum protein, lactose and citric acid contents or by the content and particle size of the fat (Weisberg et al., 1933). These authors also showed that there is a sharp drop in curd tension as the pH value is increased from 6 to 7.2, and it is also reduced by additions of NaCl (Sirry and Shipe, 1958).

Recent studies of the syneresis of casein gel have been made by Whitehead and Harkness (1954), Lawrence (1959), Cheeseman (1962) and Stoll and Morris (1966). In summary, the rate increases as the pH value is

17. MILK PROTEIN RESEARCH AND MILK TECHNOLOGY

reduced from 7 to 6; it increases with temperature, with agitation of the curd in the whey and with dry-stirring, while the addition of salts such as NaCl and KCl to the milk represses syneresis. The observation by Conochie and Sutherland (1965) that addition of $CaCl_2$ (at the salting stage) to the curd particles improves their fusion, suggests that the process of curd fusion is similar to those of coagulation and syneresis.

2. Role of Functional Groups in Coagulation and Syneresis

The requirement for calcium ion in coagulation led to the suggestion that coagulation might occur because of the formation of intermolecular calcium bridges between phosphate groups (McFarlane, 1938). The results of Hsu et al. (1958) support this idea but also show that other functional groups are likely to be involved, as enzymatically dephosphorylated casein could be coagulated by rennin if the calcium concentration were sufficiently increased. These functional groups could be carboxyl groups, which are known to form complexes with calcium(II) (Reisfeld, 1957). Hill and Laing (1965a) showed that coagulation was not affected by the modification of tryptophan, methionine and (possibly) tryosine groups of the casein but that modification of histidine (by photooxidation) did interfere with coagulation. This result, coupled with the physical changes in the structure of the rennet curd which occur at pH values near the pK of imidazole, makes it probable that histidine side chains take part in the coagulation. More recently it has been shown that the coagulation of rennin-altered whole casein can be inhibited by blocking a few lysine ϵ-amino groups (Hill and Craker, 1968) and 1–2 arginine side chains (Hill, 1970), the important groups being located on the κ-casein fraction. Hill (1970) presented evidence that the histidine, lysine and arginine residues form a positively charged cluster which plays an important part in the coagulation. Although Christ (1956) has suggested that SH groups take part in the later stages of syneresis, there is at present no definite evidence that the functional groups causing the syneresis of the curd are different from those responsible for the coagulation (however, see Cheeseman, 1962). It is suggested on the basis of the limited evidence available that both coagulation and syneresis are the result of complex interactions in which a number of different groups—imidazole, guanidine and amine, ester phosphate and colloidal phosphate, and ionic calcium—play a part.

D. NATURAL VARIATIONS IN MILK COMPOSITION AND THEIR EFFECT ON CHEESE MANUFACTURE

Milk is a rather variable raw material, as most cheesemakers know. A very thorough study by White and Davies (1958) of some 40 compositional factors in bulk and individual milks showed that seasonal changes occurred in all of them. In their opinion the factor best correlated with increased time of coagulation was increased pH value, while weaker correlations existed with lessened content of ionized calcium, lessened soluble inorganic phosphate and increased chloride content. They could not find any correlation between time of coagulation and casein content. Pyne and McGann (1962) found that low contents of colloidal calcium phosphate were associated with slow coagulation in natural milks. Changes in composition leading to slower coagulation may occur naturally as lactation progresses or may be the result of mastitic infection (White and Davies, 1958). These authors noted that mastitic milk tended to be more alkaline than normal milks (compare Feagan et al., 1966) and that slower coagulation was also associated with reduced potassium contents in these milks, although this is probably fortuitous (compare Cheeseman, 1962) since the salt (NaCl) concentration increases considerably in such milks (Kisza et al., 1964). Significant reductions in the casein-total protein ratio may also occur as a result of mastitis (Ashworth, 1965).

The firmness of the coagulum, like the rate of coagulation, is affected by the composition of the milk. Fontana et al. (1962) found weak curd to be related to high pH value and lowered content of calcium ion in poorly coagulating milks; surprisingly this condition was not remedied by the addition of calcium(II). Probst (1964) reported that the content of casein in milk was lower in hot weather, which is of interest in view of the finding that lowered contents of casein are associated with weakened curd strength (Weisberg et al., 1933). The increased salt content and higher pH value of mastitic milk must also cause a weakening of the curd and repress its syneresis (Kisza et al., 1964; Cheeseman, 1962). The increased content of bovine serum albumin in such milk is likewise associated with reductions in curd strength (Feagan et al., 1969). In addition to the foregoing changes in composition, it has been suggested by Fontana et al. (1962) and Mocquot et al. (1954) that the relative proportions of the various casein fractions may change, with possible deleterious effects on coagulation. However Hill et al. (1965), in studying seasonal variations in the sialic acid content and amino acid contents of the casein in poorly coagulating milks, were unable to find any evidence of a significant change in the relative proportions of α_s-, β-, and κ-caseins.

E. Effects of Milk Processing on the Cheesemaking Properties of Milk

The main variable factors encountered in the processing of milk are the temperature and time of holding the milk in cool storage, and the temperature and time of pasteurizing. Although there is a widespread opinion that cold storage is quite harmful to the cheesemaking properties of milk, the work of Rapp and Calbert (1954) does not fully support this. In their experiments, milk held one day at 2°C showed only a 4% increase in the time of coagulation, while storage for 2 and 3 days caused increases of 7.2% and 12.5%, respectively. These effects could be partly offset by tempering the milk (holding at 30°C) before the addition of rennet, so that the changes caused by cooling were reversible. Similar results were obtained by Vassal and Auclair (1966). Scott-Blair and Burnett (1959) reported that when milk was held at 2°C and then heated rapidly to 21°–38°C just before adding the rennet, there was a considerable delay before the normal course of coagulation followed. These effects may be caused by changes in the structure of the casein micelle resulting from the known increase of solubility of β-casein at low temperatures (Payens and Markwijk, 1963). From the practical point of view, storage overnight at 2°C followed by adequate tempering appears to be a safe procedure, although more prolonged cold storage may be harmful (Stadhouders et al., 1962).

In pasteurizing milk for cheesemaking the time and temperature of the process must be accurately controlled. Kannan and Jenness (1956) first showed that the increased coagulation times and weak curd (obtained on rennet treatment of excessively heated milks) were the results of an interaction between β-lactoglobulin and casein. They also considered that a change occurred in the form of the colloidal calcium phosphate during the heating (Kannan and Jenness, 1961). The coagulation time of the heated milk could be reduced by dialyzing it against unheated skim milk. If the conditions of pasteurization were sufficiently mild (e.g., 73°C for 15 sec), these effects did not occur. In this connection an important observation was made by Mauk and Demott (1959). They heated casein solutions and mixtures of whole acid casein and β-lactoglobulin to 62° and 71°C for 30 min and found that the reduction of curd tension in the heated lactoglobulin–casein mixtures could be prevented to a large extent by the addition of as little as 0.15% of NaCl to their calcium phosphate buffers. If they are applicable to milk, their results would indicate a means by which more drastic conditions of pasteurizing could be employed when necessary, without the usual harmful effects on the curd strength.

Another process which is widely used for market milk but which is not commonly applied to milk for cheesemaking is that of homogenization. There has been a number of investigations on the effects of homogenization of milk on the quality of different types of cheese; this subject has been reviewed by Peters (1964). The effects depend on the pressure used but, briefly, the results are that (a) fat losses are reduced, particularly at high temperature, because of better trapping of fat by the protein, (b) coagulation time is shortened as compared with unhomogenized whole milks, (c) soft cheeses are not impaired in quality but hard cheeses are affected unfavorably because the curd tends to be softer, presumably because of the dispersal of some casein on fat globule surfaces, (d) flavor is not adversely affected, and (e) the curd tends to retain more water.

F. NEW CHEESEMAKING PROCESSES

In common with other industrialists, cheese manufacturers in developed countries are under a constant pressure to reduce labor costs. This is accentuated by the shortage of skilled labor for seasonal work, particularly in isolated areas. In these circumstances the advantages of successful mechanization and automation of all or part of the cheesemaking process are evident, and Irvine (1967) has reviewed a number of systems proposed for the mechanization of cheese manufacture. An example of these is the system proposed by Czulak (1958) for mechanizing cheddar cheese production.

The rennetting and cutting are left as manual operations, in part because of their relatively low labor requirement, but cheddaring, milling, salting and hooping are mechanized. The cheddaring process, which normally occupies up to two hours and which uses labor in proportion, was speeded up, in an earlier version of the method, by the use of higher holding temperatures, such as 42°C (Czulak *et al.*, 1954), made possible by the use of thermoduric starter cultures. Apart from the saving in labor costs the advantages of this system are the following: (a) Its close relation to existing processes means that the operators require little additional training; (b) the less skilled operations with a high labor content are mechanized, freeing the skilled personnel to concentrate on the treatment of the curd in the vat; and (c) mechanization can be introduced in stages, e.g., equipment for the milling, salting, and hooping stage exists at present in a number of Australian and overseas factories. In this system cheddaring takes place on a moving segmented conveyor on which the curd fuses and is turned over and spreads.

Alternative arrangements in which cheddaring is performed in an inverted truncated cone (Budd and Chapman, 1962) or a tower (Bysouth *et al.*, 1968) have been developed. Although the capital costs of mechanization are not great by many industrial standards the costs would be too high for small factories. It is therefore probable that mechanization will hasten the concentration of cheese production in larger factories. This may have side benefits such as a better possibility of the economic use of whey.

Czulak's system is closely related to existing cheesemaking practice and owes little to the results of milk protein research. Several other proposals for mechanized cheesemaking, however, depend upon a single research finding, for example the observation by Berridge (1942), referred to earlier, that the enzymic stage of rennin action and the coagulation of the altered casein are separable at low temperatures because of the high temperature coefficient of the latter reaction. On the basis of this observation, Berridge (1963) proposed a scheme in which rennet is added to the milk at 12°C and held up to 1 hr at pH 5.8, after which the milk is coagulated by pouring it onto a hot rotating drum. The temperature attained (approximately 55°C) is sufficient to set the curd and to cause extensive syneresis in the time the curd is on the drum. As the drum rotates, whey drains off and the ribbon of curd is peeled off onto a slide for further treatment. Difficulties of the process are the inactivation of the bacterial starter at the high curdling temperature, and the slow drainage of the curd prior to pressing which also results from this treatment. It is nevertheless an attractive process mechanically. Berridge (1968) has also proposed the use of a cellulose tube for forming the curd.

A similar system for continuous manufacture of cheese is that of Ubbels and van der Linde (1962). Rennet is added in the cold (5 hr at 2°C), following which the milk is heated to 32°C in a plate heat exchanger. Coagulation occurs rapidly thereafter in a vertical cylinder in which the coagulating milk rises smoothly, and the curd is cut on emerging. Syneresis occurs in a rotating inclined cylinder, and the cheese is pressed in a vertical cylinder. An Edam type of cheese is produced.

In addition to using the cold rennetting process described, advantage is taken in the Stenne-Hutin system (Stenne, 1965) of the considerable increase in the rate of coagulation which occurs in concentrated milk (Odagiri and Nickerson, 1964). These authors found that the time of coagulation at 30°C fell from 13.5 min to 2 min as the nonfat solids content increased from 10 to 30%. In the Stenne-Hutin process, milk is concentrated to a suitable solids content (up to 30%) and treated at 10°C with rennet for periods up to 30–40 min. This treatment is designed to produce an instantaneous coagulation on warming the milk to 30°–32°C. The warming is achieved by mixing the milk with a suitable volume of water at higher temperature. After this dilution the total solids content still remains higher than that of normal milk. The curd coagulates in fine particles which synerese, agglomerate, and are sieved from the whey. Depending on the type of cheese to be made, the curd may then be washed with water or cooked in whey, after which it is treated as in normal cheesemaking. The characteristics of the curd may be altered by altering the temperature of the mixing water. The advantages of the process are its rapidity (45 min from rennetting to milling), its higher yield because of lower fat losses (vs. Mabbitt and Cheeseman, 1969) and greater retention of lactose and milk salts, its flexibility in the production of curd suitable for different types of cheese, the relatively small size of the plant, and the fact that the whey is more concentrated than normal cheese whey.

Similar principles are employed in a process for the manufacture of cottage cheese from acid curd (Ernstrom, 1964). Reconstituted skim milk of 20% total solids is acidified in the cold with either HCl or with lactic acid developed by a starter culture; on warming the milk to 38°C in a curd-forming apparatus it coagulates, and the curd is cut on emerging.

All these processes retain some treatment in a vat, result in the loss of most or all of the whey solids, and require lengthy ripening periods. Their advantages over traditional processes are therefore likely to be marginal. Any process which could successfully retain all the whey solids in the cheese would have a much greater advantage, and it is therefore of interest to read of trials of such a process in the U.S.S.R. An emulsion of 25% skim milk powder, 25% fat, 43.75% water, 2.5% salt, 2.5% sodium phosphate and 1.25% sodium citrate by weight is treated with 5% of high quality mature cheese as a starter and with rennet at the rate of 0.03% of the dried milk powder. Coagulation takes place in molds, and the cheese is kept at 40°C for 2 days and then held for about 15 days at 18–20°C. Thereafter it is stored at 4°C to delay further maturing. The mature cheese is reported to have acidity and taste comparable to a Dutch type of cheese (Kozin and Rodionova, 1962). The interesting features of this process are (a) elimination of treatment in vats—a lengthy stage in all other processes, (b) increased yield due to complete use of all milk constituents, (c) elimination of the problem of whey disposal, (d) considerable shortening of the ripening period.

Successful trials of similar processes have been reported by Peters (1965) and Denkov (1967), the latter using concentrated milk of 36–55% total solids.

G. Research and Technology

Most of the techniques and methods of manufacture used in the cheese industry have been developed empirically during its long history, while the possibility of gaining a deeper understanding of the problems of the industry through scientific research is of comparatively recent origin. Many of the most effective techniques for the study of milk proteins, for example, have been developed in the last decade. The protein system in milk is a most complex one, and although many advances have been made, our understanding of the behavior of milk proteins in cheese manufacture is far from complete. The knowledge so far gained, however, does provide opportunities for better control of existing processes and, as indicated in the preceding section, is leading to the development of new processes. It also appears that there may be possibilities for the use of this knowledge that have not yet been exploited.

The main needs of the cheese industry today are the reduction of the labor in cheesemaking and the time taken in the various stages of cheese manufacture, and the improvement of working conditions and hygiene. Ideally, these objectives would be achieved by a continuous, mechanized process. There is also a need to make effective use of the whey, much of which now appears to be wasted. If this could be done by incorporating part or all of the whey solids in an acceptable cheese, the yields of cheese would be increased considerably and costs should be reduced.

Partial incorporation of whey solids can be obtained by pasteurizing the milk in conditions that lead to interaction of the casein and the β-lactoglobulin (Kannan and Jenness, 1956; Buchanan et al., 1965). This treatment has been used for the manufacture of a cottage cheese (Durrant et al., 1961) but is not favored for other types of cheese because of the slower coagulation, weakened curd and impaired syneresis which it causes. However, some recent studies indicate ways to remedy these defects. Curd strength can be improved and the rate of coagulation increased by concentrating the milk (Odagiri and Nickerson, 1964; Stenne, 1965). Kannan and Jenness (1961) considered that one effect of heating milk was to render colloidal phosphate unavailable. Their results and those of Pyne and McGann (1962) suggest that further improvement of curd strength could be made by additions of calcium and phosphoric acid to the milk; such an improvement has been observed with natural soft-curd milks (Hill et al., 1965). Alternatively, the addition of denatured whey protein to the milk before rennetting (Genvrain, 1967) might cause less impairment of curd strength.

If all the whey solids are to be incorporated in the curd, a process similar in principle to that of Kozin and Rodionova (1962) is indicated. Instead of using a concentrated milk reconstituted from powder, it seems more feasible to employ whole milk concentrated to about 50% solids content. The aim in such a process—opposite to that in normal cheesemaking—would be to avoid any syneresis of the curd. Means available for repressing syneresis and controlling hardness of the curd are the addition of salt to the milk (Sirry and Shipe, 1958), pasteurizing at higher than normal temperature (Kannan and Jenness, 1956) and initial homogenization of the milk (Peters, 1964); if necessary, the activity of calcium(II) could be reduced by the addition of sequestering agents such as citrate. The high lactose content of the product obviously could lead to difficulties in developing an acceptable flavor. However, the rapid-ripening slurry techniques developed by Kristoffersen et al. (1967) offer the possibility of finding solutions to such problems much more quickly than heretofore, and the prospects of a better use of the milk solids seem more than sufficient to justify the investigation.

The speculative processes outlined above indicate the extent to which technology might be changed by applying a small number of research findings, each of relatively minor importance. This is also true of the changes which are taking place in the industry at present. When more detailed information is available on such topics as the mechanism of the action of rennin on casein and the formation of the casein gel, the role of colloidal phosphate and the organization of the proteins in the casein micelle, then the opportunities for obtaining better methods of production should be greatly increased.

III. Concentrated Milk and Milk Powder

A. Processes and Problems

Evaporated milk, sweetened condensed milk and milk powder are the most common forms in which milk is preserved for lengthy periods. Evaporated milk is prepared by removing part of the water from milk by evaporation under reduced pressure and then sterilizing the product by heating to a high temperature. Sweetened condensed milk contains a high level of sucrose and is manufactured by condensing under reduced pressure. The final product contains 44–46% sucrose, 28–31% milk solids and 25–26% water. Milk powder is milk from which almost all the water has been removed. For reasons of economy, as much water as possible is removed in a vacuum evaporator before the final dehydration step. In addition to these methods, concentrated milk is preserved by freezing and storing in the frozen state.

Ideally these products should have a long storage life and, on reconstitution, be capable of yielding milk corresponding to the original in all of its desired properties. In practice, however, they fall short of this ideal, and a great deal of research effort has been and is still being directed towards their improvement. Many problems in which the proteins are involved, such as gelation during the manufacture and storage of evaporated and sweetened condensed milk, insolubility or lack of dispersibility of milk powder and protein destabilization in frozen concentrated milk, are being studied. A large measure of control over these problems has already been achieved by empirical methods resulting from experience gained during the processing of milk. However, a complete understanding of the phenomena and the way in which the control measures operate is lacking.

B. Effect of Heat on Milk

The changes that occur as milk is heated during manufacture and the factors affecting its heat stability have been widely studied. Despite the great volume of data reported in the literature over the past 30 years concerning the heat stability of milk, there is as yet no satisfactory explanation for all facets of the phenomenon. This is in no way surprising when one considers the inherent complexity of the milk system and the possible interplay of its various constituents.

1. *Influence of pH*

One of the most dramatic steps forward in our understanding of this phenomenon has been derived from the discovery by Rose (1961a) that the heat stability of milk from an individual cow varies greatly over the narrow range of pH from 6.5–6.9, in many instances passing through a distinct maximum at about pH 6.6–6.7. This work indicated that a major factor determining the heat stability of a particular milk is the position of the pH value of maximum stability relative to the natural pH of the milk. Subsequent work (Rose, 1961b) implicated an interaction between the casein micelles and the noncasein protein fraction, in particular β-lactoglobulin, as a prime factor causing this effect. Systems of casein micelles suspended in milk serum devoid of other proteins did not possess this sensitivity of heat stability to pH whereas the addition of β-lactoglobulin restored the phenomenon. Heating the milk to 95°C for 10 min to denature the serum proteins had little effect on the sensitivity of heat stability to pH apart from a slight shift of the maximum to a lower pH value, and this was interpreted as indicative of the formation of a complex between β-lactoglobulin and casein which enhanced stability.

Later Tessier and Rose (1964) reported that milk samples from individual cows could be classified in two groups according to the way in which the heat stability varied within the pH range just described. In type A a maximum and minimum in the heat stability–pH plot was observed while in type B the minimum was absent. The addition of κ-casein to a type A milk caused the type B response in heat stability, and the addition of β-lactoglobulin to a type B milk resulted in conversion to type A. Feagan et al. (1969) found four distinct heat-stability patterns which were associated with the genetic variants of β-lactoglobulin present. Thus the important influence of these two protein fractions on heat stability was clearly demonstrated, although the mechanism by which they affect it is not known. Morrissey (1969) has shown that the milk serum salts, in particular calcium and phosphate, also play a part in determining the type of heat-stability response in milk. It is thought that the tendency for colloidal calcium phosphate to deposit on the caseinate–β-lactoglobulin complex renders this more sensitive to calcium ions.

Also of interest in this regard is the work of Feagan et al. (1966) demonstrating the pronounced deleterious effect on the natural heat stability of milk caused by subclinical mastitis, one manifestation of which is a shift in the pH value of the milk to a higher than normal value. Another factor thought to be involved is the higher level of serum albumin in the subclinical mastitic milk (Feagan et al., 1969).

2. Effect on Inorganic Constituents

Many changes have been observed in the inorganic composition of milk when it is heated, and these alterations doubtless have an influence on the heat stability. The concentration of calcium and magnesium ions decreases (van Kreveld and van Minnen, 1955; Davies and White, 1959; Ismael and Grimbleby, 1959) as does the amount of calcium apparently associated with the casein and the concentration of un-ionized soluble calcium. The amount of colloidal inorganic calcium increases, indicating an increase in this form of calcium at the expense of the others (Davies and White, 1959). The concentrations of the corresponding phosphates show changes parallel to those of calcium except that the casein-bound phosphate does not change until the temperature reaches 110°C (Davies and White, 1959), above which dephosphorylation of the casein commences (Howat and Wright, 1934a). The importance of colloidal calcium phosphate with regard to heat stability has been described by Pyne (1958) and Pyne and McGann (1960). The latter authors showed that the removal of the colloidal calcium phosphate from milk resulted in a considerable increase in the heat stability of the system. The variations described above in the concentrations of inorganic constituents in the protein-free serum of milk were determined in ultrafiltrate prepared after cooling the heated milk. Such variations appear to be much more pronounced at the actual temperature of heating (Rose and Tessier, 1959a). For example, the pH value of ultrafiltrate obtained at 93°C is as much as 0.5 of a pH unit lower than that prepared from the corresponding unheated milk, and since it is the casein fraction that provides the matrix of the gel that forms, these considerable alterations in the environment of the casein micelles during heating are bound to be important.

3. Interaction of Casein with the Noncasein Proteins of Milk

Physical nature. The nonreversible whitening of skim milk heated to temperatures above 60°C was interpreted by Burton (1955) as indicative of a permanent change in the size and/or number of light-scattering particles in the system as the noncasein proteins are denatured. Sullivan et al. (1957), working with milk in which the whey proteins were labeled with ^{35}S, showed that after heating 15 min at 95°C, some 52% of the noncasein proteins sedimented with the casein. On the basis of electrophoretic evidence, McGugan et al. (1954) concluded that a complex was formed between β-lactoglobulin and α-casein when the two were heated together at 85°C. Other workers have subsequently shown that κ-casein appears to be the casein fraction involved in this interaction (Fox, 1956; Long, 1959). The interaction apparently proceeds differently and to different extents depending upon the heating conditions and the system studied. Rao (1961) reported the isolation by ion-exchange chromatography of a well-defined complex with a sedimentation coefficient of 28 S from heated mixtures of α-casein and β-lactoglobulin. Zittle et al. (1962) and Long et al. (1963) found the complex formed by heating together κ-casein and β-lactoglobulin to be considerably larger (s_{20} = 45 and 44–48 S, respectively). On the other hand, Morr (1965a) showed that the largest protein component in heated skim milk (88°C for 10 min and held 18–24 hr at 5°C) has a sedimentation coefficient of only 6–7 S, and Tessier et al. (1969) were unable to demonstrate the formation in milk of a complex corresponding to that found in heated mixtures of β-lactoglobulin and κ-casein. In some instances, apparently no interaction between β-lactoglobulin and casein components occurs in heated systems of the isolated fractions, although the β-lactoglobulin is denatured (DellaMonica et al., 1958). It is clear that caution is necessary in extrapolating results obtained in simple systems to the more complex one of milk.

The work of Morr (1965b) indicated that the serum proteins react preferentially with the small casein micelles, which accords with their greater content of κ-casein (Sullivan et al., 1959) and that after heating the micelles are more uniform in size and less sensitive to calcium. Hostettler et al. (1965) also reported considerable changes in the size distribution of casein micelles when milk is heated. With increasing severity of heating there occurs an increase in the number of caseinate particles with a diameter of less than 50 mμ but there is no great change in the electrophoretic pattern of the casein prepared from them. At temperatures of around 115°C the size of the larger micelles becomes greater and they tend to aggregate.

Chemical nature. Since the sulfhydryl groups of β-lactoglobulin are exposed by heating (Larson and Jenness, 1950; Boyd and Gould, 1957) and the formation of a complex between β-lactoglobulin and casein is prevented by SH-blocking agents (Trautman and Swanson, 1959), it has been postulated that the interaction is the result of disulfide bridges formed between κ-casein and β-lactoglobulin (Sawyer et al., 1963). Although κ-casein contains sulfhydryl groups (Beeby, 1964) they are strongly masked in the casein micelle and are unavailable for reaction with methyl mercury iodide unless the protein is extensively degraded and the micellar structure destroyed (Hill, 1964). In view of this it is difficult to see how a molecule as large as β-lactoglobulin would react more readily with the masked SH groups of casein in milk than would the comparatively much smaller mercurial, although the groups may be more accessible at the high temperatures employed.

The formation of a complex involving the whey proteins and a sialic acid-rich fraction obtained from κ-casein and which is free of cysteine or cystine has been reported in unheated systems (Beeby, 1966a). Also Kenkare et al. (1965) have indicated that the complex formed between α_s-casein and κ-casein dissociates reversibly at 85°C. These results suggest the possibility of the interaction occurring as the casein components are rearranging following perturbation of the micellar structure by the heat treatment. Thus, SH groups may not be essential for complex formation, although they may be involved in preliminary self-aggregation of the β-lactoglobulin (Sawyer, 1968).

C. EVAPORATED MILK

1. *Forewarming in Relation to the Stability during Sterilization*

With milk of the same composition and bacterial quality, the more it is concentrated the less stable it is to heat (Holm et al., 1923; Wright, 1932; Howat and Wright, 1934b). In view of this, great difficulty might be expected in maintaining stability during the manufacture of evaporated milk in which the concentrated product is subjected to high-temperature sterilization. Fortunately it is well known that heating the milk to a temperature of around 90°C for 10 min (forewarming or preheating) induces in the concentrated milk a considerable degree of heat stability (Deysher et al., 1929; Webb and Holm 1932; Eilers, 1947). The efficacy of such heat treatments in conferring stability on the concentrated product varies with different milk samples (Belec and Jenness, 1960). These authors found that while preheating pooled milk increased the heat stability when it was concentrated, no such response was observed in the individual milk samples. If forewarmed before mixing the stability was not affected. Feagan et al. (1966) have discussed the possible importance of subclinical udder infections in the widely varying response of different milk samples to forewarming. Regardless of the mechanism involved in its formation, a complex between a component or components of the whey proteins and the casein fraction appears to be necessary if the concentrated milk is to be sufficiently stable during the sterilizing process (Trautman and Swanson, 1959).

2. Effect of Heat on Concentrated Milk

Influence of pH on heat stability. The stability to heat of concentrated milk varies with small changes of pH value in the same manner as the nonconcentrated material (Rose, 1961c) so that control of pH is important for stability. Schmidt and Koops (1965) have shown that the pH value of maximum stability of preheated, concentrated milk from cows homozygous with respect to κ-casein A is 0.1 of a pH unit higher than that of milk containing only κ-casein B, and this may well be a factor affecting gelation during sterilization. Citrates and phosphates are often employed as stabilizing agents during the manufacture of evaporated milk (Webb and Holm, 1932; Eilers, 1947), although as shown by Webb and Holm (1932) the heat stability of some concentrated milk samples does not increase with additions of phosphate but instead responds markedly to additions of calcium chloride or acid. This phenomenon would seem to be explained by the natural pH of the concentrated milk being higher than the pH of maximum stability, the calcium chloride producing a shift in pH to the acid and phosphate a shift to the alkaline side (Rose, 1961c).

Changes in viscosity. When skim milk is concentrated, its viscosity increases markedly with the concentration of milk solids, and the effect is greatly accentuated by high forewarming temperatures (Greenbank et al., 1927; Eilers, 1947). Further increases in viscosity, again depending upon concentration and forewarming conditions, are observed when the concentrated milk is heated (Eilers, 1947; Beeby and Loftus Hills, 1962; Beeby, 1966b). Because the viscosity of concentrated skim milk, and particularly in the heated samples, is decreased by shearing forces (Eilers, 1947), it is clear that weak interactions between the dispersed components, presumably including the casein micelles, contribute appreciably to the viscosity.

This tendency toward interaction increases greatly on heating, particularly when the level of milk solids is high, although it can be mitigated to some extent by the choice of suitable conditions of forewarming. The beneficial effect of preheating is possibly due to the blocking of reactive sites on the casein micelles through complex formation with the noncasein proteins (Trautman and Swanson, 1959). The manufacturing procedure involving sterilization in cans for approximately 15 min at 118°C (long-hold method) is, however, limited to milk of about 31% total solids. Even at this level of solids, while sterilization can be achieved without excessive increases in viscosity, age gelation (see below) may prove troublesome (Beeby and Loftus Hills, 1962). An increase in viscosity and, what is perhaps more important, the development of a definite yield value is desirable to control fat separation during storage (Bell et al., 1944; Eilers, 1947). In practice this is achieved by adjustment of the conditions of forewarming which must be varied to suit the characteristics of different milks.

Little is known regarding either the mechanism by which heating increases the net attraction between the casein micelles or the nature of the groups and forces involved. Some of the interactions appear to be effected through calcium(II) since citrate lowers the viscosity of heated concentrated milk and lessens the deviation from Newtonian behavior (Eilers, 1947). Other interactions are of a more permanent nature. When forewarmed milk is concentrated some increase in particle size is observed (Morr, 1965b), but by far the greatest change takes place during sterilization when the casein micelles expand and aggregate into clusters (Hostettler and Imhof, 1951; Wilson and Herreid, 1961).

Sterilization by heating at temperatures of 127°–146°C for 0.6 sec to 2 min (high temperature-short time, H.T.S.T., method) results in lower levels of both heated flavor and browning and smaller increases in viscosity, thus enabling more highly concentrated milk to be processed. Leviton et al. (1963) proposed a method in which the milk was sterilized before concentrating to avoid the harmful effects of heating the concentrated milk. The obvious advantages of a continuous as compared to a batch-type process are also achieved by such H.T.S.T. treatments.

A great deal of the work concerning the heat stability of milk after concentration has been done with skim milk and the data obtained have been extrapolated to commercial evaporated milk, which in general contains fat. However, the fat phase itself can have an important influence as shown by Leviton and Pallansch, (1961) who exchanged the cream of heat-labile milk with that of heat-stable milk and thereby produced a stable concentrate.

In general it may be stated that the problem of gelation during the sterilizing process in evaporated milk is adequately controlled by the forewarming treatment and the addition of stabilizers, which is an example of how a technological process may be manipulated without knowledge of the mechanism involved.

3. Age Gelation

Influence of method of manufacture. Gelation of evaporated milk during storage as distinct from gelation during manufacture is a well-known phenomenon although its occurrence appears to be rather sporadic in milk sterilized by the long-hold method. Evaporated milk sterilized under H.T.S.T. conditions is much more prone to develop this type of defect despite a lower initial viscosity (Bell et al., 1944; Tarassuk and Tamsma, 1956; Leviton et al., 1963). The defect occurs sooner with increasing concentration of milk solids; this is a serious drawback as it nullifies the advantage of a product containing less water. Beeby and Loftus Hills (1962) have suggested that the ratio of protein to water is of importance and that variations in this ratio arising from differences in the protein content of the nonfat solids fraction in the milk might explain the sporadic nature of the defect in samples sterilized by the long-hold method.

Bell et al. (1944) retarded age gelation in H.T.S.T. evaporated milk by heating after sterilization, while Tarassuk and Tamsma (1956) achieved a similar effect by heating the concentrated milk prior to sterilization. Thus it seems likely that when milk is heated the casein micelles are altered so as to be more likely to interact with each other but that other reactions, such as complex formation with noncasein proteins or reaction with lactose,* effectively block many of the sites through which micelle–micelle interactions might occur. The size of the casein micelles does not appear to be a factor in this phenomenon (Schmidt, 1969).

Leviton et al. (1963) very effectively stabilized H.T.S.T. evaporated milk containing 37% total solids by the addition of polyphosphate, which suggests that calcium may play a role in age gelation. However, as well as complexing calcium, polyphosphates react strongly with proteins (Herrmann and Perlmann, 1936), particularly with κ-casein (Melnychyn and Wolcott, 1967), and prevents aggregation of this protein after treatment with rennin (Hill, 1970). Thus, the presence of such large charged groups on the surface of the micelles may well prevent their interaction.

The possible role of enzymes. Hostettler et al. (1957) investigating the problem of age gelation in uperized (150°C for 75 sec) nonconcentrated milk found the casein in the thickened milk to be more sensitive to calcium than that of the unthickened samples and concluded that the defect was due to enzymic destabilization of the casein. Reactivation of some phosphatases in milk pasteurized at 73°C for 15 sec (Wright and Tramer, 1953) and in whole milk sterilized at 141°C for 7 sec (Edmondson et al., 1966) is known, and a similar reactivation of naturally occurring proteases may be possible. However, the conditions generally employed for preheating and sterilizing would seem more than adequate to inactivate such enzymes. Indeed, Nakai et al. (1964) found no evidence of enzymic proteolysis in samples of concentrated sterilized milk although they gelled quite quickly

* Lysine is known to decrease in milk during heating (Kisza et al., 1966).

(4–5 weeks). It is of interest to note that Downey and Andrews (1965) have reported that all the major milk proteins possess a slight esterase activity which is inherent in the proteins themselves and not due to contamination with enzymes. This activity, which is presumably not destroyed by heating unless it significantly alters the protein, may be a factor in age gelation, particularly if the esterase activity is accompanied by proteolytic activity. The ability of κ-casein to stabilize α_s-casein in the presence of calcium(II) is greatly reduced by heat treatment in some preparations while others are unaffected, and partial degradation by a naturally occurring protease in milk has been suggested as a possible cause (Zittle, 1961). This too may be a factor in age gelation. However, since the severe heat treatment to which the milk is subjected during the manufacture of evaporated milk would be expected to prevent the type of coagulation induced by rennin, it is probable that age gelation in this product involves different mechanisms.

Aged preparations of casein resemble rennin-treated casein in their sensitivity to calcium, although the amount of nonprotein nitrogen released from them by the enzyme remains unchanged (Alais et al., 1953). Further study of this aging effect may be worthwhile.

D. Sweetened Condensed Milk

1. *Viscosity*

As with evaporated milk, the viscosity of sweetened condensed milk increases markedly with the concentration of milk solids and with the temperature of forewarming (Rogers et al., 1920; Eilers, 1947) and declines with increasing shearing stress (Eilers, 1947). Although not subjected in the concentrated form to the high temperatures employed in the manufacture of evaporated milk, sweetened condensed milk contains much less water, and thus the proteins are much more susceptible to heat-induced changes which favor their interaction. The casein micelles in this product are in fact aggregated in clusters (Imhof, 1952).

Excessively high viscosity or gelation do not appear to be serious problems during normal manufacture although variations in the constituents of the milk may require slight adjustment of procedure to maintain a uniform product. These defects may be observed, however, when sweetened condensed milk is prepared from skim milk powder, butter oil, sugar and water (Pont, 1960).

2. *Age Thickening*

Age thickening is a problem limiting the shelf life of condensed milk and in this regard the conditions of forewarming are important. Increasing the temperature from 75° to 96°C results in higher initial viscosities and a greater tendency toward age thickening, the effect being very pronounced at high levels of solids (Rogers et al., 1920). Higher temperatures (110°–120°C) diminish this tendency, suggesting, as with evaporated milk, the blocking of reactive sites under these conditions, perhaps by means of a Maillard type of reaction involving casein and lactose. Reaction with lactose would be more extensive at these higher temperatures.

Samel and Muers (1962a) found that the casein fraction and not the noncasein proteins was the main factor in age thickening, which they suggested is due to slow irreversible changes in the size or shape of the casein micelles and a concomitant interaction between the altered micelles. The rate at which it proceeds is quite low at 5° and 18°C but is greatly accelerated at 39°C. The addition of urea to concentrated skim milk apparently leads to the expansion and interaction of the casein micelles (Beeby and Kumetat, 1959), and a similar expansion of the micelles due to the high concentration of sucrose may occur in sweetened condensed milk and lead to their more extensive interaction. Although it is the casein micelles which interact to cause the defect, the inorganic constituents are important in determining whether such an interaction takes place. Thus reduction of the level of milk salts by dialysis largely eliminates age thickening while the addition of various anions and cations can retard or accelerate the condition depending upon whether they are added before or after condensing. In general those ions that stabilize when added prior to condensing destabilize the system when added after concentrating and conversely (Samel and Muers, 1962b).

An indication of the subtleness of the changes in the system needed to induce the defect is given by the work of Samel and Muers (1962c), who found the temperature at which the milk is condensed to be of profound importance. Samples condensed at 23°C underwent age thickening very rapidly while samples condensed at 55°C were comparatively stable. Milk condensed at 39°C thickened at a moderate rate but when diluted and recondensed at either 45° or 29°C, yielded in the first instance a very stable product, while in the second case the condensed milk was much less stable than the original. The deleterious effect of low condensing temperatures was offset by forewarming the milk at 85°C. All this work was done with skim milk, however, and it is perhaps unwise to neglect the possible influence of fat.

E. Frozen Concentrated Milk

Although it is possible to control the development of flavor defects by storing concentrated milk at temperatures below freezing, the storage life of the product is limited by the tendency of casein to form a precipitate which does not disperse on thawing the frozen milk. This phenomenon also limits the storage life of frozen nonconcentrated milk and may exist in ice cream, although in that case it presents no technical or commercial difficulties. The precipitation has been ascribed to the increase of protein and ionic concentrations (particularly calcium) which is most marked after the (slow) transformation of the bulk of the lactose to the α-form (Desai et al., 1961). The fault is accentuated by additions of calcium and phosphate ions, while storage life is lengthened by additions of sodium chloride (Rose and Tessier, 1959b). The latter authors consider that the precipitation of the protein is mediated by calcium phosphate which forms links between the micelles. On the other hand Desai and Nickerson (1964) interpreted changes in the starch-gel electrophoresis patterns as indicating damage to the casein and whey proteins. Other factors could also bear on the problem. The binding forces of the casein micelle, which are weakened at temperatures of 0°–2°C (Waugh, 1961), are likely to become still weaker at storage temperatures of $-10°C$ or lower, and in such conditions the increased concentration of calcium ions which builds up as freezing progresses might cause a partial fractionation of the casein, forming a precipitate depleted in κ- and β-caseins. Whatever the reason for the precipitation, practical measures for control seem to be limited to reducing the activity of the calcium ions. This may be done by adding sequestering agents or by adding sugars (which may also sequester calcium ions) to increase the equilibrium volume of the liquid phase, thus limiting the increase in ionic concentration. Hydrolysis of some of the lactose by means of lactase is also effective in improving storage life (Tumerman et al., 1954). When additions to the milk are not permitted, the fault can be controlled by storing the concentrate at a temperature low enough ($-23°C$) to freeze all the liquid (Tumerman et al., 1954). Economically, the product is a marginal one and elaborate treatments do not seem to be justified.

F. Milk Powder

The most common method used to produce milk in a powder form is that of spray-drying, although the older method of roller-drying is still in use and new techniques such as foam spray-drying (Hanrahan et al., 1962) and foam-drying (Sinnamon et al., 1957) are being studied. The aim is to remove the water efficiently and economically with as little damage to the milk constituents as possible, the latter condition being of particular importance where the dried milk is to be reconstituted. Prior to drying, the milk is forewarmed as required and concentrated to approximately

40–55% total solids, a higher level of solids being used for whole milk than for skim milk. The degree of forewarming depends to a large extent upon the end use of the powder. During the production of whole milk powder, the milk is subjected to a high preheat treatment to inactivate lipases and to increase the reducing capacity of the milk in order to lessen oxidative defects in the fat. Greenbank et al., (1927) showed that forewarming to high temperatures results in an improvement in the baking quality of skim milk used in the production of bread. When skim milk powder is to be used for reconstitution for beverage purposes or cottage cheese production a low preheat treatment is preferred.

Milk is heated at three stages in the manufacture of spray-dried milk powder—during (a) forewarming, (b) concentration and (c) drying. It is during drying that the greatest increase in groups which reduce acid ferricyanide is observed in the proteins (Kumetat and Beeby, 1957), together with the greatest decrease in lysine (Kisza et al., 1966). Such reducing groups are associated with the reaction between lactose and proteins (Lea, 1947; Coulter et al., 1948), and their formation is indicative of alterations in the properties of the proteins. It is known that extensive protein–sugar interaction in milk powder renders the protein insoluble (Henry et al., 1946). The extent to which these reactions occur during processing depends upon the type of evaporator and spray-drier used, and they proceed more rapidly after forewarming at high temperatures.

As mentioned earlier, forewarming milk to a high temperature stabilizes it against heat after it is concentrated, and this stabilization also extends to concentrations of milk solids higher than those attained in the manufacture of evaporated milk (Greenbank et al., 1927). However, at approximately 45% total solids, skim milk which has been given a high preheat treatment is less stable to heat than the corresponding milk forewarmed at lower temperatures (Beeby, 1966b).

Although initially the temperature of the particles of concentrated milk in the spray-drier will be kept low through evaporation of moisture, this will become slower as dehydration proceeds; the temperature will therefore tend to rise when the concentration of solids is at its highest level and, as a consequence, the proteins are most sensitive to heat-induced changes. The casein micelles of spray-dried milk are aggregated permanently into clusters (Imhof, 1952), the extent to which this occurs presumably depending upon the combined effects of the various stages in the process. How much the milk proteins are altered will depend upon the conditions of manufacture and most likely upon seasonal and genetic variations in the protein and other constituents of the milk. Changes in the proteins brought about during processing and in particular during spray-drying may largely explain variations in the performance of skim milk powders used to prepare recombined evaporated milk and sweetened condensed milk (Muller, 1963; Muller and Kieseker, 1965).

IV. The Manufacture of Casein, Coprecipitate, and Whey Proteins

The decline in the per capita consumption of butterfat in most countries since World War II and the decrease in the industrial usage of casein have encouraged research into finding new ways of utilizing milk in addition to the traditional milk products—butter, cheese, condensed and powdered milks. Casein, for example, is no longer regarded merely as a by-product of butter manufacture and is being increasingly valued as a desirable human food. Attempts to increase the overall efficiency of dairy production by finding uses for what are at present waste products, such as cheese whey and casein whey, and efforts to develop more effective ways of using surplus dairy products to help relieve food shortage in developing countries have also led to research into new uses of milk products in food.

Changes in the pattern of milk utilization have been accompanied by improvements in manufacturing methods. Continuous processes are replacing the earlier batch methods which were based on cheesemaking equipment and practices. Large volumes of milk can now be handled in smaller equipment and greater control over the process can be maintained.

A. CASEIN MANUFACTURE

The principle stages in the manufacture of casein are temperature adjustment of the skim milk (which may be preceded by pasteurization), addition of coagulating agent, and a period during which coagulation or precipitation of the casein occurs. The casein is then separated from the whey, washed to remove minerals, lactose and whey proteins, and dried.

Acid is the most commonly used precipitating agent. In the manufacture of lactic casein, coagulation is brought about by the addition of a starter culture of lactic acid bacteria to the milk in order to ferment the lactose to lactic acid. Alternatively, lactic, hydrochloric or sulfuric acids may be added to the milk to give rapid precipitation and reduce the storage capacity which is required for methods based on the fermentation of lactose. A comprehensive account of the earlier production methods and the development of continuous processes is given by Spellacy (1953). The production of lactic casein in New Zealand is described by Oetiker (1960), and Neff (1966) has surveyed current production methods in Australia. Rennin is used as a coagulating agent to give a product which is particularly suitable for casein plastics, and a continuous process for the manufacture of rennet casein has been developed by Berridge (1963).

Acid casein can be dispersed for spray- or roller-drying by grinding the wet curd in a colloid mill or by dissolving it in alkali at pH 7 to form sodium caseinate. Alternatively, the precipitating conditions can be modified to produce fine granules which can be dried as a suspension instead of a solution. Casein is not precipitated by acid at low temperatures, and Trexler (1952) has produced fine particles by the addition of acid to milk at 2°–10°C and then heating the mixture to 70°C. This technique permits thorough mixing of the acid with the milk before precipitation at the higher temperature, and the casein precipitates as fine granules which can be separated from the whey, washed and spray-dried.

The older production methods often resulted in products of uneven ash content and quality. These arose in part from variations in the size of the curd particles produced by inefficient mixing of acid and milk, by the use of excessive temperatures during precipitation, and from inadequate washing of the curd. Better quality casein was required for high quality paper coating and for some edible purposes, and a continuous process in which the precipitating and washing conditions could be more readily controlled was developed by Muller and Hayes (1962). A wider range of precipitating conditions can, however, be used to produce casein for purposes which do not require such low levels of lactose and mineral content, and Smart and Nelson (1957) produced edible casein by precipitation with HCl at pH 4.6 in a holding tube at 41°C. The protein coagulated in a ropelike curd. Casein recovery by this process is claimed to be 98.5%.

B. COPRECIPITATE MANUFACTURE

The increasing emphasis on the food use of milk proteins and the desire to increase the overall efficiency of milk utilization by recovering all its constituents have led to the development of methods whereby whey proteins are "coprecipitated" with casein. The procedures are similar to those employed in the manufacture of casein, except that an extra stage is provided during which the whey proteins are denatured by heating the milk above 70°C prior to the addition of the coagulating agent. Some of the denatured whey proteins combine with the casein and are later precipitated with it.

Scott (1952), Howard et al. (1954) and Bernhart et al. (1955) have described batch methods of making coprecipitate. Methods for the continuous production of coprecipitate have been developed by D'yachenko et al. (1953), D'yachenko (1957), Arbatskaya et al. (1962) and Buchanan *et al.* (1965) in which 0.2–0.3% calcium chloride is the precipitating agent. Recoveries of 97% of the total milk proteins are claimed when calcium chloride is used as the precipitating agent.

As with casein, the properties of coprecipitate are modified by altering its calcium content. Coprecipitate with a high calcium content is insoluble in water, but it can be dispersed or dissolved by the addition of 4–6% (w/w) sodium tripolyphosphate (Buchanan *et al.*, 1965). Lowenstein (1965) found that the water-binding capacity of coprecipitate low in calcium is decreased by the addition of calcium before it is dried. A continuous process which uses a combination of both calcium chloride and acid to control the calcium content of the product to suit a wide range of end usage has been described by Muller *et al.* (1967).

The solubility of spray-dried coprecipitate powders can be increased by the addition of alginates (Lowenstein, 1961a) before drying or by increasing the pH above 7 with ammonia (Scott, 1958) which is removed during spray-drying. The solubility and dispersibility of the powder can also be increased by incorporating emulsifying agents before drying (Lowenstein, 1961b).

C. Research Aspects of Casein and Coprecipitate Manufacture

1. *Increasing Yields*

Improvements in manufacturing techniques have resulted in more uniform and chemically better-defined products and have also led to increased recovery of milk proteins. McDowall (1961) estimated that in New Zealand factories, losses at some periods of the year were as large as 10% of the potential yield of casein, and it is claimed that the introduction of improved methods in Australian factories has reduced losses from 9% to 1–2% (Muller and Hayes, 1962). These losses are attributed to the incomplete recovery of fine particles, especially at the low pH of precipitation required for casein with low ash content. Attempts have been made to use continuous centrifuges to reduce losses of fines (King, 1964), but it is a question of economics whether the capital cost justifies the extra returns on recovering the last 1 or 2%. Research on the coagulation process may result in a simpler way of reducing these losses. Other losses arise from the fact that some of the protein in the casein micelles is not precipitated at pH 4.5 (Hill and Hansen, 1964). If rennin is used to initiate coagulation, the loss of the soluble glycopeptide and other reaction products which are not recovered with the paracasein reduces the potential recovery of casein.

The reactions leading to the interaction between casein and whey proteins are not well understood. Only a part of the denatured whey proteins combines with or can be precipitated with the casein, and the manufacturing processes for coprecipitate rely on conditions whereby the coagulum acts as a filter to entrain as much as possible of the remaining denatured whey proteins. Some improvement in the recovery of whey proteins results from the addition of a small quantity of calcium chloride to the milk before it is heated at the start of the process (D'yachenko et al., 1953; D'yachenko, 1957; Muller et al., 1967). Research is therefore needed to find improved precipitating agents. Higher recoveries may be achieved if the casein micelles can be aggregated without the disruption which occurs during acid precipitation. A process has been described recently in which long-chain anionic polyelectrolytes are used to flocculate denatured whey proteins (Rodgers and Palmer, 1966). It may be possible to use similar reagents in the production of casein and coprecipitate. For example, Hansen (1966) has found that some of the casein is precipitated from milk by carrageenin in the concentration range 0.01–0.1%, and Cluskey et al. (1969) have shown that a complex formed between calcium caseinate and carboxymethylcellulose can be sedimented optimally at pH 7.5.

2. Improvements in Flavor

Another problem that is proving difficult to solve is the ready development of a "gluey" flavor in casein, and this discourages its use in some foods. Acid casein normally contains 1–2% lipids, although Cerbulis and Zittle (1965) found that this could be as high as 4–7%. It is not inconceivable that the oxidation of these compounds may contribute to the development of undesirable flavors. Storing casein at high temperature is known to accelerate the development of "gluey" flavor (Ramshaw, 1966). All commercial casein driers expose casein particles to a stream of hot air, and there are opportunities for undesirable chemical reactions to be initiated during the drying process. Whatever the cause of this defect, production methods may need to be modified to restrict its development. Drying casein in a vacuum at reduced temperature could favor the production of a blander product.

D. Whey Protein Manufacture

Since approximately 90% by weight of the original milk used for cheesemaking appears as whey, its economic use or mere disposal as effluent can pose a considerable problem, especially for larger factories. Wix and Woodbine (1958), reviewing this subject, point out that the production of whey in 1949 was estimated to total about 12 million metric tons in Western Europe, America, Australia and New Zealand. This figure has since increased to an estimated 25 to 30 million tons ("Dairy Produce," 1969), most of which is disposed of as effluent. The common wastage of whey, a valuable nutrient material, arises mainly from the high cost of recovering the whey proteins because of their low concentration (approximately 0.6%). On the other hand, increasing difficulties in the disposal of whey are now confronting industry with a choice of investing either in a plant for whey disposal or in one in which whey proteins and lactose can be recovered, thereby helping to offset the cost of the extra equipment. Whey solids may be recovered by

(1) *Concentrating and drying whole whey.* Concentrated or dry whey has been used for some time in the manufacture of stock food and as an ingredient in human food. However, modern techniques of water treatment developed for large scale desalination (e.g., reverse osmosis) may conceivably be applicable to whey and thus make it possible to handle economically large volumes of whey and perhaps to prepare undenatured whey protein in quantity. This, at the moment, is a matter of engineering technology rather than protein chemistry.

(2) *Fermentation.* The application of large-scale, continuous fermentation processes to whey has made it possible to use most of the lactose in whey as a nutrient for yeast cells, which are then harvested and used as a basis for food (Wasserman et al., 1961). A source of nitrogen, such as urea or ammonium salts, must of course be supplied. In this way carbohydrate that is now wasted can be converted into protein and fat.

(3) *Removal of whey proteins.* A number of methods have been developed for the removal of proteins from whey as a preliminary step in the recovery of lactose. These methods generally involve the denaturation of the whey proteins since they depend on heat coagulation. The coagulum disperses on agitation and is therefore difficult to remove using equipment designed for casein manufacture (Josh and Hull, 1950; Henika et al., 1958). Mention has already been made of a method (Rodgers and Palmer, 1966) in which long-chain polyelectrolytes are used to increase the size of the particles so that they can be retained on a screen. Another promising approach is that of Hidalgo and Hansen (1969), who have shown that complexes of carboxymethylcellulose with undenatured whey proteins can be readily precipitated from whey with good yield at their isoelectric points.

E. Research Aspects of Whey Protein Manufacture

1. Precipitation Conditions

An interesting claim in the patent of Rodgers and Palmer (1966) is that the best precipitation conditions result from carrying out the heating process in several stages: (a) preheating the whey at pH 4.6 to the stage just before appreciable coagulation takes place, (b) heating the whey to about 95°C under conditions of turbulence and in which flocs of precipitated whey proteins are present, and (c) after 1–10 min, the addition of polyelectrolytes, which precipitates and stabilizes the flocs to a size at which they can be removed on a screen, or by centrifugation. The denaturation of β-lactoglobulin proceeds in two stages as has been shown by Tanford et al. (1959) and Tanford and Taggart (1961). They demonstrated that near pH 7.5, β-lactoglobulin undergoes a reversible change in which 2 carboxyl groups, thought to be normally buried in the hydrophobic interior of the molecule, become titratable. Ionization of imidazole groups accompanies this change. Pantaloni (1963) confirmed that structural changes in the region pH 6–9 increase the dissociation of carboxyl groups and involve tyrosine side chains. Pantaloni (1964) also showed that the changes which accompany heating increase the reactivity of SH groups and alter the optical rotation. It appears therefore that in the preheating stage the whey proteins undergo some subtle change such that they interact before unfolding more extensively at the higher temperatures. In this way larger flocs are built up than would otherwise occur.

The amount of whey protein recovered in the continuous process for the manufacture of coprecipitate depends on an optimum combination of pH, temperature, and holding stages to give a casein curd of suitable stickiness onto which the denatured whey proteins can be adsorbed. This observation, together with the work on whey protein aggregation, suggests that the interactions between denatured whey protein molecules and between denatured whey proteins and casein may not depend on covalent bonds only but may also occur through surface or charge effects.

2. Protein Modification

The large-scale recovery of undenatured whey proteins is more difficult to achieve than the recovery of denatured protein. Although continuous gel filtration, ultrafiltration, or electroconcentration, in conjunction with a technique such as that of Hidalgo and Hansen (1969), might be used for this purpose, another approach to the problem might be to modify the solubility of the native protein by treatment with enzymes. The work of

Greenberg and Kalan (1965) on the modification of β-lactoglobulin with carboxypeptidase, which removes two C-terminal amino acids and produces a modified protein which is much easier to crystallize from solution than the original, indicates one approach to the problem. It is necessary only to consider the profound alterations in behavior which can occur as a result of the alteration of just two residues in hemoglobin (Ingram, 1958) in order to appreciate the possibilities of investigation along these lines.

V. Specialized Products

Casein has for some time been used as a relatively cheap form of protein, being added as a protein supplement in a wide range of foods, particularly in invalid foods and "low-calorie" diets. A better understanding of the effects of heat and minerals on the properties of milk powder and casein and the availability of chemically better-defined products have resulted in a variety of special-purpose high-protein powders. For example, Lowenstein (1965) found it preferable to use coprecipitate with a high calcium content to increase the protein level in baked goods. The more highly hydrated coprecipitate with a low calcium content tended to retain water in the dough and prolonged the baking time. Conversely, other products, such as ice cream, require a protein which is more highly hydrated. Lee (1952) found that skim milk powder with a low preheat treatment reduced the loaf volume and quality of bread and that a high preheat powder was required to give a satisfactory product. The loaf was further improved by the addition of small quantities of potassium bromate, hydrogenated stearin, and glycerol monostearate to the milk before drying.

Present trends suggest that the pattern of future development of some milk products may be in the direction of greater diversification and in the development of specialized products. This implies that an increasing feedback from user to manufacturer will be necessary to ensure that these specialized products will not only be produced efficiently but that they will also be used in the most effective way. Some of the recently published research in this area will be dealt with in the remainder of this section.

A. Use of Milk Products in Other Foods and as Special-Purpose Foods

In supplementing protein-deficient diets it seems preferable to provide the extra protein not as a high-protein additive, but rather in the form of food in which a balance is maintained between protein and the other nutritional requirements (Milner, 1964). The tendency of casein to form stable intermolecular bonds and its ability to immobilize water by hydration suggest that it could be used to produce desirable textures in a variety of foods in addition to improving their nutritional status. For example, Subrahmanyan et al. (1961), Chandrasekhara et al. (1962), and Claydon (1957) have increased the nutritive value of macaroni products and biscuits by incorporating milk protein. These products are nutritionally better and are more attractive as food than reconstituted skim milk powder. For special diets and in areas where lactase deficiency is common, the use of coprecipitate instead of skim milk powder in protein-enriched biscuits may be beneficial (Bolin and Davis, 1970).

Some of the techniques used in protein chemistry have been scaled up for the treatment of milk for special purposes. For example, human milk has less protein and minerals than cow milk (Platt and Moncrieff, 1947) and it consequently produces a softer curd than cow milk when coagulated in the infant's stomach. Formulations based on cow milk and whey demineralized by electrodialysis (Al and Wiechers, 1952) approximate the protein, mineral, and amino acid content of human milk (McLoughlin et al., 1963). Ion-exchange resins have been used to deionize whey prior to the recovery of lactose (McGlasson and Boyd, 1951) and for the removal of strontium-90 from milk (Heinemann and Baldi, 1965). Morr et al. (1963, 1964) and Hill and Hansen (1964) have used Sephadex to separate milk into its constituents. It may be possible to prepare undenatured lactalbumin and lactoglobulin in quantity by large-scale column methods (Samuelsson et al., 1967) or by a centrifugal Sephadex procedure (Morr et al., 1967), if a need for these products arises.

B. Use of Milk Products to Replace Other Foods: Modification of Proteins

The modification of casein to confer on it desirable properties for specific applications has a long history. For example, formaldehyde is used in the manufacture of casein plastics as a means of cross-linking casein molecules and decreasing their solubility in water (Sutermeister and Browne, 1939). Dicyandiamide is used in the paper-coating industry to reduce the viscosity of casein solutions, and Salzberg et al. (1961) and Muller and Hayes (1963) have used proteolytic enzymes for the same purpose.

While casein in industrial use is modified in its physical properties by various chemical treatments, these, with the exception of enzyme treatments, are unsuited for use in food.

1. Foaming Properties of Milk Protein—Egg Substitutes

Many uses of eggs and egg white depend upon their foaming capacity, and many attempts have been made to duplicate these properties in products containing milk proteins. Kumetat and Beeby (1954) have reviewed various methods which have been used to increase the foaming properties of milk proteins. These include treatment at high pH with calcium hydroxide, partial proteolysis to give a polypeptide content of 5–40%, the addition of pyrophosphates, polyphosphates and neutral salts of strong acids, and the lowering of the coagulation temperature of the whey proteins by treatment with sulfite.

The two foam requirements in foods for which egg substitutes from milk have been sought are (a) as a support for sugar, for example in meringues, and (b) as a means of strengthening the walls around the gas bubbles during the cooking of cake batter so that it will rise and develop a suitable texture. Kumetat and Beeby (1954) have developed a powder for use in meringues by treating concentrated skim milk with calcium hydroxide at low temperature, and an egg substitute for cakes by treating concentrated skim milk with sodium hexametaphosphate, trisodium phosphate and sodium hydroxide (Kumetat and Beeby, 1956). All the egg required in sponge cake cannot, however, be replaced by the powder.

Beeby and Kumetat (1959) and Beeby and Lee (1959) have suggested, on the basis of viscosity measurements on concentrated skim milk, that following treatment with alkali or calcium sequestering agents, the casein micelles at first swell and then disintegrate. The molecular transformations which subsequently give rise to the foaming properties are however not known.

2. Gel- and Fiber-Forming Properties—Meat Substitutes

To simulate the texture of meat, a substitute in dried form must be able to hydrate and soften when cooked but not to the extent that it disintegrates. It must also have chewing properties similar to those of meat. Granules which have a chewy texture after cooking can be made by mixing milk proteins with soybean or cereal flours (Wrenshall, 1951; Kende and Ketting, 1959; MacAllister and Finucane, 1963). The κ-casein-depleted precipitate formed by treating sodium caseinate with calcium chloride readily forms a tough rubbery gel which can be used as the basis of a meat substitute (Anson and Pader, 1957, 1958, 1959).

An interesting development in meat substitutes has been the application of the techniques used in making and spinning artificial fibers. The protein solutions, based on soybean protein and casein, are extruded through a spinneret and spun into fibers. A texture similar to the structure of muscle can then be built up by orientating the fibers (Boyer, 1956; Szczesniak and Engel, 1960). Progress in textured foods using this technique has been reviewed by Hartman (1966) and Ziemba (1966).

C. INVESTIGATIONS RELATED TO FAT–PROTEIN INTERACTIONS

1. *Naturally Occurring Complexes and Homogenization*

The naturally occurring fat–protein interactions are important in milk because of the stability conferred on the fat globules by the membrane protein. The probable structure of the fat globule and its associated membrane is described in the classical work of King (1955) and the isolation and characteristics of the membrane protein by Herald and Brunner (1957) and Thompson and Brunner (1959). Also important in milk technology are the forcibly induced fat–protein interactions that occur during homogenization. It was recognized some years ago that proteins other than the normal membrane protein must be involved in the latter interactions (Brunner *et al.*, 1953; Sasaki *et al.*, 1956). Jackson and Brunner (1960) eventually demonstrated that the membrane material in homogenized milk included casein and whey proteins.

The proportions in which casein and lipid materials may combine is quite variable, and the complex may be less dense than the serum (Jackson and Brunner, 1960) or more dense (Fox *et al.*, 1960) depending on the conditions of homogenization and the composition of the sample. Fox *et al.* (1960) considered that the complex is formed because of interaction between the fat surface and hydrophobic regions on the casein made momentarily accessible by the turbulent conditions at the homogenizing valve. They also showed that the proportion of the fat which sedimented as a lipid–protein complex was increased by relatively small additions of calcium ion; they interpreted this to mean that calcium ion promoted formation of the complex. An alternative explanation is that the additional calcium ion causes an increase in the proportion of denser fat–protein particles because of the binding of additional casein to the layer(s) already adsorbed.

These results are important for milk technology because, apart from homogenized market milk, other milk products may pass through a homogenizing stage during manufacture. Examples are reduced or sterilized cream, evaporated milk, ice cream, recombined milk and filled milk, and high-fat cream for the manufacture of butter powder. In a product such as a canned sterilized cream of 25% fat content, the initial purpose of homogenization is to improve the consistency of the product. It does however have an incidental undesirable effect because the complex formed is somewhat lighter than the serum and therefore separates from it. The cream plug so formed results in an unsatisfactory product. It was found that this tendency could be eliminated or reduced by increasing the density of the lipid–protein complex. A complex of higher density was formed by introducing additional casein into the system prior to homogenization, and the stability of the cream was greatly improved (Hill and Hay, 1963; Swanson, 1964).

2. Emulsifying Properties of Milk Proteins

Casein forms complexes with other lipids as well as with milk fats and acts as an emulsifier by forming a stable coating around the fat globules (Pearson et al., 1965). When added to sausages, casein assists the retention of fat during cooking by dispersing and stabilizing the fat as an emulsion. It also serves as a water-binding agent, particularly when treated with pyrophosphate (Freund and Danes, 1960).

Lipoproteins appear to act as emulsifiers and have other desirable properties in manufactured foods. For example, Waldt et al. (1963) have shown that a lipoprotein containing 33% lipid and 65% protein enhances the action of eggs and emulsifiers in a wide range of foods and that it can replace two-thirds of the eggs in cake. It may prove possible to make complexes with milk fat and proteins which will behave in a similar way and improve the quality of certain manufactured foods.

3. High-Fat Powders

Research on the manufacture of high-fat powders has been stimulated by the possibility of developing powdered creams and shortenings and as a means of storing milk fat without refrigeration.

A number of approaches to the problem of manufacturing powdered shortenings were reviewed by Hansen (1963a). Requirements for the manufacture of a satisfactory powder include an oil-in-water emulsion that will be stable enough to withstand drying. Sufficient fat must be released from the powder to give the action required in the product in which the powder is used. Some de-emulsification of the reconstituted powder may be necessary to achieve this. The crystalline state of the fat should also be controlled to give a structure commensurate with the requirements of the mixture in which it is being used. For example in making cakes, Hansen (1963b) and Hoerr et al. (1966) found that the density of the batter and cake volume are related to the crystalline state of the shortening.

Hansen (1963a) found that a butter powder meeting these requirements could be made by homogenizing cream (or butter fat and skim milk), mixing with casein and emulsifier, and spray-drying. Sodium citrate was added to assist the emulsification of the fat by sequestering calcium ion (Fox, 1960). The baking performance of butter powder is improved by reducing its protein content and increasing the emulsifier level (Snow et al., 1967).

Boudreau et al. (1966) have described the preparation of a free-flowing, spray-dried butter powder containing 80% fat but without the edible additives used by Hansen. Preliminary baking tests show that this powder can be used as a powdered shortening, although comparative data on the volume and texture of cakes made from butter powder incorporating emulsifier are not given. Spray-drying the emulsion at pH 9 increases the recovery of volatile fatty acids, and this finding could be of importance in improving the butter flavor of the powder.

The nature of the coating around the fat globule is critical in determining stable conditions for emulsification and drying and for the partial de-emulsification of the powder when it is reconstituted. At the present time little is known about the effect of different types of coating materials in relation to the properties of the emulsion or about the minimum thickness, molecular weight and functional groupings that are required to produce a stable emulsion. A start in this area has been made by Tripp et al. (1966a, b) in their investigation of the effects of different types of carbohydrates on the amount of fat extracted from high-fat powders by petroleum and ethyl ethers. In most powders, 80–90% of the fat was extracted, but the addition of 9–10% dextrose or sucrose reduced this value to 50 and 68%, respectively. Further research on suitable coating materials and on the colloid chemistry of these systems is needed to find a range of emulsification and de-emulsification conditions. This information is required before a systematic approach can be made to the problem of manufacturing high-fat powders for specific purposes in foods.

VI. Conclusion

In considering the problems which occur during the manufacture of various milk products, it is possible to select factors that are common to a number of them. All manufactured milk products are heated at some stage of production, and this heating may strongly influence the properties of the final product. Thus the technique of forewarming is used to impart stability to evaporated milks during sterilization and storage; this may depend in part for its effectiveness on the interaction of serum protein and casein micelles at high temperatures. The same reaction is normally to be avoided in pasteurizing milk for cheese manufacture, although it is an important stage in the production of milk protein coprecipitates, and as suggested earlier it may be possible to make use of the reaction to produce cheese containing whey protein as well as casein. While there is evidence that the interaction depends upon –SS–SH exchange (Sawyer et al., 1963), this is not yet certain, and information on the detailed course of the reaction and the conditions of pH, temperature, ionic strength and composition which promote or hinder it is far from complete (see DellaMonica et al., 1958; Mauk and Demott, 1959; Sawyer, 1969). In view of its industrial importance, a thorough study of the phenomenon is desirable.

When milk is subjected to high temperatures, changes in conformation of the proteins and association-dissociation reactions occur. Present knowledge of these processes is limited and has for the most part been obtained by studying at ordinary temperatures the products of the high-temperature reactions. Pertinent information might be gained from studies made at the actual reaction temperature. In spite of the difficulties of working at these temperatures, the results of Kenkare et al. (1965) on the α_s-κ-casein dissociation at 85°C and of Rose and Tessier (1959a) on changes in the milk serum at 93°C indicate the potential value of this approach. Studies of the changes in ionic composition which occur at high temperatures and the effect of heat on the organization of the casein micelle could help our understanding not only of serum protein–casein interactions in heated milk, but also of matters such as the heat stability of concentrated milks and their gelation on storage. At this stage, it is probable that enough is known about the composition of the micelle for such studies to be useful.

There are also some similarities between the coagulation of milk following rennin action and the gelation of concentrated milks (as regards the effect of various ions and complexing agents for example), and detailed knowledge of the mechanism of the clotting reaction might help in dealing with the gelation problem.

Although considerable advances in our knowledge of the action of rennin have recently been made, many puzzling features remain. The κ-casein substrate of the rennin has been isolated, but its distribution in the casein micelle is not known. The reason for the high reactivity of the rennin-sensitive bond and the detailed course of the enzymic and clotting reactions are likewise not fully known. It is frequently not possible to determine the reason for the slow setting of a particular milk, which illustrates the state of ignorance of the factors involved. On the practical level, our inability to control adequately the rates of clotting, syneresis and curd fusion is a limiting factor in the design of more efficient cheese-making processes. The tendency of concentrated sterilized milk to coagulate is at present controlled by the technique of forewarming and by limiting the degree of concentration of the milk. If the nature of the functional groups responsible for the gelation were known, it might be possible to find an alternative to the forewarming treatment for imparting stability and also to make a more concentrated product of improved flavor.

Techniques for the separation of milk proteins on an industrial scale have changed little in the recent past. However, the method of Hidalgo and Hansen (1969) for the separation of complexes of whey proteins and carboxymethylcellulose at their respective isoelectric points not only points to a general means for obtaining pure proteins for research purposes, but it also has potential as an industrial method for the recovery of protein and peptide materials. The usefulness of the method may be increased by alteration of the pK of the ionic groups on the precipitating agent, by altering its hydrophobicity or size, or by selectively modifying the protein partner to improve the selectivity of the separation.

It is evident from this short survey that some fundamental problems in the field of milk proteins are closely related to problems of milk technology. In addition the research problems themselves are interrelated. Thus, studies of the organization of the proteins and the role of calcium phosphate in the casein micelle can provide information relevant to the studies of the action of rennin in milk, the coagulation of milk and the nature of the interaction between the serum proteins and the casein micelles. In this situation the results gained in one field of investigation are likely to shed light on the problems in another, and we can expect that the results themselves will have a greater chance of being usefully applied.

Acknowledgments

The authors wish to record their thanks to their colleagues, Messrs. G. Loftus Hills, J. Czulak and L. L. Muller, for the helpful discussions during the preparation of the manuscript.

References

Al, J., and Wiechers, S. G. (1952). *Research (London)* **5**, 256.
Alais, C., Mocquot, G., Nitschmann, Hs., and Zahler, P. (1953). *Helv. Chim. Acta* **36**, 1955.
Anson, M. L., and Pader, M. (1957). U.S. Patent 2,813,794.
Anson, M. L., and Pader, M. (1958). U.S. Patent 2,830,902.
Anson, M. L., and Pader, M. (1959). U.S. Patent 2,879, 163.
Arbatskaya, N., Panasenkov, N., Sokolov, A., and Fialkov, M. (1962). *Moloch. Prom.* **23**, 9.
Ashworth, U. S. (1965). *J. Dairy Sci.* **48**, 537.
Beeby, R. (1964). *Biochim. Biophys. Acta* **82**, 418.
Beeby, R. (1965). *J. Dairy Res.* **32**, 57.
Beeby, R. (1966a). *Proc. Int. Dairy Congr., 17th, 1966* **B2**, 95.
Beeby, R. (1966b). *Proc. Int. Dairy Congr., 17th, 1966* **E3**, 115.
Beeby, R., and Kumetat, K. (1959). *J. Dairy Res.* **26**, 248.
Beeby, R., and Lee, J. W. (1959). *J. Dairy Res.* **26**, 258.
Beeby, R., and Loftus Hills, G. (1962). *Proc. Int. Dairy Congr., 16th, 1962* **2**, 1019.

Belec, J., and Jenness, R. (1960). *J. Dairy Sci.* **43**, 849.
Bell, K. W., Curran, H. R., and Evans, F. R. (1944). *J. Dairy Sci.* **27**, 913.
Bernhart, F. W., Eckhardt, E. R., Janson, M. H., and Tinkler, F. H. (1955). U.S. Patent 2,714,068.
Berridge, N. J. (1942). *Nature* **149**, 194.
Berridge, N. J. (1963). *Dairy Eng.* **80**, 130.
Berridge, N. J. (1968). *J. Soc. Dairy Technol.* **21**, 52.
Bolin, T. D., and Davis, A. E. (1970). *Aust. J. Dairy Technol.* **25**, 119.
Boudreau, A., Richardson, T., and Amundson, C. H. (1966). *Food Technol. (Chicago)* **20**, 668.
Boyd, E. N., and Gould, I. A. (1957). *J. Dairy Sci.* **40**, 1294.
Boyer, R. A. (1956). U.S. Patent 2,730,447.
Brunner, J. R., Duncan, C. W., and Trout, G. M. (1953). *Food Res.* **18**, 454.
Buchanan, R. A., Snow, N. S., and Hayes, J. F. (1965). *Aust. J. Dairy Technol.* **20**, 139.
Budd, R. T., and Chapman, H. R. (1962). *Proc. Int. Dairy Congr., 16th, 1962* **C**, 194.
Burton, H. (1955). *J. Dairy Res.* **22**, 74.
Bysouth, R., Gillies, J., Harkness, W. L., McGillivray, W. A., and Robertson, P. S. (1968) *N.Z. J. Dairy Technol.* **3**, 21.
Cerbulis, J., and Zittle, C. A. (1965). *J. Dairy Sci.* **48**, 1154.
Chandrasekhara, M. R., Soma, K., Indiramma, K., Bhatia, D. S., Swaminathan, M., Sreenivasan, A., and Subrahmanyan, V. (1962). *Food Sci.* **11**, 27.
Cheeseman, G. C. (1962). *Proc. Int. Dairy Congr., 16th, 1962* **B**, 465.
Cheeseman, G. C. (1963). "Annual Report of the National Institute of Research in Dairying," p. 98. Shinfield, Reading.
Christ, W. (1956). *Milchwissenschaft* **11**, 381.
Claydon, T. J. (1957). *Milk Prod. J.* **48**, 14.
Cluskey, F. J., Thomas, E. L., and Coulter, S. T. (1969). *J. Dairy Sci.* **52**, 1181.
Conochie, J., and Sutherland, B. J. (1965). *J. Dairy Res.* **32**, 35.
Coulter, S. T., Harland, H., and Jenness, R. (1948). *J. Dairy Sci.* **39**, 699.
Czulak, J. (1958). *Dairy Eng.* **75**, 67.
Czulak, J., Hammond, L. A., and Meharry, H. J. (1954). *Aust. Dairy Rev.* **22**, 18.
"Dairy Produce." (1969). The Commonwealth Secretariat, London.
Davies, D. T. and White, J. C. D. (1959). *Proc. Int. Dairy Congr., 15th, 1959* **3**, 1677.
Delfour, A., Jollès, J., Alais, C., and Jollès, P. (1965). *Biochem. Biophys. Res. Commun.* **19**, 452.
DellaMonica, E. S., Custer, J. H., and Zittle, C. A. (1958). *J. Dairy Sci.* **41**, 465.
Dennis, E. S., and Wake, R. G. (1965). *Biochim. Biophys. Acta* **97**, 159.
Denkov, T. (1967). *Izv. Nauchnoizsled. Inst. Mlech. Prom. Vidin* **2**, 15.
Desai, I. D., and Nickerson, T. A. (1964). *Nature* **202**, 183.
Desai, I. D., Nickerson, T. A., and Jennings, W. G. (1961). *J. Dairy Sci.* **44**, 215.
Deysher, E. F., Webb, B. H., and Holm, G. E. (1929). *J. Dairy Sci.* **12**, 80.
Downey, W. K., and Andrews, P. (1965). *Biochem. J.* **96**, 21C.
Durrant, N. W., Stone, W. K., and Large, P. M. (1961). *J. Dairy Sci.* **44**, 1171.
D'yachenko, P. (1957). *Moloch. Prom.* **18**, 5.
D'yachenko, P., Vlodavets, I., and Bogomolova, E. (1953). *Moloch. Prom.* **14**, 33.
Edmondson, L. F., Avants, J. K., Douglas, F. W., and Easterly, D. G. (1966). *J. Dairy Sci.* **49**, 708.
Eilers, H. (1947). *In* "Chemical and Physical Investigations on Dairy Products. Monographs on the Progress of Research in Holland during the War" (R. Houwink and J. A. Ketelsar, eds.), pp. 1–114. Elsevier, Amsterdam.

Ernstrom, C. A. (1964). *Milk Dealer* **54,** 50.
Feagan, J. T., Griffin, A. T., and Lloyd, G. T. (1966). *J. Dairy Sci.* **49,** 933.
Feagan, J. T., Hehir, A. F., and Bailey, L. F. (1969). *J. Dairy Sci.* **52,** 887.
Fish, J. C. (1957). *Nature* **180,** 345.
Foltmann, B. (1966). *C. R. Trav. Lab. Carlsberg* **35,** 143.
Fontana, P., Colagrande, O., and Corrandini, C. (1962). *Proc. Int. Dairy Congr., 16th, 1962* **B,** 901.
Fox, K. K. (1956). *Diss. Abstr.* **16,** 2129; *Dairy Sci. Abstr.* **20,** 705 (1958).
Fox, K. K., Holsinger, V., Caha, J., and Pallansch, M. J. (1960). *J. Dairy Sci.* **43,** 1396.
Freund, E. H., and Danes, E. N. (1960). U.S. Patent 2,957,770.
Genvrain, S. A. (1967). British Patent 1 079 604.
Greenbank, G. R., Steinbarger, M. C., Deysher, E. F., and Holm, G. E. (1927). *J. Dairy Sci.* **10,** 335.
Greenberg, R., and Kalan, E. B. (1965). *Biochemistry* **4,** 1660.
Hanrahan, F. P., Tamsma, A., Fox, K. K., and Pallansch, M. J. (1962). *J. Dairy Sci.* **45,** 37.
Hansen, P. M. T. (1963a). *Aust. J. Dairy Technol.* **18,** 79.
Hansen, P. M. T. (1963b). *Aust. J. Dairy Technol.* **18,** 86.
Hansen, P. M. T. (1966). *J. Dairy Sci.* **49,** 698.
Harland, H. A., and Ashworth, U.S. (1945). *J. Dairy Sci.* **28,** 15.
Hartman, W. E. (1966). *Food Technol. (Chicago)* **20,** 39.
Heinemann, B., and Baldi, E. J. (1965). *J. Dairy Sci.* **48,** 781.
Henika, R. C., Rodgers, N. E., and Miersch, R. E. (1958). U.S. Patent 2,826,571.
Henry, K. M., Kon, S. K., Lea, C. H., Smith, J. A. B., and White, J. C. D. (1946). *Nature* **158,** 348.
Herald, C. T., and Brunner, J. R. (1957). *J. Dairy Sci.* **40,** 948.
Herrmann, J., and Perlmann, G. (1936). *Nature* **140,** 807.
Hidalgo, J., and Hansen, P. M. T. (1969). *J. Agr. Food Chem.* **17,** 1089.
Hill, R. D. (1964). *J. Dairy Res.* **31,** 185.
Hill, R. D. (1968). *Biochem. Biophys. Res. Commun.* **33,** 659.
Hill, R. D. (1970). *J. Dairy Res.* **37,** 187.
Hill, R. D., and Craker, B. A. (1968). *J. Dairy Res.* **35,** 13.
Hill, R. D., and Hansen, R. R. (1964). *J. Dairy Res.* **31,** 291.
Hill, R. D., and Hay, A. K. (1963). *Aust. J. Dairy Technol.* **18,** 97.
Hill, R. D., and Laing, R. R. (1965a). *J. Dairy Res.* **32,** 193.
Hill, R. D., and Laing, R. R. (1965b). *Biochim. Biophys. Acta* **99,** 352.
Hill, R. D., and Laing, R. R. (1966). *Nature* **210,** 1160.
Hill, R. D., and Laing, R. R. (1967). *Biochim. Biophys. Acta* **132,** 188.
Hill. R. D., Laing, R. R., Snow, N. S., and Hammond, L. A. (1965). *Aust. J. Dairy Technol.* **20,** 122.
Hoerr, C. W., Moncrieff, J., and Paulicka, F. R. (1966). *Baker's Dig.* **40,** 38.
Holm, G. E., Deysher, E. F., and Evans, F. R. (1923). *J. Dairy Sci.* **6,** 556.
Hostettler, H., and Imhof, K. (1951). *Milchwissenschaft* **6,** 351.
Hostettler, H., Stein, J., and Bruderer, G. (1957). *Landwirtsch. Jahrb. Schweiz* **6,** 143; *Dairy Sci. Abstr.* **20,** 1209 (1958).
Hostettler, H., Imhof, K., and Stein, J. (1965). *Milchwissenschaft* **20,** 189.
Howard, H. W., Block, R. J., and Sevall, H. E. (1954). U.S. Patent 2,665,989.
Howat, G. R., and Wright, N. C. (1934a). *Biochem. J.* **28,** 1336.
Howat, G. R., and Wright, N. C. (1934b). *J. Dairy Res.* **5,** 236.

Hsu, R. Y. H., Anderson, L., Baldwin, R. L., Ernstrom, C. A., and Swanson, A. M. (1958). *Nature* **182**, 798.
Imhof, K. (1952). Dissertation. University of Bern, Bern, Switzerland.
Ingram, V. M. (1958). *Biochim. Biophys. Acta* **28**, 539.
Irvine, D. M. (1967). *Dairy Sci. Abstr.* **29**, 243, 317.
Ismael, A. A., and Grimbleby, F. H. (1959). *Proc. Int. Dairy Congr., 15th, 1959* **3**, 1873.
Jackson, R. H., and Brunner, J. R. (1960). *J. Dairy Sci.* **43**, 912.
Josh, G., and Hull, M. E. (1950). U.S. Patent 2,521,853.
Kannan, A., and Jenness, R. (1956). *J. Dairy Sci.* **39**, 911.
Kannan, A., and Jenness, R. (1961). *J. Dairy Sci.* **44**, 808.
Kende, S., and Ketting, F. (1959). *Proc. Int. Dairy Congr., 15th, 1959* **2**, 1211.
Kenkare, D. B., Hansen, P. M. T., and Gould, I. A. (1965). *J. Dairy Sci.* **48**, 779.
King, D. W. (1964). "36th Annual Report of the Dairy Research Institute," p. 29. Palmerston North, New Zealand.
King, N. (1955). "The Milk Fat Globule Membrane and Some Associated Phenomena." Commonwealth Agricultural Bureaux, Farnham Royal, England.
Kisza, J., Karwowicz, E., and Sobina, A. (1964). *Milchwissenschaft* **19**, 437.
Kisza, J., Sobina, A., and Zbikowski, Z. (1966). *Proc. Int. Dairy Congr., 17th, 1966* **E3**, 85.
Kozin, N. I., and Rodionova, I. R. (1962). *Izv. Vyssh. Ucheb. Zaved., Pisch. Tekhnol.* **2**, 61, as quoted in *Dairy Eng.* **80**, 62 (1963).
Kristoffersen, T., Mikolajcik, E. M., and Gould, I. A. (1967). *J. Dairy Sci.* **50**, 292.
Kumetat, K., and Beeby, R. (1954). *Dairy Ind.* **19**, 730.
Kumetat, K., and Beeby, R. (1956). *Dairy Ind.* **21**, 287.
Kumetat, K., and Beeby, R. (1957). *Dairy Ind.* **22**, 642.
Larson, B. L., and Jenness, R. (1950). *J. Dairy Sci.* **33**, 896.
Lawrence, A. J. (1959). *Aust. J. Dairy Technol.* **14**, 166.
Lea, C. H. (1947). *Analyst (London)* **72**, 336.
Lee, J. W. (1952). *Aust. J. Dairy Technol.* **7**, 118.
Leviton, A., and Pallansch, M. J. (1961). *J. Dairy Sci.* **44**, 633.
Leviton, A., Anderson, H. A., Vettel, H. E., and Vestal, J. H. (1963). *J. Dairy Sci.* **46**, 310.
Long, J. E. (1959). *Diss. Abstr.* **19**, 2242; *Dairy Sci. Abstr.* **21**, 1712 (1959).
Long, J. E., van Winkle, Q. and Gould, I. A. (1963). *J. Dairy Sci.* **46**, 1329.
Lowenstein, M. (1961a). U.S. Patent 3,001,876.
Lowenstein, M. (1961b). U.S. Patent 2,970,913.
Lowenstein, M. (1965). U.S. Patent 3,218,173.
Mabbitt, L. A., and Cheeseman, G. C. (1969). *J. Dairy Res.* **34**, 73.
MacAllister, R. V., and Finucane, T. P. (1963). U.S. Patent 3,102,031.
McDowall, F. H. (1961). *Aust. J. Dairy Technol.* **16**, 143.
McFarlane, A. S. (1938). *Nature* **142**, 1023.
McGann, T. C. A., and Pyne, G. T. (1960). *J. Dairy Res.* **27**, 403.
McGlasson, E. D., and Boyd, J. C. (1951). *J. Dairy Sci.* **34**, 119.
McGugan, W. A., Zehren, V. F., Zehren, V. L., and Swanson, A. M. (1954). *Science* **120**, 435.
Mackinlay, A. G., and Wake, R. G. (1965). *Biochim. Biophys. Acta* **104**, 167.
McLoughlin, P. T., Bernhart, F. W., and Tomarelli, R. M. (1963). *Clin. Pediat.* **2**, 302.
Mauk, B. R., and Demott, B. J. (1959). *J. Dairy Sci.* **42**, 39.
Melnychyn, P., and Wolcott, J. M. (1967). *J. Dairy Sci.* **48**, 780.
Milner, M. (1964). *WHO/FAO/UNICEF, Protein Advisory Group News Bulletin* No. 4, 2.

Mocquot, G., and Garnier, J. (1965). *J. Agr. Food Chem.* **13,** 414.
Mocquot, G., Alais, C., and Chevalier, R. (1954). *Ann. Technol. Agr.* **3,** 1.
Morr, C. V. (1965a). *J. Dairy Sci.* **48,** 8.
Morr, C. V. (1965b). *J. Dairy Sci.* **48,** 29.
Morr, C. V., Kenkare, D. B., and Gould, I. A. (1963). *J. Dairy Sci.* **46,** 593.
Morr, C. V., Kenkare, D. B., and Gould, I. A. (1964). *J. Dairy Sci.* **47,** 621.
Morr, C. V., Nielsen, M. A., and Coulter, S. T. (1967). *J. Dairy Sci.* **50,** 305.
Morrissey, P. A. (1969). *J. Dairy Res.* **36,** 343.
Muller, L. L. (1963). *Aust. Soc. Dairy Technol., Tech. Pub.* **12,** 25.
Muller, L. L., and Hayes, J. F. (1962). *Aust. J. Dairy Technol.* **17,** 189.
Muller, L. L., and Hayes, J. F. (1963). *Aust. J. Dairy Technol.* **18,** 184.
Muller, L. L., and Kieseker, F. G., (1965). *Aust. J. Dairy Technol.* **20,** 130.
Muller, L. L., Hayes, J. F., Buchanan, R. A., and Snow, N. S. (1967). *Aust. J. Dairy Technol.* **22,** 12.
Nakai, S., Wilson, H. K., and Herreid, E. O. (1964). *J. Dairy Sci.* **47,** 754.
Neff, E. (1966). *J. Agr. (Victoria)* **64,** 157, 269.
Nitschmann, Hs., and Beeby, R. (1960). *Chimia* **14,** 318.
Nitschmann, Hs., and Bohren, Hr. (1955). *Helv. Chim. Acta* **38,** 942.
Odagiri, S., and Nickerson, T. A. (1964). *J. Dairy Sci.* **47,** 1306.
Oetiker, N. (1960). *Aust. J. Dairy Technol.* **15,** 69.
Pantaloni, D. (1963). *Compt. Rend.* **256,** 4994.
Pantaloni, D. (1964). *Compt. Rend.* **258,** 5753.
Payens, T. A. J., and van Markwijk, B. W. (1963). *Biochim. Biophys. Acta* **71,** 517.
Pearson, A. M., Spooner, M. E., Hegarty, G. R., and Bratzler, L. J. (1965). *Food Technol. (Chicago)* **19,** 1841.
Peters, I. I. (1964). *Dairy Sci. Abstr.* **26,** 458.
Peters, I. I. (1965). *J. Dairy Sci.* **48,** 764.
Platt, B. S., and Moncrieff, A. (1947). *Brit. Med. Bull.* **5,** 177.
Pont, E. G. (1960). *Aust. J. Dairy Technol.* **15,** 17.
Probst, A. (1964). *Molkerei–Kaserei–Z.* **15,** 428.
Pyne, G. T. (1958). *J. Dairy Res.* **25,** 467.
Pyne, G. T., and McGann, T. C. A. (1960). *J. Dairy Res.* **27,** 9.
Pyne, G. T., and McGann, T. C. A. (1962). *Proc. Int. Dairy Congr., 16th, 1962* **B,** 611.
Ramshaw, E. H. (1966). Personal communication.
Rao, P. S. (1961). *Diss. Abstr.* **21,** 3251; *Dairy Sci. Abstr.* **23,** 3027 (1961).
Rapp, H., and Calbert, H. E. (1954). *J. Dairy Sci.* **37,** 637.
Reisfeld, R. A. (1957). *Diss. Abstr.* **17,** 1204.
Rodgers, N. E., and Palmer, G. M. (1966). U.S. Patent 3,252,961.
Rogers, L. A., Deysher, E. F., and Evans, F. R. (1920). *J. Dairy Sci.* **3,** 468.
Rose, D. (1961a). *J. Dairy Sci.* **44,** 430.
Rose, D. (1961b). *J. Dairy Sci.* **44,** 1405.
Rose, D. (1961c). *J. Dairy Sci.* **44,** 1763.
Rose, D., and Tessier, H. (1959a). *J. Dairy Sci.* **42,** 969.
Rose, D., and Tessier, H. (1959b). *J. Dairy Sci.* **42,** 989.
Salzberg, H. K., Georgevits, L. E., and Cobb, R. M. K. (1961). T.A.P.P.I. Monograph 22, p. 103. New York.
Samel, R., and Muers, M. M. (1962a). *J. Dairy Res.* **29,** 249.
Samel, R., and Muers, M. M. (1962b). *J. Dairy Res.* **29,** 269.
Samel, R., and Muers, M. M. (1962c). *J. Dairy Res.* **29,** 259.
Samuelsson, E. G., Emneus, A., and Hallström, B. (1967). *Sv. Mejeritidn.* **59,** 89.

Sasaki, R., Tsugo, T., and Miyazawa, K. (1956). *Proc. Int. Dairy Congr., 14th, 1956* **1,** 233.
Sawyer, W. H. (1968). *J. Dairy Sci.* **51,** 323.
Sawyer, W. H. (1969). *J. Dairy Sci.* **52,** 1347.
Sawyer, W. H., Coulter, S. T., and Jenness, R. (1963). *J. Dairy Sci.* **46,** 564.
Schmidt, D. G. (1969). *Neth. Milk Dairy J.* **23,** 128.
Schmidt, D. G., and Koops, J. (1965). *Neth. Milk Dairy J.* **19,** 63.
Scott, E. C. (1952). U.S. Patent 2,623,038.
Scott, E. C. (1958). U.S. Patent 2,832,685.
Scott-Blair, G. W., and Burnett, J. (1959). *J. Dairy Res.* **26,** 144.
Sinnamon, H. I., Aceto, N. C., Eskew, R. K., and Schoppet, E. F. (1957). *J. Dairy Sci.* **40,** 1036.
Sirry, I., and Shipe, W. F. (1958). *J. Dairy Sci.* **41,** 204.
Smart, P. H., and Nelson, C. E. (1957). U.S. Patent 2,807,608.
Snow, N. S., Townsend, F. R., Bready, P. J., and Shimmin, P. D. (1967). *Aust. J. Dairy Technol.* **22,** 125.
Spellacy, J. R. (1953). "Casein, Dried and Condensed Whey." Lithotype Process, San Francisco.
Stadhouders, J., Blauw, J., and Badings, H. T. (1962). *Officëel Organ vande koninglijke nederlandsche Zuivelbond* **54,** 337.
Stenne, P. (1965). *Lait* **45,** 143.
Stepanov, V. M., Lobareva, L. S., and Mal'tsev, N. I. (1968). *Biochim. Biophys. Acta* **151,** 719.
Stoll, W. F., and Morris, H. A. (1966). *J. Dairy Sci.* **49,** 698.
Subrahmanyan, V., Gopalakrishna Rao, N., Venkata Rao, S., Bains, G. S., Bhatia, D. S., Swaminathan, M., and Sreenivasan, A. (1961). *Food Sci. (Mysore)* **10,** 379.
Sullivan, R. A., Hollis, R. A., and Stanton, E. K. (1957). *J. Dairy Sci.* **40,** 830.
Sullivan, R. A., Fitzpatrick, M. M., and Stanton, E. K. (1959). *Nature* **183,** 616.
Sutermeister, E., and Browne, F. L. (1939). "Casein and its Industrial Applications," 2nd ed. Reinhold, New York.
Swanson, A. M. (1964). U.S. Patent 3,117,879.
Szczesniak, A. S., and Engel, E. (1960). U.S. Patent 2,952,543.
Tanford, C., and Taggart, V. G. (1961). *J. Amer. Chem. Soc.* **83,** 1634.
Tanford, C., Bunville, L. G., and Nozaki, Y. (1959). *J. Amer. Chem. Soc.* **81,** 4032.
Tarassuk, N. P., and Tamsma, A. F. (1956). *Agr. Food Chem.* **4,** 1033.
Tessier, H., and Rose, D. (1964). *J. Dairy Sci.* **47,** 1047.
Tessier, H., Yaguchi, M., and Rose, D. (1969). *J. Dairy Sci.* **52,** 139.
Thompson, M. P., and Brunner, J. R. (1959). *J. Dairy Sci.* **42,** 369.
Trautman, J. C., and Swanson, A. M. (1959). *J. Dairy Sci.* **42,** 895.
Trexler, P. C. (1952). U.S. Patent 2,618,629.
Tripp, R. C., Amundson, C. H., and Richardson, T. (1966a). *J. Dairy Sci.* **49,** 695.
Tripp, R. C., Amundson, C. H., and Richardson, T. (1966b). *Mfd. Milk Prod. J.* **57,** 6.
Tumerman, L., Fram, H., and Cornelly, K. W. (1954). *J. Dairy Sci.* **37,** 830.
Ubbels, J., and van der Linde, J. T. (1962). *Proc. Int. Dairy Congr., 16th, 1962* **C,** 185.
van Kreveld, A., and van Minnen, G. (1955). *Neth. Milk Dairy J.* **9,** 1.
Vassal, L., and Auclair, J. (1966). *Ind. Lait.* **237,** 666.
Wake, R. G. (1959). *Aust. J. Biol. Sci.* **12,** 479.
Waldt, L. M., Debreczeni, E. J., Schwarcz, M., and O'Keefe, T. (1963). *Food Technol. (Chicago)* **17,** 927.

Wasserman, A. E., Hampson, J., Alvare, N. F., and Alvare, N. J. (1961). *J. Dairy Sci.* **44,** 387.
Waugh, D. F. (1961). *J. Phys. Chem.* **65,** 1793.
Waugh, D. F., and von Hippel, P. H. (1956). *J. Amer. Chem. Soc.* **78,** 4576.
Webb, B. H., and Holm, G. E. (1932). *J. Dairy Sci.* **15,** 345.
Weisberg, S. M., Johnson, A. H., and McCollum, E. V. (1933). *J. Dairy Sci.* **16,** 225.
White, J. C. D., and Davies, D. T. (1958). *J. Dairy Res.* **25,** 267.
Whitehead, H. R., and Harkness, W. L. (1954). *Aust. J. Dairy Technol.* **9,** 103.
Wilson, H. K., and Herreid, E. O. (1961). *J. Dairy Sci.* **44,** 552.
Wix, P., and Woodbine, M. (1958). *Dairy Sci. Abstr.* **20,** 539, 623.
Wrenshall, C. L. (1951). U.S. Patent 2,560,621.
Wright, N. C. (1932). *J. Dairy Res.* **4,** 122.
Wright, R. C., and Tramer, J. (1953). *J. Dairy Res.* **20,** 177.
Ziemba, J. V. (1966). *Food Eng.* **38,** 82.
Zittle, C. A. (1961). *J. Dairy Sci.* **44,** 2100.
Zittle, C. A., Thompson, M. P., Custer, J. H., and Cerbulis, J. (1962). *J. Dairy Sci.* **45,** 807.

Part G
The Future

General Introduction

In the preface of Volume I we promised the reader a detailed presentation of our knowledge of milk proteins and of important methods for their investigation. In some areas progress has been rapid and a detailed picture has been revealed. In others progress has been slower and the picture revealed is analogous to a bird's eye view, the details not yet being in sharp focus. Nevertheless the picture emerging in both cases is one of great beauty. We have stressed the great tendency for milk proteins to associate and dissociate and the need to consider this property carefully in the interpretation of physical and other types of investigations. In addition, the caseins interact extensively to form polymers of large size. The most impressive of these aggregates are the micelles of milk itself. Because of the range of caseins and ions involved in micelles, the elucidation of their structure and properties is a complex problem. An important example of the effect of micelle formation on the properties of the individual caseins can be seen in the work of Fox (1970). In normal milk, where a large part of the $\alpha_{s,1}$- and β-caseins is in micellar form at 20°C, the susceptibility of these two caseins to proteolysis is very slight. Fox has shown that on disaggregation of the micelle by removal of calcium phosphate, $\alpha_{s,1}$- and β-caseins become accessible to proteolysis. Furthermore, on lowering the temperature of casein micelles sufficient β-casein dissociates to make it susceptible to proteolysis.

Problems such as these are of great interest in protein chemistry, but some of them also have a profound significance in milk technology. Members of the XVIII International Dairy Congress in 1970 stressed the need for the peoples of the world and their international agencies to take action, however radical, to see that malnutrition is banished from the earth.

They believe that, because of their high nutritional value and the advances in manufacture, milk proteins can play a leading role in overcoming protein malnutrition. Resolutions such as these lend a greater urgency to the solution of problems in milk protein chemistry, technology, and allergenicity reactions.

It is considered appropriate at this stage to assess some of the outstanding problems in the chemistry of milk proteins and to speculate on some of the possible courses of action for their solution.

H. A. McKenzie

18 □ Milk Proteins in Prospect

H. A. McKENZIE

I.	Introduction	469
II.	Prospects	470
	A. Caseins and Micelles	470
	B. β-Lactoglobulin	473
	C. α-Lactalbumin and Lysozyme	474
	D. Minor Proteins and Enzymes	477
	E. Allergenicity of Milk Proteins	478
	F. Immunoglobulins in Milk	478
	G. Some Important Technological Problems	479
	References	480
	Note Added in Proof: Some Further Recent Advances	482

I. Introduction

Progress was made in the early period of milk protein studies mainly because of increasing knowledge of (a) protein structure, function and methodology in general and (b) milk proteins in particular. In the first area we recall the basic contributions made by chemists such as the Danes to whom this book is dedicated. Concomitant with this work, information was gradually accumulating on the protein composition of milk and of the isolation, characterization and function of individual milk proteins. Some of the eminent workers of this early period of development (examples being Tiselius, Linderstrøm-Lang, and McMeekin) overlapped into the second period. The dramatic progress in this second period, commencing about 1950, has arisen out of several major developments. These are (a) increasing knowledge of the covalent structure of proteins and the location and behavior of functional groups (given tremendous impetus by the work of Sanger and later by that of Moore and Stein); (b) the determination of the conformation of proteins in the crystalline state (given particular impetus by the studies of Corey and Pauling); (c) improved knowledge of the properties of proteins in solution, exemplified by the work of Edsall, Kauzmann, Doty, Gilbert and others; (d) the application of new methods for the isolation of proteins (for example, the use of Sephadex gel filtration by Flodin and Porath); (e) the development of physico-chemical methods, such as electrophoresis, ultracentrifugation, circular dichroism, optical rotatory dispersion; (f) the first demonstration of genetic variants in milk proteins (β-lactoglobulin) by Aschaffenburg and Drewry and the discovery of κ-casein by Waugh and von Hippel.

We have reached a stage at which we are familiar with the majority of proteins and enzymes in cow milk and appreciably less familiar with those of the milk of other species. Knowledge of the properties and structure of most of them is far from complete. However, in standing at the threshold of the future we are armed with an unprecedented range of techniques and an understanding of the properties and interactions of protein molecules. Let us look at several prospects for some of the various classes of milk proteins.

II. Prospects

A. Caseins and Micelles

The caseins are the major proteins of milk, but in no area of milk protein research, despite tremendous progress, are we more aware of deficiencies in our knowledge. It is the author's view that the most promising line of attack in gaining information on the individual caseins is to carry out comparative studies on their chemical evolution in the various mammals. Aschaffenburg (1968) has clearly demonstrated the formidable amount of data we now have on the occurrence of the genetic variants of α_s-, β-, and κ-caseins in the various breeds of cattle. Thompson has shown in his own work and in Chapter 11 how valuable chemical studies on the variants of α_s- and β-caseins can be. It is important to characterize these caseins further: to ascertain which genetic variants may occur, the physical interactions of these variants and the noncovalent forces responsible for the interactions.

We have stressed in Part C, Volume I, that the basis for the interpretation of experiments on interacting systems, using existing physical methods, is firmly established although not widely recognized. Nevertheless new methods must be developed in the future in order to carry out effective and comprehensive studies on interacting systems such as the caseins.

Annan and Manson (1969) have isolated several minor α_s-caseins (see Chapter 11). They introduced the following terminology for the minor α_s-caseins: $\alpha_{s,0}$-, $\alpha_{s,2}$-, $\alpha_{s,3}$-, $\alpha_{s,4}$-, and $\alpha_{s,5}$-casein. This notation distinguishes them from the major α_s- caseins, still designated as $\alpha_{s,1}$-caseins. Hoagland et al. (1970) have shown that $\alpha_{s,5}$-casein can be converted, in the presence of reducing agents, to $\alpha_{s,3}$- and $\alpha_{s,4}$-caseins. The amino acid compositions of $\alpha_{s,3}$- and $\alpha_{s,4}$-caseins differ markedly from those of the $\alpha_{s,1}$ variants. There is either one SS bond or 2 SH groups per molecule. Thus these α_s-caseins differ from all the known $\alpha_{s,1}$-caseins in containing cystine (or cysteine) but resemble κ-casein in this regard. However, unlike κ-casein, they are calcium sensitive. It is important to determine the role of these minor α_s-caseins, their physical properties, their genetic variation, and their occurrence in the milk of other species.

Comparatively little has been done to determine whether $\alpha_{s,1}$-caseins are present in species other than the ruminant, and, if they are present, to characterize them. Malpress and Seid-Akhaven (1966) reported that human casein is composed mainly of κ- and α_s-caseins. On the other hand, Nagasawa et al. (1970a) have concluded that β- and κ-caseins are the major caseins of human milk. They believe that the component they call β-casein is similar to the fraction designated as an α_s-casein by Malpress' group. Obviously, further investigations are needed, not only of the human calcium-sensitive caseins, but also of those of other species.

No complete sequences of any of the α_s- or β-caseins have yet been determined. It is important that we obtain this information in order to understand better the nature of the interactions involved in the interesting associations of these caseins. The temperature dependence of the association of β-caseins is rather unusual, and it is gratifying to note the further attempts of Payens to elucidate this system. Payens and Heremans (1969) have studied the effect of pressure on the association of β-casein and suggested possible mechanisms for the effects observed. Manson and Annan (1970) have found that four of the five phosphate groups of β-casein A^1 are located in an uninterrupted sequence of phosphoseryl residues close to the N-terminal end of the polypeptide chain, supporting the contention that the phosphate groups have a special physical significance.

Creamer et al. (1970) have found that a single bond in bovine β-casein is more susceptible to the action of rennin than any of its other bonds and that the peptide released is near the C-terminal end. Rennin was found to be more specific than pepsin in its action on β-casein. We look forward to further results from this work.

Our knowledge of κ-casein is gradually expanding, as can be seen from Fig. 10, Chapter 12, where our present knowledge is summarized of the structure of this protein in its various forms. There is still a real need to develop gentler methods for the isolation of κ-casein and to obtain it in higher purity. It is evident that urea should be avoided in new isolation procedures (see Chapter 12). (Of interest in the overall picture of κ-casein heterogeneity is the pH 4.7 or acid-soluble fraction of Beeby, 1970.)

There is now general agreement that the bond split when rennin acts on κ-casein is a phenylalanine-methionine linkage. However this bond does not seem to be intrinsically sensitive to rennin. Hill (1968) has carried out important studies of the action of rennin on model peptides in an attempt to elucidate the pertinent structures around the rennin-sensitive linkage (for example, nearby serine appears to be important). The effect of modification of various side chains in κ-casein on the action of rennin has also been studied by Hill, the more recent studies being concerned with the role of arginine residues (Hill, 1969). Investigations such as these are of importance in elucidataing the mechanism of the primary action of rennin, but they will also be of value in finding model substrates for evaluating new "rennet-like" enzymes (see, for example, Humme, 1970a,b,c).

Information is gradually being obtained on the κ-caseins from animals other than the cow. Alais and Jollès (1967) studied the κ-casein of sheep milk and found that whole ovine casein contains 0.85% P, 0.09% sialic acid, 0.24% galactosamine and 0.33% galactose and that ovine κ-casein has 0.35–0.40% P, 0.26–0.36% sialic acid, 0.32% galactosamine and 0.56% galactose. Fiat et al. (1970) have studied the rennin-sensitive sequence in ovine κ-casein A; they found that the bond concerned is Phe-Met and that it is situated in a tryptic tridecapeptide similar to that of the bovine κ-casein. Zittle and Custer (1966) isolated a caprid κ-casein. They found this variant to be much less stable than the bovine protein. Knoop and Wortmann (1967) made the very interesting observation in electron microscope studies that while casein micelles become apparent about 1 day after parturition in cow milk and sow milk, they do not become evident in human milk until about 10 days *post partum*. Malpress and Seid-Akhavan (1966) isolated a κ-casein-like protein from mature human milk and examined the nature of the carbohydrate linkage; they found that it was similar, in some respects, to the bovine carbohydrate linkage. Groves and Gordon (1969) have made a comparison of human and bovine caseins, finding some similarity in amino acid composition but marked differences in leucine and alanine content. Studies of human κ-casein have been made by Alais and Jollès (1969). Woychik and Wondolowski (1969) found that porcine κ-caseins yielded a single para-κ-casein on rennin treatment, the polymorphic amino acid substitutions residing in the macropeptide portion of the molecules, as in the bovine proteins. There are marked differences in the micelles in mature milk from various species, for example the cow and echidna, and it is necessary to determine whether these differences reside in the κ-caseins and/or in the other caseins.

The basic problem of the role of sulfhydryl groups in the behavior of κ-casein in micelle formation and in milk still remains to be solved. The important work of Beeby in this regard (see Chapters 10 and 12) needs reassessment and extension.

The presence of minor casein components and other phosphoproteins and glycoproteins in milk has been discussed in Chapter 10. We still do not know the role of these components in micelle formation. Groves (1969) has done a valuable service in attempting to reassess the importance and genetics of the γ-caseins. It is surprising that the γ-caseins have been so long neglected. Further study of them is necessary, especially as Kirchmeier (1970) has shown that they play an important part in micelle structure.

Lawrence and Creamer (1969) have shown that the lag period in the aggregation of para-κ-casein is increased by a high κ-casein/para-κ-casein ratio, suggesting that the aggregation of newly formed para-κ-casein is inhibited by the unchanged κ-casein. They also noted that the presence of small amounts of $\alpha_{s,1}$- or β-casein in the κ-casein increase the time of aggregation of κ-casein. It is surprising to the present author that this effect does not appear to be generally recognized among those who have worked with the caseins. Lawrence and Creamer have done a valuable service in stressing this.

Despite all of the work that has been carried out on micelle structure we still do not know the basic structure of the casein micelle in detail. Fundamental problems, such as the roles of water and calcium phosphate, remain to be elucidated. It will be interesting to see if there is any evolutionary change in micelle structure and in the mechanism of limited proteolysis and coagulation of various caseins as in the case of fibrinogen of blood (see the excellent review of Lorand, 1969).

B. β-Lactoglobulin

Tremendous progress has been made in our understanding of the β-lactoglobulins, as has been shown in Chapter 14. The important problem of the detailed X-ray crystal structure is not yet solved. We look forward to the further developments in this area from D. Green and his collaborators. At the same time there is a wealth of solution chemistry remaining to be done, including the completion of the amino acid sequence. The nature of the carbohydrate linkage in the Droughtmaster variant is being elucidated.

For many years it was assumed that β-lactoglobulin was the predominant whey protein in all milk. However Bell and McKenzie (1964) suggested that it was confined to ruminant milk. It has been found in cow, goat, and sheep milk but not in human or guinea pig milk. It does not seem to be present in kangaroo milk and there is, as yet, no evidence of it in platypus or echidna milk. In early studies of pig milk Bell and McKenzie noted a major whey protein of high mobility in starch-gel electrophoresis at alkaline pH. Neither this nor any other whey protein appeared to be immunologically similar to β-lactoglobulin. However there has been reinvestigation of this protein in Australia and Great Britain. Kessler and Brew (1970) found that this major whey protein has a molecular weight of 18,500 daltons and an $s_{20,w}$ of 1.85 S at pH 7.0. It has two cystine residues per monomer but no cysteine. The N-terminal residue is valine (leucine for the bovine β-lactoglobulins) and the C-terminal is also valine (isoleucine for the bovine). These differences involve a single base exchange in the messenger RNA codons. The amino acid composition has considerable homology with that of the ruminant proteins. The occurrence of the protein as a 18,500 dalton monomer unit is of special significance in view of the role postulated for the cysteine residue in the ruminant proteins by the author in Chapter 14. Bell et al. (1970) have reported further studies of their major protein of pig milk. They found that this protein can occur as several variants that are genetically determined. The two major ones have been designated A and B and have been isolated from homozygotes. It would appear that Kessler and Brew's protein is similar to the A variant. Studies of the optical rotatory dispersion, circular dichroism, urea denaturation, interaction with casein micelles, etc., are being carried out.

Kessler and Brew (1970) report that they have found that another artiodactyl, the camel, is devoid of β-lactoglobulin.

C. α-Lactalbumin and Lysozyme

One of the most startling discoveries concerning milk proteins during the 1960s was that made by Ebner and his colleagues of the biological function of α-lactalbumin. Current views on this have undergone some modification since the original paper (Ebner et al., 1966), mainly as a result of the work of Brew, Hill and others, and are discussed in the review of Ebner (1970) and in Chapter 15. At present it is generally believed that the "A protein" of the lactose synthetase system is a galactosyl transferase and that α-lactalbumin (the "B protein") modifies its enzymic function to enhance markedly its lactose synthetase activity.

The reactions involved in lactose synthesis and related reactions are shown in Scheme I.

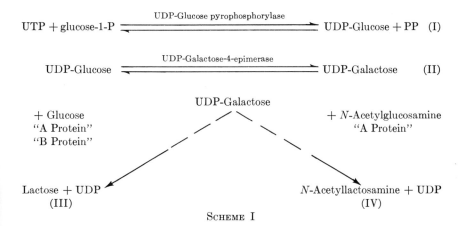

Scheme I

Outstanding problems in connection with the lactose synthetase system include the following:

(1) The need for a reasonably simple method of preparing pure "A protein."
(2) The need for improved methods of assay for the "A protein" and for lactose synthetase ("AB protein").
(3) The nature of the complex (if any) formed between A and B proteins.
(4) The possible broader role of the "A protein" in glycoprotein synthesis.
(5) The nature of evolution of lysozyme and α-lactalbumin.

Several new methods for preparing "A protein" have been published in full or in abstract during 1970, and several of these are mentioned in Chapter 15. It seems that one of the most promising approaches is to use human milk as a starting point. Nagasawa et al. (1970b) have isolated the human "A protein" by a new method, and they report that it is a glycoprotein (24.1% hexose, 22.1% hexosamine, 9.1% sialic acid).

The use of simple radioactivity assays of "A" and "B protein" activities in mammary gland tissues is complicated by the appreciable amount of nonspecific UDP-galactose hydrolases found in the tissue homogenates. Before reaction IV (Scheme I) was used to measure total "A protein" activity and excess purified "A protein" and glucose were used to measure "B protein" activity (reaction III), lactose synthetase activity was measured in the presence of glucose as total "AB protein" activity. Uridine triphosphate (UTP) was shown, by Watkins and Hassid (1962) and Coffey and Reithel (1968), to enhance the enzymic activity in these assays, presumably by protecting the UDP-galactose from nonspecific hydrolysis. McGuire (1969) has developed assays for "A protein" and "B protein" activity in mammary tissue in which he demonstrates reduction of the interfering UDP-galactose hydrolase activity by the addition of low concentrations of UTP. Fitzgerald et al. (1970) have indicated that their group has developed improved assays for the lactose synthetase system. Their work resulted from the following observations: The specific activity of α-lactalbumin increases as the "A protein" becomes more purified; under certain conditions there is a nonlinear response of enzymic activity to protein concentration. In the assay for the "A protein," high levels of α-lactalbumin are inhibiting in a linear manner, an optimum level of α-lactalbumin being required for maximum activity of a limiting amount of "A protein"; also, there is an inverse relationship between concentration of glucose and α-lactalbumin required for optimum activity. During the development of an assay for α-lactalbumin Fitzgerald et al. noted that although the "A protein" has a very low ability to synthesize lactose in the absence of α-lactalbumin, it can be stimulated to synthesize appreciable amounts of lactose in the presence of high concentrations of glucose. This effect appears to be specific for glucose since mannose, galactose and sucrose do not give it.

Fitzgerald et al. (1970) define the following activities: (a) LS_A denotes lactose synthetase activity (reaction III) for limiting amounts of "A protein" and saturating amounts of α-lactalbumin; (b) $LS_{(\alpha-1a)}$, lactose synthetase activity with limiting amounts of α-lactalbumin and saturating amounts of "A protein"; (c) LS_{end}, endogenous lactose synthetase activity for "A protein" in the absence of α-lactalbumin, at a high level of glucose; (d) LacNAc reaction, activity of A protein in the absence of α-lactalbumin but with N-acetylglucosamine instead of glucose as substrate (reaction IV). The revised assay conditions are summarized by Fitzgerald et al. (1970) and full details will be published by them.

The nature of the complex between the "A protein" and α-lactalbumin remains to be elucidated. It is obvious that studies of this interaction will be meaningful only if pure proteins are used. Hopper and McKenzie (1970) have attempted to purify the proteins for such a study. They found that each genetic variant (A and B) of bovine α-lactalbumin is associated with three minor α-lactalbumins, at least two of which are glycoproteins. All of them have been separated and they are active in the lactose synthetase system. The "A protein" of lactose synthetase is, in the opinion of Hopper and McKenzie, probably still not completely pure. Their optical rotatory dispersion and circular dichroism studies of the proteins are being continued.

Brew (1969) has attempted to rationalize the apparent inefficiency of the large amounts of α-lactalbumin involved in the lactose synthetase system; however his hypothesis needs more experimental evidence.

K. E. Hopper and H. A. McKenzie are studying the evolution of α-lactalbumin and lysozyme in a variety of species, and Hopper, McKenzie and G. B. Treacy have examined the whey proteins of kangaroo (for a preliminary report of both studies see Hopper et al., 1970). It is now well known that α-lactalbumin is a major whey protein in bovine and human milk, but there is much more lysozyme in human milk than in bovine milk. α-Lactalbumin is also a major whey protein in the pig. The situation in the marsupial, the kangaroo, is more complex. The whey protein composition changes appreciably during the lactation cycle. Furthermore, at a given time, two glands may give milk with protein compositions that differ from one another. Two of the proteins concerned are synthesized in the mammary gland and are among the three fastest-moving proteins in starch-gel electrophoresis at pH 7.8. The amounts of the fastest-moving protein (a) and the protein (c) of third-highest mobility change dramatically during lactation, whereas the intermediate one does not show this change. These effects were first reported by Bailey and Lemon (1966), who related them to the resumption of development of a quiescent blastocyst resulting from *post partum* mating. This explanation has now been shown to be incorrect in this author's laboratory. Protein (b) has been shown to be an α-lactalbumin in amino acid composition and behavior in the lactose synthetase system. It has not been possible to relate proteins (a) and (c) to other known whey proteins.

In an examination of the milk of the monotremes, the echidna and platypus, Hopper and McKenzie have shown that there is little lactose, lysozyme or α-lactalbumin in the platypus. The echidna has lactose and high lysozyme activity. Its lysozyme has an unusual amino acid composition and also has lactose synthetase activity. These studies are obviously of importance in tracing the evolutionary change in lysozyme and α-lactalbumin.

D. Minor Proteins and Enzymes

It can be seen readily from Chapter 16 that there has been a tremendous upsurge of interest in the "minor" proteins and enzymes of milk. We can expect this interest to continue at a high level.

While there is a considerable amount of knowledge being accumulated on the composition and behavior of individual enzymes, there is still little knowledge of the distribution of specific activities. Kitchen et al. (1970)

are making a valuable survey of the distribution of alkaline phosphatase, acid phosphatase, catalase, xanthine oxidase, aldolase, ribonuclease and carbonic anhydrase. The range of activities of some of the enzymes has been determined for both normal and mastitic milk. No carbonic anhydrase activity has been detected in any samples examined, although one might anticipate it to be present in mastitic milk due to the high amount of red blood cells present in severe cases of mastitis. In general, the milk enzymes have a greater proportion of their total activity located in skim milk. However, in terms of specific activity, only one enzyme, ribonuclease, is principally associated with the nonlipid fraction of milk. The other enzymes are distributed between the skim milk and fat to various extents.

The iron-binding protein, lactoferrin (red protein, ekkrinosiderophilin), of humans appears to be present not only in milk, but also in saliva, sweat, cervical mucin, sperm sheaths, bronchial mucin, urine, and the gastrointestinal tract. Its normal function in mucins is thought to be bacteriostatic. In the intestinal tract it appears to be involved also in iron absorption. It increases in concentration in the urine in certain pathological conditions, for example, some cases of kidney dysfunction. In diseases of the mammary gland (both cancerous and noncancerous) the autoantigen responsible for the production of antibodies has been found to be lactoferrin, bound to nucleic acid and to an enzyme that is nonantigenic. Some of this work has been discussed by, *inter alia*, de Laey et al. (1969), Masson et al. (1969), Loisillier et al. (1969) and Schade et al. (1969). It is of great importance that further work on the structure and function of the lactoferrins (in a variety of species) be carried out.

E. ALLERGENICITY OF MILK PROTEINS

With the availability of methods for the isolation of the important whey and casein proteins in high purity, it is now possible to reinvestigate the allergenic properties of milk proteins. This is important not only to the study of allergenicity reactions in general, but also because of the increasing use of milk, and processed milk in particular, by ethnic groups whose possible allergic reactions are not well known or understood. Close attention in planning such work should be given to the important studies of Bleumink and his collaborators on β-lactoglobulin (see Chapter 3, Volume I).

F. IMMUNOGLOBULINS IN MILK

It is apparent from a perusal of the comprehensive review on immunoglobulins by Edelman and Gall (1969) that we now have a reasonable grasp of the structure of these proteins. Nevertheless our knowledge of their structure and interactions is far from complete. One of the most important problems in this area is the relationship between the secretory immunoglobulins and the immunoglobulins of blood. Several workers found in the period 1958–1961 that the dominant immunoglobulin in human milk is not IgG as in blood serum, but IgA. Hanson found that it had an antigenic determinant not present in blood serum IgA (see Chapter 3, Volume 1). It is now known that IgA is the predominant immunoglobulin in many human secretions and that parotid duct saliva contains the same additional determinant as colostral IgA. On the basis of such findings it has been postulated that the role of the IgA of secretions is part of an immune system protecting the mucous membranes. The present knowledge and implications of this work are critically evaluated in an important review by Hanson et al. (1970).

G. Some Important Technological Problems

The greatest technical problems involving milk proteins that confront the dairy industry today are, in the authors opinion, the problems of heat stability, the role of subclinical mastitis in heat stability and curd firmness, and lactose intolerance. In devising milk protein products for areas where there is a high level of lactose intolerance in the population, close attention will have to be given to producing products that are relatively free of lactose. Cognizance must be taken of the important studies of Davis and his collaborators (Bolin et al., 1968) on lactose intolerance and their conclusion that it is an adaptive phenomenon.

The nature of the interaction between κ-casein and β-lactoglobulin is still not well understood; for example, we do not yet know for certain whether the interaction involves –SH/–SS exchange between subtly altered β-lactoglobulin monomers and κ-casein on the surface of the micelle or between β-lactoglobulin aggregates and the κ-casein.

It is apparent from the important studies of Feagan and his collaborators that subclinical mastitis plays a very important role in the heat stability curves (and renneting properties) that are obtained for different milk samples. The knowledge derived from this work of the role of β-lactoglobulin variant (amount and type), pH, level of subclinical mastitis, serum albumin and immunoglobulin A is reviewed by McKenzie (1970) (see also the original papers of Feagan et al., 1966a,b, 1969, 1970).

In planning future studies, workers will need to consider carefully the basic work on the structure of the casein micelle by Waugh and his collaborators (see Chapter 9) and the views of Rose (1965, 1969) on protein stability and the structure of the micelle.

There is a considerable amount of attention being given to the development of protease preparations, other than rennet, for cheesemaking. In 1967 Arima *et al.* reported the identification of a fungus, *Mucor pusillis* var. Lindt, that produces a powerful milk-clotting acid protease. The active enzyme involved has been purified and studied by several groups of workers. A similar type of protease, originating from *Mucor miehei*, has become available commercially. The acid protease involved has been isolated and partially characterized by Ottesen and Rickert (1970). They have compared its properties with those of the protease of Arima *et al.* (1967). It is important that such studies be improved to enable us not only to better understand the mechanism of limited proteolysis, but also to produce better proteases for cheesemaking. A micromethod developed by Lawrence and Sanderson (1969) for the quantitative estimation of rennets should prove valuable in such studies.

In a previous review the author stated that "In our greater knowledge of the chemistry of individual proteins isolated from milk, we all run the real danger of losing sight of their ultimate natural environment: milk itself. Only when we can understand the many reactions of whole milk will we be able to say that we understand the proteins." If this is accepted (and even if it is not) the road to success in milk protein research will be hard and long in the future, but the rewards will be rich.

REFERENCES

Alais, C., and Jollès, P. (1967). *J. Dairy Sci.* **50**, 1555.
Alais, C., and Jollès, P. (1969). *J. Chromatog.* **44**, 573.
Annan, W. D., and Manson, W. (1969). *J. Dairy Res.* **36**, 259.
Arima, K., Iwasaki, S., and Tamura, G. (1967). *Agr. Biol. Chem.* **31**, 540.
Aschaffenburg, R. (1968). *J. Dairy Res.* **35**, 447.
Bailey, L. F., and Lemon, M. (1966). *J. Reprod. Fert.* **11**, 473.
Beeby, R. (1970). *Biochem. Biophys. Acta.* **214**, 364.
Bell, K., and McKenzie, H. A. (1964). *Nature* **204**, 1275.
Bell, K., McKenzie, H. A., and Ralston, G. B. (1970). *Proc. Aust. Biochem. Soc.* **3**, 82.
Bolin, T. D., Crane, G. C., and Davis, A. E. (1968). *Australas. Ann. Med.* **17**, 300.
Brew, K. (1969). *Nature* **222**, 671.
Coffey, R. G., and Reithel, F. J. (1968). *Biochem. J.* **109**, 169.
Creamer, L. K., Mills, O. E., and Richards, H. L. (1970). *Proc. Int. Dairy Congr., 18th, 1970.* **1E**, 45.
de Laey, P., Masson, P. L., and Heremans, J. F. (1969). *In* "Protides of the Biological Fluids: Proceedings of the 16th Colloquium, 1968" (H. Peeters, ed.), p. 627. Pergamon, Oxford.
Ebner, K. E. (1970). *Accounts Chem. Res.* **3**, 41.
Ebner, K. E., Denton, W. L., and Brodbeck, U. (1966). *Biochem. Biophys. Res. Commun.* **24**, 232.
Edelman, G., and Gall, W. E. (1969). *Ann. Rev. Biochem.* **38**, 415.
Feagan, J. T., Griffin, A. T., and Lloyd, G. T. (1966a). *J. Dairy Sci.* **49**, 933.
Feagan, J. T., Griffin, A. T., and Lloyd, G. T. (1966b). *J. Dairy Sci.* **49**, 1010.
Feagan, J. T., Hehir, A. F., and Bailey, L. F. (1969). *J. Dairy Sci.* **52**, 887.

Feagan, J. T., Hehir, A. F., and Bailey, L. F. (1970). *Proc. Int. Dairy Congr. 18th, 1970.* IE, 525.
Fiat, A. M., Alais, C., and Jollès, J. (1970). *Chimia* **24**, 220
Fitzgerald, D. K., Brodbeck, U., Kiyosawa I., Mawal, R., Colvin, B., and Ebner, K. E. 1970). *J. Biol. Chem.* **245**, 2103.
Fox, P. F. (1970). *J. Dairy Sci.* **37**, 173.
Groves, M. L. (1969). *J. Dairy Sci.* **52**, 1155.
Groves, M. L., and Gordon, W. G. (1969). *J. Dairy Sci.* **52**, 906.
Hanson, L. Å., Borssen, R., Holmgrem, J., Jodal, U., Johansson, B. G., and Kaijser, B. (1970). "Proceedings of the Symposium on Immunological Incompetence, Florida, March 1970."
Hill, R. D. (1968). *Biochem. Biophys. Res. Commun.* **33**, 659.
Hill, R. D. (1969). *J. Dairy Sci.* **52**, 902.
Hoagland, P. D., Thompson, M. P., and Kalan, E. B. (1970). To be published.
Hopper, K. E., and McKenzie, H. A. (1970). To be published.
Hopper, K. E., McKenzie, H. A., and Treacy, G. B. (1970). *Proc. Aust. Biochem. Soc.* **3**, 86.
Humme, H. E. (1970a). *Neth. Milk Dairy J.* **24**, 3.
Humme, H. E. (1970b). *Neth. Milk Dairy J.* 24, 10.
Humme, H. E. (1970c). In press.
Kessler, E., and Brew, K. (1970). *Biochim. Biophys. Acta* **200**, 449.
Kirchmeier, O. (1970). *Proc. Int. Dairy Congr., 18th, 1970,* IE, 25.
Kitchen, B. J., Taylor, G. C., and White, I. C. (1970). *J. Dairy Res.* **37**, 279.
Knoop, V. E., and Wortmann, A. (1967). *Milchwissenschaft* **22**, 198.
Lawrence, R. C., and Creamer, L. K. (1969). *J. Dairy Res.* **36**, 11.
Lawrence, R. C., and Sanderson, W. B. (1969). *J. Dairy Res.* **36**, 21.
Loisillier, F., Buxtin, P., and Grabar, P. (1969). *In* "Protides of the Biological Fluids: Proceedings of the 16th Colloquium, 1968" (H. Peeters, ed.), p. 647. Pergamon, Oxford.
Lorand, L. (1969). *In* "Dynamics of Thrombus Formation and Dissolution" (S. A. Johnson and M. M. Guest, eds.), p. 212. Lippincott, Philadelphia.
McGuire, W. L. (1969). *Anal. Biochem.* **31**, 39.
McKenzie, H. A. (1970). *Proc. Int. Dairy Congr., 18th, 1970,* **2**. In press.
Malpress, F. H., and Seid-Akhavan, M. (1966). *Biochem. J.* **101**, 764.
Manson, W., and Annan, W. D. (1970). *Proc. Int. Dairy Congr., 18th, 1970,* IE, 33.
Masson, P. L., Heremans, J. F., Schonne, E., and Crabbe, P. (1969). *In* "Protides of the Biological Fluids: Proceedings of the 16th Colloquium, 1968" (H. Peeters, ed.), p. 633. Pergamon, Oxford.
Nagasawa, T., Kiyosawa, I., and Kuwahara, K. (1970a). *J. Dairy Sci.* **53**, 136.
Nagasawa, T., Kiyosawa, I., and Tanahashi, N. (1970b). *Proc. Int. Dairy Congr., 18th, 1970.* IE, 63.
Ottesen, M., and Ricket, W. (1970). *C. R. Trav. Lab. Carlsberg* **37**, 301.
Payens, T. A. J., and Heremans, K. (1969). *Biopolymers* **8**, 335.
Rose, D. (1965). *J. Dairy Sci.* **48**, 139.
Rose, D. (1969). *Dairy Sci. Abstr.* **31**, 171.
Schade, A. L., Pallavicini, C., and Wiesmann, U. (1969). *In* "Protides of the Biological Fluids: Proceedings of the 16th Colloquium, 1968" (H. Peeters, ed.), p. 619. Pergamon, Oxford.
Watkins, W. M., and Hassid, W. Z. (1962). *J. Biol. Chem.* **237**, 1432.
Woychik, J. H. and Wondolowski, M. V. (1969). *J. Dairy Sci.* **52**, 901.
Zittle, C. A., and Custer, J. H. (1966). *J. Dairy Sci.* **49**, 788.

Note Added in Proof: Some Further Recent Advances

In Chapter 10, Fig. 12, our knowledge of the chemical structure of κ-casein was summarized at the time the chapter was prepared. However, appreciable further advances have recently been made by Jollès et al. (1970), who have now established the sequences for approximately one-half the amino acid residues in bovine κ-casein A. They point out that, although κ-casein consists of a hydrophilic part (the glycopeptide) and a hydrophobic part (para-κ-casein), the latter does contain several hydrophilic sequences. The protein also exhibits frequent duplication or triplication of certain residues. The reader is referred to Table VIII in their paper for further details.

A comparison of the amino acid sequences of rennin and pepsin has been presented in Fig. 9 of Chapter 13. Tang (1970) has now studied the homology in the N-terminal region of the two enzymes. When the homologous residues are aligned rennin contains two more residues at the N-terminus than pepsin:

Rennin: ↓ Gly-Glu-Val-Ala-Ser-Val-Pro-Leu-Thr-Asn-Tyr-

Pepsin: Ala-Glu-Ile-Gly-Asp-Glu-Pro-Leu-Glu-Asn-Tyr

This is as a result of the difference in cleavage sites (marked with arrows) during activation of the respective zymogens. It is somewhat surprising that this difference in cleavage sites exists since the specificities of the two enzymes in proteolysis are similar.

References

Jollès, J., Alais, C., and Jollès, P. (1970). *Helv. Chim. Acta* **53**, 1918.
Tang, J. (1970). *Biochem. Biophys. Res. Commun.* **11**, 697.

☐ Appendix

H. A. McKENZIE

Methods for Zone Electrophoresis of Milk Proteins 483
A. Gel and Buffer Concentrations 484
B. Resolution 484
C. Hydrolyzed Starch 484
D. Urea 485
E. Gel Dimensions and Preparation of Thin Starch Gels 485
F. Apparatus and Gel Preparation of Acrylamide Electrophoresis . . . 486
G. Staining 486
H. Abbreviations 486
Summary of Zone Electrophoresis Methods 487
References 508

Methods for Zone Electrophoresis of Milk Proteins

A bewildering array of procedures has been developed for the zone electrophoretic study of milk proteins. In early methods filter paper or cellulose acetate strips were employed as supporting medium, but more recent methods involve the use of hydrolyzed starch or acrylamide gels (and occasionally agarose). Most of these methods have been developed in an empirical fashion, and preference for one method over another is sometimes largely subjective. It has been pointed out in Chapter 7, Volume I, that there is a considerable need for the development of methods that have a sound theoretical basis. The relevant principles are discussed in Chapter 7, Section II.

In order to simplify the lot of the worker in selecting methods for zone electrophoretic examination of milk proteins, a selection of methods is summarized in this appendix. This should be read in conjunction with the discussion of zone electrophoresis of whey proteins in Chapter 14 and of caseins in Chapters 10 and 11. The reader is also referred to the review of methods of phenotyping casein polymorphs by Thompson (1970).

A. Gel and Buffer Concentrations

It is to be emphasized that concentrations slightly different from those given in Table I will sometimes be necessary to get optimum resolution. Although most workers use starch and acrylamide from similar sources, slight variations in properties from lot to lot mean that the concentration for optimum resolution may vary.

In order to obtain low conductivity and, hence, to keep the heating effects low during electrophoresis, most of the buffer systems used are of low ionic strength. In some cases the buffering capacity is inadequate to overwhelm effects of residual acid in the starch *and/or* to maintain sufficient pH control during the electrophoresis. The concentration of the buffer system in such cases will need to be increased, even if this means using a lower voltage gradient in the electrophoresis.

The pH given for the gel in the column "Buffers and Medium" is an average value when the gel contains the concentration of the starch or acrylamide given. The pH of a sample of the *gel* used in each run should always be checked. Do *not* rely alone on the pH of the buffer solution. Care must also be given to the concentration of buffer in the electrode vessels to make sure that electroosmosis is minimized (see Smithies, 1959).

B. Resolution

In Table I some indication of resolution is given. However with new genetic variants being continuously discovered, resolution of new variants will have to be checked by the worker with a given method.

C. Hydrolyzed Starch

Partially hydrolyzed starch made by the Connaught Laboratories, Toronto, Canada, is the hydrolyzed material most often used in starch-gel electrophoresis. If no brand of starch is given in Table I, the starch used is Connaught. In the methods of Larsen and Thymann (1966) hydrolyzed Danish potato starch is used. In those of Bell (1962, 1967) involving resolution in borate buffers of bovine β-lactoglobulins A, B, C, and $_{\text{Droughtmaster}}$, hydrolyzed Australian potato starch is used; it is prepared from the potato starch of Drug Houses of Australia Ltd., Brisbane, Queensland.

D. UREA

In gels involving the use of urea, care must be taken to use urea free of cyanate and not to heat the urea too strongly during preparation of the gel (see Cole and Mecham, 1966; Melamed, 1967).

E. GEL DIMENSIONS AND PREPARATION OF THIN STARCH GELS

In many of the methods of starch-gel electrophoresis, gels $\sim$0.6 cm in thickness are used. In general these gels must be sliced prior to staining. The use of thinner gels for electrophoresis of milk proteins was introduced by Aschaffenburg and Thymann (1965). It has the advantages that (a) slicing is unnecessary and (b) heating due to the electric current is lower during electrophoresis.

The following method of preparing thin (0.15 cm) gels is used in the author's laboratory (McKenzie and Treacy, 1967). A tray with a cross section of $\sim$17 × 20 cm is used. The base of the tray is plate glass since it does not warp like Plexiglas (Perspex) with frequent pouring of hot gel solutions on it. This tray is supported in a modified LKB filter-paper electrophoresis apparatus (LKB-Produkter, Stockholm, Sweden). A Beckman Model RD-2 Duostat power supply (Spinco Division, Beckman Instruments, Palo Alto, California) is used, either at constant voltage *or* constant current according to the method.

The glass plate is greased on one side with Dow Corning silicone stopcock grease. The hot gel solution is poured on to the assembled tray so that it is slightly overfull. A polythene sheet lubricated with silicone grease is laid over the gel and a piece of plate glass placed on top of it. Two iron weights are placed on top of the glass and the gel is allowed to set. The weights and glass may be removed when the gel is set but the polythene is left in position until the gel is used.

At this time the polythene is carefully peeled back near one end of the gel. The gel is cut $\sim$5 cm from this end and the samples inserted on filter-paper strips (6 strips, 1 × 0.6 cm for each gel). The gel is re-covered with the polythene sheet during the electrophoresis.

At the end of the electrophoresis the polythene sheet is removed and a piece of nylon fishing line, monofilament, is drawn across the plate between the bottom of the gel and the plate. The plate and gel are inverted into the dye bath and after 3 min the plate is slowly lifted, the gel remaining behind in the bath. It is then stained.

F. Apparatus and Gel Preparation for Acrylamide Electrophoresis

A full description is given by Davis (1964) for disc-acrylamide electrophoresis. Apparatus and procedures for vertical acrylamide-gel slab electrophoresis are given by Raymond (1962, 1964).

G. Staining

Each worker has a favorite dye system. The following method is recommended and is referred to as the "standard method" in *starch-gel electrophoresis*. (It has been found to be superior, in the author's laboratory for milk proteins, to Procion Brilliant Blue RS and amido black.) Dissolve Nigrosine (water soluble, CI 50420) (1.1 g) in methanol (750 ml), water (750 ml), 17 M acetic acid (100 ml) and filter (Ashton, 1957). Stain gels for 15 min and use only three times. Remove excess dye with the following solvent: methanol (1000 vol) plus water (1000 vol) plus 17 M acetic acid (200 vol) *or* methanol (1000 vol) plus water (780 vol) plus 17 M acetic acid (200 vol) plus glycerol (220 vol). (The glycerol is used for drying the gel for photography; see Wake and Baldwin, 1961.)

Some workers prefer amido black (Amido Schwarz 10B, Naphthol Blue Black, CI 20470) (Grassman *et al.*, 1951): Make a saturated solution of amido black in methanol (500 vol) plus water (500 vol) plus 17 M acetic acid (100 vol).

El-Negoumy (1966) prefers a mixture of amido black and Nigrosine: Dissolve Amido Schwarz 10B (0.125 g), Nigrosine (0.250 g) in ethanol (500 ml) plus water (400 ml) plus 17 M acetic acid (100 ml).

In *acrylamide-gel electrophoresis* the following system is almost universal (although some workers may prefer Coomassie Blue, introduced by Fazekas de St. Groth *et al.*, 1963, for cellulose acetate): Dissolve 1 g amido black (Amido Schwarz 10B, Naphthol Blue Black, CI 20470) in 100 ml (7% v/v) CH_3COOH. Filter the solution. Stain for at least 1 hr. Destaining solvent: 17 M CH_3COOH (70 ml), water to 1 liter. Destain by immersion *or* electrolytically (see Davis, 1964).

H. Abbreviations

BIS	N,N'-Methylenebisacrylamide
Cyanogum	Cyanogum-41, a mixture of acrylamide and BIS made by American Cyanamid Co.
DMAPN	Dimethylaminopropionitrile
Na_2EDTA	Disodium salt of ethylenediaminetetraacetic acid
TEMED	N,N,N',N'-Tetramethylenediamine
Tris	Tris(hydroxymethyl)aminomethane
Me	2-Mercaptoethanol

APPENDIX

TABLE I
Summary of Zone Electrophoretic Methods

Reference	Samples	Resolution	Buffers and medium	Method
Wake and Baldwin (1961)	Whole casein or casein fractions	Many bands but κ-casein variants not resolved	Electrode: ~0.3 M H_3BO_3 plus ~0.08 M NaOH, pH 8.6 Gel: ~0.076 M Tris plus ~0.005 M citric acid 7 M urea, pH 8.5 with hydrolyzed starch (11.4 g/dl)	Electrode buffer: titrate 0.3 M H_3BO_3 to pH 8.6 with NaOH Gel-buffer stock solution: titrate 0.76 M Tris (92.0 g/liter) to pH 8.6 with citric acid Gel preparation: heat hydrolyzed starch (40 g), gel buffer (35 ml) and water (205 ml) to boiling; add urea (147 g) slowly, stir and heat; deaerate, pour Gel dimensions: 12 × 24 × 0.6 cm, horizontal tray Sample preparation: casein solution made 7 M in urea and 2% in unheated starch Electrophoresis: apply sample solutions in slots; 170 V, 16 hr, 2°C room Stain: Buffalo Black

(Continued)

TABLE I—(*Continued*)

SUMMARY OF ZONE ELECTROPHORETIC METHODS

Reference	Samples	Resolution	Buffers and medium	Method
Schmidt (1964)	Casein solutions	Caseins; designed for κ-casein but resolution also $\alpha_{s,1}$-, β-, γ-caseins	Electrode: ~0.3 M H_3BO_3 plus ~0.08 M NaOH, pH 8.6 Gel: ~0.076 M Tris plus 0.005 M citric acid, 7 M urea, 0.022 M ME, pH 8.5 with hydrolyzed starch (12.2 g/dl)	Electrode buffer: similar to Wake and Baldwin (1961) Gel-buffer stock solution: similar to Wake and Baldwin (1961) Gel preparation: similar to Wake and Baldwin (1961) but containing 12.2 g/dl hydrolyzed starch, and ME added at 50°C while gel cooling to give 0.022 M ME Gel dimensions: similar to Wake and Baldwin (1961). Sample preparation: casein solutions (1.5–3.0 g/dl) in 0.076 M Tris–citrate buffer, 7 M in urea, 0.022 M in ME, 2% in hydrolyzed starch Electrophoresis: room temperature Stain: same as Wake and Baldwin (1961)

| Aschaffenburg and Thymann (1965) | Milk | α_s-, β-, κ-Caseins; β-lactoglobulin A, B | Electrode: 0.3 M H_3BO_3 plus 0.1 M NaOH, pH 8.7
Gel: 0.36 M Tris, 0.054 M H_3BO_3, 0.0158 M Na_2EDTA, 6 M urea, ~0.02 M ME, pH 9.3, hydrolyzed starch (20 g/dl) | Electrode buffer: H_3BO_3 (37.1 g), 1 M NaOH (200 ml) to 2 liter with water
Gel buffer: Tris (60 g), Na_2EDTA (8 g), H_3BO_3 (4.6 g) to 1 liter with water
Gel preparation: add hydrolyzed starch (20 g) to gel buffer (80 ml), heat to boiling point; add urea (40 g), stir, reheat to boiling point; deaerate, add ME (0.15 ml) and pour gel
Gel dimensions: 17.5 × 9 × 0.15 cm
Sample preparation: add ME (0.05 ml) to milk sample (0.5 ml) for 15 min (do not hold >24 hr)
Electrophoresis: apply samples on Whatman 3 MM filter paper (0.8 × 0.15 cm); 300 V, 17.5 mA, 5 hr, in 4°C room
Stain: amido black, 10 min |

(Continued)

TABLE I—(Continued)

SUMMARY OF ZONE ELECTROPHORETIC METHODS

Reference	Samples	Resolution	Buffers and medium	Method
Larsen and Thymann (1966)	Milk	α_{s-1}-, β-, κ-Caseins; β-lactoglobulin A, B	Electrode: 0.076 M Tris plus 0.0032 M Na$_2$EDTA plus 0.0113 M H$_3$BO$_3$, pH 9.3[a] Gel: 0.076 M Tris plus 0.0032 M Na$_2$EDTA plus 0.0113 M H$_3$BO$_3$, 5.5 M urea, ~0.02 M ME, pH 9.3, hydrolyzed Danish potato starch (12 g/dl)	Electrode buffer: see footnote a Gel-buffer stock solution: Tris (12.2 g), Na$_2$EDTA (1.56 g), H$_3$BO$_3$ (0.92 g) to 1 liter with water (0.1 M Tris plus 0.00419 M Na$_2$EDTA plus 0.0149 M H$_3$BO$_3$) Gel preparation: prepare gel from hydrolyzed Danish potato starch (90 g), urea (236 g), gel buffer (540 ml) and ME (1 ml); pour the hot mixture directly on the cooling plate of a high-voltage electrophoresis apparatus capable of taking 100 samples (Technik A. Hölzel, Munich, Germany); apparatus divided to form 3 trays (10 cm wide) Sample preparation: reduce by method of Aschaffenburg and Thymann (1965) Electrophoresis: apply samples on Whatman No. 1 filter-paper strips; 1000 V (12–15 V/cm), 70–100 mA, 6 hr, cooling plate at 4°C Stain: use method of Aschaffenburg and Thymann (1965)

Aschaffenburg and Michalak (1968); see also Michalak (1967), Aschaffenburg and Thymann (1965)	Skim milk	$\alpha_{s,1}$-, β-, κ-Caseins; β-lactoglobulin A, B, C, D	Electrode: 0.05 M Tris plus 0.385 M glycine, pH 8.5 Gel: ~0.09 M Tris plus ~0.01 M citric acid, 6.4 M urea, 0.04 M ME, pH ~8.6 with hydrolyzed starch (15 g/dl)	Electrode buffer: Tris (6.0 g), glycine (28.8 g) to 1 liter with water (pH 8.5) Gel-buffer stock: Tris (9.0 g), citric acid (1.8 g) to 1 dl with water; adjust pH to 8.6 Gel preparation: Hydrolyzed starch (10 g) in gel buffer (8 ml) plus water (40 ml); boil until high viscosity lost (at least 1 min); stir in urea (26 g); proceed as in Aschaffenburg-Thymann method (but add 4 drops ME); allow gel to set for 30 min, transfer to cold room for at least 5 hr before use Gel dimensions: 17.5 × 9.0 × 0.15 cm Sample preparation: similar to Aschaffenburg and Thyman (1965) Electrophoresis: 210 V, ~10 mA, 16 hr, 4°C room Stain: (20 min) with amido black (0.125 g) plus Nigrosine (0.250 g), ethanol (500 ml), water (400 ml), 17 M acetic acid (100 ml)

(Continued)

TABLE I—(Continued)
SUMMARY OF ZONE ELECTROPHORETIC METHODS

Reference	Samples	Resolution	Buffers and medium	Method
El-Negoumy (1966)	Whole casein or casein fractions or whey protein fractions	$\alpha_{s,1}$-, β-, κ-, γ-Caseins; whey proteins: α-lactalbumin, β-lactoglobulin, serum albumin	Electrode: $\sim$0.3 M NaOH plus $\sim$0.08 M H$_3$BO$_3$, $\sim$0.03 M ME, pH 8.6 Gel: $\sim$0.12 M Tris plus $\sim$0.008 M citric acid, $\sim$5.6 M urea, $\sim$0.07 M ME, pH 8.5 with hydrolyzed starch (15 g/dl)	Electrode buffer: similar to Wake and Baldwin (1961) but 0.2% in ME Gel-buffer stock solution: similar to Wake and Baldwin (1961) Gel preparation: mix deionized distilled water (205 ml) with Tris–citrate buffer (55 ml) in a flask; add hydrolyzed starch (50 g) gradually with shaking; heat until mixture thickens and subsequently clarifies; add urea (115 g), then ME (1.4 ml); cool to 30°C and pour; stand 12–24 hr at 25°–30°C before use Gel dimensions: horizontal, 24 × 12 × 0.5 cm (Research Specialities Corp.); vertical, 20 × 14 × 0.30 cm (E. C. Apparatus Corp.) Sample preparation: mix whole milk, 10 ml, and pH 4.6 buffer, 10 ml (1 M CH$_3$COOH plus 1 M CH$_3$COONa) at 50°C; centrifuge; disperse casein in 7 M urea (6 ml) → casein ($\sim$5 g/dl); samples may be dried on pieces of Whatman 3 MM paper and kept in desiccator for application to gel Electrophoresis: horizontal—250 V, 35 mA; vertical—150 V, 15 mA; 15–22 hr in 2°C room or 20°C circulating water Stain: Buffalo Black

McKenzie and Treacy (1968) (continuous system) [In a modified version (1971) improved resolution can be obtained. Buffer concentrations are 5× those given here. Stock: Tris (30.0 g), glycine (144.0 g) per liter. Use same volumes for dilution.]	Skim milk or casein	κ- and γ- Caseins but useful for $\alpha_{s,1}$- and β-caseins (some whey proteins resolved)	Electrode: 0.01 M Tris plus 0.077 M glycine, pH 8.5 Gel: 0.01 M Tris plus 0.077 M glycine, 7 M urea, ~0.028 M ME; pH 8.4 with hydrolyzed starch (14 g/dl)	Electrode and gel stock solution: Tris (6.0 g), glycine (28.8 g), to 1 liter with water (0.05 M Tris plus 0.385 M glycine) Electrode solution: dilute 300 ml to 1500 ml with water Gel solution: dilute 30.4 ml to 100 ml with water Gel preparation: make a slurry of gel buffer (30 ml) and starch (21 g); warm buffer (70 ml) to 75°C, pour onto urea (64 g), warm to 65°C; pour urea buffer onto slurry, heat to 65°C, deaerate, add ME (0.2 ml), stir, pour gel, cover and stand overnight before use Gel dimensions and tray: standard Sample preparation: dissolve urea (6 g) in stock buffer solution (4 ml), water (1 ml), ME (0.2 ml) and make up to 10 ml with water (solution is 10 M urea plus 0.28 M ME, 0.02 M Tris, 0.154 M glycine); to sample solution (0.25 ml), add water (0.05 ml) or rennin solution (0.05 ml); add urea–ME–buffer solution (0.30 ml) immediately or after rennin action; stand mixture overnight at 2°C Electrophoresis: Insert samples in Whatman 3 MM paper; 340 V, 12–14 hr, 2°C room Stain: standard

(Continued)

TABLE I—(Continued)

SUMMARY OF ZONE ELECTROPHORETIC METHODS

Reference	Samples	Resolution	Buffers and medium	Method
McKenzie and Treacy (1968) [modification of method of McKenzie and Murphy (1965); discontinuous system]	Skim milk or caseins	$\alpha_{s,1}$-, β-, κ-Caseins	Electrode: 0.3 M H_3BO_3 plus 0.08 M NaOH, pH 8.6 Gel: 0.038 M Tris plus 0.0025 M citric acid,[b] 7 M urea, ~0.02 M ME, pH 8.7 with hydrolyzed starch (14 g/dl)	Electrode buffer: H_3BO_3 (37.1 g) plus 1 M NaOH (160 ml) to 2 liter with water Gel-buffer stock solution: Tris (92.04 g), citric acid (10.55 g) to 1 liter with water (→0.76 M Tris plus 0.05 M citric acid) Gel preparation: dilute 7.6 ml stock buffer to 100 ml with H_2O (→0.0578 M Tris plus 0.0038 M citric acid), pH 8.6; make a slurry of this buffer (30 ml) and hydrolyzed starch (21 g); heat 70 ml buffer to 75°C, pour onto urea (64 g), stir, and warm to 65°C; pour the urea-buffer mixture on to the slurry, heat to 65°C, deaerate; add ME (0.2 ml), stir and pour onto gel tray Gel dimensions and tray: standard Sample preparation: dissolve urea (6 g) in gel stock buffer solution (2 ml), water (3 ml), ME (0.2 ml); make up to 10 ml with water; treat sample solution as in continuous system of McKenzie and Treacy (1968) Electrophoresis: 300 V, ~10 mA, 12 hr, 2°C room Stain: standard

| Thompson et al. (1964) | Skim milk or caseins | β-Casein A, B, C | Electrode: 0.033 M Tris plus 0.0014 M Na₂EDTA plus 0.005 M H₃BO₃, pH 9.2 Gel: 0.033 M Tris plus 0.0014 M Na₂EDTA plus 0.005 M H₃BO₃, 7 M urea; pH 9 with Cyanogum (7 g/dl) | Stock buffer: H₃BO₃ (2.3 g), Na₂EDTA (3.9 g), Tris (30.25 g), water to 1 liter pH 9.2 buffer: 1 vol stock plus 2 vol water Stock gel solution: dissolve Cyanogum (70 g) in pH 9.2 buffer (400 ml), add urea (270 g), DMAPN (1 ml); dilute to 1 liter with pH 9.2 buffer Gel preparation: add (NH₄)₂S₂O₈ (0.3 g) to stock gel solution (150 ml), pour solution (gelling time, 20 min; aging time, 45 min) Gel dimensions: vertical, 15 × 23 cm (E. C. Apparatus Corp.) Sample insertion: 30 μl 1% casein solution or 30 μl skim milk plus pH 9.2 buffer solution layered in gel slot (or 30 μl skim milk made 8% in sucrose) Electrophoresis: 100 V, 55 mA for 15 min; then 150 V, 60 mA for 15 min; then slowly increased to 250 V, ≫ 70 mA; after 5 hr, current is 25–30 mA Stain: amido black, 3 min; wash with acetic acid (7%) and destain electrolytically |

(Continued)

TABLE I—(Continued)

SUMMARY OF ZONE ELECTROPHORETIC METHODS

Reference	Samples	Resolution	Buffers and medium	Method
Peterson (1963)	α_s- and β-Caseins	α_s- and β-Caseins	Electrode: 0.085 M Tris plus 0.0036 M Na$_2$EDTA plus 0.0128 M H$_3$BO$_3$, pH 9.1 Gel: ~0.085 M Tris plus 0.0036 M Na$_2$EDTA plus 0.0128 M H$_3$BO$_3$, 4.5 M urea; pH 9.0 with Cyanogum (7 g/dl)	Stock buffer solution: mix water (2700 ml), Tris (270 g), Na$_2$EDTA (35.1 g), H$_3$BO$_3$ (20.7 g) Electrode buffer: dilute 1 vol of stock buffer to 9 vol with water Gel preparation: mix stock buffer (55 ml), water (300 ml), Cyanogum (35 g), urea (135 g), dilute with water to 500 ml; add DMAPN (0.5 ml) to stock gel solution and store at 2°C (<3 weeks); immediately before use add solid (NH$_4$)$_2$S$_2$O$_8$ (0.3 g/125 ml); pour gel solution (125 ml) into mold Gel dimensions: E. C. Apparatus Corp. vertical, 15 × 23 × 0.35 cm (see Raymond, 1962) Sample preparation: dilute 1 vol stock buffer to 3 vol with water, adding sufficient urea → 7 M urea; dissolve casein sample (1 mg) in this solution (0.1) and 0.05 ml methyl red indicator (1 g/dl) Electrophoresis: 100 V, 15 min then 250 V, 75 mA, 5 hr (or 190 V, 16 hr for β-caseins) Stain: amido black, 1 min

Peterson et al. (1966)	Whole casein or β-casein	β-Casein A^{2-1}, A^1, A^{2-3}, A^{2-2}	Same as Peterson (1963); pH 9.2 but Cyanogum (10 g/dl)	Same as Peterson (1963)
		β-Casein A^{2-1}, A^{2-2}, A^{2-3}, A^1	Electrode: 1 M CH_3COOH Gel: 1 M CH_3COOH, 4.5 M urea, pH 3.0 with Cyanogum (10 g/dl)	Same as Peterson and Kopfler (1966) but pre-run 40 mA, 2 hr

(Continued)

TABLE I—(Continued)

SUMMARY OF ZONE ELECTROPHORETIC METHODS

Reference	Samples	Resolution	Buffers and medium	Method
Peterson and Kopfler (1966)	Whole casein or β-casein	β-Casein A^3, A^2, A^1, B, C	Electrode: CH_3COOH (7.7% v/v) pH 2.4 Gel: ~1.25 M CH_3COOH plus ~1.7 M HCOOH, 4.5 M urea, pH 3.5 with Cyanogum (10 g/dl)	Stock buffer: 17 M CH_3COOH (86 ml) plus 90% HCOOH (25 ml) to 1 liter with water → 1.46 M CH_3COOH plus 2 M HCOOH Electrode buffer: CH_3COOH (7.7% v/v) Gel preparation: Cyanogum (15 g), urea (40.5 g) dissolved in stock buffer → 150 ml solution; warm to 25°C, add TEMED (1.0 ml) and $(NH_4)_2S_2O_8$ (0.35 g) Gel dimensions: vertical, 15 × 23 cm; same as Peterson (1963) Sample preparation: whole casein (1 g/dl), β-casein (0.25 g/dl) in buffer made 10 M in urea Electrophoresis prerun: 95 mA, 2 hr Electrophoresis: add each sample solution (0.015 ml) to slots; run at 150 V, 25 mA for 20 min; then 100 mA; increase voltage to 300 V to keep at 100 mA; then 300 V, ~20 hr (final current 60 mA) Stain: saturated solution Naphthol Blue Black (CI 20470) in mixture 50 parts by vol CH_3OH, 50 parts H_2O, 10 parts 17 M CH_3COOH; wash with this solvent or destain electrolytically

Aschaffenburg (1966)	Whole milk or casein	β-Casein A³, A², A¹, B, D, C	Electrode: 0.87 M HCOOH plus 2.04 M CH₃COOH, pH 1.7 Gel: 0.7 M HCOOH plus 1.6 M CH₃COOH, ~4 M urea, pH ~1.7 with hydrolyzed starch (~16 g/dl)	Stock buffer: dilute HCOOH (30 ml 98%) and 17 M CH₃COOH (120 ml) to 1 liter with water Electrode buffer: use stock buffer Gel preparation: disperse hydrolyzed starch (20 g) in stock buffer (80 ml), heat until grains disrupt; add urea (24 g); heat to boiling, deaerate and pour Gel dimensions: 17.5 × 9.0 × 0.15 cm, horizontal tray Sample preparation: dilute whole milk with equal-volume urea solution (100 g urea added to 100 ml H₂O) *or* dissolve casein (10 mg) in pH 1.7 buffer (1 ml) containing 6 g urea/10 ml buffer Electrophoresis: 220 V, 16 mA, 16 hr, 4°C room (methyl red marker moves 11 cm) Stain: amido black

(Continued)

TABLE I—(Continued)

SUMMARY OF ZONE ELECTROPHORETIC METHODS

Reference	Samples	Resolution	Buffers and medium	Method
Bell (1966)	Skim milk	β-Caseins	Electrode: 0.158 M HCOOH plus 0.08 M NaOH, pH 3.7 Gel: 0.052 M HCOOH plus 0.0104 M NaOH, ~8 M urea, pH ~3.0 with hydrolyzed starch (~15 g/dl)	Electrode buffer: HCOOH (5.95 ml 98%), 1 M NaOH (80 ml) to 1 liter with water Gel buffer stock: HCOOH (38.5 ml 98%), 1 M NaOH (200 ml) to 1 liter with water Gel buffer: dilute 8.33 ml to 100 ml with water (→0.0833 M HCOOH plus 0.0167 M NaOH) Gel preparation: similar to Aschaffenburg (1966) using hydrolyzed starch (24 g), gel buffer (100 ml) and urea (80 g) Gel dimensions: 17.5 × 9 × 0.15 cm, horizontal tray Sample preparation: use skim milk samples or samples prepared for alkaline gel containing ME or without ME Electrophoresis: insert samples on Whatman No. 1 paper (1.0 × 0.15 cm). 185 V, 15 mA, 16 hr, 4°C room or higher V, 23 mA, 8 hr, 4°C room Stain: amido black

| Groves and Kiddy (1968) | Whole casein or γ-casein fractions | γ-Casein variants; *also* β- and α_s-caseins and TS (temperature sensitive) fractions | Electrode: 0.025 M Tris plus 0.192 M glycine, pH 8.3
Spacer gel: 0.06 M HCl, 0.062 M Tris, 4 M urea, ~pH 6.7 with acrylamide (2.5 g/dl), BIS (0.63 g/dl), TEMED (0.058 ml/dl), riboflavin (0.0005 g/dl)
Running gel: 0.06 M HCl, 0.377 M Tris, 4 M urea, ~pH 8.9 with acrylamide (7.0 g/dl), BIS (0.184 g/dl), TEMED (0.029 ml/dl), $(NH_4)_2S_2O_8$ (0.07 g/dl) | Electrode buffer: Tris (6 g), glycine (28.8 g), with water to 2 liter
Spacer gel solutions:^{c} (large pore)
B: 1 M HCl (~48 ml to give pH 6.7), Tris (5.98 g), TEMED (0.46 ml), urea (24 g), water to 100 ml, pH 6.7
D: acrylamide (10.0 g), BIS (2.5 g), urea (24 g), water to 100 ml
E: riboflavin (0.004 g), urea (24 g), water to 100 ml
F: urea (24 g), water to 100 ml
Mix: 1 vol B, 2 vol D, 1 vol E, 4 vol F (→B/8, D/4, E/8, F/2)
Running gel solutions: (small pore)
A: 1 M HCl (24 ml), Tris (18.3 g), TEMED (0.115 ml), urea (24 g), water to 100 ml, pH 8.9
C: acrylamide (28.0 g), BIS (0.735 g), urea (24 g), water to 100 ml
G: $(NH_4)_2S_2O_8$ (0.14 g), urea (24 g), water to 100 ml
Mix: 4 vol A, ¼ vol C, 8 vol G (→A/4, C/4, G/2)
Gel dimensions: tubes (6.3 cm long × 0.5 cm i.d.) of Model 12 Apparatus of Canalco
Sample preparation: dissolve casein sample (~0.2 mg) in large-pore gel solution (0.1 ml)
Electrophoresis: "discontinuous method" of Ornstein (1964) and Davis (1964); ~5 mA/tube; runs at pH 9.6
Stain: amido black |

(Continued)

TABLE I—(Continued)

SUMMARY OF ZONE ELECTROPHORETIC METHODS

Reference	Samples	Resolution	Buffers and medium	Method
Groves and Gordon (1969)	γ- and β-Casein fractions	γ- and β-Casein variants A^1, A^2, A^3	Electrode: 0.35 M β-alanine plus $\sim$0.025 M CH_3COOH, pH 5.0 Spacer gel: 0.06 M KOH, $\sim$ 0.06 M CH_3COOH, 8 M urea, pH 6.7 with acrylamide (2.5 g/dl), BIS (0.63 g/dl), TEMED (0.058 ml/dl), riboflavin (0.0005 g/dl) Running gel: 0.06 M KOH, $\sim$ 0.37 M CH_3COOH, 8 M urea, pH 4.5 with acrylamide (5.6 g/dl), BIS (0.15 g/dl), TEMED (0.5 ml/100 ml), $(NH_4)_2S_2O_8$ (0.14 g/100 ml)	Electrode buffer stock: β-alanine (62.4 g), 17 M CH_3COOH ($\sim$3.0 ml to give pH 5.0), water to 2 liter Spacer gel solutions (large pore):[d] B: 2 M KOH (24 ml), 17 M CH_3COOH ($\sim$2.87 ml to give pH 6.7), TEMED (0.46 ml), urea (48 g), water to 100 ml D: acrylamide (10 g), BIS (2.5 g), urea (48 g), water to 100 ml E: riboflavin (0.004 g), urea (48 g), water to 100 ml F: urea (48 g), water to 100 ml Mix: 1 vol B, 2 vol D, 1 vol E, 4 vol F ($\rightarrow$B/8, D/4, E/8, F/2) Running gel solutions (small pore): A: 2 M KOH (24 ml), 17 M CH_3COOH (17.2 ml), TEMED (4.0 ml), urea (48 g), water to 100 ml C: acrylamide (30 g), BIS (0.8 g), urea (48 g), water to 100 ml G: $(NH_4)_2S_2O_8$ (0.28 g), urea (48 g), water to 100 ml Mix: 2 vol A, 3 vol C, 3 vol F, 8 vol G ($\rightarrow$A/8, 3C/16, 3F/16, G/2) Gel dimensions: similar to Groves and Kiddy (1968) Sample preparation: dissolved in large-pore gel solution

Bell (1962, 1967)	Skim milk or whey protein	Bovine β-lactoglobulin A, B, C, Droughtmaster; α-lactalbumin A, B (A coincides with β-lactoglobulin Droughtmaster); serum albumin; Fe proteins; immunoglobulin	Electrode: 0.3 M H_3BO_3 plus 0.075 M NaOH, pH 8.5 Gel: 0.028 M H_3BO_3 plus 0.0112 M NaOH, pH 8.5 with hydrolyzed starch, D.H.A. (15 g/dl)	Electrode buffer: H_3BO_3 (37.1 g), 1 M NaOH (150 ml) to 2 liter with water Gel buffer stock: H_3BO_3 (7.73 g), 1 M NaOH (64 ml) to 250 ml with water; dilute 64 ml to 500 ml with water Gel preparation: add hydrolyzed starch to gel buffer (15 g/dl), heat until viscosity decreases, deaerate, pour, cover, leave overnight Gel dimensions: 25 × 12 × 0.6 cm (or five gel compartments 25 × 4 × 0.6 cm) Electrophoresis: insert samples[e] on Whatman No. 17 paper (1 cm × 0.6 cm); 6.5–7.5 V/cm, 5 hr, 20°C room; remove inserts after 1 hr Stain: standard Nigrosine, 1 hr Wash: standard

(Continued)

TABLE I—(Continued)

SUMMARY OF ZONE ELECTROPHORETIC METHODS

Reference	Samples	Resolution	Buffers and medium	Method
Bell and McKenzie (1964, 1967), McKenzie and Treacy (1967), Bell et al. (1970)	Ovine milk and ovine whey proteins; porcine milk and porcine whey proteins	Whey proteins: β-lactoglobulin, α-lactalbumin; serum albumin	Electrode: 0.275 M Na$_2$HPO$_4$ plus 0.031 M KH$_2$PO$_4$ (or ½ this conc), pH 7.7 Gel: 0.011 M Na$_2$HPO$_4$ plus 0.00124 M (or ½ this conc), pH 7.2 with hydrolyzed starch (14.5 g/dl)	Stock buffer solution (version A): Na$_2$HPO$_4$ (42 g) in water to 1 liter; titrate to pH 7.8 with saturated KH$_2$PO$_4$ solution (~14 ml) Electrode buffer: use stock undiluted Gel buffer: dilute 1 vol to 25 vol with water (this is twice conc given by Bell and McKenzie, 1964, 1967) Stock buffer solution (version B): Na$_2$HPO$_4$ (39.0 g), KH$_2$PO$_4$ (4.2 g), water to 1 liter (this is twice conc used by Bell, 1968) Electrode buffer: use stock undiluted Gel buffer: dilute 1 vol to 25 vol with water Gel preparation: make a slurry of hydrolyzed starch (14.5 g) with gel buffer (25 ml); heat gel buffer (75 ml) to 100°C, add to starch suspension, shake vigorously for 15 sec, deaerate and pour; cover, stand overnight Gel dimensions: thick gel, 0.6 cm (Bell, 1962, 1967); thin gel, 0.15 cm (standard) Sample preparation: skim milk or whey protein dialyzed vs. 0.05 M NaCl; insert samples on Whatman No. 3 MM paper Electrophoresis: 160 V (6 V/cm) 5 hr, 20°C room; remove inserts after 2 min Stain: standard

Bell (1965)	Ovine skim milk or whey protein fractions	Ovine β-lactoglobulin A, B; α-lactalbumin; serum albumin	Electrode:ᵍ 0.1 M NaOH plus 0.3 M H₃BO₃, pH 8.7 Gel: 0.014 M Tris plus 0.004 M citric acid, pH 7.5 with hydrolyzed starch (14.5 g/dl)	Electrode buffer: NaOH (200 ml, 1 M), H₃BO₃ (37.08 g) to 2 liter with water Gel buffer: Tris (1.70 g), citric acid (0.768 g) to 1 liter with water Gel preparation: make a slurry of hydrolyzed starch (72.5 g) in gel buffer (110 ml); heat buffer (390 ml) to 100°C, add rapidly to starch slurry; shake 15 sec, deaerate (1 min), pour into form; cover gel with Saran wrap 15 min after pouring, allow to stand overnight at 21°C, or cool 1 hr in refrigerator; then 1 hr at 20°C Gel dimensions: 25 × 12 × 0.6 cm Sample preparation: skim milk or dialyzed whey protein; apply samples on Whatman 3 MM paper inserts (1 × 0.6 cm) Electrophoresis: 160 V, 30 mA, 20°C, 15 min; remove filter-paper inserts after 1 min; increase to 210 V, ~40 mA; run until brown "borate" (~3.5 hr) boundary 10 cm past insertion line Stain: standard

(Continued)

TABLE I—(Continued)

SUMMARY OF ZONE ELECTROPHORETIC METHODS

Reference	Samples	Resolution	Buffers and medium	Method
Bailey and Lemon (1966), Larsen and Thymann (1966), McKenzie and Treacy (1966), Hopper et al. (1970)	Bovine skim milk, kangaroo whole milk (1:3), echidna skim milk or whey protein fractions	Whey proteins Bovine: β-lactoglobulin A, B, C, D (Droughtmaster) α-lactalbumin A, B Kangaroo: α-lactalbumin transferrins, serum albumin, etc.	Electrode:^{h} 0.1 M LiOH plus 0.38 M H_3BO_3, pH 8.4 Gel: 0.0144 M Tris plus 0.00297 M citric acid, 0.002 M LiOH, 0.0076 M H_3BO_3, pH 7.7 with hydrolyzed starch (~13 g/dl)	Electrode buffer solution: $LiOH \cdot H_2O$ (8.4 g), H_3BO_3 (47.0 g) to 2 liter with water Gel buffer solutions: A: Tris (1.94 g), citric acid (0.7 g), to 1 liter with water (0.016 M Tris, 0.0033 M citric) B: $LiOH \cdot H_2O$ (0.84 g), H_3BO_3 (4.7 g), to 1 liter with water (0.02 M LiOH, 0.076 M H_3BO_3) Mix: 90 vol A with 10 vol B (pH 7.9) Gel preparation: make a slurry of hydrolyzed starch (13 g) with gel-buffer mixture (25 ml); heat gel buffer (75 ml) to 100°C, add to the starch suspension, shake vigorously for 15 sec, heat for 1 min; deaerate and pour; cover, stand overnight Gel dimensions: thin gel (0.15 cm); standard Sample preparation: skim milk *or* whey protein fraction dialyzed vs 0.05 M NaCl Electrophoresis: 160 V, 13 mA (decreasing with time) 6 hr, 20°C room Stain: standard

| Peterson (1963) | Whey proteins | Bovine whey proteins | Electrode: similar to Bell (1962, 1967)
Gel: similar to Bell (1962, 1967), pH 8.5 with Cyanogum (5 g/dl) | Buffer solutions: similar to Bell (1962, 1967)
Gel preparation: similar to Peterson (1963) for caseins, but urea omitted and Cyanogum conc of 8 g/dl
Gel dimensions: similar to Peterson (1963) for caseins
Sample preparation: no details given
Electrophoresis: no details given
Stain: similar to Peterson (1963) for caseins |

---

a Authors state that their buffer system is continuous but give their electrode buffer as same as gel-buffer stock solution. This would not give a continuous system. Electrode buffer concentrations given here are for a continuous system. (Dilute 75 vol stock buffer to 100 vol with water.)

b Final buffer concentration is one-half concentration of buffer of Poulik (1957) or of Wake and Baldwin (1961).

c Modifications of solutions of Ornstein (1964) and Davis (1964).

d Modified from Reisfeld et al. (1962).

e See also McKenzie and Sawyer (1966).

f Version B is simpler than version A.

g Based on buffers of Kristjansson (1963).

h Based on semidiscontinuous buffer system of Ferguson and Wallace (1963).

References

Aschaffenburg, R. (1966). *J. Dairy Sci.* **49,** 1284.
Aschaffenburg, R., and Michalak, W. (1968). *J. Dairy Sci.* **51,** 1849.
Aschaffenburg, R., and Thymann, M. (1965). *J. Dairy Sci.* **48,** 1524.
Ashton, G. C. (1957). *Nature* **180,** 917.
Bailey, L. F., and Lemon, M. (1966). *J. Reprod. Fert.* **11,** 473.
Bell, K. (1962). *Nature* **195,** 705.
Bell, K. (1965). Private communication.
Bell, K. (1966). Private communication.
Bell, K. (1967). *Biochim. Biophys. Acta* **147,** 100.
Bell, K. (1968). Private communication.
Bell, K., and McKenzie, H. A. (1964). *Nature* **204,** 1275.
Bell, K., and McKenzie, H. A. (1967). *Biochim. Biophys. Acta* **147,** 123.
Bell, K., McKenzie, H. A., and Ralston, G. B. (1970). *Proc. Aust. Biochem. Soc.* **3,** 82.
Cole, E. G., and Mecham, D. K. (1966). *Anal. Biochem.* **14,** 215.
Davis, B. J. (1964). *Ann. N. Y. Acad. Sci.* **121,** 404.
El-Negoumy, A. M. (1966). *Anal. Biochem.* **15,** 437.
Fazekas de St. Groth, S., Webster, R. G., and Datyner, A. (1963). *Biochim. Biophys. Acta* **71,** 377.
Ferguson, K. A., and Wallace, A. L. C. (1963). *Recent Progr. Horm. Res.* **19,** 1.
Grassman, W., Hannig, K., and Knedel, M. (1951). *Deut. Med. Wochschr.* **76,** 333.
Groves, M. L., and Gordon, W. G. (1969). *Biochim. Biophys. Acta* **194,** 421.
Groves, M. L., and Kiddy, C. A. (1968). *Arch. Biochem. Biophys.* **126,** 188.
Hopper, K. E., McKenzie, H. A., and Treacy, G. B. (1970). *Proc. Aust. Biochem. Soc.* **3,** 86.
Kristjansson, F. K. (1963). *Genetics* **48,** 1059.
Larsen, B., and Thymann, M. (1966). *Acta Vet. Scand.* **7,** 189.
McKenzie, H. A., and Murphy, W. H. (1965). Unpublished data.
McKenzie, H. A., and Sawyer, W. H. (1966). *Nature* **212,** 161.
McKenzie, H. A., and Treacy, G. B. (1966). Unpublished data.
McKenzie, H. A., and Treacy, G. B. (1967). Private communication.
McKenzie, H. A., and Treacy, G. B. (1968). Private communication.
Melamed, M. D. (1967). *Anal. Biochem.* **19,** 187.
Michalak, W. (1967). *J. Dairy Sci.* **50,** 1319.
Ornstein, L. (1964). *Ann. N.Y. Acad. Sci.* **121,** 321.
Peterson, R. F. (1963). *J. Dairy Sci.* **46,** 1136.
Peterson, R. F., and Kopfler, F. C. (1966). *Biochem. Biophys. Res. Commun.* **22,** 388.
Peterson, R. F., Nauman, L. W., and Hamilton, D. F. (1966). *J. Dairy Sci.* **49,** 601.
Poulik, M. D. (1957). *Nature* **180,** 1477.
Raymond, S. (1962). *Clin. Chem.* **8,** 455.
Raymond, S. (1964). *Ann. N.Y. Acad. Sci.* **121,** 350.
Reisfeld, R. A., Lewis, V. J., and Williams, D. E. (1962). *Nature* **195,** 281.
Schmidt, D. G. (1964). *Biochim. Biophys. Acta* **90,** 411.
Smithies, O. (1959). *Advan. Protein Chem.* **14,** 65.
Thompson, M. P. (1970). *J. Dairy Sci.* In press.
Thompson, M. P., Kiddy, C. A., Johnston, J. O., and Weinberg, R. M. (1964). *J. Dairy Sci.* **47,** 378.
Wake, R. G., and Baldwin, R. L. (1961). *Biochim. Biophys. Acta* **47,** 225.

Author Index

Numbers in italics refer to the pages on which the complete references are listed.

A

Aasa, R., 373, 374, 375, *411*
Abe, M., 178, *215*
Aceto, N. C., 444, *464*
Adachi, S., 11, *79*
Adam, A., 185, 208, *213*
Adams, E. T., Jr., 307, *325*
Affsprung, H. E., 63, *79*
Agren, G., 405, *413*
Aisen, P., 373, 375, *411*
Åkeson, Å., 387, *418*
Al, J., 453, *459*
Alais, C., 16, 17, 20, 21, 76, *79*, *81*, 110, *115*, 176, 185, 186, 187, 189, 197, 198, 199, 201, 202, 203, 204, 207, 208, 209, 210, 211, *212*, *213*, 225, 229, 249, 250, *251*, *253*, 424, 425, 428, 442, *459*, *460*, *463*, 472, *480*, *481*, *482*
Alberty, R. A., 44, *79*, *83*
Albizati, L. D., 227, 231, 242, *251*
Albright, D. A., 305, 306, *325*
Alderton, G., 368, *411*
Allen, P. Z., 373, 387, 395, 396, *411*, *416*
Alvare, N. F., 450, *465*
Alvare, N. J., 450, *465*
Ambler, R. P., 242, *251*
Amundson, C. H., 457, *460*, *464*
Anacker, E. W., 8, *80*
Andersen, B., 250, *253*
Anderson, H. A., 440, 441, *462*
Anderson, L., 18, *79*, 427, *462*
Anderson, W., 303, *328*
Andreotti, R. E., 341, 346, 347, 348, 349, *363*, *364*

Andrews, P., 229, 238, 239, *251*, 357, *361*, *362*, 397, 401, 402, 403, *411*, *413*, 442, *460*
Annan, W. D., 18, *82*, 145, *169*, 471, *480*, *481*
Annino, R., 14, 16, 37, *83*, 99, *115*, 166, 167, *172*
an Piette, L. H., 396, 397, *418*
Anson, M. L., 454, *459*
Arave, C. W., *170*, 273, *325*
Arbatskaya, N., 447, *459*
Ariga, H., 271, *327*
Arima, K., *480*
Arima, S., 208, *213*
Armstrong, C. E., 194, 205, 207, 208, *212*
Armstrong, J. McD., 262, 265, 260, 267, 268, 269, 308, *325*, 337, *361*, *362*
Arneson, R. M., 399, *411*
Aschaffenburg, R., 105, 108, 110, 111, 112, *114*, 118, 119, 122, 123, 125, 126, 128, 129, 133, 138, 145, 147, 155, 157, 160, *169*, *170*, *171*, 191, 192, *212*, 258, 259, 260, 261, 262, 271, 273, 274, 275, 291, 304, 307, 323, *325*, *327*, 336, 337, 341, 345, 351, 352, 353, 355, 356, 357, *362*, *363*, 370, 403, 404, *411*, *415*, 470, *480*, 485, 489, 490, 491, 499, *508*
Ashton, G. C., 376, *411*, 486, *508*
Ashworth, U. S., 14, 19, *79*, *84*, 88, 89, 101, *116*, 120, 121, 145, *172*, 428, *459*, *461*
Askonas, B. A., 266, 276, *325*

Atassi, M. Z., 358, *362*
Auclair, J. E., 395, *417*, 429, *464*
Avants, J. K., 441, *460*
Avis, P. G., 396, 397, 398, *411*

B

Babad, H., 357, 361, *362*, 409, *411*
Badings, H. T., 429, *464*
Bailey, L. F., 272, 275, 323, *325*, 428, 435, 436, *461*, 477, 479, *480*, *481*, 506, *508*
Bailie, M. J., 386, *411*
Bains, G. S., 453, *464*
Baker, E., 371, 373, 376, *411*
Baker, H. P., 303, *325*
Baker, J. M., 63, *79*
Baldi, E. J., 453, *461*
Baldwin, R. L., 13, *84*, 98, 101, 109, *116*, 118, 120, 121, 128, 157, *173*, 179, 190, 197, 200, *214*, 238, *251*, 427, *462*, 486, 487, 488, 492, 507, *508*
Ball, E. G., 396, *411*
Bang-Jensen, V., 247, 248, *251*
Bargmann, W., 381, *411*
Barlow, G., 291, *330*
Barman, T. E., 404, 405, *411*
Barthel, H., 317, *329*
Basch, J. J., 17, *80*, *81*, 139, 141, 143, 155, *170*, *171*, 266, 271, 276, 278, 279, 283, 294, 295, 297, 298, 300, 301, 302, 303, 310, 313, *325*, *326*, *327*, *330*, 370, 371, 372, 374, 375, 376, 379, *413*, *414*
Baud, C. A., 77, *79*
Baudet, P., 5, *79*, *80*, 101, *114*, 176, *212*, 225, *252*
Baumber, M. E., 244, *251*
Bayer, E., 397, *411*
Becker, C. A., 224, 227, 231, 237, *251*
Beeby, R., 16, 20, 21, 76, *79*, *82*, 99, *114*, 184, 189, 200, 201, 204, 208, *212*, *214*, 425, 438, 439, 440, 441, 443, 445, 454, *459*, *462*, *463*, 472, *480*
Beekes, H. W., 404, 405, *413*
Beinert, H., 399, *411*, *416*
Belec, J., 438, *460*
Belitz, H. D., 154, *172*

Bell, K., 258, 262, 265, 266, 271, 272, 273, 274, 275, 276, 278, 279, 281, 282, 283, 284, 299, 300, 310, 313, 316, *325*, *326*, 352, 354, *362*, 474, *480*, 500, 503, 504, 507, *508*
Bell, K. W., 440, 441, *460*
Bellinzona, G., 405, *411*
Bendixen, H. A., 402, *413*
Bengtsson, C., 338, *362*
Bennich, J., 15, *79*
Bergel, F., 396, 397, 398, *411*, *413*
Berger, A., 248, *251*
Bergmann, M., 248, *251*
Bernfield, M., *214*
Bernhart, F. W., 447, 453, *460*, *462*
Berridge, N. J., 75, *79*, 198, 199, *212*, 218, 219, 220, 224, 225, 226, 239, 246, *251*, 424, 426, 431, 447, *460*
Bertagnolio, G., 403, *413*
Bezkorovainy, A., 107, 108, 109, *114*, 373, 384, *411*
Bhalerao, V. R., 147, *170*
Bhatia, D. S., 453 *460*, *464*
Bhattacharya, S. D., 351, *362*
Bingham, E. W., 154, *170*, *173*, 380, 385, 386, 404, 405, 406, *412*, *418*
Binon, N., 288, *326*, *329*
Biserte, G., 337, 342, *362*, *363*, 373, *416*
Bishop, W. H., 324, *326*
Blake, C. C. F., 358, *362*, 407, *416*
Blalock, C. R., 384, *415*
Blanc, B., 372, 374, 375, *412*
Blauw, J., 429, *464*
Bleumink, E., 337, *362*
Block, R. J., 21, *85*, 335, 339, *362*, *365*, 447, *461*
Bloemmen, J., 289, *326*, *327*
Blondel-Queroix, J., 207, 208, *212*
Blum, R., 348, 349, *363*, *364*
Blumberg, B. S., 351, *362*
Bock, R. M., 44, *79*
Bodanszky, A., 316, *326*
Bogomolova, E., 447, 449, *460*
Bohren, H. U., 76, *82*, 92, 93, 95, 98, *114*, 199, *214*, 424, *463*
Bolin, T. D., 453, *460*, 479, *480*
Bolling, D., 335, *362*

Borgstrom, B., 7, *79*
Borssen, R., 479, *481*
Bosc, J., 250, *251*
Bosworth, A. W., 59, *84*
Both, P., 16, 17, 20, 47, *83*, 151, 152, *172*, 191, 192, 193, 194, 195, *214*
Boudreau, A., 457, *460*
Boulet, M., 61, 63, 65, *79*
Bovey, F. A., 245, *251*
Boyd, E. N., 437, *460*
Boyd, J. C., 453, *462*
Boyer, R. A., 455, *460*
Boyer, S. H., 127, *170*
Bradley, T. B., 145, *170*
Brand, E., 285, *326*
Bratzler, L. J., 456, *463*
Braunitzer, G., 280, 281, 282, 283, 284, 291, *327*
Bray, R. C., 372, 396, 397, 398, 399, *411*, *412*, *413*, *414*, *417*
Bready, P. J., 457, *464*
Brew, K., 276, *326*, 339, 340, 343, 344, 345, 347, 350, 354, 358, 359, 360, 361, *362*, *363*, *364*, 407, *412*, 474, 477, *480*, *481*
Brewer, J. M., 391, *412*
Briggs, D. R., 316, 317, *326*
Bright, J., 395, *416*
Brignon, G., *80*, 186, 202, *213*, 278, 279, 283, 294, 300, *326*
Brignot, J., 373, *416*
Brimacombe, B., *214*
Brinkhuis, J. A., 99, *115*, 150, *172*
Brodbeck, U., 339, 340, 357, 360, 361, *362*, *363*, *364*, 410, *412*, *413*, 474, 476, *480*, *481*
Brown, J. R., 242, *251*, 291, *326*
Browne, F. L., 453, *464*
Browne, W. J., 350, 358, *362*
Bruderer, G., 441, *461*
Brum, E. W., *170*
Brumby, P. E., 397, 399, *412*, *416*
Brunner, J. R., 14, 16, 17, 19, 21, 32, 48, 49, *83*, 89, 99, 101, 107, 108, *114*, *115*, *116*, 119, 120, 139, *170*, *172*, 180, 182, 186, 188, 189, *214*, 380, 381, 382, 383, 391, *412*, *413*, *414*, *417*, *418*, 455, *460*, *461*, *462*, *464*
Bryan, R. F., 356, *363*

Buchanan, R. A., 433, 448, 449, *460*, *463*
Buchet, J.-P., 288, 315, *326*, *328*
Budd, R. T., 430, *460*
Bujard, E., 372, *412*
Bull, H. B., 305, *326*
Bundy, H. F., 224, 227, 231, 237, 242, *251*
Bunn, C. W., 244, *251*
Bunville, L. G., 294, 303, *328*, *329*, 451, *464*
Burk, N. F., 14, *79*, 150, *170*
Burnett, G., 22, *79*
Burnett, J., 429, *464*
Burtin, P., 373, *415*
Burton, H., 437, *460*
Buruiana, L. M., 405, *412*
Butler, J. E., 384, *412*
Butler, L. G., 285, 319, *328*
Buvanendran, V., 127, *170*
Buxtin, P., 478, *481*
Bysouth, R., 430, *460*

C

Cable, R. S., 16, 17, *80*
Caha, J., 455, *461*
Calbert, H. E., 429, *463*
Calvin, M., 315, *326*
Camerman, N., 244, *251*
Campanella, S., 408, *415*
Campbell, J. E., 404, *417*
Campbell, P. N., 276 *326*, 340, 343, 354, *362*
Canfield, R. E., 344, *362*
Cannan, R. K., 258, 294, *326*, 332, *362*
Caputto, R., 248, *254*
Carey, F. G., 396, *412*
Carlson, D. M., 360, *364*
Carlström, A., 388, 391, 393, 394, 395, *412*
Caroline, L., 368, *417*
Carr, C. W., 39, *79*
Carroll, R. J., 34, *82*, 381, *412*
Carver, B. R., 285, *327*
Caspar, D. L. D., 8, *79*
Castellino, F. J., 343, 344, 351, 358, *362*
Cechova, D., 409, *412*
Cerankowski, L., 348, *363*

Cerbulis, J., 21, 32, *85*, 136, *173*, 246, *251*, 321, *330*, 437, 449, *460*, *465*
Chambers, D. C., 339, *364*
Chandan, R. C., 399, 400, 401, 402, 406, 407, *412*, *416*, *417*, *418*
Chandrasekhara, M. R., 453, *460*
Changeux, J., 8, *82*
Chanutin, A., 38, *79*
Chapman, H. R., 396, 398, *414*, 430, *460*
Charlemagne, D., 406, 407, *412*, *414*
Chatterjee, R., 303, *329*
Chaudhuri, S., 266, 276, 294, *327*, *329*, 340, 353, *362*, *364*
Cheeseman, G. C., 14, *79*, 181, 184, 186, *212*, 219, 238, 239, 241, 245, *251*, 383, *412*, 426, 427, 428, 432, *460*, *462*
Chen, A. H., 37, *81*
Cherbuliez, E., 5, *79*, *80*, 101, *114*, 176, *212*, 225, *252*
Chevalier, R., 428, *463*
Chien, H. C., 382, 383, *412*
Choate, W. L., 13, 38, 72, *80*
Christ, W., 427, *460*
Christensen, L. K., 285, 319, *326*, *327*
Christianson, G., 63, *80*
Chun, P., 137, *170*
Clark, G., 38, *84*
Clark, J. R., 374, *418*
Clark, W. R., 382, *414*
Claydon, T. J., 453, *460*
Clezy, P. S., 395, *416*
Cluskey, F. J., 449, *460*
Cobb, R. M. K., 453, *463*
Coffey, R. G., 361, *363*, 476, *480*
Cohen, C., 22, *80*
Cohn, E. J., 10, *80*
Colagrande, O., 428, *461*
Cole, E. G., 485, *508*
Cole, S. J., 407, *416*
Coleman, J. C., 43, *83*
Colvin, B., 476, *481*
Colvin, J. R., 11, 38, *83*
Connors, W. M., 225, 246, *252*
Conochie, J., 427, *460*
Cook, B. B., 285, *327*
Copius Peereboom, J. W., 383, 404, 405, *413*
Corbin, E. A., 368, *413*
Cornelly, K. W., 444, *464*
Corran, H. S., 396, *413*

Corrandini, C., 428, *461*
Coughlan, M. P., 398, *413*
Coulson, E. J., 378, 382, 383, 384, 386, *412*, *413*, *414*
Coulter, S. T., 63, *80*, 321, *329*, 438, 445, 449, 453, 458, *460*, *463*, *464*
Cox, A. C., 373, *415*
Crabbe, P. A., 373, *416*, 478, *481*
Craker, B. A., 212, *213*, 427, *461*
Crane, G. C., 479, *480*
Craven, D. A., 181, 184, *212*
Creamer, L. K., 35, 37, 42, 43, *84*, 471, 473, *480*, *481*
Crestfield, A. M., 203, *212*
Crick, F. H. C., 283, *326*
Crouwy, F., 337, 342, *362*, *363*
Crutchfield, G., 245, *252*
Cullis, A. F., 8, *80*
Cunningham, L. W., 291, *326*
Cuperlovic, M., 122, 125, *170*
Curran, H. R., 440, 441, *460*
Custer, J. H., 14, 16, 17, 21, 32, 39, *81*, *85*, 101, *115*, 118, 135, 136, 137, 157, 158, 160, *170*, *173*, 180, 182, 186, 187, *213*, *215*, 246, *251*, 318, 321, 322, *326*, *330*, 333, *364*, 378, 408, *417*, *418*, 437, 458, *460*, *465*, 472, *481*
Czulak, J., 430, *460*

D

Dalaly, B. K., 386, *413*
Damodaran, G., 15, *80*
Danes, E. N., 456, *461*
Datyner, A., 486, *508*
Dautrevaux, M., 337, 342, *362*, *363*, 373, *416*
Davie, E. W., 277, 278, 279, 285, *326*, *328*, *329*, 341, *363*
Davies, C. W., 44, 61, 65, *80*
Davies, D. T., 59, 60, 61, *80*, *84*, 181, 185, *215*, 427, 428, 436, *460*, *465*
Davis, A. E., 453, *460*, 479, *480*
Davis, B. J., 486, 501, 507, *508*
De, S. K., 293, *328*
de Baun, R. M., 225, 246, *252*
Debreczeni, E. J., 456, *464*

AUTHOR INDEX 513

Debye, P., 8, *80*
de Koning, P. J., 16, 17, 20, 45, 47. *80
83*, 121, 139, 151, 152, 155, 165, *170,
172*, 187, 191, 192, 193, 194, 195,
203, 205, 207, 208, 209,*213, 214*,
238, 239, 241, *252*
de Laey, P., 478, *480*
Delfour, A., 185, 187, 197, 203, 208, 209,
213, 425, *460*
DellaMonica, E. S., 39, *85*, 136, *173*,
318, 321, 322, *326, 330*, 346, *365*, 403,
418, 437, 458, *460*
Demott, B. J., 429, 458, *462*
Denkov, T., 432, *460*
Dennis, E. S., 203, *213*, 425, *460*
Denton, W. L., 339, 340, 357, *362, 363*,
410, *413*, 474, *480*
Desai, I. D., 444, *460*
Descamps, J., 373, *416*
Deschamps, 2, *80*, 218, *252*
Desnuelle, P., 401, *413*
de Spain Smith, L., 14, *81*
Dewan, J. G., 396, *413*
Deysher, E. F. 438, 439, 442, 443, 445,
460, 461, 463
Dianoux, A.-C., 406, 407, *414*
Dickinson, W. L., 2, *81*, 218, *253*
Dickson, I. R., 39, *80*
Dietrich, J. W., 77, *83*
Dive, C., 373, 374, *416*
Dixon, G. H., 249, *252*
Dixon, M., 89, *114*, 262, *326*
Djurtoft, R., 229, 237, 238, 239, *252*
Dolmans M., 351, *364*
Donohue, J., 293, *326*
Dopheide, T. A. A., 243, 249, *252*
Douglas, F. W., 441, *460*
Dowben, R. M., 381, *413*
Downey, W. K., 178, *213*, 401, 402,
413, 442, *460*
Dreizen, I. R., 39, *80*
Dreizen, P., 150, 151, *170*
Dresdner, G. W., 14, 17, 22, 35, 37, 42,
43, *80, 84*
Drewry, J., 105, *114*, 118, *169*, 258, 259,
260, 271, 274, 304, *325*, 336, 345,
351, 355, 356, *362*, 370, *411*
Dumas, J. B., *252*
Dummel, B. M., 224, 227, 231, 237, *251*

Duncan, C. W., 108, *116*, 455, *460*
Dunnill, P., 292, 323, *326*
Dupont, M., 317, 318, *326*
Durrant, N. W., 433, *460*
Dutheil, H., 250, *251*
D'yachenko, P. F., 203, *213*, 447, 449,
460

E

Easterly, D. G., 441, *460*
Ebner, K. E., 339, 340, 357, 358, 360,
361, *362, 363, 364*, 410, *412, 413*,
474, 476, *480, 481*
Eckhardt, E. R., 447, *460*
Edelman, G., 478, *480*
Edmondson, L. F., 441, *460*
Edsall, J. T., 10, *80*, 293, *326*
Edwards, P., 397, *411*
Ege, R., 231, 234, *252*
Eiler, J. J., 44, 46, *81*
Eilers, H., 10, *80* 438, 439, 440, 442,
460
Ellman, L., 287, *326*
El-Negoumy, A. M., 88, 94, 102, *114*,
137, *170*, 492, *508*
Emneus, A., 453, *463*
Engel, E., 455, *464*
Engelstadt, W. P., 39, *79*
Engström, L., 405, *413*
Enkelmawn, D., 271, *329*
Erickson, D. R., 382, *414*
Ernstrom, C. A., 119, 120, *170*, 218, 224,
225, 229, 231, 234, 244, *252, 253*,
254, 427, 432, *461, 462*
Eskew, R. K., 444, *464*
Evans, F. R., 438, 440, 441, 442, 443,
460, 461, 463
Ezekiel, E., 371, 375, *413*

F

Fahrney, D., 249, *253*
Fama Cambria, M., 408, *415*
Fantes, K. H., 391, *413*
Farrell, H. M., Jr., 124, 145, *170, 172*

Fazekas de St. Groth, S., 486, *508*
Feagan, J. T., 322, 323, *326*, 428, 435, 436, 438, *461*, 479, *480*, *481*
Feeney, R. E., 373, 374, 375, *413*, *414*, *418*
Ferguson, K. A., 273, *326*, 507, *508*
Fevold, H. L., 368, *411*
Fialkov, M., 447, *459*
Fiat, A. M., 208, *213*, 472, *481*
Filmer, D., 8, *81*
Finucane, T. P., 454, *462*
Fish, J. C., 199, 203, *213*, 246, 247, 248, *252* 425, *461*
Fish, W. W., 373, *415*
Fitzgerald, D. K., 360, 361, *363*, 476, *481*
Fitzpatrick, M. M., 13, 14, 16, 21, 29, 37, 73, *83*, 99, *115*, 166, 167, *172*, 437, *464*
Flory, P. J., 78, *80*
Fölsch, G., 46, *80*
Foley, M., 6, 10, 14, 16, 17, 21 22, *84*, 92, *116*, 120, 133, 135, 147, 148, 155, *173*
Folk, J. E., 278, 279, 285, *329*
Foltmann, B., 2, *80*, 218, 219, 220, 221, 222, 223, 224, 225, 226, 227, 228, 229, 230, 231, 232, 233, 234, 235, 237, 238, 239, 240, 241, 242, 243, 244, 245, 246, 247, 248, *251*, *252*, 424, 425, 426, *461*
Fontana, P., 428, *461*
Ford, T. F., 13, 38, 72, *80*
Forster, T. L., 402, *413*, *414*, *415*, *416*
Fowkes, F. M., 381, *417*
Fox, K. K., 261, 262, *326*, 338, *363*, 437, 444, 455, 457, *461*
Fox, P. F., 178, *213*, *214*, 246, *252*, 400, 401, 402, *413*, *416*, *481*
Fraenkel-Conrat, H., 277, 285, *327*, *328*, 374, *413*
Fram, H., 444, *464*
Frank, G., 280, 281, 282, 283, 284, 291, *327*
Frankel, E. N., 399, 401, *413*, *418*
Franklin, J. G., 286, *327*
Freimuth, V., 322, *327*
Freund, E. H., 456, *461*
Fridovich, I., 396, 398, *412*, *413*

Friedman, L., 238, 248, *252*
Frot-Coutaz, J., 403, *413*
Fruton, J. S., 248, *251*
Fuld, E., 218, *252*
Furminger, I. G. S., 391, *413*

G

Gaffney, P. J., Jr., 178, *213*, 400, 401, *413*
Gahne, B., 375, 376, *413*
Gall, W. E., 478, *480*
Gander, G. W., 402, *414*
Ganguli, N. C., 147, *170*
Garnier, J., 14, 16, 17, 19, 20, 21, 32, 33, 48, 77, *80*, *83*, *84*, 121, 124, 125, 126, 127, 150, 151, 157, 160, 165, *170*, *172*, 181, 184, 186, 187, 189, 194, 195, 200, 202, *213*, *214*, 218, *252*, 275, 278, 279, 283, 294, 300, *326*, *327*, 368, *413*, 426, *463*
Gehrke, C. W., 63, *79*, 137, *170*, 181, 184, *212*
Genvrain, S. A., 433, *461*
Georges, C., 313, 317, 318, *327*
Georgevits, L. E., 453, *463*
Ghose, A. C., 266, 276, 294, *327*
Ghosh, S. K., 341, 352, 353, *363*
Gibson, J. F., 399, *413*
Gilbert, D. A., 396, *413*
Gilbert, G. A., 307, *327*
Gillespie, D. C., 378, *415*
Gillespie, J. M., 6, 10, 14, 16, 17, 19, 21, 22, *84*, 92, *116*, 120, 133, 135, 147, 148, 155, *173*
Gillies, J., 430, *460*
Gladner, J. A., 277, 278, 279, 285, *328*, *329*
Glasnak, V., 128, *170*
Glazer, A. N., 374, *413*
Goldwater, W. H., 285, *326*
Gopalakrishna Rao, N., 453, *464*
Gorbunoff, M. J., 291, 292, 293, *330*, 349, *363*
Gordon, A. H., 396, *413*
Gordon, W. G., 16, 17, *80*, *81*, 101, 102, 103, 104, 107, 109, *114*, 135, 139, 141, 143, 155, 157, 162, 166, *170*, *171*, *172*, 279, *327*, 333, 334, 335, 339, 340, 341, 345, 347, 352, 353,

356, *363*, 370, 371, 372, 374, 375, 376, 379, 383, *413*, *414*, *481*, 502, *508*
Gorguraki, V., 227, 231, 242, *253*
Gorin, G., 285, 319, *328*
Got, R., 276, *327*, 373, 403, *413*, *414*, *415*
Gough, P., 318, *327*
Gould, I. A., 14, 21, 48, *81*, *82*, 108, *115*, 135, *171*, 178, *213*, 271, 321, 322, *327*, *328*, 380, 400, 401, 408, *413*, *414*, *417*, *418*, 434, 437, 438, 453, 458, *460*, *462*, *463*
Goussault, Y., 276, *327*, 403, *414*
Grabar, P., 373, *415*, 478, *481*
Graham, E. R. B., 90, 95, 97, 98, 100, 113, *114*
Grassman, W., 486, *508*
Gray, W. R., 242, *252*
Green, D. E., 396, *413*, *415*
Green, D. W., 291, 292, 307, 308, 323, *325*, *326*, *327*, 356, *363*
Green, M.-L., 245, *252*
Green, R. C., 399, *414*
Greenbank, G. R., 439, 445, *461*
Greenberg, D. M., 14, *79*, 150, *170*
Greenberg, R., 16, 17, *81*, 143, 145, 147, 148, 155, 160, 162, 164, 165,168, *171*, *172*, 277, 278, 279, 288, *327*, 452, *461*
Greene, F. C., 373, *414*
Griffin, A. T., 322, *326*, 428, 436, 438, *461*, 479, *480*
Grimbleby, F. H., 436, *462*
Grohlich, D., 373, 384, *411*
Grosclaude, F., 16, 18, *80*, *82*, 121, 124, 125, 126, 127, 143, 148, 155, *170*, *171*, 275, *327*
Grosjean, N., 282, 310, *329*
Groves, M. L., 14, 15, 16, 17, *80*, *81*, 101 102, 103, 104, 107, 109, *114*, *115*, 118, 135, 137, 154, 156, 157, 158, 160, *170*, *171*, 180, 183, 186, *213*, 271, 315, 318, 323, *327*, *328*, 338, *363*, 368, 369, 370, 371, 372, 375, 376, 377, 378, 379, 383, 384, 386, 388, 389, 390, 408, *412*, *413*, *415*, 473, *481*, 501, 502, *508*

Gruber, M., *84*
Grundig, E., 15, *82*
Güll, J., 410, *415*
Guinand, S., 278, 279, 283, 294, 300, 313, 317, 318, *326*, *327*
Gutfreund, H., 404, 405, *411*
Guy, E. J., 403, *414*

H

Habeeb, A. F. S. A., 246, *252*, 285, *327*, 358, *362*
Habermann, W., 203, *213*
Hageman, E. C., 335, 352, *363*, *364*
Hallström, B., 453, *463*
Halwer, M., 14, *80*, 99, *115*
Hamilton, D. F., 111, 112, *115*, 162, 165, *171*, 497, *508*
Hamilton, H. B., 368, 369, 370, 387, 394, *416*
Hammarsten, O., 88, *115*, 222, *252*
Hammond, L. A., 428, 430, 433, *460*, *461*
Hampson, J., 450, *465*
Han, K., 342, *363*
Handler, P., 396, 397, 398, *412*, *413*, *416*, *417*
Hankinson, C. L., 224, 239, *252*, 402, *417*
Hannig, K., 486, *508*
Hanrahan, F. P., 444, *461*
Hans, R., 154, *172*
Hansen, P. M. T., 438, 449, 450, 451, 456, 458, 459, *461*, *462*
Hansen, R. R., 17, *80*, 89, *115*, 184, *213*, 338, *363*, 448, 453, *461*
Hanson, L. A., 338, *362*, 410, *413*, 479, *481*
Hansson, E., 406, *414*
Hardy, W. B., 317, *327*
Harkness, W. L., 426, 430, *460*, *465*
Harland, H. A., 445, *460*, *461*
Harley, B. S., 291, *326*
Harper, W. J., 178, *213*, 400, 401, 404, 408, *413*, *414*, *415*, *417*
Harris, F. E., 43, *80*, *83*
Hart, L. I., 396, 398, *414*

Hartley, B. S., 242, 243, 249, *251*, *252*, *254*
Hartman, W. E., 455, *461*
Harwalkar, V. R., 382, 410, *414*, *416*
Hasselle, C., 310, *329*
Hassid, W. Z., 357, 361, *362*, *364*, 409, *411*, 475, *481*
Hata, T., 380, 386, *414*, *416*
Hawes, R. O., *171*
Hay, A. K., 456, *461*
Hayashi, S., 381, 382, *414*
Hayashikawa, R., 396, 397, *418*
Hayes, J. F. 433, 447, 448, 449, 453, *460*, *463*,
Heckman, F. A., 13, 38, 72, *80*
Hegarty, G. R., 456, *463*
Hehir, A. F., 323, *326*, 428, 435, 436, *461*, 479, *480*, *481*
Heimburger, N., 229, 245, *253*, 271, *329*
Heinemann, B., 453, *461*
Hellung-Larsen, P., *414*
Henika, R. C., 450, *461*
Henry, K. M., 445, *461*
Henschel, M. J., 250, *252*
Henzi, R., 76, *82*, 200, 201, 202, 208, 209, 210, *214*
Herald, C. T., 108, *115*, 382, 383, *414*, 455, *461*
Heremans, J. F., 373, 374, 375, *416*, 478, *480*, *481*
Heremans, K., 471, *481*
Hermann, J., 407, *414*
Hermans, J. J., 43, *80*
Herreid, E. O., 21, 50, *82*, 186, 187, 189, *214*, 440, 441, *463*, *465*
Herriott, R. M., 250, *252*
Herrmann, J., 441, *461*
Herskovits, T. T., 22, *80*, 185, *213*, 291, 292, 293, *330*, 349, *363*
Heyndrickx, G. V., 409, *414*
Hickman, C. G., 376, *417*
Hidalgo, J., 450, 451, 459, *461*
Hill, R. D., 11, 17, 49, *80*, *83*, 89, *115*, 181, 184, 197, 212, *213*, 245, 249, *253*, 338, *363*, 426, 427, 428, 433, 438, 441, 453, 456, *461*, 472, *481*
Hill, R. J., 20, 48, *80*, *82*, 185, 190, 191, 194, 196, 197, 205, 206, 207, 208, 209, 210, *212*, *213*, *214*, 292, *327*

Hill, R. L., 339, 340, 343, 344, 345, 347, 350, 351, 358, 359, 360, 361, *362*, *363*, *364*, 407, *412*
Hill, W. B., 250, *252*
Hines, H. C., *170*
Hipp, N. J., 14, 15, 16, 17, *80*, *81*, 101, *114*, *115*, 118, 135, 137, 154, 156, 157, 158, 160, *170*, *171*, 180, 183, 186, *213*, 315, 318, 323, *327*, *328*, 408, *415*
Hirs, C. H. W., 386, *417*
Ho, C., 17, 18, 37, 39, 52, *81*, 135, 154, 155, *170*
Hoagland, P. D., 137, 138, 145, 168, *170*, *171*, 471, *481*
Hodgkin, D. C., 258, 323, *327*
Hoerr, C. W., 456, *461*
Hofman, T., 18, *81*, 154, *171*, 203, *213*
Hogancamp, D. M., 227, 231, 242, *251*
Hollis, R. A., 119, 120, *170*, 437, *464*
Holly, R. G., 406, *412*
Holm, G. E., 438, 439, 445, *460*, *461*, *465*
Holmes, L. G., 348, 349, *363*, *364*
Holmgren, J., 410, *414*, 479, *481*
Holsinger, V. H., 261, 262, *326*, 338, *363*, 455, *461*
Holter, H., 203, *213*, 219, 220, 250, *253*
Holwerda, B. J., 231, *253*
Hood, L. F., 381, *414*
Hoogendoorn, M. P., *171*
Hoover, S. R., 17, *82*, 147, 160, *171*
Hopper, K. E., 266, 267, 268, 269, *325*, 337, 352, *362*, 476, 477, *481*
Horst, M. G., 68, *81*
Hostettler, H., 10, 77, *81*, 219, 225, *253*, 437, 440, 441, *461*
Howard, H. W., 447, *461*
Howat, G. R., 436, 438, *461*
Hoyle, B. E., 44, 61, 65, *80*
Hsu, R. Y. H., 427, *462*
Hull, M. E., 450, *462*
Hull, R., 316, 317, *326*
Hultquist, D. E., 369, 387, 391, 393, 395, *414*, *416*
Humme, H. E., 472, *481*
Hunziker, H. G., 271, 322, *327*, *330*
Hutchinson, E., 8, *83*
Hutton, J. T., 285, *327*
Hutton, T. J., 21, *81*

AUTHOR INDEX 517

I

Ibuki, F., 380, 386, *414*, *416*
Ikenaka, T., 227, 231, 242, *253*
Imhof, K., 10, 77, *81*, 437, 440, 442, 445, *461*, *462*
Indiramma, K., 453, *460*
Ingram, V. M., 452, *462*
Inman, J. K., 356, *363*, 375, *414*
Irvine, D. M., 381, *417*, 430, *462*
Isemura, T., 7, 8, *83*
Isliker, H., 372, 375, 384, *412*, *415*
Ismael, A. A., 436, *462*
Iwasaki, S., *480*

J

Jackson, R. H., 382, 383, *413*, *414*, 455, *462*
Jacobsen, C. F., 259, 315, 319, *327*, *328*
Jaenike, R., 317, *329*
Jago, G. R., 395, *414*
James, D. W. F., 397, 398, *411*
Janson, M. H., 447, *460*
Jasewicz, L. B., 51, *85*, 406, *412*
Jayasinghe, W., 395, *416*
Jenard, R., 310, *329*
Jenness, R., 19, 59, 60, 63, 68, *80*, *81*, *84*, 88, 89, 101, 105, 107, 108, *115*, *116*, 120, 121, 145, *171*, *172*, 276, 278, 279, 285, 286, 287, 288, 292, 317, 318, 321, *327*, *328*, *329*, 335, 352, 354, *364*, 368, 380, 396, 403, 408, *414*, 429, 433, 437, 438, 445, 458, *460*, *462*, *464*
Jennings, R. K., 14, *81*
Jennings, W. G., 444, *460*
Jensen, R. G., 399, 402, *414*
Jeppsson, J.-O., 373, *414*
Jeunet, R., 125, 126, *170*
Jirgensons, B., 227, 231, 242, *253*
Jodal, U., 479, *481*
Johansen, A., 229, 237, 238, 239, *252*
Johansen, G., 319, *327*
Johansson, B. G., 15, *79*, 338, *362*, *363*, 368, 373, 374, 375, *414*, 479, *481*
Johke, T., 352, *363*

Johnson, A. H., 426, 428, *465*
Johnson, P., 271, *329*, 338, 346, *364*, 370, 371, 372, 373, 375, 376, *418*
Johnston, J. O., 118, 122, 124, 125, 128, 130, 155, *171*, *172*, 495, *508*
Jollès, J., 16, 17, 20, 21, *81*, 110, *115*, 185, 186, 187, 189, 198, 201, 202, 203, 207, 208, 209, 210, 211, 212, *213*, 249, *253*, 406, 407, *414*, *416*, 425, *460*, 472, *481*, *482*
Jollès, P., 16, 17, 20, 21, 75, 76, *79*, *81*, 110, *115*, 120, *171*, 185, 186, 187, 189, 197, 198, 201, 202, 203, 204, 207, 208, 209, 210, 211, *212*, *213*, 249, *253*, 344, *363*, 406, 407, *412*, *414*, *417*, 425, *460*, 472, *480*, *482*
Jones, F. S., 395, *415*
Joniau, M., 289, *326*, *327*
Jordan, S. M., 371, 374, 376, *411*, *415*
Josephson, R. V., 322, *328*
Josh, G., 450, *462*
Jourdian, G. W., 360, *364*
Jukes, T. B., 283, *327*

K

Kaijser, B., 479, *481*
Kalan, E. B., 16, 17, 18, 20, 21, 47, *81*, *84*, 89, 101, 107, *115*, 120, 137, 138, 139, 141, 145, 147, 148, 155, 160, 162, 164, *171*, *172*, 187, 189, 191, 193, 194, 195, 196, 201, 204, 207, 208, *213*, *215*, 271, 276, 277, 278, 279, 286, 287, 288, 292, *325*, *327*, *328*, 386, *412*, 452, *461*, 471, *481*
Kaldor, I., 371, *415*
Kaminogawa, S., 409, *418*
Kamiyama, S., 21, *81*
Kannan, A., 429, 433, *462*
Karlsson, B. W., 411, *415*
Karwowicz, E., 428, *462*
Kassel, B., 285, *326*
Kauzmann, W., 316, 319, *326*, *327*
Kay, H. D., 404, *415*
Kayser, C., 394, *415*
Keenan, T. W., 381, *415*
Keil, B., 409, *412*
Kekwick, R. A., 332, *363*
Keller, W., 199, *214*

Kelley, J. J., 18, *79*
Kelly, P. L., 406, *417*
Kendall, R. A., 378, *415*
Kende, S., 454, *462*
Kenkare, D. B., 271, *328*, 438, 453, 458, *462*, *463*
Kennedy, E. P., 22, *79*
Kessler, E., 474, *481*
Ketting, F., 454, *462*
Kibrick, A. C., 258, 294, *326*, 332, *362*
Kiddy, C. A., 16, 17, *84*, 102, 111, 112, *114*, *115*, 118, 120, 121, 122, 124, 125, 126, 128, 130, 135, 136, 137, 138, 145, 152, 155, 164, *170*, *171*, *172*, 278, 279, 292, *330*, 370, 371, 388, 389, 390, *414*, 495, 501, 502, *508*
Kielland, J., 44, *81*
Kiermeier, F., 21, *81*, 394, 406, 408, 409, 410, 411, *415*
Kieseker, F. G., 446, *463*
Kim, Y. K., 20, *81*, 178, 181, 185, 193, 196, 206, 207, *213*, *215*
King, C. W., 219, *253*
King, D. W., 448, *462*
King, J. W. B., 125, 126, 128, 145, *171*, 276, *327*
King, N., 380, *415*, 455, *462*
Kirchmeier, O., 473, *481*
Kissel, G., 14, 16, 37, *83*, 99, *115*, 166, 167, *172*
Kisza, J., 428, 441, 445, *462*
Kitchen, B. J., 477, *481*
Kiyosawa, I., 354, 360, 361, *363*, *364*, 471, 475, 476, *481*
Kjellberg, B., 411, *415*
Klebanoff, S. J., 395, *415*
Kleiner, E. S., 6, 10, 14, 16, 17, 21, 22, *84*, 92, *116*, 120, 133, 135, 147, 148, 155, *173*
Kleiner, I. S., 222, 231, *253*
Kleyn, D. H., 404, *415*
Klostergaard, H., 345, *363*
Klotz, I. M., 285, 291, *327*, *330*
Klug, A., *79*
Knedel, M., 486, *508*
Knoop, A., 381, *411*
Knoop, V. E., 472, *481*
Kodama, S., 4, 81, 117, 118, *171*

Koike, K., 271, *327*
Kok, A., 17, 20, *80*, 187, 193, 194, 203, 207, 208, 209, *213*
Kolar, C. W., Jr., 107, 108, *115*, 391, *418*
Komai, H., 399, *415*, *416*
Komatusu, S. K., 375, *413*
Kon, S. K., 445, *461*
Koops, J., 59, 63, 68, *81*, *172*, 425, 439, *464*
Kopfler, F. C., 111, 112, *115*, 133, 134, 155, 164, *171*, 497, 498, *508*
Korn, A. H., 17, *82*, 147, 160, *171*
Korn, E. D., 402, *415*
Koshland, D. E., *81*
Kovacs, G., 122, 125, *170*
Kozin, N. I., 432, 433, *462*
Krause, W., 322, *327*
Krecji, L. E., 14, *81*
Krekel, R., 317, *329*
Kresheck, G. C., 14, 22, *81*, 321, *327*, 403, 404, *415*
Krigbaum, W. R., 351, *363*
Kristjannson, F. K., 273, *327*, 507, *508*
Kristoffersen, T., 434, *462*
Kronman, M. J., 265, *329*, 336, 337, 341, 346, 347, *363*, *364*
Krukovsky, V. N., 399, *415*
Kügler, F. R., 351, *363*
Kumetat, K., 443, 445, 454, *459*, *462*
Kumler, W. D., 44, 46, *81*
Kumosinski, T. F., 310, 311, 312, *327*, *330*
Kurland, R. J., 18, *81*, 154, 155, *170*
Kuwahara, K., 471, *481*
Kuwatu, T., 208, *213*
Kuyper, A. C., 62, *81*

L

Lahav, E., 88, 94, *115*
Laing, R. R., 49, *80*, 212, *213*, *245*, *253*, 426, 427, 428, 433, *461*
Lanzani, G. A., 405, *411*
Large, P. M., 433, *460*
Larsen, B., 124, *170*, 191, 192, *213*, 273, 275, *328*, 484, 490, 506, *508*

Larsen, E., 119, 122, 124, 127, 130, *172*
Larson, B. L., 89, 95, 101, 105, 107, *115*, 119, 120, 139, *170, 171, 172*, 285, 317, *327*, 335, 352, *363, 364*, 378, *415, 417*, 437, *462*
Lascelles, A. K., 96, *115*
Laskowski, M., 409, *415*
Laskowski, M., Jr., 409, *415*
Latour, N. J., 39, *83*
Laurell, C. B., 370, *415*
Laurent, T. C., 7, *81*
Lawrence, A. J., 426, *462*
Lawrence, R. C., 219, *253*, 473, 480, *481*
Lea, A. S., 2, *81*, 218, *253*
Lea, C. H., 445, *461, 462*
Leach, B. E., 384, *415*
Leach, S. J., 285, *328*
Le Baron, J. L., 406, *414*
Leder, P., *214*
Lee, J. W., 452, 454, *459, 462*
Lehmann, W., 176, *214*
Leibman, A., 373, *411*
Lemon, M., 272, *325*, 477, *480*, 506, *508*
Léonis, J., 351, *364*
Leslie, J., 285, 286, 319, *327, 328*
Leujeune, N., 288, *329*
Levin, Y., 249, *253*
Leviton, A., 440, 441, *462*
Levy, H., 375, *417*
Lewis, M. S., 307, *325*
Lewis, V. J., 507, *508*
Li, S. O., 203, *213*
Liang Tung T'sai, 244, *251*
Lillevik, H. A., 14, 16, 17, 19, 21, 48, 49, *82, 83, 84*, 88, 89, 101, *116*, 120, 121, 145, 150, *171, 172*, 186, 188, *214*
Lin, S. H. C., 404, *415*
Linderstrøm-Lang, K. V., 4, *81*, 117, 118, *171*, 175, *214*, 315, 319, *328*
Lindqvist, B., 75, *82*, 120, *171*, 218, 246, *253*
Linklater, P. M., 234, *253*
Lipmann, F., 405, *417*
Liü, A. K., 344, *362*
Livrea, G., 408, *415*
Lloyd, G. T., 322, *326*, 428, 436, 438, *461*, 479, *480*

Lobareva, L. S., 426, *464*
Locquet, J.-P., 406, *417*
Loftus Hills, G., 439, 440, 441, *459*
Loisillier, F., 373, *415*, 478, *481*
Long, J. E., 21, 48, *82*, 108, *115*, 135, *171*, 321, 322, *328*, 437, *462*
Lontie, R., 267, 282, 288, 289, 290, 310, 313, 315, *326, 327, 328, 329*, 338, *364*
Lorand, L., 473, *481*
Lovrien, R., 303, *328*
Lowenstein, M., 448, 452, *462*
Lowndes, J., 15, *82*
Ludwig, M. L., 6, 10, 14, 16, 17, 19, 21, 22, *84*, 92, *116*, 120, 133, 135, 147, 148, 155, *173*
Ludwig, S., 38, *79*
Luebke, R. G., 395, *415*
Lundsteen, E., 231, 234, *252*
Lyster, R. L. J., 352, 354, *364*, 403, 404, *415*

M

Mabbitt, L. A., 383, *412*, 432, *462*
MacAllister, R. V., 454, *462*
Macara, T. J. R., 15, *82*
McBain, J. W., 7, *82*
McCabe, E. M., *214*
McCollum, E. V., 426, 428, *465*
Macdonald, C. A., 203, *214*
McDowall, F. H., 448, *462*
McFarlane, A. S., 427, *462*
McFarren, E. F., 404, *417*
McGann, T. C. A., 13, 62, 68, 69, 71, *82, 83*, 426, 428, 433, 436, *462, 463*
McGartoll, M. A., 396, 398, *414*
McGillivray, W. A., 430, *460*
McGlasson, E. D., 453, *462*
McGugan, W. A., 437, *462*
McGuire, E. J., 360, *364*
McGuire, W. L., 360, *364*, 476, *481*
Mach, J. P., 384, *415*
Macheboeuf, M., 315, *328*
Mackenzie, D. D. S., 96, *115*

McKenzie, H. A., 14, 16, 21, 33, 47, 48, *82*, 89, 90, 95, 97, 98, 100, 113, *114*, *115*, *171*, 175, 177, 179, 180, *214*, 258, 259, 262, 263, 264, 265, 266, 267, 268, 269, 271, 272, 273, 274, 275, 276, 278, 279, 280, 281, 282, 283, 284, 288, 299, 304, 308, 309, 310, 311, 313, 315, 316, 319, 320, *325*, *326*, 337, 352, 354, *361*, *362*, *364*, 374, *413*, 474, 476, 477, 479, *480*, *481*, 485, 493, 494, 504, 506, 507, *508*

Mackinlay, A. G., 16, 20, 21, 47, 48, *82*, 99, 110, *115*, 194, 205, 207, 208, *212*, 425, *462*

Mackler, B., 396, *415*

McLoughlin, P. T., 453, *462*

McMeekin, T. L., 14, 15, 16, 17, 18, *80*, *81*, *83*, 101, 111, *114*, *115*, 118, 120, 135, 137, 154, 156, 157, 158, 160, *170*, *171*, 180, 183, 186, *213*, 226, *253*, 302, 315, 318, 323, *327*, *328*, *330*, 368, 408, *415*

Macquiddy, E. L., Sr., 406, *417*

MacRae, H. F., *171*

Maeno, M., 354, *364*

Maes, E. D., 351, *364*

Magnusson, J. A., 18, *81*, 155, *170*

Magnusson, N. S., 18, *81*, 155, *170*

Mahler, H. R., 396, *415*

Mair, G. A., 358, *362*

Makey, D. G., 371, *417*

Malmström, B. G., 372, 373, 374, 375, *411*, *412*

Malpress, F. H., 208, *214*, 250, *253*, 471, 472, *481*

Mal'tsev, N. I., 426, *464*

Mann, K. G., 373, *415*

Manson, W., 16, 17, 18, 22, *82*, 145, 147, *169*, *171*, 471, *480*, *481*

Marier, J. R., 21, 61, 63, 65, *79*, *82*, 186, *214*

Marin, M., 405, *412*

Marnay, A., 276, *327*, 403, *414*

Marquardt, R. R., 402, *415*

Mars, P. H., 409, *415*

Masket, N., 38, *79*

Massey, V., 397, 399, *412*, *415*, *416*

Masson, P. L., 373, 374, 375, *416*, 478, *480*, *481*

Masters, C. J., 178, *214*

Matsen, H., 219, *254*

Matsushita, S., 380, 386, *414*, *416*

Mattenheimer, H., 203, *213*

Mattock, P., 358, 361, *363*, *364*

Maubois, J.-L., *172*, 181, 184, *214*, 218, *252*

Mauk, B. R., 429, 458, *462*

Mauron, J., 372, *412*

Mawal, R. B., 271, *328*, 360, 361, *363*, 476, *481*

Mecham, D. K., 485, *508*

Meharry, H. J., 430, *460*

Meinl, E., 406, *415*

Melamed, M. D., *508*

Mellander, O., 5, *82*, 101, *115*, 117, *171*

Mellon, E. F., 17, *82*, 147, 160, *171*

Mellors, A., 410, *416*

Melnychyn, P., 89, 101, 107, *115*, 120, 139, 150, *171*, *172*, 441, *462*

Melton, B., 92, *116*, 120, 133, 135, 147, 148, 155, *173*

Melville, E. M., 219, *253*

Mercier, J.-C., 16, 18, *80*, *82*, 143, 148, 155, *170*, *171*

Mescanti, L., 22, *80*, 310, 311, 313, 316, *330*, 349, *363*

Metton, B., 6, 10, 14, 16, 17, 21, 22, *84*

Meyer, H., 275, *328*

Michalak, W., 110, *114*, 130, 131, 147, *171*, 275, *328*, 485, 491, *508*

Mickelsen, R., 244, *253*

Miersch, R. E., 450, *461*

Mihályi, E., 203, *214*

Mikolajcik, E. M., 434, *462*

Miller, R. W., 397, *412*

Mills, O. E., 471, *480*

Milner, M., 453, *462*

Mitchell, W. M., 391, *416*

Miyazawa, K., 455, *464*

Mocquot, G., 14, 32, 48, 76, *79*, *80*, 151, 157, 160, *170*, *172*, 176, 181, 184, 186, 199, 202, *212*, *213*, *214*, 218, *252*, 424, 426, 428, 442, *459*, *463*

Moews, P. C., 244, *251*

Moncrieff, A., 453, 456, *461*, *463*

Monod, J., 8, *82*

Monsigny, M., 373, *416*, *417*

Montgomery, M. W., 402, *413*, *416*

AUTHOR INDEX   521

Montoure, J. E., 402, *413*
Montreuil, J., 368, 373, 374, 375, *416*, *417*
Moore, S., 22, *83*, 203, *212*, 243, 249, *252*, 285, *329*
Morard, J. C., 77, *79*
Morell, D. B., 395, *416*
Morelle, A., 282, 310, *329*
Morgan, A. F., 285, *327*
Morgan, E. H., 371, 373, 374, 376, *411*, *415*
Mori, T., 380, 386, *414*, *416*
Morr, C. V., 271, 319, 322, *328*, 437, 440, 453, *463*
Morre, D. J., 381, *415*
Morris, H. A., 426, *464*
Morris, M., 16, 17, *80*
Morrison, M., 368, 369, 370, 372, 373, 387, 389, 391, 393, 394, 395, 396, *411*, *414*, *416*, *417*
Morrissey, P. A., 436, *463*
Morrow, B. H., 397, 398, *417*
Morton, R. K., 368, 380, 386, 403, 404, *411*, *416*
Moschetto, Y., 337, 342, *362*, *363*
Moustgaard, J., *173*
Mouton, A., 407, *416*
Moxley, J. E., *171*
Mülder, G. J., 88, *115*
Muers, M. M., 443, *463*
Muirhead, H., 8, *80*
Mukherjee, S., 293, *328*
Mulder, H., 218, *253*
Mullen, J. E. C., 404, 405, *415*, *416*
Muller, L. L., 446, 447, 448, 449, 453, *463*
Mullet, S., 368, 374, 375, *416*
Murphy, R. F., 178, *213*, 401, *413*
Murphy, W. H., 90, 95, 97, 98, 100, 113, *114*, 266, 267, 268, 269, 271, 272, 275, 278, 279, 284, 310, *325*, *326*, 337, 352, *362*, 494, *508*
Murthy, G. K., 102, *115*

Nakagawa, T., 7, 8, *83*
Nakai, S., 21, 50, *82*, 186, 187, 189, *214*, 441, *463*
Nakamura, S., 399, *416*
Naughton, M. A., 209, *213*
Nauman, L. W., 15, 18, *83*, 111, 112, *115*, 154, 162, 163, 165, *171*, 497, *508*
Neelin, J. M., 13, 20, *82*, 110, *115*, 118, *171*, 179, 180, 182, 184, 190, *214*
Neff, E., 446, *463*
Nelson, C. A., 398, *416*
Nelson, C. E., 447, *464*
Némethy, G., 8, *81*
Neuman, M. W., 62, *82*
Neuman, W. F., 62, *82*
Neumann, H., 248, *251*
Neurath, H., 277, *328*
Newman, C. R., 277, *326*
Ng, W. S., 107, 108, *115*
Nickerson, T. A., 399, *418*, 431, 433, 444, *460*, *463*
Nielsen, H. C., 14, *82*, 150, *171*
Nielsen, M. A., 453, *463*
Niki, R., 208, *213*
Nirenberg, M., *214*
Nitschmann, Hs., 10, 14, 20, 21, 76, *79*, *82*, 147, *173*, 176, 199, 200, 201, 202, 204, 208, 209, 210, *212*, *214*, 238, 239, 241, 246, *253*, 424, 425, 442, *459*, *463*
Niu, C. L., 277, *328*
Noble, R. W., 7, 10, 12, 14, 16, 17, 19, 21, 23, 25, 28, 30, 33, 34, 48, 50, 51, 71, 74, 77, *80*, *82*, *84*, 92, 100, *115*, 150, 151, *170*, *171*, 178, 184, 197, *214*, *215*
Noelken, M. E., 16, 17, 20, 21, 37, 47, *82*, *84*, 150, 151, *171*, 189, 191, 193, 194, 195, 204, 207, *215*
North, A. C. T., 350, 358, *362*
Nozaki, Y., 294, 303, *328*, *329*, 451, *464*
Nuenke, B. J., 291, *326*

N

Nagant, D., 288, *328*, *329*
Nagasawa, T., 471, 475, *481*

O

O'Brien, P. J., 399, *414*
Odagiri, S., 431, 433, *463*

Oeda, M., 225, *253*
Österberg, R., 15, 18, 36, 46, *79*, *80*, *82*
Oetiker, N., 446, *463*
Ogston, A. G., 258, 304, *328*
Oh, Y. H., 137, *170*
O'Keefe, T., 456, *464*
Olson, D. E., 381, *415*
O'Neal, C., *214*
Oosthuizen, J. C., 76, *83*
Oppenheimer, C., 218, *253*
Oram, J. D., 374, 395, *416*, *417*
Orme-Johnson, W. H., 399, *416*
Ornstein, L., 501, 507, *508*
Osborne, T. B., 105, *115*
Osserman, E. F., 407, *416*
Osterhoff, D. R., 274, *328*
O'Sullivan, A. C., 404, *416*
Osumi, K., 271, *327*
Ottesen, M., 480, *481*
Outteridge, P. M., 96, *115*
Overbeek, J. ThG., 7, 42, 43, *80*, *82*, *84*
Overcast, W. W., 400, *417*

P

Pace, N. C., 319, *328*
Pader, M., 454, *459*
Pahud, J. J., 384, *415*
Palermiti, F., 14, 16, 37, *83*, 99, *115*, 166, 167, *172*
Palit, S. R., 293, *328*
Pallansch, M. J., 261, 262, *326*, 338, *363*, 384, *415*, 440, 444, 455, *461*, *462*
Pallavicini, C., 373, *417*, 478, *481*
Palmer, A. H., 258, 294, *326*, *328*, 332, *362*, *364*
Palmer, G., 399, *411*, *415*, *416*
Palmer, G. M., 449, 451, *463*
Palmer, L. S., 239, *252*, 380, *414*, *416*
Palmiter, R. D., 361, *364*
Panasenkov, N., 447, *459*
Pantaloni, D., 258, 278, 279, 283, 289, 294, 300, 313, 315, 317, *326*, *328*, 451, *463*
Pantlitschko, M., 15, *82*
Parry, R. M., Jr., 34, *82*, 406, 407, *412*, *416*, *418*

Pasternak, R. A., 345, *363*
Patel, C. V., 178, *214*, 400, 401, 402, *416*
Patton, S., 21, 60, *81*, 285, *327*, 368, 380, 381, 396, 403, 408, *414*, *416*, *417*
Paul, K. G., 387, 393, 394, *417*, *418*
Paulicka, F. R., 456, *461*
Payens, T. A. J., 14, 16, 17, 31, 33, 38, *82*, *83*, 99, *115*, 135, 136, 147, 148, 149, 150, 167, 168, *171*, *172*, 229, *253*, 429, *463*, 471, *481*
Pearson, A. M., 456, *463*
Pedersen, K. O., 4, *82*, 258, 304, *328*, 332, 347, *364*, 393, 394, *418*
Pepper, L., 14, 16, 17, 32, 50, *81*, *83*, *84*, 118, 121, 128, 136, 151, 152, 154, 157, 158, 160, 162, 164, 165, *171*, *172*, *173*, 186, 198, *214*
Perkins, D. J., 39, *80*
Perlmann, G. E., 18, *83*, 153, 154, *171*, 203, *214*, 250, *253*, 441, *461*
Pernoux, E., 77, *79*
Persson, H., 7, *81*
Perutz, M. F., 8, *80*
Peters, I. E., 77, *83*
Peters, I. I., 429, 432, 433, *463*
Peters, R., 217, *253*
Peterson, R. F., 15, 18, *83*, 111, 112, *115*, 130, 133, 134, 154, 155, 162, 163, 164, 165, *171*, 273, *328*, 370, 371, *414*, 496, 497, 498, 507, *508*
Petz, E., 21, *81*, 411, *415*
Phillips, D. C., 350, 358, *362*
Phillips, N. I., 276, 278, 279, 286, 287, 288, 292, *328*, 352, 354, *364*
Philpott, D. E., 381, *413*
Pick, F. M., 399, *417*
Piette, L. H., 396, 397, *418*
Piez, K. A., 278, 279, 285, *329*
Pin, P., 153, *172*
Pinder, T. W., 303, *330*
Pion, R., 16, 17, 20, *83*, 165, *172*, 187, 189, 194, 195, *214*
Platt, B. S., 453, *463*
Plimmer, R. H. A., 15, *82*
Plummer, T. H., Jr., 386, *417*
Polis, B. D., 333, *364*, 368, 378, 387, 388, 393, 394, 407, *417*
Polis, E., 408, *418*

Pont, E. G., 442, *463*
Porcher, C., 218, 251, *253*
Porter, J. W. G., 250, *252*
Porter, R. R., 247, 248, *253*
Portmann, A., 395, *417*
Posati, L. P., 261, 262, *326*, 338, *363*
Poulik, M. D., 110, *115*, 507, *508*
Pradal, M. B., 403, *413*
Préaux, G., 267, 282, 288, 289, 290, 310, 313, 315, *326*, *328*, *329*, 338, *364*
Prentice, J. H., 380, *417*
Pretorius, A. M. G., 274, *328*
Prignot, J. J., 374, *416*
Printz, I., 229, 245, *253*
Probst, A., 428, *463*
Pujolle, J., 121, 124, 127, *170*, 187, 189, 194, 195, *214*, 275, *327*
Purkayastha, R., *329*
Pyne, G. T., 13, 62, 68, 69, 71, 75, *82*, *83*, 426, 428, 433, 436, *462*, *463*

R

Racker, E., 154, *172*
Radema, L., 218, *253*
Rajagopalan, K. V., 397, 398, *413*, *417*
Ralston, G. B., 281, 283, 319, 320, *328*, *329*, 474, *480*, 504, *508*
Ramachandran, B. V., 15, *80*
Ramaekers, C., 288, *328*; *329*
Ramshaw, E. H., 449, *463*
Rand, A. G., 224, 231, 233, 234, *253*
Rao, P. S., 437, *463*
Rapp, H., 429, *463*
Rasmusson, Y., 406, *414*
Ray, A., 303, *329*
Raymond, S., 486, 496, *508*
Reeves, R. E., 39, *83*
Reibstein, M., 37, *82*
Reich, H. A., 373, *411*
Reid, T. W., 249, *253*
Reinhart, R. W., 375, *417*
Reisfeld, R. A., 427, *463*, 507, *508*
Reiter, B., 374, 395, *416*, *417*
Reithel, F. J., 361, *363*, 476, *480*
Rendel, J., 376, *413*
Revel, H. R., 154, *172*
Rhee, K. C., 383, *418*

Ribadeau-Dumas, B., 16, 17, 18, 20, 33, *80*, *82*, *83*, 121, 124, 125, 126, 127, 143, 148, 155, 157, 160, 165, *170*, *171*, *172*, 181, 184, 187, 189, 194, 195, *214*, 218, *252*, 275, 278, 279, 283, 294, 300, *326*, *327*
Rice, S. A., 43, *80*, *83*
Richards, F. M., 324, *326*
Richards, H. L., 471, *480*
Richardson, G. A., 402, *417*
Richardson, L. A., 404, *417*
Richardson, T., 382, *412*, 457, *460*, *464*
Ricket, W., 480, *481*
Rieder, R. F., 145, *170*
Riley, D. P., 258, 323, *327*
Rimington, C., 15, *83*
Robbins, F. M., 265, *329*, 336, 337, 348, 349, *364*
Robert, B., 315, *328*
Roberts, R. C., 371, *417*
Robertson, J. A., Jr., 401, 408, *414*, *417*
Robertson, P. S., 430, *460*
Roche, H., 38, *84*
Roche, J., 153, *172*
Rodgers, N. E., 449, 450, 451, *461*, *463*
Rodionova, I. R., 432, 433, *462*
Roels, H., 288, 289, 290, 315, *329*
Rogers, H. M., 62, *83*
Rogers, L. A., 442, 443, *463*
Rolleri, G. D., 95, 105, 107, *115*, 335, *364*, 378, *417*
Rombauts, W., 247, 248, *251*
Rombauts, W. A., 372, 387, 389, 391, 393, 394, 395, *417*
Roop, W. E., 375, *417*
Rose, D., 11, 13, 19, 20, 21, 31, 33, 38, 61, 63, 67, 68, 73, *81*, *82*, *83*, *84*, 88, 89, 101, 107, 110, *115*, *116*, 120, 121, 139, 145, *172*, 179, 180, 184, 186, 193, 196, 206, 207, *213*, *214*, 322, *329*, *330*, 435, 436, 439, 444, 458, *463*, *464*, 479, *481*
Roseman, S., 360, *364*
Ross, P. D., 44, *83*
Rossman, M. G., 8, *80*
Rottman, F., *214*
Roussos, G. G., 397, 398, *417*
Rout, T. P., 178, *214*

Rowland, S. J., 95, 105, 108, *115*
Roychoudhury, A. K., 351, *362*
Rucknagel, D. L., 127, *170*
Rudd, R. K., 318, *330*
Rüegger, H. R., 219, *253*
Rupley, J. A., 385, *417*
Rutz, W. D., 10, *84*
Ryan, F. J., 285, *326*
Rydstedt, L., 358, *362*
Ryle, A. P., 247, 248, *253*

S

Saal, R. N. J., 10, *80*
Saidel, L. J., 285, *326*
Saint-Blancard, J., 406, *414*, *417*
Salmon, C. S., 7, *82*
Saltman, P., 373, 374, *411*
Salzberg, H. K., 453, *463*
Samel, R., 443, *463*
Sampath Kumar, K. S. V., 153, *172*
Sampugna, J., 402, *414*
Samuelsson, E. G., 410, *414*, 453, *463*
Sandberg, K., 122, *172*
Sanderson, W. B., 219, *253*, 480, *481*
Sanger, F., 247, *253*
Sarma, P. S., 153, *172*
Sarma, V. R., 358, *362*
Saroff, H. A., 303, *325*
Sasaki, R., 455, *464*
Sawyer, W. H., 262, 265, 266, 269, 272, 281, 288, 309, 310, 311, 313, 315, 317, 321, 322, *325*, *328*, *329*, 337, *361*, 438, 458, *464*, 507, *508*
Scatchard, G., 43, *83*
Schade, A. L., 368, 373, 375, *417*, 478, *481*
Schanbacher, F. L., 361, *363*
Schardinger, F., 396, *417*
Schellman, J. A., 319, *329*
Scheraga, H. A., 385, *417*
Schmid, K., 21, *81*
Schmidt, D. G., 14, 16, 17, 20, 47, *82*, *83*, 99, 110, *115*, 119, 135, 136, 147, 148, 149, 150, 151, 152, *171*, *172*, 190, 191, 192, 193, 194, 195, *214*, 425, 439, 441, *464*, 488, *508*

Schober, R., 229, 245, *253*, 271, *329*
Schonne, E., 373, *416*, 478, *481*
Schoppet, E. F., 444, *464*
Schormuller, J., 154, *172*
Schroeder, W. A., 372, 387, 389, 391, 393, 394, 395, *417*
Schwander, H., 238, 239, 241, *253*
Schwarcz, M., 456, *464*
Schwartz, J. H., 405, *417*
Schwendener, S., 7, *84*
Scott, E. C., 447, 448, *464*
Scott-Blair, G. W., 76, *83*, 429, *464*
Seal, U. S., 371, *417*
Sebelien, J., 332, *364*
Segelcke, T., 220, *253*, *254*
Seibles, T. S., 303, *329*, 339, 341, 342, 356, *365*
Seid-Akhavan, M., 208, *214*, 471, 472, *481*
Sela, M., 248, *251*
Semmett, W. F., 16, 17, *80*, 333, 334, 339, 345, 347, *363*
Semper, G., 408, 409, *415*
Sen, A., *114*, 119, 122, 123, 125, 126, 128, 147, 155, *170*, 266, 276, 294, *327*, *329*, 340, 341, 351, 352, 353, *362*, *363*, *364*
Sevall, H. E., 447, *461*
Shahani, K. M., 368, 385, 399, 400, 401, 402, 403, 406, 407, *412*, *413*, *416*, *417*, *418*
Sharp, P. F., 10, *84*
Shaw, D. C., 266, 271, 272, 275, 276, 278, 279, 281, 282, 283, 284, 299, 310, *326*, 352, *362*, 371, 373, 376, *411*
Sheinson, R. S., 302, *330*
Shen, A. L., 43, *83*
Sherman, J. M., 10, *84*
Shimmin, P. D., 11, *83*, 457, *464*
Shinoda, K., 7, 8, *83*
Shipe, W. F., 404, *416*, 426, 433, *464*
Shmukler, H. W., 333, *364*, 368, 378, 387, 388, 393, 394, 407, *417*
Shooter, K. V., 397, 398, *411*
Shukri, N. A., 234, *254*
Shulman, J. H., 9, *83*
Simmons, R. M., 323, *326*
Simms, H. S., 395, *415*
Simons, R. M., 323, *325*

Simpson, R. B., 319, *327*
Sinha, N. K., 351, 353, *362*, *364*
Sinkinson, G., 185, 208, *215*
Sinnamon, H. I., 444, *464*
Sinohara, H. 203, *213*
Sirry, I., 426, 433, *464*
Sjögren, B., 332, *364*
Skean, J. D., 400, *417*
Sky-Peck, H., 203, *213*
Slatter, W. L., 14, *83*, 321, *329*
Slattery, C. W., 35, 37, 42, 43, *83*, *84*
Sloan, R. E., 352, 354, *364*
Smart, P. H., 447, *464*
Smeets, G. M., 61, 63, *83*
Smith, E. L., 383, *417*
Smith, J. A. B., 445, *461*
Smith, L. M., 381, 382, *414*
Smith, M. B., 266, 281, 309, 310, 311, *328*
Smith, R. M., 44, *79*, *83*
Smithies, O., 110, *115*, 271, *329*, 376, *417*, 484, *508*
Snow, N. S., 428, 433, 448, 449, 457, *460*, *461*, *463*, *464*
Sobina, A., 428, 441, 445, *462*
Sørensen, M., 259, *329*, 333, 334, *364*, 368, *417*
Sørensen, S. P. L., 259, *329*, 333, 334, *364*, 368, *417*
Sokolov, A., 447, *459*
Soma, K., 453, *460*
Sommer, H. H., 219, *254*
Sorm, F., 409, *412*
Soxhlet, F., 220, *254*
Spellacy, J. R., 446, *464*
Spies, J. R., 141, *172*, 193, *214*, 339, *364*
Spik, G., 373, *416*, *417*
Spooner, M. E., 456, *463*
Sreenivasan, A., 453, *460*, *464*
Stadhouders, J., 395, *417*, 429, *464*
Stadtman, T. C., 403, *417*
Stanton, E. K., 13, 14, 16, 21, 29, 37, 73, *83*, 99, *115*, 166, 167, *172*, 437, *464*
Stark, G. R., 22, *83*, 285, *329*
Stauff, J., 317, *329*
Steele, W., 395, *417*
Stein, J., 77, *81*, 219, 225, *253*, 437, 441, *461*

Stein, W. H., 22, *83*, 203, *212*, 243, 249, *252*, 285, *329*
Steinbarger, M. C., 439, 445, *461*
Stenne, P., 431, 433, *464*
Stepanov, V. M., 243, *254*, 426, *464*
Stevens, H., 378, 386, *413*
Stewart, P. S., 381, *417*
Stoll, W. F., 426, *464*
Stone, W. K., 433, *460*
Storch, V., 220, *253*, *254*
Storgårds, T., 246, *253*
Stormont, C., 276, *326*
Storrs, F. C., 219, *254*
Stotz, E., 368, 369, 370, 387, 394, *416*
Strauss, U. P., 44, *83*
Subrahmanyan, V., 453, *460*, *464*
Sud, S. K., 147, *170*
Sullivan, R. A., 13, 14, 16, 21, 29, 37, 73, *83*, 99, *115*, 166, 167, *172*, 225, 246, *252*, 437, *464*
Sundararajan, T. A., 153, *172*
Susi, H., 311, *330*
Sutermeister, E., 453, *464*
Sutherland, B. J., 427, *460*
Svedberg, T., 258, *329*, 332, 347, *364*
Svestkova, V., 409, *412*
Swaisgood, H. E., 14, 16, 17, 19, 21, 32, 48, 49, *83*, 89, 99, 101, 107, *115*, *116*, 120, 139, 151, *172*, 180, 182, 186, 188, 189, *214*
Swaminathan, M., 453, *460*, *464*
Swan, I. D. A., 407, *416*
Swanson, A. M., 120, *171*, 321, *330*, 427, 437, 438, 439, 456, *462*, *464*
Swope, F. C., 380, 381, 383, 391, *412*, *417*, *418*
Symons, L., 90, 95, 97, 98, 100, 113, *114*
Szczesniak, A. S., 455, *464*
Szent-Györgyi, A. G., 22, *80*
Szuchet-Derechin, S., 271, *329*, 338, 346, *364*, 370, 371, 372, 373, 375, 376, *418*

T

Taggart, V. G., 296, 313, *329*, 451, *464*
Takahashi, K., 250, *254*
Takemoto, S., 152, *173*
Talbot, B., 47, 49, *84*, 92, *116*, 182, *214*

Tamamushi, B., 7, 8, *83*
Tamsma, A., 441, 444, *461*, *464*
Tamura, G., *480*
Tanahashi, N., 339, 340, 357, *362*, *364*, 475, *481*
Tanford, C., 43, 44, *84*, 294, 296, 297, 303, 313, 319, *328*, *329*, 373, *415*, 451, *464*
Tang, J., 243, 248, 249, *254*, *482*
Tang, K. I., 248, *254*
Tarassuk, N. P., 19, *84*, 88, 89, 101, *116*, 120, 121, 145, *172*, 178, *213*, *214*, *215*, 271, 322, *327*, *329*, *330*, 399, 400, 401, 402, *413*, *416*, *418*, 441, *464*
Tauber, H., 222, 231, *253*
Taylor, G. C., 477, *481*
Taylor, W. H., 248, *254*
Telka, M., 18, *81*, 154, *171*, 342, *365*
Tessier, H., 13, 21, 61, 63, *82*, *83*, *84*, 110, *115*, 179, 180, 184, 186, *214*, 322, *329*, *330*, 435, 436, 437, 444, 458, *463*, *464*
Theorell, H., 387, 393, 394, *418*
Thoai, N., 153, *172*
Thomas, E. L., 449, *460*
Thomas, M. A. W., 203, *214*
Thompson, M. P., 14, 16, 17, 19, 21, 32, 50, *80*, *81*, *83*, *84*, *85*, 88, 89, 101, 108, *114*, *116*, 118, 119, 120, 121, 122, 123, 124, 125, 126, 128, 130, 133, 135, 136, 137, 138, 139, 141, 143, 145, 147, 148, 151, 152, 154, 155, 157, 158, 160, 162, 164, 166, 168, *169*, *170*, *171*, *172*, 186, 198, *214*, 271, 321, *325*, *330*, 338, 364, 437, 455, *464*, *465*, 471, *481*, 483, 495, *508*
Thymann, M., 110, *114*, 119, 122, 124, 125, 127, 129, 130, *170*, *172*, *173*, 191, 192, *212*, *213*, 273, 275, *328*, 484, 485, 489, 490, 491, 506, *508*
Tilley, J. M. A., 258, 259, 304, *328*, *330*
Timasheff, S. N., 151, *172*, 278, 279, 283, 291, 292, 293, 294, 295, 297, 298, 300, 302, 303, 305, 306, 307, 308, 309, 310, 311, 312, 313, 316, *325*, *326*, *327*, *330*, 378, *418*
Tinkler, F. H., 447, *460*
Tobias, J., 321, *330*

Tomarelli, R. M., 453, *462*
Tombs, M. P., 304, *328*, *330*, 351, *362*
Tonnelat, J., 313, 317, *327*, 368, 374, 375, *416*
Topol, L., 39, *79*
Topper, Y. J., 22, *84*
Touchberry, R. W., 95, *115*, 378, *417*
Townend, R., 141, *171*, 266, 276, 278, 279, 281, 282, 291, 292, 293, 294, 301, 305, 306, 307, 308, 309, 310, 311, 312, 313, 316, *330*
Townley, R. C., 380, *418*
Townsend, F. R., 457, *464*
Tracy, P., 321, *330*
Tramer, J., 395, *418*, 441, *465*
Trautman, J. C., 321, *330*, 438, 439, *464*
Trayer, I. P., 358, 361, *363*, *364*
Treacy, G. B., 98, *115*, 266, 272, *328*, *330*, 477, *481*, 485, 493, 494, 504, 506, *508*
Treece, J. M., 302, *330*, 396, *418*
Trexler, P. C., 447, *464*
Tripp, R. C., 457, *464*
Trout, G. M., 108, *116*, 455, *460*
Trucco, R. E., 248, *254*
Trupin, J., *214*
Tsugo, T., 152, *173*, 409, *418*, 455, *464*
Tuckey, S., 38, *84*
Tuffery, A. A., 376, *411*
Tumerman, L., 444, *464*
Tuppy, H., 247, *253*
Turkington, R. W., 22, *84*

U

Ubbels, J., 431, *464*
Ühlein, E., 317, *329*
Uozumi, M., 396, 397, *418*

V

Vakil, J. R., 386, 406, *413*, *418*
Vanaman, T. C., 339, 340, 343, 344, 347, 350, 358, 360, *362*, *363*, *364*, 407, *412*
van Bruggen, E. F. J., 8, *84*
van Dam, W., 218, *254*
van den Berghe-van Orshoven, M., 288, *329*

AUTHOR INDEX

van der Burg, B., 222, 225, *254*
van der Linde, J. T., 431, *464*
van der Scheer, A. E., 222, 225, *254*
van der Waarden, M., 10, *80*
van Kreveld, A., 63, *84*, 436, *464*
van Maele, A., 408, *418*
van Markwijk, B. W., 14, 16, *82*, 99, *115*, 150, 167, 168, *171*, *172*, 429, *463*
van Minnen, G., 63, *84*, 436, *464*
Vånngård, T., 373, 374, 375, 399, *411*, *412*
van Rooijen, P. J., 16, 17, 20, 45, *80*, *83*, 121, 139, 155, 165, *170*, *172*, 187, 193, 194, 203, 207, 208, 209, *213*
van Slyke, L., 59, *84*
van Wazer, J. R., 62, *84*
van Winkle, Q., 14, 21, 48, *81*, *82*, *83*, 108, *115*, 135, *171*, 321, 322, *327*, *328*, *329*, 437, *462*
Varin, R., 246, *253*
Varrichio, F., 285, *328*
Vassal, L., 429, *464*
Venge, O., 376, *413*
Venkata Rao, S., 453, *464*
Vercauteren, R., 408, *418*
Veringa, H. A., 395, *417*
Verwey, E. J. W., 7, *84*
Vestal, J. H., 440, 441, *462*
Vesterberg, O., 388, 391, *412*
Vettel, H. E., 440, 441, *462*
Vincentelli, J. B., 351, *364*
Vitols, R., 346, 347, *363*
Vlodavets, I., 447, 449, *460*
Voelter, W., 397, *411*
von Hippel, P. H., 5, 6, 10, 14, 21, 32, 33, 37, 48, *84*, 89, 91, 92, 98, 100, *116*, 118, 119, 151, 167, 169, *173*, 176, 200, *214*, *215*, 425, *465*
von Nägeli, C., 7, *84*

W

Wake, R. G., 13, 14, 16, 20, 21, 47, 48, 76, *80*, *82*, *84*, 98, 99, 101, 109, 110, *115*, *116*, 118, 120, 121, 128, 157, *171*, *173*, 177, 179, 180, 182, 184, 188, 189, 190, 191, 194, 195, 196, 197, 200, 202, 203, 204, 205, 206, 207, 208, 209, *212*, *213*, *214*, 238, *251*, 425, *460*, *462*, *464*, 486, 487, 488, 492, 507, *508*
Wakeman, A. J., 105, *115*
Waldt, L. M., 456, *464*
Wallace, A. L. C., 273, *326*, 507, *508*
Walter, M., 34, *85*, *173*, 277, 278, 279, *327*
Ward, W. H., 368, *411*
Warner, R. C., 5, 6, 11, 14, 16, *84*, 118, 156, 157, 160, *173*, 374, 375, 408, *418*
Wasserman, A. E., 450, *465*
Watkins, W. M., 357, *364*, 475, *481*
Watson-Williams, E. J., 127, *170*
Watts, D. C., 398, *412*
Waugh, D. F., 5, 6, 7, 10, 12, 14, 16, 17, 19, 20, 21, 22, 23, 25, 28, 30, 32, 33, 34, 35, 37, 39, 42, 43, 47, 48, 49, 50, 51, 52, 71, 74, 76, 77, *80* *81*, *82*, *83*, *84*, 89, 91, 92, 98, 100, 101, 107, 110, *115*, *116*, 118, 119, 120, 133, 135, 139, 147, 148, 150, 151, 154, 155, 162, 167, 169, *170*, *171*, *172*, *173*, 176, 178, 182, 184, 188, 197, 200, *214*, *215*, 425, 444, *465*
Wauters, G., 374, *416*
Weatherall, D. J., 127, *170*
Webb, B. H., 438, 439, *460*, *465*
Webb, E. C., 89, *114*, 178, *214*, 262, *326*
Weber, I., 374, 375, *418*
Webster, R. G., 486, *508*
Weil, L., 339, 341, 342, 356, *365*
Weinberg, R. M., 118, 125, 128, 130, 155, *172*, 495, *508*
Weinstein, B. R., 108, *116*
Weisberg, S. M., 426, 428, *465*
Weiss, K. W., 335, 339, *362*
Wenner, V. R., 92, 93, 95, 98, *114*
Westberg, N. J., 224, 227, 231, 237, *251*
Wetlaufer, D. B., 341, 345, 346, 347, *365*
Wheelcock, J. V., 185, 208, *215*
Whitaker, R., 10, *84*
White, I. C., 477, *481*
White, J. C. D., 59, 60, 61, *80*, *84*, 427, 428, 436, 445, *460*, *461*, *465*

Whitehead, H. R., 426, *465*
Whitnah, C. H., 10, *84*
Whitney, R. McL., 13, *84*, 102, *115*, 119, 120, *170*, *171*, 321, *330*, 368, 403, *418*
Whittier, E. O., 368, *413*
Wichmann, A., 332, *365*
Wiebenga, E. H., *84*
Wiechers, S. G., 453, *459*
Wiersema, A. K., 374, *418*
Wiesmann, U., 373, *417*, 478, *481*
Wilcox, P. E., 277, *326*, 341, *365*
Williams, D. E., 507, *508*
Williams, D. L., 285, *328*
Williams, J. W., 305, 306, *325*
Wilson, H. K., 21, 50, *82*, 186, 187, 189, *214*, 440, 441, *463*, *465*
Wilson, J. B., 18, *81*, 155, *170*
Windle, J. J., 374, *418*
Wishnia, A., 303, *330*, 374, *418*
Wissmann, H., 20, *82*, 147, *173*, 200, 202, 208, 209, *214*
Witnah, C. H., 120, *171*
Wix, P., 450, *465*
Woerner, F., 394, *418*
Wohl, R. C., 145, *170*
Wolcott, J. M., 150, *171*, 441, *460*
Wolf, S., 248, *254*
Wondolowski, M. V., 472, *481*
Woodbine, M., 450, *465*
Woodward, C., 225, *251*
Wortmann, A., 472, *481*
Woychik, J. H., 16, 17, 20, 21, 47, *84*, 119, *173*, 183, 187, 188, 189, 190, 191, 193, 194, 195, 196, 201, 204, 207, 208, *213*, *215*, 472, *481*
Wrenshall, C. L., 454, *465*
Wright, N. C., 436, 438, *461*, *465*
Wright, R. C., 395, *418*, 441, *465*
Wyman, J., 8, *82*

Y

Yaguchi, M., 20, 48, *81*, *84*, 178, 181, 185, 193, 196, 206, 207, *213*, *215*, 271, 322, *329*, *330*, 399, 400, 402, *413*, *418*, 437, *464*
Yamamoto, T., 250, *254*
Yamauchi, K., 152, *173*
Yamazaki, I., 399, *416*
Yanari, S. S., 245, *251*
Yap, W. T., 43, *83*
Yasunobu, K. T., 341, *365*
Yon, J., 151, *170*

Z

Zahler, P., 76, *79*, 176, 199, *212*, 238, 239, 241, *253*, 424, 442, *459*
Zbikowski, Z., 441, 445, *462*
Zehren, V. F., 437, *462*
Zehren, V. L., 437, *462*
Ziegler, J., 335, 339, 340, 347, *363*, 370, 371, 374, 375, *413*
Ziemba, J. V., 455, *465*
Zikakis, J. P., 396, *418*
Zimmerman, J. K., 291, *330*
Zittle, C. A., 16, 17, 21, 25, 32, 34, 39, 50, 51, *84*, *85*, 114, *116*, 118, 119, 120, 121, 128, 135, 136, 151, 152, 154, *170*, *172*, *173*, 180, 182, 186, 187, 212, *215*, 246, *251*, 318, 321, 322,,*326*, *330*, 345, 346, *365*, 380, 385, 386, 396, 403, 404, 405, 406, 408, *412*, *418*, 437, 442, 449, 458, *460*, *465*, 472, *481*
Zunz, E., 218, *254*
Zurcher, H., 14, *82*
Zweig, G., 21, *85*, 335, *362*, *365*

SUBJECT INDEX

This index is to be used in conjunction with the Subject Index of Volume I. Further information on a given subject can be ascertained by checking the Subject Index for Volume I.

A

"A protein" of lactose synthetase, 269, 360, 361, 474–476
 N-acetyl lactosamine synthetase, 360
 bovine milk, 361
 chromatography of whey proteins and, 269
 galactosyl acceptor specificity, 361
 human milk, 361, 475
 isolation, 269, 361, 475
 nature of complex with α-lactalbumin, 476
 partial purification, 269
 preparation, 269, 361, 475
β-N-Acetylglycosaminase, 410
 bovine milk, 410
 milk leucocytes, 410
Acid phosphatase, 405–406
 effect of stages of lactation, 406
 properties, 406
Acrylamide gel electrophoresis
 α-casein, 111, 496
 $\alpha_{s,1}$-casein, 130, 496
 β-casein, 111, 130, 495–498, 501, 502
 γ-casein, 103, 111, 501–502
 α-lactalbumin, 273
 β-lactoglobulin, 273
 methods for milk proteins, 486, 495–498, 501–502
 whey proteins, 273
Age gelation of milk, role of enzymes, 441

Aggregation, see also Association
 β-casein, 99, 168, 471
 κ-casein, 441, 473
 α-lactalbumin, 347
 β-lactoglobulin δ, 315, 317, 318
 para-κ-casein, 473
Aldolase, fat globule, 407
Alkaline phosphatase, 403–406
 association with lipids, 403
 chromatography, 405
 composition, 404
 effect of stages of lactation, 406
 fat globule membrane and, 403
 inactivation, 404
  test for efficiency of pasteurization, 405
 isozymes, 405
 molecular weight, 403
 reactivation process, 404
  β-glycerophosphate and, 404
  β-lactoglobulin and, 404
  magnesium and, 404
  sulfhydryl groups and, 404
Alkaline phosphatase assay test, 405
 efficiency of pasteurization and, 405
Allergenicity of milk proteins, in prospect, 478
Amino acid composition
 "B protein" of lactose synthetase, 339, 357
 casein, bovine compared to human, 472
 $\alpha_{s,0}$-casein A, bovine, 471

Amino acid composition (Cont.)
    $\alpha_{s,1}$-casein A, bovine, 139, 161
    $\alpha_{s,1}$-casein B, bovine, 139, 140
    $\alpha_{s,3}$-casein, bovine, 143, 146, 471
    $\alpha_{s,4}$-casein, bovine, 145, 146, 471
    β-casein A, bovine, 103, 104, 164, 165
    β-casein C, bovine, 103, 104, 164, 165
    γ-casein, bovine, 104
    κ-casein, bovine, 185
    κ-casein A, bovine, 192, 193
    κ-casein B, bovine, 192, 193
    fat globule membrane protein, 382
    M_1-glycoproteins, 107, 109, 384
    α-lactalbumin
        A and B of Droughtmaster, 352
        bovine, 339, 340
        guinea pig, 340
        water buffalo, 353
        Zebu, 351
    lactoferrin, 372
        bovine, 372
        human, 372
    β-lactoglobulin
        bovine variants, 277–294
        caprid, 280
        ovine variants, 284
    lactollin, 372
    lactoperoxidase, 372, 394, 395
    lipase, 400
    lysozyme, human milk, 407
    major whey proteins, pig, 474
    prorennin A, 230, 231, 241
    prorennin B, 230
    rennin, 230, 231, 241
    ribonuclease, 386
    serum transferrin, 376
serum transferrin (milk), 376
    trypsin inhibitor, 409
    xanthine oxidase, 372, 397
Amino acid sequence
    $\alpha_{s,1}$-casein, 18, 143
    β-casein, 162, 166
    κ-casein, 187, 211
    α-lactalbumin, 344, 349, 358
    β-lactoglobulin, 208–284
    lysozyme, hen egg white, 344
    rennin, 242, 243
α-Amylase, 402–403
    human colostrum, 403
    properties, 403
    purification, 403
β-Amylase, 402
Apatite-like structures in milk phosphates, 62
Aqueous phase of milk, 58–75
    constituents of, 60–61
Assay centrifugation in study of micelle structure, 24
Association
    α-casein, 99
    α_s-casein, 36, 148
    β-casein, 99, 471
    κ-casein, 99, 177–178
    α-lactalbumin, 347
    β-lactoglobulin, 308, 309, 313
    whole casein, 99

B

"B protein" of lactose synthetase
    amino acid composition, 339, 357
    immunology, 357
    molecular weight, 357
    relation to α-lactalbumin, 339, 357, 474
    UV spectra, 357
Binding
    calcium, 39–47
    α_s-casein, 39
    β-casein, 39
    α-lactalbumin, 346
    β-lactoglobulin, 302–303, 315
    micelle core, 38–47
    whole casein, 40, 43–45
Bovine serum albumin
    from crude "lactalbumin," 333
    whey protein chromatography and, 269
Bovine whey proteins, see also individual proteins
    zone electrophoresis, 489, 503–507
Buffer systems for zone electrophoresis, 273, 478–508
Buffers simulating milk environment, 63, 64, 69

C

Calcium binding
    acid charge reversal and, 47

to casein
  acidic peptide, 45
  electrostatic free energy and, 43–44
  proton release and, 43
  α_s-casein, 39
  β-casein, 39
  solvation of casein and, 42
  whole casein and, 40
Canned sterilized cream, 455
  stability, 456
Carboxyl group
  abnormal
    in β-lactoglobulin, 295, 298, 299, 301, 313
    in casein, 427
  role in β-lactoglobulin polymerization, 308, 313
Carboxypeptidase action
  on $\alpha_{s,1}$-casein, 147, 148
  on α-lactalbumin, 341
  on β-lactoglobulin, 252, 292
Casein, see also Caseins
  bovine
    amino acid composition, comparison with human, 472
    linkage of genes controlling synthesis, 127
    synthesis, linkage of genes, 127
  components of, 13–14
  definition, 88, 119
    solubility in $CaCl_2$, 119
    in urea, 119
  early work, 117–118
  nature of organic phosphate in, 153
  nomenclature, 89, 120, 153
  phosphorylation of, in synthesis, 22
  technology and
    action of rennin, 424–426
    coagulation and syneresis, 427
      calcium ion, 427
      carboxyl groups, 427
      factors affecting rate and extent, 426
      role of functional groups, 427
      sulfhydryl groups, 427
    emulsifying properties, 456
      waterbinding agent, 456
    modification for industrial use, 453
      dicyandiamide and, 453
      formaldehyde and, 453
      proteolytic enzymes and, 453
    reaction with lactose, 443
    rennin-sensitive bond, 426
    syneresis, 425–427
      variation in rate of, 426
    "whole," see "Whole" casein
    zone electrophoresis methods, 587–502
α-Casein
  bovine, see also α_s-Casein
    absorbance, 135
    association, 99
    nature of organic phosphate in, 153
    nitrogen content, 135
    phosphorus content, 135
    sedimentation coefficient, 135
  buffalo, end group analysis, 147
α_3-Casein, bovine
  isolation, 180, 183
  relation to κ-casein, 183
  sedimentation coefficient, 183
  sialic acid content, 183
  solubility, 183
  stabilizing ability, 183
α_s-Casein
  bovine, 177–155
    acidic peptide titration curve, 46
    association, model for, 36
    binding of calcium, 40, 41
    complexes with κ-casein, 32, 177–178
    composition, 139–148
      amino acid analysis, 139–141
      molecular weight, 141, 150
      peptide mapping, 139
    genetic variants, 470
    interaction with κ-casein, 32, 99, 438
    introduction, 117
    isolation, 133–138
      by column chromatography, 137
      method of Schmidt and Payens, 136
        of Thompson and Kiddy, 137
        of Waugh et al., 133
        of Zittle and Custer, 136
    minor $\alpha_{s,0^-}$, $\alpha_{s\ 2^-}$, $\alpha_{s,3^-}$, $\alpha_{s,4^-}$, $\alpha_{s,5^-}$, 145–146, 471
      amino acid composition, 146, 471
      effect of reducing agents, 471
      molecular weight, 147, 150
    nomenclature, 120
      zone electrophoresis numbering system, 120

α_s-Casein (Cont.)
  primary structure
    carboxypeptidase A, action of, 143
    chymotryptic peptides, 143
    properties, 15–17, 148–153
    reaction with κ-casein, 32, 99, 438, 473
    sulfhydryl group and, 438
    solvation at boundary, 42
    stabilization by κ-casein, effect of heat, 442
    zone electrophoresis, 138
  caprid
    genetic polymorphism, 128
    mode of inheritance, 128
  human, 471, 472
  ovine
    genetic polymorphism, 128
    mode of inheritance, 128
$\alpha_{s,1}$-Casein
  bovine, see also α_s-Casein
    breed specificity, 122
    carboxypeptidase A digest, 148
    chymotryptic digest, 142
    electrophoretic mobilities, 121
    gene frequencies and genetic polymorphism, 118, 121–133
    general considerations, 168
    inheritance pattern, 122, 123
    linkages of genes controlling $\alpha_{s,1}$- and β-, 127
    methods of phenotyping, 128–133, 488–490, 492–494
    mode of inheritance, 496, 501
    peptide digest, 121–125, 144
    relationship to α-casein, 144
    separation from κ-casein, 177
    stabilization, 177
    tryptic digest, 140
    zone electrophoresis, 128–133, 488–490, 492–494, 496, 501
  nonmicellar, susceptibility to proteolysis, 467
$\alpha_{s,1}$-Casein A, bovine
  amino acid composition, 141
  characteristics, 152–153
  end group analysis, 147
  evolution, 145
  molecular weight, 148, 151

  occurrence and genetics, 121–125
  peptide maps, 141
    chymotryptic, 141
    peptic, 141
    tryptic, 141
  solubility, 153
$\alpha_{s,1}$-Casein B, bovine
  amino acid composition, 140
  association, 148
  end group analysis, 147
  evolution, 145
  molecular weight, 148, 150
  occurrence and genetics, 121–125
  peptide maps, 140
    chymotryptic, 140
    peptic, 140
    tryptic, 140
  solubility, 153
  zone electrophoresis, 136
$\alpha_{s,1}$-Casein C, bovine
  association, 148
    thermodynamics, 150
  end group analysis, 147
  molecular weight, 148–151
  occurrence and genetics, 121–125
  peptide maps, 140
    chymotryptic, 140
    peptic, 140
    tryptic, 140
$\alpha_{s,1}$-Casein D, bovine
  amino acid composition, 141
  occurrence and genetics, 121–125
$\alpha_{s,2}$-Casein, bovine, method of isolation, 138
$\alpha_{s,3}$-Casein, bovine
  amino acid composition, 143, 146, 471
  method of isolation, 138
$\alpha_{s,4}$-Casein, bovine, amino acid composition, 145, 146, 471
β-Casein
  bovine, 104, 112, 117–155, 471
    action of pepsin, 471
      of rennin, 471–472
    amino acid composition, 103, 104, 165
    association, 99, 471
      effect of pressure, 471
      temperature dependence of, 471
    breed specificity, 125
    calcium(II) binding, 40

composition, 160–166
electrophoretic mobilities, 121
end group analysis, 160–164
gene frequencies, 123, 125
general considerations, 168
genetic polymorphism, 121–133, 181, 470
genetically linked with γ-casein, 102
inheritance pattern, 123
introduction, 117
in κ-casein preparations, 184
linkage of genes controlling synthesis $\alpha_{s,1}$- and β-, 127
methods of isolation, 156–160
  alcohol fractionation, 156
  chromatography, 158
  Ribadeau–Dumas method, 157
  Thompson method, 158
  urea fractionation, 156
  Warner method, 156
methods of phenotyping, 128–133
  paper electrophoresis, 128
  polyacrylamide gel electrophoresis, 130
  starch-gel electrophoresis, 128, 129
micelles from mixtures with κ-casein, 34
nature of organic phosphate in, 153
nitrogen content, 157
nomenclature, 155
nonmicellar, susceptibility to proteolysis, 467
occurrence of phenotypes, 126, 127
phosphate groups, 471
phosphorus content, 157
physical properties, 166–168
  aggregation behavior, 168
  molecular weight, 18, 167
properties, brief summary, 16, 17, 18
sedimentation coefficient, 167
sedimentation patterns, 168
solubility, 19, 153
structure, 160–166
zone electrophoresis, 112, 161, 163, 488–502
buffalo
  end group analysis, 147
  genetic polymorphism, 128
  mode of inheritance, 128
human, 471

ovine
  polymorphism, 128
  mode of inheritance, 128
β-Casein A, bovine
  aggregation behavior, 165
  amino acid composition, 164
  discovery, 155
  end group analysis, 162
  molecular weight, 162
  occurrence, 125
  zone electrophoresis, 129
  methods, 495
  variants A^1–A^3, 497–499, 502
β-Casein A^1, bovine, 102, 104, 111, 133, 497–499, 502
β-Casein A^2, bovine, 102, 104, 111, 133, 497–499, 502
β-Casein A^3, bovine, 102, 111, 133, 497–499, 502
β-Casein B, bovine
  aggregation behavior, 166
  amino acid composition, 166
  comparison of Jersey and Zebu, 166
  discovery, 155
  end group analysis, 162
  molecular weight, 167
  occurrence, 125
  zone electrophoresis, 128–133, 495, 499
β-Casein C, bovine
  aggregation behavior, 168
  discovery, 155
  end group analysis, 162
  molecular weight, 162–163
  occurrence, 125
  zone electrophoresis, 129, 495, 499
β-Casein D
  bovine
    occurrence, 125
    phenotyping, 155
    zone electrophoresis, 133, 499
  Zebu, amino acid composition, 166
γ-Casein
  bovine, 101–104, 473
  amino acid composition, 104
  compared with IgG, 102
  definition, 101
  genetic variants, 102
  genetically linked with β-casein, 102
  genetics, effect in micelle structure, 473

γ-Casein (Cont.)
  bovine (Cont.)
    isolation, 101
    phosphorus content, 101
    zone electrophoresis methods, 488, 493, 501–502
  buffalo, end group analysis, 147
κ-Casein
  bovine
    action of rennin (chymosin) on, 198–212
    aggregation, 414, 472, 473
    amino acid sequence (partial), 187, 211
    association, 99
      with $α_s$-casein, 33, 100, 177–178, 438
    attacks by rennin (chymosin), 175–215
    carbohydrate content, 194–195
    chemical composition, 185–188
      amino acid, 185
      carbohydrate, 185
      galactose, in, 185
      galactosamine in, 185
      phosphorus, 186
      sialic acid, 186
    complexes with α-casein, 32, 100, 177–178, 438
    conclusion, 197
    cystine content, 188
    dissociation, 188
    distribution
      between breeds, 191, 193
      in casein micelle, 458
    disulfide bonding, 99, 188
    early work, 175–178
    effects of disulfide cleavage, 188–189
      of modification on action of rennin, 472
      of presence of $α_{s,1}$-casein, 473
      of presence of β-casein, 473
    fractionation of reduced, 194
    general properties, 175–198
    genetic polymorphism, 118, 122, 185, 470
    genetic variants, 118, 122, 190, 192, 470
    genetics, 123, 127, 192
    heterogeneity, 189–197
      from individual cows, 190
      variation in carbohydrate, 194–195
      in para-κ-casein, 196
    homogeneity check, 187
    interaction with β-lactoglobulin, 321–322, 437
    isolation, 178–185
      alcohol fractionation, 179, 180
      Cheeseman method, 181, 184
      Craven and Gehrke method, 181, 184
      gel filtration, 181, 184
      ion-exchange chromatography, 181, 184
      Morr method, 180, 184
      trichloracetic acid–urea, 180, 182
      urea–sulfuric acid, 180, 182
        artifacts, 183
        stabilizing ability, 183
    macropeptide
      amino acid composition, 192, 193
      heterogeneity, 195
    method of phenotyping, gel electrophoresis, 128–133, 488, 490, 492–494
    micelle formation, 128–133
      effect of SH groups, 473
    micelle from mixtures with β-casein, 34
    micelle stabilizing ability, 177, 187
    molecular weight, 187
    nature of rennin-sensitive linkage, 201–204, 426, 472
      lithium borohydride and, 202
    para-κ-casein variation, 196, 205–207
    penetration into casein micelle, 57
    phosphorus content, 185, 186
    position on micelle, 197
    primary effect of rennin (chymosin), 201
    primary structure, 197
    products of action by rennin, 204–209
      insoluble, 205–207
      soluble, 205, 207–209
        amino acid composition, 208–209
        carbohydrate composition, 208
        molecular weight, 209
    properties, 16–17, 19–20, 175–198
    "protective colloid" theory, 176
    reaction with $α_s$-casein, 32, 100, 177–178, 438

SUBJECT INDEX 535

sulfhydryl groups, 438
relation to α_3-casein, 183
  sialic acid content, 183
  sedimentation coefficient, 183
  solubility, 183
  stabilizing ability, 183
rennin-sensitive bond, 201–204, 426, 472
resolved from α-casein, 177
role
  as lipase, 178
  as micelle stabilizer, 178
  in micelle formation, 49
  of sulfhydryl groups and micelle formation, 473
sialic acid content, 185, 186
solubility, 153
solubilization of $\alpha_{s,1}$-, 151
stability of micelles from, 51
stabilization of α_s-casein, effect of heat, 442
starch-gel electrophoresis, 179
structure, 189, 471
sulfhydryl groups, 189, 427, 473
summary of properties of, 19–20
variation between individual cows, 190
zone electrophoresis methods, 128–133, 488–490, 492–494
caprid
  compared to bovine, 472
  occurrence, 178
colostrum, carbohydrate content, 195
human, 471
  carbohydrate linkage, 472
  occurrence, 178
ovine
  composition, 472
  occurrence, 178
  rennin-sensitive bond, 472
porcine, rennin action, 472
reduced
  dissociation, 188
  molecular weight, 189
κ-Casein A, bovine, 192, 193
  amino acid composition, 192, 193
  mode of inheritance, 123
κ-Casein B, bovine
  amino acid composition, 192, 193
  mode of inheritance, 123

Casein manufacture, 446–447
  acid precipitation, 446
  rennin coagulation, 446–447
  research aspects, 448–449
    improving flavor, 449
    increasing yield, 448
Casein micelles
  abbreviations, 6
  age thickening in condensed milk, 443
  bovine
    electron microscopy, 10–12, 472
    ionic environment and inorganic colloid, 61
    species differences, 472
  from α_s- and κ-casein, 51
    effect of ionic strength, 51
    stability, 51
  changes with heat, 437
  charged groups, 441
  echidna, species differences, 472
  effect of heat, 440–441
  equilibrium systems, 50–58
    free energy, 55–56
    size transformations, 54
  formation and structure, 3–85
  hysteresis effects, 69–73
  introduction, 4
  natural, *see* Natural casein micelles
  path dependencies in formation, 51–52
  penetration of κ-casein, 57
  pig, electron microscopy, 472
  in prospect, 470–473
  protective effect and
    Cherbuliez and Baudet, 5
    Linderstrøm-Lang, 4
    Von Hippel and Waugh, 5
  rennin coagulation and, 75–79
  role in minor caseins in, 74
  sketch of model, 30
  structure, 23–25
    micelles from mixtures of α_s- and κ-casein, 23
  terms and nonstandard abbreviations, 6
α_s-κ-Casein micelles
  calcium requirements, 31
  coat and core formation, 31
Casein monomers
  interaction of, 14
  properties of, 13–14

Casein typing
  acrylamide-gel electrophoresis, 111, 501–502
  zone electrophoresis
    in urea, 109–112, 487, 490–502
α_s-κ-Caseinate micelle systems
  conclusions, 29–34
  developing model of micelle, formation and structure, 29
  experimental results, 29–34
  stability
    to close approach, 29
    of final states, 29
Caseins, *see also* Casein; Casein, bovine
  controversal components, 101–109
  effect micelles on properties, 467
  general remarks, 21
  minor, role in casein micelles, 74
  nomenclature, 88, 89, 117
    ADSA, 88
  primary structure, 120
  properties, general summary, 15–17
  in prospect, 470–473
  relation to gastric proteolytic enzymes, 251
  rennin (chymosin) and, 1
Catalase, distribution, 408
Ceruloplasmin, 410
Cheese
  incorporating whey solids, 433
  manufacturing processes, 422–434
    association problems, 424
  variety, 423
    difference in state of proteins, 424
Cheese curd, properties influencing mature cheese, 426
Cheese manufacture
  addition of salt, 422–424
  adjustment of composition, 423
  behavior of milk proteins, 423
  coagulation, 432
  coagulum firmness, 423
    calcium ions, 428
    casein content, 428
    casein fractions, 428
    mastitic milk, 428
    pH, 428
  continuous, Edam type, work of Ubbels and Van der Linde, 431
  cottage cheese, 432

Dutch type, 432
  work of Kozen and Rodionova, 432
  effects of casein content, 428
    of chloride content, 428
    of colloidal calcium phosphate, 428
    of homogenization, 429
    of interaction between β-lactoglobulin and casein, 429
    of ionized calcium, 428
    of mastitic milk, 428
    of milk processing, 428
    of pasteurizing, 429
    of phosphate, 428
    of potassium, 428
    of temperature, 428
    of time of coagulation, 428
    of variation in milk composition, 427
  expulsion of whey, 423
  fusion of curd particles, 423
  maturing, 423
  mechanization of cheddar cheese, 430
    Berridge method, 431
    Czulak method, 430–431
  new processes, 430–432
  pasteurization, 423
  problems associated with, 424
  research and technology, 430–434
Chymosin, 218, *see also* Rennin
  action on κ-casein, 198–212
  attack on κ-casein, 175–215
  nomenclature, 218
  origin of term, 2
Chymotrypsin
  action on α_s-caseins, 143
  on $\alpha_{s,1}$-casein, 143
  on $\alpha_{s,1}$-casein A, bovine, 141
  on $\alpha_{s,1}$-casein B, bovine, 140
  on $\alpha_{s,1}$-casein C, bovine, 140
  on β-lactoglobulin B, 283
Circular dichroism (CD)
  α-lactalbumin, 350
  β-lactoglobulin, 310, 312, 316
Citrate, content in milk, 61
Coagulum (curd tension), 426
  firmness, 426
Colloidal association products, examples of, 8
Colloidal particles of milk, 58–75
  constituents of, 59–60
  inorganic, 62

Colloidal phosphate-free milk, comparison with skimmilk, 69
Colloidal systems
  metastability of, 9
  particle composition changes, in going from surface to center, 9
Colostrum
  cow
    ceruloplasmin, 410
    β-glucuronidase, 410
    glycoprotein-a, 384
    lactollin, 376–378
    serum transferrin (milk), 376
  human, α-amylase, 403
Complexes of α_s- and κ-caseins, 32, 100, 177–178, 438
"Component 3," 106–108
  properties, 106–107
    composition, 106–107
    mobility, 106
    molecular weight, 107
"Component 5," 106–108
  properties, 106, 107
    composition, 106–107
    mobility, 106
    molecular weight, 107
"Component 8," 106–108
  properties, 106, 107
    composition, 106, 107
    mobility, 106
    molecular weight, 107
Concentrated milk
  evaporated, see Evaporated milk
  frozen, 443, 444
    effect of addition of salts, 444
  interaction of casein with β-lactoglobulin
    chemical nature of, 437
    physical nature of, 437
  milk powder, 434–438, see also Milk powder
  sweetened condensed milk, 434–438, see also Sweetened condensed milk
Conformation
  β-lactoglobulin
    effect of pH, 305–315
    sulfhydryl groups blocked, 291
    in urea, 319
    without urea, 319
  lysozyme, hen egg white, 358

β-Conformation, β-lactoglobulin, 310, 311, 316
Conformational changes
  α-lactalbumin, 348–351
  β-lactoglobulin, 297, 313, 320
  Tanford theory, 297
Constituent levels in milk, total and environmental, 60, 61
Cooperative interactions, 68
Coprecipitate manufacture, 447–448
  addition of alginates, 448
  improving flavor, 449
  increasing yield of, 449
  optimum conditions, 451
  properties, 448
  research aspects, 448–449
Covalent bonds, role in whey protein manufacture, 451
Crystalline albumins, isolation from cow milk, 332
"Crystalline insoluble substance"
  early α-lactalbumin studies, 333
  equivalence to α-lactalbumin, 333
Cyanogum, in zone electrophoresis, 495–498, 507

D

Dairy produce, statistics, 422
Denaturation of β-lactoglobulin, 316–321
  kinetics, 320–321
Dissociation
  reduced κ-casein, 188–189
  β-lactoglobulin, 305–315
Disulfide bridges
  in κ-casein, 93, 188
  in β-lactoglobulin, 285–293, 322
  location in α-lactalbumin, 359
  in β-lactoglobulin, 285–293
  rennin, 243
Disulfide cleavage, effects of, in κ-casein, 188–189

E

Electron microscopy
  casein micelles, 10–12, 472

Electron microscopy (Cont.)
  casein micelles (Cont.)
    bovine, 10–12, 472
    pig, 472
  fat globule membrane, 381
  rennin coagulation, 77
Electrophoresis, see specific types, and individual proteins
Enzymes
  effect on flavor of cheese, 367
  on stability of milk, 367
  in milk, 385–411, 477
    β-N-acetylglucosaminase, 410
    aldolase, 407
    amylases, 402–403
    catalase, 408
    esterases, 399–402
    β-glucuronidase, 410
    lactic dehydrogenase, 411
    lactoperoxidase, 387–395
    lactose synthetase, 356–361, 409–410, 474–476
    lipases, 399–402
    lysozyme, 406–407, 477
    lysylacrylamidase, 410
    malic dehydrogenase, 411
    α-mannosidase, 410
    nucleases, 385–386
    phosphatases, 403–406
    protease inhibitor, 408–409
    trypsin inhibitor, 408–409
    sulfhydryl oxidase, 410
    xanthine oxidase, 396–399
  role in age gelation, 441
Esterases, cow
  aliphatic esterase (lipase), 399–402
  aryl esterase, 402
  cholinesterase, 402
Evaporated milk, 438–442
  age gelation, influence of method of manufacture, 440–441
  effects of heat, 439–441
    casein micelles and, 440, 441
    citrate and, 439
    milk solid level and, 439
    pH and, 439
    phosphate and, 439
    viscosity and, 439
  forewarming, relation to stability, 438
  stability, complex between whey proteins and casein, 437
Evolutionary significance
  caseins, 145
  lactalbumin, 477
  lysozyme, 477
  rennin, 251

F

Fat globule membrane proteins, 380–383, 455
  amino acid composition, 382
  content in cream, 380
  electron microscopy, 381
  electrophoresis, 382
  enzymes, 380
  fractionation, 382
  glycoprotein content, 383
    composition, 383
  hexosamine content, 382
  iron distribution, 382
  isolation, 378, 383, 455
  milk homogenization and, 455
    characteristics, related to, 455
  mineral composition, 383
  model, 382
  molecular weights, 382
  morphology, 381
  mucoprotein, 383
    composition, 383
    electron microscopy, 381
    electrophoresis, 383
    immunology, 383
    molecular weight, 383
  properties, 380
  ribonucleic acid content, 382
Fat-protein interactions, 455
  homogenization, 455

G

Gastric juice, mammals, proteolytic activity, 250
Gene linkage
  $\alpha_{s,1}$- and β-caseins, 125–127
  β- and γ-caseins, 102
  κ-casein, 127

SUBJECT INDEX

Genetic polymorphism
  bovine, 118, 121, 124–127
  buffalo, 128
  caprid, 128
  α-casein, 470
  α_s-casein, 127
  $\alpha_{s,1}$-casein, bovine, 118, 121
  $\alpha_{s,1}$-casein A, B, C, 121
  β-casein, 470
  κ-casein, 118, 122, 185, 190, 192, 470
  heat stability and, 435
  α-lactalbumin, 351–352
  β-lactoglobulin, 271–278, 435
  ovine, 128
  prorennin, 227
  serum transferrin (milk), 376
Genetics
  caseins, 102, 121–128, 190–191, 470
  γ-caseins in micelle structure, 473
β-Glucuronidase, 410
  occurrence, 410
Glycoprotein,
  minor, 104–109
  M-1 type, 106–109, 384
    amino acid composition, 107, 109, 384
    properties, 106–108
      carbohydrate content, 384
      comparison with from milk and serum, 384
      composition, 106, 107
      electrophoresis, 384
      mobility, 106
      molecular weight, 107, 384
  M-2 type, 108
    properties, 108
Glycoprotein-a, bovine, 109, 383–384
  isolation, 383
  properties, 106, 107, 109, 383
    composition, 106, 107, 109
    electrophoresis, 106, 109, 383
    hexose content, 383
    molecular weight, 107, 109, 383
  "secretory piece," 384

H

Heat
  effect on casein micelles, 439
  inorganic constituents of milk and, 436, 439
  interaction β-lactoglobulin and κ-casein and, 321–323, 437
  pH and, 435, 439
Heat-stability of milk
  effect of calcium(II), 436
    of genetic variants of β-lactoglobulin and, 435
    of magnesium(II) and, 436
    of phosphate, 436
  serum albumin and, 436
  subclinical mastitis and, 436
α-Helix
  in α-lactalbumin, 349
  in β-lactoglobulin, 310, 316
High-fat powders, 456
  butter powder, 456–457
  powdered shortening, 456
Historical notes
  casein, 117–118
  α-lactalbumin, 332
  β-lactoglobulin, 258, 332
  micelles, 7
  rennin, 217–218
Homogenization, 455
  casein and whey proteins, 455
  effect of calcium ion, 455
  fat-protein interactions, 455
  relation to fat globule membrane protein, 455

I

Imidazole group, β-lactoglobulin, 313
Immunodiffusion analysis
  lactoferrin, 373
  lactoperoxidase, 373
Immunoglobulins
  in milk, 479
  relation of secretory proteins to blood proteins, 479
Immunology
  "B protein" of lactose synthetase, 357
  α-lactalbumin, 352, 357
  lactoferrin
    bovine, 372
    human, 372
  lactoperoxidase, 388, 391, 396
  ribonuclease, 386
  serum transferrin (milk), 375

Infrared spectroscopy, β-lactoglobulin, 311
Ion binding, micelle core, 38–47
Ionic calcium, concentration in milk, 63
Iron-binding protein, *see also* Lactoferrin, Transferrin
  from cow milk, 371
  iron-free, 370
  from quokka, 375
  from rabbit milk, 371
  from rat milk, 370

K

Kininogen, chromatography, 384

L

"Lactalbumin," early studies, 333
α-Lactalbumin
  caprid
    amino acid composition, 340
      comparison with bovine B and guinea pig, 340
    properties, 353
  general and bovine, 331–365
    absorptivity (bovine), 341
    amino acid composition, 339–340
      comparison of bovine, goat, and guinea pig, 340
      comparison with "B protein" of lactose synthetase, 339
    amino acid sequence, 344
      homology with hen egg white lysozyme, 343, 344, 359
    association
      aggregation and, 347
      in acid solution, 347
      in alkaline solution, 347
      Kronman hypothesis, 348
      with minor lactalbumins, 476
    "B protein" of lactose synthetase and, 339, 410, 474
    binding of anions, 346
    biological function, 356–361, 474
      enzymatic, 356–361
    comparison with lysozyme, 358
      conformation, 358
      immunology, 358
    composition, 339–344
    conformation and conformational changes, 348–351
      circular dichroism, 350
      comparison of modified and native protein, 349
      comparison with lysozyme, 350
      electrophoretic behavior, 349
      ORD, 349
      sedimentation velocity, 349
      solvent perturbation difference spectrophotometry, 349
      UV difference spectra, 349
      UV fluorescence, 349
    crystals and crystallization, 334–338, 355–356
    early studies, 332
    electrophoretic behavior, 345
      heterogeneity, 345
    end group analysis, 341
      action of carboxypeptidase, 341
    equivalence with "crystalline insoluble substance," 333
      crystal form, 334
      electrophoretic mobility, 334
      moving-boundary electrophoresis, 333
      sedimentation, 333
      ultraviolet spectrum, 334
    evolutionary changes, 358, 477
    genetic polymorphism, 351–352
    α-helix, 349
    immunological comparison of various animals, 352
    isoelectric point, 346
    isolation, 259–271, 334–338
      chromatographic methods, 266, 338
      method of Armstrong, McKenzie, and Sawyer, 337
      of Aschaffenburg, 337
      of Aschaffenburg and Drewry, 336
      of Dautrevaux and Bleumink, 337
      of Gordon, Semmett, and Zeigler, 334–335
      of Robbins and Kronman, 336
      of Sørensen and Sørensen, 334
      of Zweig and Block, 335

milk
  bovine, occurrence in, 477
  human, occurrence in, 477
  pig, occurrence in, 477
molecular weight, 339, 346, 347
  comparison of methods, 347
nature of complex with "A protein," 476
nomenclature, 332
primary structure, 342–346
  action of carboxypeptidase, 342
  preparation of derivatives, 342–346
in prospect, 474–477
purification, 269, 334–338
"satellite α-lactalbumin," 345
  amino acid composition, 345
  electrophoretic behavior, 345
  hexosamine content, 345
similarity to "B protein" of lactose synthetase, 357
  amino acid composition, 357
  immunology, 357
  molecular weight, 357
  UV spectra, 357
solubility, 346
structure, 342–345
  homology with hen egg white lysozyme, 343
  X-ray diffraction, 343, 356
  three-dimensional model, 350
ultracentrifuge behavior, 346
X-ray diffraction, 343, 356
zone electrophoresis methods, 492, 503, 506
guinea pig, 352
  amino acid composition, comparison with bovine B and caprid, 340
  immunology, 354
  isolation, 354
  molecular weight, 354
human, 477
  properties, 354
kangaroo, 477
  zone electrophoresis method, 506
ovine, 352, 354
  zone electrophoresis methods, 504, 505
water buffalo, 352, 353
  amino acid composition, 353
  molecular weight, 353

α-Lactalbumin A$_{\text{Droughtmaster}}$
  amino acid composition, 352
  genetic variants, 352
α-Lactalbumin A, Zebu, amino acid analysis, 351–352
α-Lactalbumin B, bovine
  amino acid composition, 340
  comparison with goat and guinea pig, 340
Lactic dehydrogenase, 411
Lactoferrin, 368–375, *see also* Iron-binding protein, "Red protein"
  amino acid composition, 372
  bacterial inhibitor, 374
  bovine
    amino acid composition, 372
    comparison with bovine transferrin (serum), 372
    with human lactoferrin, 372
    electrophoretic pattern, 371
    genetic polymorphism, 371
    iron affinity, 375
      comparison with human lactoferrin and human serum transferrin, 375
    iron complex, 375
    iron content, 371
    molecular weight, 371
  diseased mammary glands, in autoantigen, 478
  distribution in milk, 371
  function, 478
  human, 372–375
    content
      amino acid composition, 372
      carbohydrate, 372
      iron, 373
    immunology, 372
    iron affinity, 375
      comparison with human serum transferrin and bovine lactoferrin, 375
    iron complex, 375
    molecular weight, 373
    peptide maps, 372
    physical properties, 373
    zone electrophoresis method, 503
  human milk, 368
  iron content, 374

Lactoferrin *(Cont.)*
  isolation
    from acid precipitated casein, 370
    from whey fraction, 368
  nomenclature, 368
  occurrence, 373, 478
    in bronchial secretions, 373
    in colon tissues, 373
    in genital tract, 373
    in intestine, 373
    in kidneys, 373
    in leucocytes, 373
    in mammary glands, 373, 478
    in saliva, 373
    in semen, 373
    in stomach, 373
  structure, 374
  synthesis, 373
    in mammary gland, 374
β-Lactoglobulin
  caprid, 474
    amino acid composition, 280
    detection, 276
    isolation, 266
    ORD, 302
    pH and conformational changes, 302
    pH titration, 301
      ε-amino group, 301
      carboxyl group, 301
    pH titration curve, 301
    tyrosine residues in, 293
  general and bovine, 257–330, 474
    aggregation, 315
      binding of ions, 315
      chromatography, 315
      electrophoresis, 315
      ORD, 315
      oxidation of SH groups, 315
      ultracentrifuge measurements, 315
    amino acid composition, 277–294
    amino acid sequence, 280–284
    association (polymerization), role of carboxyl groups, 308, 313
    binding of ions and molecules, 302–303
    chromatography, 266–271
    circular dichroism, 310, 312, 316
    comparison of amino acid composition of bovine, caprid, and ovine, 280

    of sulfhydryl groups in bovine A, B, C variants, 292
    conformation, 296, 299, 302, 304–316
      blockage of sulfhydryl groups, 288–291
      effects of pH, 305–316
      ORD, 310, 315–317
      pH titration curves, 310
    conformation changes, Tanford theory, 296–297
    denaturation, 316–321
      effect of detergents, 285–288, 318
        of guanidine hydrochloride, 285, 319
        of temperature, 316–318
        of urea, 285–286, 319
    dissociation
      molecular weight, 305–316
      pH effect, 305–315
    disulfide groups, 289
    early physico-chemical studies, 258
    early studies, 258, 332
    effect of heat (temperature), 316–318
      on gel electrophoresis, 317
      on light scattering, 317
      on moving boundary electrophoresis, 316
      on optical rotatory dispersion, 317
      on sulfhydryl groups, 437, 451
      on UV difference spectra, 317
    of interaction with casein on cheese manufacture, 429
    electrochemical properties, 294–304
    electrophoresis, 271–274
    electrophoretic patterns, 268, 272
    end group analysis, 277
    genetic polymorphism, 274–277
    heat stability patterns of milk associated with genetic variants of, 435
    interaction with κ-casein, 321–322, 437
      effect of calcium(II), 322
      heat stability, 322
      light scattering, 321
      mastitis and, 321

SUBJECT INDEX 543

moving boundary electrophoresis
   and, 322
rennin clotting, 321
sulfhydryl groups, 321
zonal ultracentrifugation, 322
introduction, 258
isoionic points, 303
isolation, 260, 263
isolation methods,
   Aschaffenburg and Drewry, 259
   early methods, 259
   Fox et al., 262
   McKenzie et al., 262–266
location of tryptophan groups, 293
   of tyrosine groups, 293
modification with carboxypeptidase,
   451–452
molecular size, 304–309
   pH 1.8–3.5, 305–307
   pH 3.8–5.4, 307–310
   pH 5.4–9.2, 313–316
moving boundary electrophoresis, 304
   Gilbert's theory, 304
nomenclature, 258
nonruminant milk, 276
occurrence, 258
ORD, 310, 316
partial specific volume, 323
peptide studies, 280–284
pH and conformational change, 296,
   299
   Tanford theory of ionization, 296
   linked transitions, 296
pH titration curves, 294–303
in prospects, 474–477
purification, 270
reaction with DTNB, 286, 287
role of denaturation in whey protein
   manufacture, 451
ruminant, 276, 474
species differences, 274–277
structure, 315
   IR spectroscopy, 311
   ORD, CD, 310, 315
sulfhydryl content, 285
sulfhydryl groups, 321
   with cystamine hydrochloride, 289
   with DTNB, 287–288
   with ferricyanide, 285
   with N-ethyl maleimide, 285

with PHMB, 289
reactivity, 285–293
   effect of carboxypeptidase, 292
   of guanidine hydrochloride,
      285
   of sodium dodecylsulfate, 285,
      286, 288
   of sodium tetrathionate, 289
   of urea, 285, 319
summary, 324
titration of sulfhydryl groups, 289
tryptophan residues, solvent
   perturbation spectroscopy, 293
tyrosine residues, spectrophotometric
   titration, 293
water in crystals, 323
X-ray diffraction, 323
ovine, zone electrophoresis methods,
   474, 504, 505
porcine, monogastric analog, 277
β-Lactoglobulin A
   bovine
      amino acid analysis, 279
      amino acid sequence, 280–282
      chymotryptic peptides, 282
      conformation, effect of pH, 305–315
      conformational transitions, 311
      detection, 276
      dissociation and pH, 305–310
      effect of pH, 290
         of urea, 319
      kinetics of denaturation, 312
      octamer
         association constants, 309
         thermodynamic parameters, 309
      pH titration curve, 294–296
      separation from
         β-lactoglobulin$_{Droughtmaster}$, 270
      solubility, 264
      structure, 308
      sulfhydryl group reactivity, 314
      tryptic peptides, 281
      zone electrophoresis, 271–273, 489–
         491, 503, 506
      zone electrophoretic pattern, effect
         of pH, 314
   ovine
      amino acid composition, 284
      detection, 276

β-Lactoglobulin A (*Cont.*)
  ovine (*Cont.*)
    isolation, 266
    occurrence, 276
    tryptic peptides, 284
    zone electrophoresis, 273
β-Lactoglobulin AB, bovine, sulfhydryl groups, 317
β-Lactoglobulin B
  bovine
    amino acid composition, 279
      compared with other species, 278
      with other variants, 278
    amino acid sequence, 280–283
    CD, 312
    chymotryptic peptides, 283
    conformation, effect of pH, 305–315
    conformational transitions, 311
    detection, 276
    dissociation and pH, 305, 313
    effect of pH, 290
    occurrence, 274
    ORD, 320
    pH titration curve, 294–296
    solubility, 264
    sulfhydryl group reactivity, 314
    zone electrophoresis, 273, 489–491, 503, 506
    zone electrophoretic pattern, effect of pH, 314
  ovine
    amino acid composition, 284
    detection, 276
    isolation, 266
    occurrence, 274
    tryptic peptides, 284
    zone electrophoresis, 273, 504, 505
β-Lactoglobulin C, bovine
  amino acid composition, 283
  conformations, effect of pH, 310–315
  conformational transitions, 311
  detection, 274
  occurrence, 275
  pH titration, 297–301
    α-amino groups, 297–301
    carboxyl groups, 297–301
    imidazole groups, 297–301
  pH titration curve, 295
  sulfhydryl group reactivity, 314
  tryptic peptides, 283
  zone electrophoresis, 273, 491, 503, 506
  zone electrophoretic patterns, effect of pH, 314
β-Lactoglobulin D, bovine
  amino acid analysis, 283
  chymotryptic peptides, 283
  detection, 275
  gel electrophoresis, 130
  occurrence, 273
  pH titration, 300
  zone electrophoresis, 273, 491, 503, 506
β-Lactoglobulin $_{Droughtmaster}$, bovine
  amino acid composition, 279
  conformation, 308
  detection, 275
  occurrence, 275
  purification, 271
  separation from β-lactoglobulin A, 270
  tryptic peptides, 284
  zone electrophoresis methods, 273, 503, 506
Lactollin, 376–378
  from bovine milk, 376
  amino acid composition, 372, 378
  electrophoresis, 378
  isolation, 378
  molecular weight, 378
Lactoperoxidase, 387–395
  amino acid composition, 272, 394, 395
  bovine, immunodiffusion analysis
    harderian and lacrimal glands, 395
    sublingual, submaxillary, and parotid glands, 395
  bovine lacrimal glands, in, 373
  bovine salivary glands, in, 373
  caprid, immunological comparison to sheep and cow, 396
  carbohydrate content, 394, 395
  chromatography, 388
  content in milk, 394
  electrophoresis, 389, 390
  function, 395

heterogeneity, 388, 391
immunology, 387, 396
iron content, 388, 392, 394
isolation, 369, 387, 388
molecular weight, 392, 394
ovine, immunological comparison to goat and cow, 396
prosthetic group, 395
variation in activity, 394

Lactose
biosynthesis, 356
intolerance, adaptive phenomenon, 479
reaction with casein, 443

Lactose synthetase, 356–357, 474–476
assay, 476
partial purification in chromatography, 269
properties, 409
relation to α-lactalbumin, 410
subunits, 410
"A protein," 269, 360, 410, 474–476
"B protein," 339, 357, 410, 474

Lipase, 399–402
activation, 399
amino acid composition, 400
association with α-casein, 400
with κ-casein, 400
cheese flavors, 399
clarifier slime and, 400
composition, 400
electrophoresis, 400
fractionation, 400
inhibitors, 401
lipoprotein lipase, 402
milk flavors and, 399
molecular weight, 400
physico-chemical measurements, 400
properties
heat stability, 401
inhibitors, 401
sulfhydryl groups, 401
sulfur content, 402

Lysozyme
bovine milk, 406–407
homogeneity, 407
properties, 407
purification, 406
role, antibacterial activity, 406
content in cow milk, 406
in echidna milk, 477
in goat milk, 406
in human milk, 406
in pig milk, 406
evolutionary changes, 477
hen egg white
amino acid sequence, 344
homology with α-lactalbumin, 344, 349, 358
human milk, 477
amino acid composition, 407
biological function, 407
comparison with tissue lysozyme, 407
relation between chemical structure and biological function, 407
X-ray structure, 407
in prospect, 474–477

Lysyl anylamidase, 410

M

Maillard reaction, 443
reaction of casein and lactose, 443
Malic dehydrogenase, 411
α-Mannosidase, 410
Mastitic milk
effect on cheese manufacture, 428
electrophoretic pattern, 100
enzymes, 478
sedimentation patterns, 100
whole casein, 96
chemical composition, 96
comparison with normal, 96
Mastitis
complex between whey proteins and casein and, 437
effect on heat stability of milk, 436, 438, 479
bovine serum albumin and, 479
IgA and, 479
β-lactoglobulin and, 479
pH and, 479
Metastable colloid, 54, 73
stabilization, 54
transformation, 54
Micelle
antibodies in study of, 34
assay centrifugation, 24.
casein, electron microscopy, 12, 472

$\alpha_{s,1}$-casein, susceptibility to
  proteolysis and, 467
β-caseins, susceptibility to proteolysis
  and, 467
effect on properties of caseins and, 467
electron microscopy of, 12, 472
formed from molecules of biological
  origin, 7
    from mixtures of α_s- and κ-caseins,
      23, 24
        assay centrifugation, 24
        studies of Waugh and Noble, 23
    from small polar organic molecules,
      7
McBain definition, 7
types of, 6
Micelle coat, 47–50
  association products of SH-κ-casein
    and, 48
  dephosphorylation of κ-casein and, 50
  κ-casein and formation of, 49
  modification of κ-casein and, 49
  surface activity of monomeric and
    polymeric caseins, 47
Micelle core
  binding of ions, 38–47
  nature of core polymers
    views of Payens, 38
    of Waugh, 35
Micelle distribution in milk simulating
  buffers, 63, 66
Micelle formation
  effect of ionic strength, 52
  with several buffers, 64
Micelle stability
  test of Noble and Waugh, 25
  of Zittle, 25
Micelle stabilization, role of κ-casein, 177
Micelle structure, γ-casein, effect of
  genetics on, 473
Milk
  camel, 474
  cow
    "A protein," 360–361, 474–476
    β-N-acetylglycosaminase, 410
    aldolase, 407
    α-amylase, 402–403
    β-amylase, 403
    catalase, 408
    ceruloplasmin, 410

enzyme distribution, 477–478
esterases, 402
glycoprotein-a, 383–384
kininogen, 384
α-lactalbumin, 331–365
lactoferrin, 368, 371–372
β-lactoglobulin, 257–330
lactollin, 376, 378
lactoperoxidase, 395
lipase, 399–402
lysyl arylaminidase, 410
lysozyme, 406–407
protease inhibitor, 408–409
serum albumin, 378
  amount, 378
  isolation, 378
  properties, 378
serum transferrin (milk), 375–376
sulfhydryl oxidase, 410
trypsin inhibitor, 408–409
xanthine oxidase, 396–399
echidna, 472, 474, 477
  lactose, 477
  lactose synthetase, 477
  lysozyme, 477
    amino acid composition, 477
effects of heat, 435–438, 458
  association-dissociation reactions, 458
  changes in conformation, 458
    in ionic composition, 458
  gelation in storage, 458
  heat stability, 458
  organization of casein micelle, 458
goat
  lactoperoxidase, 396
  lysozyme, 406
guinea pig, 474
human, 471, 472, 474
  "A protein," 361
    carbohydrate composition, 475
    preparation, 475
  antigenic determinant, 479
  α_s-casein, 470–471
  β-casein, 470–471
  γ-casein, 473
  κ-casein, 178, 470–473
  β-glucuronidase, 410
  IgA, 479
  lactoferrin, 368, 372
  lysozyme, 406–407

kangaroo, 474
  whey proteins, 477
    changes during lactation, 477
pig, 472
  major whey protein
      ("β-lactoglobulin"), 474
    amino acid composition, 474
    genetic polymorphyism, 474
    molecular weight, 474
  platypus, 474, 477
    α-lactalbumin, 477
    lactose, 477
    lysozyme, 477
  rabbit
    iron binding protein, 371
    serum transferrin (milk), 376
      comparison with from serum, 376
  rat, iron binding protein, 370
  sheep, lactoperoxidase, 396
Milk-clotting test, 219–221
  Berridge technique, 219
  British Standards Institution test, 219
  experimental performance, 219
  micromethod, 219
Milk environment
  complex ion interactions, 68
  components, 60
  cooperative interactions, 63, 67, 68
    basis of, 67
Milk enzymes, 385–411
  acid phosphatase, 405–406
  β-N-acetylglucosaminase, 410
  aldolase, 407
  alkaline phosphatase, 403–405
  amylase, 402–403
  catalase, 408
  ceruloplasmin, 410
  esterase, 399–402
  β-glucuronidase, 410
  lactic dehydrogenase, 411
  lactoperoxidase, 387–396
  lactose synthetase, 409–410
  lipase, 399–402
  lysozyme, 406–407
  lysyl arylamidase, 410
  malic dehydrogenase, 411
  α-mannosidase, 410
  phosphodiesterase, 386
  protease inhibitor, 408–409
  ribonuclease, 385–386

sulfhydryl oxidase, 410
  trypsin inhibitor, 408–409
  xanthine oxidase, 396–399
Milk powder, 444–445
  alterations in milk protein, 445
  high protein, 452
    properties, 452
    uses, 452
Milk proteins
  allergenic properties, 478
  allergenicity in prospect, 478
  behavior in cheese manufacture, 432
  emulsifying properties, 456
  large scale manufacture, 459
    complex of whey proteins and
        carboxymethyl cellulose, 459
    egg substitutes, 454
    meat substitutes, 454
  minor, 367–418
    complexes with metal ions, 367
    effect of flavor of cheese, 367
    on stability of milk, 367
    in prospect, 477
  in prospect, 469–481
  research and milk technology, 421–465
Milk solids, use of, 434
Milk technology and milk protein
    research, 421–465
Minor glycoproteins, bovine milk,
    104–109, *see also* specific glycoproteins

N

Natural casein micelles
  electron microscopy of, 10–12, 472
  properties of, 10, 57–75
  stability of, 11, 57–75
  total and environmental levels of
      constituents in milk, 60
Nomenclature, 88–89
  casein, 88–89
  α_s-caseins, 120
  $\alpha_{s,0}$-casein, 120
  $\alpha_{s,1}$-casein, 117, 120
  $\alpha_{s,2}$-casein, 120
  $\alpha_{s,3}$-casein, 120
  β-caseins, 153
  chymosin, 2, 218
  α-lactalbumin, 332
  lactoferrin, 368

Nomenclature (Cont.)
  β-lactoglobulin, 258
  micelle, 7
  prochymosin, 2
  prorennin, 2
  "proteose peptone," 89
  rennin, 2, 218
Nucleases, 385–386

O

Optical rotatory dispersion (ORD), see individual proteins

P

Para-κ-casein, 196–197, 205–207
  aggregation, 473
  amino acid composition, 192, 193, 196
Parapepsin, insulin hydrolysis by, 247
Pasteurization
  effect on cheese, 423, 429
    on milk, 429
  efficiency assay test, 405
  test for effective pasteurization chromatography, 405
Pepsin
  action on $\alpha_{s,1}$-casein A, B, C, 140
    on β-casein, 471
  comparison with rennin, 243, 248–251
  configuration pepsinogen and pepsin, 250
  insulin hydrolysis by, 247
Peptide models, action of rennin on, 472
pH titration curves, β-lactoglobulin, 294–303
Phenotyping, see Zone electrophoresis and individual proteins
Phosphate levels in milk, 61
Phosphodiesterase, 386
  from fat globule membrane, 386
  isolation, 386
  properties, 386
Phosphohexose isomerase, 409
Preparation, see Isolation
Prochymosin, 217–254, see also Prorennin
  origin of term, 2
Prorennin (prochymosin), 217–254

activation, 230–236
  pH dependence, 231–234
  possible mechanism of, 234–236
  salt concentration dependence, 234
amino acid composition, 230, 231, 241
chromatography, 227
fractionation, 228
genetic variants, 227
isolation, 222–224
molecular weight, 238
  by amino acid composition, 237, 238
  by gel filtration, 237, 238
  by sedimentation, diffusion, 237, 238
nomenclature, 2
physical chemical properties, 229, 237
  diffusion coefficients, 237
  isoelectric point, 239
  partial specific volume, 237
  sedimentation coefficients, 238
  solubility, 239
sedimentation coefficients, 237
stability, 244
Prorennin B, amino acid composition, 230
Protease, preparations for cheese making, 480
Protease inhibitor
  associated with casein, 408
  properties, 408
Protein malnutrition, 468
Protein modification, special purpose foods, 453–454
  egg substitutes, 454
  meat substitutes, 454
σ-Proteose, 108
"Proteose peptone," 105–109
  isolation, 105
  nomenclature, 89

Q

Quokka, iron-binding proteins, 375

R

"Red protein," see also Lactoferrin
  composition, 371
  molecular weight, 371
  nomenclature, 368
Rennets, quantitative estimation, 480

SUBJECT INDEX

Rennin, see also Chymosin
  action on β-casein, 471
    on κ-casein, 211–212, 424–426
  activity, 221
  amino acid composition, 230, 231, 241
  amino acid sequence, 242, 243
  assay, 218–221
  attack on κ-casein, 175–215
  calculation of Rennin Unit (RU), 220
  casein soluble product analysis, 205
  chromatographic heterogeneity, 227–230
  chromatography, 228
  comparison with pepsin, 248–249
    action on ribonuclease, 248
    amino acid sequences, 243, 249
    common ancestor, 249
    hydrolysis of synthetic peptides, 248
    isoelectric point of proenzyme, 249
    primary structure, 249
  course of reaction, 424
  denaturation with urea, 245
  early studies, 217–218
  effect on κ-casein, 198–211, 472
    on modification of κ-casein, 472
  evolutionary significance, 251
    relation to casein, 251
  formation from prorennin, 230–236
  functional groups concerned, 425–426
  fractionation, 227–228
    A-, B-, and C-rennin, 228
  general proteolytic activity, 248
  insulin hydrolysis by, 247
  isolation, 224–226
  molecular weight, 238–239, 245
    by amino acid composition, 238, 239
    by Archibald method, 239
    by gel filtration, 238, 239
    by sedimentation-diffusion, 238
  nature of substrate, 425
  nomenclature, 2, 218
  pH activity curve, on denatured hemoglobin, 198
  physical chemical properties, 229, 237–239
    diffusion coefficients, 237
    isoelectric point, 239
    sedimentation coefficient, 238
    solubility, 239
  preparation, 224–226
  primary effect on κ-casein, 201
  primary structure, 243
    amino acid sequence, 243
    cystine bridges, 243
    disulfide bridges, 243
  products of action on κ-casein, 204–209
    chemical composition, 204–209
    heterogeneity, 204–209
    insoluble product, 205–207
    origin, 204–209
    soluble product, 207–209
      amino acid composition, 208–209
      carbohydrate composition, 208
      molecular weight, 209
  proteolytic specificity, 248
  rennin-pepsin ratio, 250
  rennin-sensitive bond in κ-casein, 201–204, 426, 472
  ruminants, 250–251
  sedimentation coefficients, 237
  solubility, 240
  stability, 244–245
  tertiary structure, 243–244
  X-ray diffraction, 244
Rennin coagulation
  casein micelles and, 75–79
  coat-core model and, 77
  electron microscopy, 77
  linear polymerization model, 78
Rennin-sensitive linkage, nature in κ-casein, 201–204, 426, 472
Rennin unit (RU)
  comparison, 221, 226
  definition, 220
A-Rennin
  amino acid composition, 242
  fractionation, 228
B-Rennin
  amino acid composition, 230, 241–242
  chromatography, 229
  fractionation, 228–229
Ribonuclease
  bovine, 385–386
    amino acid composition, 386
    carbohydrate content, 386
    immunology, 386
    isolation, 385, 386
  chromatography, 385
  electrophoresis, 389

Ruminants, milk clotting enzymes, 250–251

S

"Satellite α-lactalbumin," 345
  amino acid composition, 345
  hexosamine content, 345
"Secretory piece," glycoprotein-a, 384
Serum albumin
  in cow milk,
    amount, 378
    isolation, 378
    properties, 378
    zone electrophoresis method, 492, 503, 504
  ovine, zone electrophoresis methods, 504, 505
Serum transferrin (milk)
  bovine, 375–376
    amino acid composition, 376
    chemical analysis, 376
    electrophoresis, 375
    genetic polymorphism, 376
    immunology, 375
    isolation, 376
    physical measurements, 376
  rabbit, 376
    N-acetylneuraminic acid content, 376
    amino acid content, 376
    comparison with from serum, 376
    electrophoresis, 376
    molecular weight, 376
SH groups, see Sulfhydryl groups
Sialic acid
  α_3 casein, 183
  κ-casein, 183, 185, 186
  glycoproteins, 108
  β-lactoglobulin, 280
Solubility curves for calcium
  α_s-κ-caseinate mixtures, 25–26
  dip, 25
  peak, 26
  pseudoplateau, 26
Solvation of casein, 42
Specialized products from milk, 452–457
  fat-protein products, 455–457
  modified proteins, 452–455
  special purpose food, 452–455

SS groups, see Disulfide groups
Staining methods in zone electrophoresis, 486–508
Starch-gel electrophoresis, see also Zone electrophoresis
  methods, 128–129, 484–485, 487–494, 499–500, 503–506
  $\alpha_{s,1}$-casein, 128–129
  β-casein, 128–129
  γ-casein, 488, 493
  κ-casein, 128–129, 488–490, 492–494
  immunoglobulins, 503
  α-lactalbumin, 504, 506
  β-lactoglobulin, 489–491, 503–506
  method of preparation, 484–485, 487–494, 499–500, 503–506
  serum albumin, 492, 503, 504
  staining, 486–494, 499–500, 503–506
  transferrin, 503
Sulfhydryl groups
  aggregation involving, 318
  alkaline phosphatase, 404
  in casein, 427
  in α_s-casein, 438
  in κ-casein, 99, 189, 321, 473.
    role in micelle, 473
  comparison in β-lactoglobulin A, B, and C, 292
  in β-lactoglobulin, 285–293, 313, 315, 317, 318, 321
    blockage and conformation, 288–291
    effect of heat, 318, 321–323, 437–438, 451
    location, 285–293
    pH titration, 289
    reactivity, 285–293
  in β-lactoglobulin-κ-casein interaction
    moving boundary electrophoresis and, 321
  lipase, 401
  milk lipase inhibitors, 401
  reaction in β-lactoglobulin, 285–293
    effect of cystamine hydrochloride, 289
    DTNB, 287–288
    guanidine hydrochloride, 285
    N-ethyl maleimide, 285
    ferricyanide, 285
    PHMB, 289

sodium dodecylsulfate, 285, 286, 288, 318
sodium tetrathionate, 289
urea, 285
reactivity in β-lactoglobulin, 292
effect of carboxypeptidase, 292
role in κ-casein micelle formation, 473
Sulfhydryl oxidase, 410
Sweetened condensed milk, 442–443
age thickening, 442
casein fractions, 443
casein micelles, 443
inorganic constituents, 443
temperature, 443
viscosity, 442

T

Transferrin
human
comparison with lactoferrin, human, 373
iron-binding sites, 373
molecular weight, 373
subunits, 373
kangaroo, zone electrophoresis methods, 506
milk, 368–375
rabbit, physical studies, 373
serum, 368–375
isolation
from albumin fraction of milk, 370
human, iron affinity, 375
comparison with human lactoferrin and bovine lactoferrin, 375
Trypsin
action on $\alpha_{s,1}$-casein B and C, 141
on β-lactoglobulin, ovine, 284
on β-lactoglobulin A and B, 281
Trypsin inhibitor, 408–409
amino acid composition, 409
molecular weight, 409
properties, 409
stability of milk, 409
of purified caseins, 409

U

Urea
cyanate as contaminant of, 22, 485

effect of caseins, 102, 110, 151
on β-lactoglobulin, 285–286, 319, 320
on rennin, 245
heavy metals as contaminant of, 22
in zone electrophoresis, 485, 487, 489, 491–495, 499–502
UV spectra, see individual proteins

W

Whey protein(s), see individual proteins
bovine
fractionation, 264
starch-gel electrophoresis, 268, 503, 506
methods of zone electrophoresis 271–274, 489–492, 502–507
porcine, zone electrophoresis methods, 273, 504
Whey protein aggregation, role of covalent bonds and surface or charge effects, 451
Whey protein manufacture, 449–450
concentration of whole whey, 450
drying of whole whey, 450
fermentation, 450
precipitation condition, 451
protein modification, enzyme treatment, 451
removal of whey proteins, 450
research aspects, 451–452
"Whole" casein, 87–116
bovine
chemical composition, 96, 97
N-acetylneuraminic acid, 96
comparison with mastitic milk, 96
hexosamine content, 96, 97
hexose content, 96, 97
nitrogen content, 96, 97
phosphorus content, 96, 97
comparison of product of various preparations, 90
electrophoretic patterns, 98, 113
isolation, 89–94
pitfalls in, 113
isolation methods, 90–94
acid precipitation, 90–91
ammonium sulfate precipitation, 93

"Whole" casein (*Cont.*)
  bovine (*Cont.*)
    centrifugation with calcium, 91–92
      without calcium, 93
    properties, 94–101
    sedimentation, 98–99
    solubility, 94
    yield, 95–96
      mastitic milk, 96
      normal milk, 95
    zone electrophoresis, 98, 487–488, 492–495, 497, 499, 501
  calcium binding and, 39
  isolation, method of, 90–94
    acid precipitation, 90–91
    ammonium sulfate precipitation, 93
    centrifugation, 91–93
  summary, 113

X

Xanthine oxidase, 396–399
  amino acid composition, 372, 397
  associated with fat globule membrane, 396
  FAD content, 397
  heat stability, 397
  iron content, 397
  isolation, 396, 397
  molecular weight, 397
  molybdenum, 396–398
  polymorphism, 396
    molybdenum availability, 396
    nutritional, 396
  properties, 397
  thiol groups, 398
X-ray crystallography
  α-lactalbumin, 343, 356
  β-lactoglobulin, 323–324
  lysozyme, 407
  rennin, 244

Z

Zone electrophoresis
  α_s-casein, 138

$\alpha_{s,1}$-casein, 128–134, 488–496, 501
β-casein, 488–496, 499–502
γ-casein, 488, 493, 501–502
κ-casein, 488–490, 492–494
immunoglobulins, 503
α-lactalbumin
  bovine, 492, 503, 506
  kangaroo, 506
  ovine, 504, 505
lactoferrin, human, 503
β-lactoglobulins
  bovine, 271–274, 489–491, 503, 506
  ovine, 273, 504–505
methods for milk proteins, 271–274, 483–508
  acrylamide electrophoresis, 486, 495–499, 501–502, 507
  buffer concentrations, 273, 487–508
  cyanogum, 495–498, 507
  gel concentrations, 484–508
  gel dimensions, 485–508
  hydrolyzed starch, 271–274, 484–494, 499–500
  pH effects, 272–273, 314
  resolution, 484, 487–508
  staining, 486–508
    acrylamide gel, 486, 495–498, 501–502, 507
    starch gel, 486–494, 499–500, 503–506
  starch gel preparation, 484–494, 499–500
  starch reproducibility, 273
transferrin, kangaroo, 503
use of urea, 485, 487, 489, 491–495, 497–502
used for phenotyping, 271–274
whey protein, 271–274
"whole" casein, 98, 109–112, 492–495, 497, 499, 501
Zone electrophoretic methods, 483–508
  abbreviations, 486
Zone electrophoretic numbering, casein nomenclature, 120